Lecture Notes in Computer Science   14507

The series Lecture Notes in Computer Science (LNCS), including its subseries Lecture Notes in Artificial Intelligence (LNAI) and Lecture Notes in Bioinformatics (LNBI), has established itself as a medium for the publication of new developments in computer science and information technology research, teaching, and education.

LNCS enjoys close cooperation with the computer science R & D community, the series counts many renowned academics among its volume editors and paper authors, and collaborates with prestigious societies. Its mission is to serve this international community by providing an invaluable service, mainly focused on the publication of conference and workshop proceedings and postproceedings. LNCS commenced publication in 1973.

Oscar Camara · Esther Puyol-Antón ·
Maxime Sermesant · Avan Suinesiaputra ·
Qian Tao · Chengyan Wang · Alistair Young
Editors

Statistical Atlases and Computational Models of the Heart

Regular and CMRxRecon Challenge Papers

14th International Workshop, STACOM 2023
Held in Conjunction with MICCAI 2023
Vancouver, BC, Canada, October 12, 2023
Revised Selected Papers

 Springer

Editors
Oscar Camara ⓘ
Universitat Pompeu Fabra
Barcelona, Spain

Maxime Sermesant ⓘ
Inria
Sophia Antipolis, France

Qian Tao ⓘ
Technische Universiteit Delft
Delft, The Netherlands

Alistair Young ⓘ
King's College London
London, UK

Esther Puyol-Antón ⓘ
King's College London
London, UK

Avan Suinesiaputra ⓘ
King's College London
London, UK

Chengyan Wang ⓘ
Fudan University
Shanghai, China

ISSN 0302-9743 ISSN 1611-3349 (electronic)
Lecture Notes in Computer Science
ISBN 978-3-031-52447-9 ISBN 978-3-031-52448-6 (eBook)
https://doi.org/10.1007/978-3-031-52448-6

Preface

Cardiac image analysis has remained an active research field in medical imaging. Numerous studies have focused on developing statistical atlases and computational models of the heart to enhance our understanding of cardiac structure and function. These models provide valuable insights into the complex dynamics and behaviour of the heart, allowing for improved diagnosis and treatment of various cardiac conditions. Additionally, the use of artificial intelligence and machine learning algorithms in cardiac image analysis has further revolutionized the field.

However, clinical translations of these methods are still constrained by the lack of complete and rigorous technical and clinical validations, as well as benchmarking with common data. The Statistical Atlases and Computational Modelling of the Heart (STACOM) workshop aims to achieve these goals by providing a platform for researchers to exchange ideas, present their advancements, and collaborate on the development of robust and validated computational models of the heart.

The 14th edition of STACOM (https://stacom.github.io/stacom2023/), was held in conjunction with the 26th MICCAI conference in Vancouver, Canada, on 12 October 2023. It followed the thirteen successful previous editions (https://stacom.github.io/): STACOM 2010 (Beijing, China), STACOM 2011 (Toronto, Canada), STACOM 2012 (Nice, France), STACOM 2013 (Nagoya, Japan), STACOM 2014 (Boston, USA), STACOM 2015 (Munich, Germany), STACOM 2016 (Athens, Greece), STACOM 2017 (Quebec City, Canada), STACOM 2018 (Granada, Spain), STACOM 2019 (Shenzhen, China), STACOM 2020 (Lima, Peru), STACOM 2021 (Strasbourg, France), and STACOM 2022 (Singapore). Throughout these thirteen years, the STACOM workshop has provided a forum to discuss the latest developments in various areas of computational cardiac imaging, cardiac modelling, application of artificial intelligence and machine learning to cardiac image analysis, electro-mechanical modelling of the heart, novel methods in preclinical/clinical imaging for tissue characterization and image reconstruction, as well as statistical cardiac atlases.

The STACOM 2023 workshop attracted 29 paper submissions, of which 24 were accepted for presentation during the workshop. Topics range from cardiac segmentation, modelling, strain quantification, registration, statistical shape analysis, and quality control. Deep learning methods were still the predominant approach to performing automated cardiac image analysis. Left atrial image analysis and modelling gained more attention in this workshop with atrial fibrillation being the common area of interest.

The workshop awarded the best oral to **Laura Alvarez Florez** from Amsterdam University Medical Center, the Netherlands, for her paper entitled *Deep Learning for Automatic Strain Quantification in Arrhythmogenic Right Ventricular Cardiomyopathy*.

The best poster presenter was awarded to **Jiachuan Peng** from University of Oxford, UK, for his paper entitled *Generating Virtual Populations of 3D Cardiac Anatomies with Snowflake-Net*.

This year STACOM 2023 also hosted a challenge on reconstruction of cardiac MRI: CMRxRecon (https://cmrxrecon.github.io/), which aimed to establish a platform for fast CMR image reconstruction and provide a benchmark dataset that enables the broader research community to promote advances in this area of research. The challenge included two independent tasks: 1) to accelerate reconstruction of cine MRI by raw data subsampling and to address the image degradation due to motion artefacts caused by voluntary breath-hold imperfections or cardiac arrythmia, and 2) to improve the T1 and T2 mapping estimation accuracy from raw data under-sampling. The CMRxRecon challenge attracted 24 paper submissions, of which 21 are presented in this volume.

Finally, the STACOM workshop organisers would like to express our gratitude to external reviewers from King's College London (UK), University of Auckland (New Zealand), Inria (France), University of Oxford (UK), Fudan University (China), Imperial College London (UK), Technische Universiteit Delft (the Netherlands), and Universitat Pompeu Fabra (Spain). Special thanks to Tareen Dawood, Debbie Zhao, Edward Ferdian, Gabriel Bernardino, Josquin Harrison, Shuo Wang, Yidong Zhao, Chen Qin, Carlos Albors Lucas, and Vicky Wang, who spent their time to meticulously review papers for these proceedings. Ultimately, we would like to thank all authors who participated in this workshop.

November 2023

Esther Puyol-Antón
Oscar Camara
Maxime Sermesant
Avan Suinesiaputra
Qian Tao
Chengyan Wang
Alistair Young

Organization

Program Committee Chairs

Camara, Oscar — Universitat Pompeu Fabra, Spain
Puyol-Anton, Esther — King's College London, UK
Sermesant, Maxime — Inria, Sophia Antipolis, France
Suinesiaputra, Avan — King's College London, UK
Tao, Qian — Delft University of Technology, The Netherlands
Young, Alistair — King's College London, UK

Program Committee Members

Albors Lucas, Carlos — Universitat Pompeu Fabra, Spain
Bernardino, Gabriel — Universitat Pompeu Fabra, Spain
Camara, Oscar — Universitat Pompeu Fabra, Spain
Chen, Chen — University of Oxford, UK
Dawood, Tareen — King's College London, UK
Ferdian, Edward — University of Auckland, New Zealand
Harrison, Josquin — Inria, Sophia Antipolis, France
Puyol-Anton, Esther — King's College London, UK
Sermesant, Maxime — Inria, Sophia Antipolis, France
Suinesiaputra, Avan — King's College London, UK
Tao, Qian — Delft University of Technology, The Netherlands
Ugurlu, Devran — King's College London, UK
Wang, Shuo — Fudan University, China
Wang, Chengyang — Fudan University, China
Wang, Vicky — Stanford University, USA
Young, Alistair — King's College London, UK
Zhao, Debbie — University of Auckland, New Zealand
Zhao, Yidong — Delft University of Technology, The Netherlands

Reviewers

Albors Lucas, Carlos — Universitat Pompeu Fabra, Spain
Bernardino, Gabriel — Universitat Pompeu Fabra, Spain
Camara, Oscar — Universitat Pompeu Fabra, Spain

Chen, Chen	University of Oxford, UK
Dawood, Tareen	King's College London, UK
Ferdian, Edward	University of Auckland, New Zealand
Harrison, Josquin	Inria, Sophia Antipolis, France
Li, Lei	University of Oxford, UK
Puyol-Anton, Esther	King's College London, UK
Suinesiaputra, Avan	King's College London, UK
Tao, Qian	Delft University of Technology, The Netherlands
Ugurlu, Devran	King's College London, UK
Wang, Shuo	Fudan University, China
Wang, Chengyang	Fudan University, China
Wang, Vicky	Stanford University, USA
Young, Alistair	King's College London, UK
Zhao, Debbie	University of Auckland, New Zealand
Zhao, Yidong	Delft Unversity of Technology, The Netherlands

Contents

Regular Papers

CardiacSeg: Customized Pre-training Volumetric Transformer with Scaling
Pyramid for 3D Cardiac Segmentation 3
   Zhiyu Ye, Hairong Zheng, and Tong Zhang

Voxel2Hemodynamics: An End-to-End Deep Learning Method
for Predicting Coronary Artery Hemodynamics 15
   *Ziyu Ni, Linda Wei, Lijian Xu, Qing Xia, Hongsheng Li,
Shaoting Zhang, and Dimitris Metaxas*

Deep Learning for Automatic Strain Quantification in Arrhythmogenic
Right Ventricular Cardiomyopathy 25
   *Laura Alvarez-Florez, Jörg Sander, Mimount Bourfiss,
Fleur V. Y. Tjong, Birgitta K. Velthuis, and Ivana Išgum*

Patient Stratification Based on Fast Simulation of Cardiac
Electrophysiology on Digital Twins 35
   *Dolors Serra, Pau Romero, Miguel Lozano, Ignacio Garcia-Fernandez,
Diego Penela, Antonio Berruezo, Oscar Camara, Miguel Rodrigo,
Miriam Gil, and Rafael Sebastian*

Deep Conditional Shape Models for 3D Cardiac Image Segmentation 44
   Athira J. Jacob, Puneet Sharma, and Daniel Ruckert

Global Sensitivity Analysis of Thrombus Formation in the Left Atrial
Appendage of Atrial Fibrillation Patients 55
   *Zineb Smine, Paolo Melidoro, Ahmed Qureshi, Stefano Longobardi,
Steven E. Williams, Oleg Aslanidi, and Adelaide De Vecchi*

Sparse Annotation Strategies for Segmentation of Short Axis Cardiac MRI 66
   Josh Stein, Maxime Di Folco, and Julia A. Schnabel

Contrast-Agnostic Groupwise Registration by Robust PCA for Quantitative
Cardiac MRI 77
   Xinqi Li, Yi Zhang, Yidong Zhao, Jan van Gemert, and Qian Tao

FM-Net: A Fully Automatic Deep Learning Pipeline for Epicardial
Adipose Tissue Segmentation 88
   Fan Feng, Carl-Johan Carlhäll, Yongyao Tan, Shaleka Agrawal,
   Peter Lundberg, Jieyun Bai, John Zhiyong Yang, Mark Trew,
   and Jichao Zhao

Automated Quality-Controlled Left Heart Segmentation from 2D
Echocardiography 98
   Bram W. M. Geven, Debbie Zhao, Stephen A. Creamer,
   Joshua R. Dillon, Gina M. Quill, Nicola C. Edwards,
   Malcolm E. Legget, Robert N. Doughty, Alistair A. Young,
   Thiranja P. Babarenda Gamage, and Martyn P. Nash

Impact of Hypertension on Left Ventricular Pressure-Strain Loop
Characteristics and Myocardial Work 108
   Stephen A. Creamer, Debbie Zhao, Gina M. Quill,
   Abdallah I. Hasaballa, Vicky Y. Wang, Thiranja P. Babarenda Gamage,
   Nicola C. Edwards, Malcolm E. Legget, Boris S. Lowe,
   Robert N. Doughty, Satpal Arri, Peter N. Ruygrok, Alistair A. Young,
   Julian F. R. Paton, Gonzalo D. Maso Talou, and Martyn P. Nash

Automated Segmentation of the Right Ventricle from 3D Echocardiography
Using Labels from Cardiac Magnetic Resonance Imaging 119
   Joshua R. Dillon, Debbie Zhao, Thiranja P. Babarenda Gamage,
   Gina M. Quill, Vicky Y. Wang, Nicola C. Edwards, Timothy M. Sutton,
   Boris S. Lowe, Malcolm E. Legget, Robert N. Doughty,
   Alistair A. Young, and Martyn P. Nash

Neural Implicit Functions for 3D Shape Reconstruction from Standard
Cardiovascular Magnetic Resonance Views 130
   Marica Muffoletto, Hao Xu, Yiyang Xu, Steven E Williams,
   Michelle C Williams, Karl P Kunze, Radhouene Neji,
   Steven A Niederer, Daniel Rueckert, and Alistair A Young

Deep Learning-Based Pulmonary Artery Surface Mesh Generation 140
   Nina Krüger, Jan Brüning, Leonid Goubergrits, Matthias Ivantsits,
   Lars Walczak, Volkmar Falk, Henryk Dreger, Titus Kühne,
   and Anja Hennemuth

Impact of Catheter Orientation on Cardiac Radiofrequency Ablation 152
   Massimiliano Leoni, Argyrios Petras, Zoraida Moreno Weidmann,
   Jose M. Guerra, and Luca Gerardo-Giorda

Generating Virtual Populations of 3D Cardiac Anatomies
with Snowflake-Net 163
    *Jiachuan Peng, Marcel Beetz, Abhirup Banerjee, Min Chen,
    and Vicente Grau*

Effects of Fibrotic Border Zone on Drivers for Atrial Fibrillation:
An In-Silico Mechanistic Investigation 174
    *Shaheim Ogbomo-Harmitt, George Obada, Nele Vandersickel,
    Andrew P. King, and Oleg Aslanidi*

Exploring the Relationship Between Pulmonary Artery Shape and Pressure
in Pulmonary Hypertension: A Statistical Shape Analysis Study 186
    *Malak Sabry, Uxio Hermida, Ahmed Hassan, Michael Nagy,
    David Stojanovski, Irini Samuel, John Locas, Magdi H. Yacoub,
    Adelaide De Vecchi, and Pablo Lamata*

Type and Shape Disentangled Generative Modeling for Congenital Heart
Defects 196
    Fanwei Kong and Alison L. Marsden

Automated Coronary Vessels Segmentation in X-ray Angiography Using
Graph Attention Network 209
    Haorui He, Abhirup Banerjee, Robin P. Choudhury, and Vicente Grau

Inherent Atrial Fibrillation Vulnerability in the Appendages Exacerbated
in Heart Failure 220
    *Shaleka Agrawal, Joseph Ashby, Jeiyun Bai, Fan Feng, Xue J. Cai,
    Joseph Yanni, Caroline B. Jones, Sunil J. R. J. Logantha, Akbar Vohra,
    Robert C. Hutcheon, Antonio F. Corno, Halina Dobrzynski,
    Robert S. Stephenson, Mark Boyett, George Hart, Jonathan Jarvis,
    Bruce Smaill, and Jichao Zhao*

Two-Stage Deep Learning Framework for Quality Assessment of Left
Atrial Late Gadolinium Enhanced MRI Images 230
    *K M Arefeen Sultan, Benjamin Orkild, Alan Morris,
    Eugene Kholmovski, Erik Bieging, Eugene Kwan, Ravi Ranjan,
    Ed DiBella, and Shireen Elhabian*

Automatic Landing Zone Plane Detection in Contrast-Enhanced Cardiac
CT Volumes 240
    *Lisette Lockhart, Xin Yi, Nathan Cassady, Alexandra Nunn,
    Cory Swingen, and Alborz Amir-Khalili*

A Benchmarking Study of Deep Learning Approaches for Bi-Atrial
Segmentation on Late Gadolinium-Enhanced MRIs 250
   Yongyao Tan, Fan Feng, and Jichao Zhao

CMRxRecon Challenge

Fill the K-Space and Refine the Image: Prompting for Dynamic
and Multi-Contrast MRI Reconstruction 261
   Bingyu Xin, Meng Ye, Leon Axel, and Dimitris N. Metaxas

Learnable Objective Image Function for Accelerated MRI Reconstruction 274
   Artem Razumov and Dmitry V. Dylov

Accelerating Cardiac MRI via Deblurring Without Sensitivity Estimation 283
   Jin He, Weizhou Liu, Yun Tian, and Shifeng Zhao

T1/T2 Relaxation Temporal Modelling from Accelerated Acquisitions
Using a Latent Transformer 293
   Michael Tänzer, Fanwen Wang, Mengyun Qiao, Wenjia Bai,
   Daniel Rueckert, Guang Yang, and Sonia Nielles-Vallespin

T1 and T2 Mapping Reconstruction Based on Conditional DDPM 303
   Yansong Li, Lulu Zhao, Yun Tian, and Shifeng Zhao

k-t CLAIR: Self-consistency Guided Multi-prior Learning for Dynamic
Parallel MR Image Reconstruction 314
   Liping Zhang and Weitian Chen

Cardiac MRI Reconstruction from Undersampled K-Space Using
Double-Stream IFFT and a Denoising GNA-UNET Pipeline 326
   Julia Dietlmeier, Carles Garcia-Cabrera, Anam Hashmi,
   Kathleen M. Curran, and Noel E. O'Connor

Multi-scale Inter-frame Information Fusion Based Network for Cardiac
MRI Reconstruction 339
   Wenzhe Ding, Xiaohan Liu, Yong Sun, Yiming Liu, and Yanwei Pang

Relaxometry Guided Quantitative Cardiac Magnetic Resonance Image
Reconstruction 349
   Yidong Zhao, Yi Zhang, and Qian Tao

A Context-Encoders-Based Generative Adversarial Networks for Cine
Magnetic Resonance Imaging Reconstruction 359
   Weihua Zhang, Mengshi Tang, Liqin Huang, and Wei Li

Accelerated Cardiac Parametric Mapping Using Deep Learning-Refined
Subspace Models 369
Calder D. Sheagren, Brenden T. Kadota, Jaykumar H. Patel,
Mark Chiew, and Graham A. Wright

DiffCMR: Fast Cardiac MRI Reconstruction with Diffusion Probabilistic
Models 380
Tianqi Xiang, Wenjun Yue, Yiqun Lin, Jiewen Yang, Zhenkun Wang,
and Xiaomeng Li

C³-Net: Complex-Valued Cascading Cross-Domain Convolutional Neural
Network for Reconstructing Undersampled CMR Images 390
Quan Dou, Kang Yan, Sheng Chen, Zhixing Wang, Xue Feng,
and Craig H. Meyer

Space-Time Deformable Attention Parallel Imaging Reconstruction
for Highly Accelerated Cardiac MRI 400
Lifeng Mei, Kexin Yang, Yi Li, Shoujin Huang, Yilong Liu,
and Mengye Lyu

Multi-level Temporal Information Sharing Transformer-Based Feature
Reuse Network for Cardiac MRI Reconstruction 410
Guangming Wang, Jun Lyu, Fanwen Wang, Chengyan Wang, and Jing Qin

Cine Cardiac MRI Reconstruction Using a Convolutional Recurrent
Network with Refinement 421
Yuyang Xue, Yuning Du, Gianluca Carloni, Eva Pachetti,
Connor Jordan, and Sotirios A. Tsaftaris

ReconNext: A Encoder-Decoder Skip Cross Attention Based Approach
to Reconstruct Cardiac MRI 433
Ruiyi Li, Hanyuan Zheng, Weiya Sun, and Rongjun Ge

Temporal Super-Resolution for Fast T1 Mapping 443
Xunkang Zhao, Jun Lyu, Fanwen Wang, Chengyan Wang, and Jing Qin

NoSENSE: Learned Unrolled Cardiac MRI Reconstruction Without
Explicit Sensitivity Maps 454
Felix Frederik Zimmermann and Andreas Kofler

CineJENSE: Simultaneous Cine MRI Image Reconstruction
and Sensitivity Map Estimation Using Neural Representations 467
Ziad Al-Haj Hemidi, Nora Vogt, Lucile Quillien, Christian Weihsbach,
Mattias P. Heinrich, and Julien Oster

Deep Cardiac MRI Reconstruction with ADMM 479
  George Yiasemis, Nikita Moriakov, Jan-Jakob Sonke, and Jonas Teuwen

Author Index 491

Regular Papers

CardiacSeg: Customized Pre-training Volumetric Transformer with Scaling Pyramid for 3D Cardiac Segmentation

Zhiyu Ye[1,2,3], Hairong Zheng[2,3], and Tong Zhang[1(✉)]

[1] Peng Cheng Laboratory, Shenzhen, China
zhangt02@pcl.ac.cn
[2] Shenzhen Institute of Advanced Technology, Shenzhen, China
[3] University of Chinese Academy of Sciences, Beijing, China
hr.zheng@siat.ac.cn

Abstract. Congenital heart disease (CHD) is the most common type of birth defect and a leading cause of death worldwide. The volumetric segmentation of the whole heart anatomy serves as a basic step towards accurate diagnosis and treatment planning for CHD patients. Although deep learning segmentation networks can be powerful tools, it is still very challenging to apply them to CHD images due to the complex nature of the defect and the limited availability of training data and annotations. In this paper, we present CardiacSeg, a volumetric transformer for 3D cardiac image segmentation with masked image pre-training. Following the classic "U-shaped" encoder-decoder architecture, CardiacSeg is composed of a vision transformer (ViT) encoder, a scaling feature pyramid and a convolutional neural network decoder. Specifically, the scaling pyramid is generated solely from the output of the last layer of the encoder, and converts the single-scale feature map into a multi-scale representation, thereby enabling the decoder to effectively reconstruct the segmentation results. We evaluated our pre-trained ViT backbone and downstream segmentation network on the 3D Computed Tomography Image Dataset for CHD (ImageCHD) and the Multi-Modality Whole Heart Segmentation Challenge (MM-WHS) dataset. To further validate the few-shot learning ability, we conduct comparison experiments using a randomly sampled 10%-subset of the training data. Experimental results show that CardiacSeg outperforms five benchmark models, particularly in the few-shot learning scenario. The codes will be open-sourced to https://openi.pcl.ac.cn/OpenMedIA/CardiacSeg.

1 Introduction

Congenital heart disease (CHD) refers to the anatomical abnormality caused by the formation disorder or abnormal development of the heart or great vessels

---

T. Zhang and H. Zheng—Co-corresponding authors of this paper.

Supplementary Information The online version contains supplementary material available at https://doi.org/10.1007/978-3-031-52448-6_1.

during embryonic development, which affects nearly 1% newborn babies world-wide every year [17]. Complex and serious malformations may present severe and life-threatening symptoms after birth, while even mild symptoms require timely diagnosis and treatment. Cardiac computed tomography (CT) is a widely used imaging modality for assessing patients with CHDs. Creating patient-specific whole-heart segmentation models of CT images is typically a fundamental step for accurate diagnosis and surgical planning [2]. Convolutional neural networks (CNNs), such as UNet [3,14] and its improved variants, have proven to be pow-erful segmentation tools when provided with ample and unbiased training and labelling data. However, due to the significant variability in cardiac anatomy among CHD patients, such fully-supervised methods may struggle to segment long-tailed data [20]. Additionally, acquiring sufficient and unbiased training labels poses a significant challenge.

Recently, pre-training transformer-based models, such as BERT [9], GPT-3 [1] for natural language processing (NLP), Vision Transformer (ViT) [4], masked autoencoder (MAE) [8] for computer vision, and Swin UNETR [6] for 3D med-ical image analysis, have become increasingly popular due to their remarkable performance on a variety of the downstream tasks. These models are usually pre-trained on large amounts of unlabelled data in self-supervised fashion, which significantly reduces the amount of annotations required for specific downstream tasks to achieve high performance. Fang et al. [5] further proposed EVA, a general vision-centric foundation model that can scale-up ViT to one billion parameters, which was pre-trained via simple masked image modelling and achieved promis-ing results in a number of downstream tasks. Hence, our CardiacSeg strives to leverage the exceptional performance of these foundation models to enhance the performance in medical segmentation tasks.

In this work, we propose a pre-training based scaling pyramid transformer for 3D cardiac segmentation, named CardiacSeg. We first pre-train a volumet-ric asymmetric encoder-decoder transformer via masked image pre-training and freeze the encoder layers in the fine-tuning segmentation stage. Particularly, we design a scaling feature pyramid that only connects the last layer of the trans-former backbone. The feature pyramid is then combined with the CNN-based decoder for image segmentation. The main contributions of this work can be concluded as the follows:

- We propose CardiacSeg, a customized volumetric ViT that leverages the ben-efits of vision foundation models with masked image modelling for cardiac image segmentation.
- A scaling feature pyramid structure is designed to connect the ViT encoder and the CNN decoder for 3D image segmentation. This paper revisits the design of the hierarchical skip connections for ViT backbones through exten-sive ablation studies, especially in the pre-training paradigm.
- In terms of the segmentation accuracy, CardiacSeg outperformed five state-of-the-art (SOTA) models on both the ImageCHD [19] and MM-WHS [23] datasets, and demonstrated even more promising results in the few-shot learn-ing experiments.

2   Related Works

Transformers for Medical Image Segmentation. Since its introduction in 2020, ViT [4], based on a transformer architecture originally proposed for NLP, has had a significant impact on the field of computer vision. The ViT-based architecture utilises the self-attention mechanism to learn features from an image in a highly efficient way and has spurred self-supervised learning research, which has the potential to significantly reduce the need for labelled data. Numerous research efforts were put into exploring the ViT backbones for medical image segmentation tasks. CoTr [18] is a hybrid framework that combines CNNs and transformers for 3D medical image segmentation. In CoTr, CNNs extract feature representations, while transformers model the long-range dependency. UNETR [7] and Swin UNETR [6] may be the most representative transformer-based networks for medical image segmentation. Both UNETR and Swin UNETR maintain the "U-shaped" structure, and UNETR straightforwardly replaces the entire CNN encoder in UNet with a ViT backbone, while Swin UNETR uses a Swin Transformer [12] as the encoder. The hierarchical skip connection is preserved in UNETR and Swin UNETR by extracting feature maps from various transformer layers and matching them with different scales. Tang et al. [16] further investigated the self-supervised pre-training strategies for Swin UNETR.

Feature Pyramids. The Feature Pyramid Network (FPN) [11] is an advanced neural network architecture that was initially designed to solve the challenge of object recognition and detection at various scales in images. FPN leverages the inherent pyramidal feature hierarchy of CNNs to build feature pyramids with high-level semantics at all scales through the use of skip connections. It also worth noting that the hierarchical skip connections are a crucial component in the U-Net [14] architecture for achieving accurate image segmentation. They enable the network to recover spatial information that may be lost during down-sampling. Despite its successful in CNNs, the popular hierarchical skip connections settings need to be revisited for vision transformers like ViT, which do not use downsampling as CNNs, and instead use self-attention to capture global relationships between image patches. It can also be observed that the last transformer block of ViT inherits the feature representations learned by previous blocks. In ViTDet [10], it is demonstrated that a simple feature pyramid generated from a single-scale feature even outperformed the commonly used hierarchical feature pyramid if pre-trained with MAE [8]. As for the image segmentation tasks with ViT backbones, the SOTA methods as mentioned above are still based on the popular but sub-optimal hierarchical skip connections. Inspired by ViTDet, we propose CardiacSeg to incorporate the scaling feature pyramid constructed from the output of the last transformer block. To the best of our knowledge, it is the first time to explore the design of the scaling feature connections with transformer backbones for 3D cardiac image segmentation, while inheriting the advances in vision foundation models.

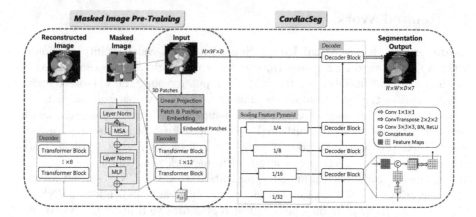

Fig. 1. Overview of the masked image pre-training and CardiacSeg architecture. For pre-training, we adopted MAE with the 3D settings. CardiacSeg contains a transformer encoder, a scaling feature pyramid and a CNN decoder. The scaling feature pyramid is set up from z_{12}. The overlap of two dashed boxes is the encoder, which remains the same in both the pre-training stage and CardiacSeg.

3   Method

As displayed in Fig. 1, the proposed CardiacSeg contains a transformer encoder, a scaling feature pyramid and a CNN decoder. It is first pre-trained with masked image modelling and then fine-tuned with cardiac segmentation tasks.

Masked Image Pre-Training. Following MAE [8], we masked 75% of the 3D patches in the unlabelled image dataset and built a 3D auto-encoder to learn the masked patches. Specifically, the original input image $x \in \mathbb{R}^{H \times W \times D}$ is first divided into $N = \frac{H}{P} \times \frac{W}{P} \times \frac{D}{P}$ non-overlapping patches p_i $(i = 1, 2, \cdots, N)$, where H, W and D represent the height, width and depth of the image, respectively, with fixed size $P \times P \times P$. 25% of these patches are randomly chosen as the input of the transformer encoder (see Sect. 3), which is exactly the same in both pre-training and fine-tuning segmentation stage. The rest patches are masked and will be later input into the pre-training decoder. The pre-training decoder is composed of eight transformer blocks. The masked pixel values need to be predicted by combining the position embeddings of visible and masked patches after decoding. The mean square error (MSE) is used as a loss function to optimize the pre-training weights.

Transformer Encoder. The encoder consists of $L = 12$ repeated transformer blocks made up of multi-head self-attention (MSA) and multi-layer perceptron (MLP) which are computed by

$$\mathbf{z_i}' = MSA\left(LN\left(\mathbf{z_{i-1}}\right)\right) + \mathbf{z_{i-1}}, \tag{1}$$

$$\mathbf{z_i} = MLP\left(LN\left(\mathbf{z_i}'\right)\right) + \mathbf{z_i}', \tag{2}$$

where z_{i-1} and z_i are the input and output feature of the i-th transformer block respectively, $i = 1, 2, \cdots, L$; LN denotes layer normalization. An MSA is a concatenation of $n = 12$ SA to combine the information learned from different attention heads.

Scaling Feature Pyramid and Decoder. A scaling feature pyramid for reconstructing segmentation results is formed by utilizing the output feature of the last transformer block, *i.e.* z_{12}. The feature pyramid in our Cardiac-Seg converts features in multiple resolutions, specifically, $\left\{ \frac{1}{4}, \frac{1}{8}, \frac{1}{16}, \frac{1}{32} \right\}$ from the original single-scale feature representations. Since the default scale of z_{12} is $\frac{1}{16}$, the feature pyramid is obtained via convolutions of stride $\left\{ 4, 2, 1, \frac{1}{2} \right\}$ where the stride $\frac{1}{2}$ implies maxpool operation of stride 2.

The decoder consists of four CNN decoder blocks. In each block, a feature representation with a relatively small resolution is first upsampled by a convolution transpose layer and then concatenated with feature maps from feature pyramids or the original input image (in the last decoder block). Next, the concatenated feature maps pass through two sets of convolution, batch normalization and ReLU layers. After the final fully connected layer, the segmentation result is obtained. We use the sum of soft Dice loss [13] and cross-entropy loss to train the model.

4   Experiments

4.1   Datasets and Evaluations

We evaluate our CardiacSeg on two public whole-heart segmentation datasets ImageCHD [19] and MM-WHS [23]. ImageCHD contains 110 cases with various diseases and is divided into training, validation and testing sets which contain 77, 11 and 22 cases respectively according to the proportion of different diseases. Detailed division information has been listed in Table 1. MM-WHS contains 20 labelled CT images, 16 of which are for training and 4 for testing. The whole hearts in these images are manually labelled into seven substructures: left ventricle, right ventricle, left atrium, right atrium, myocardium of the left ventricle, ascending aorta and pulmonary artery. Besides, another public dataset CT Lymph Nodes [15] is used for masked image pre-training stage. We use the Dice score [13] to evaluate the accuracy of segmentation results.

4.2   Implementation Details

We built our CardiacSeg framework in PyTorch[1] and MONAI[2]. Our model was pre-trained with an NVIDIA-DGX-1 server with 8 T V100 GPUs with

---

[1] https://pytorch.org/.
[2] https://monai.io/.

Table 1. The distribution of diseases in training, testing and validation sets of ImageCHD. Since some cases may be associated with several diseases, the sum of each row is not equal to the overall. ASD: atrial septal defect, AVSD: atrio-ventricular septal defect, PDA: patent ductus arteriosus, PuA: pulmonary atresia, VSD: ventricular septal defect, CA: co-arctation, TOF: tetrology of fallot, TGA: transposition of great arteries, PAS: pulmonary artery sling, DORV: double outlet right ventricle, CAT: common arterial trunk, DAA: double aortic arch, APVC: anomalous pulmonary venous drainage, AAH: aortic arch hypoplasia, IAA: interrupted aortic arch, DSVC: double superior vena cava.

Sets	Diseases																
	Overall	ASD	VSD	AVSD	ToF	TGA	DORV	CAT	CA	AAH	DAA	IAA	PuA	APVC	DSVC	PDA	PAS	Normal
Training	77	19	30	13	9	3	7	2	4	2	4	2	13	4	6	9	2	3
Testing	22	6	10	3	2	2	1	1	1	1	1	0	0	0	1	4	0	1
Validation	11	1	4	2	1	0	0	1	1	0	0	1	3	2	1	1	0	2
Overall	110	26	44	18	12	5	8	4	6	3	5	3	16	6	8	14	3	6

open-sourced implementation at OpenMedIA[3,4] [22]. All the fine-tuning and segmentation experiments were conducted on an NVIDIA A100 GPU with 80G memory. In the segmentation training stage, models were trained for 300 epochs with batch size of 2 and optimizer AdamW with learning rate 10^{-3} and weight decay 10^{-5}. Besides, a 10-epoch linear warmup and a cosine annealing learning rate scheduler were used. Our encoder is composed of 12 repeated transformer blocks with embedding size $C = 768$ and patch resolution $16 \times 16 \times 16$. All input data is first resampled to spacing $1 \times 1 \times 1$ mm^3, then the HU values are clipped into $[500, 2000]$ (for ImageCHD) or $[-200, 1500]$ (for MM-WHS) and normalized to $[0, 1]$. Next, the input image is cropped into a bounding box which only contains the foreground according to its label. This bounding box is further randomly cropped to four cubes in size $96 \times 96 \times 96$ as input samples. As for data augmentation, we randomly add Gaussian noise with a standard deviation of 0.05; randomly scale the intensity with factor 0.1; and shift intensity with randomly picked offset from $[-0.1, 0.1]$ to the images.

4.3   Results

We conducted an extensive comparative analysis of CardiacSeg against state-of-the-art (SOTA) segmentation networks, including CoTr and its variant networks [18], UNETR [7] with a ViT backbone, and Swin UNETR [6,16] with a Swin Transformer backbone. To ensure a fair comparison, we aligned the pre-training settings of UNETR with those of our CardiacSeg. The pre-training weights for Swin UNETR were obtained from the official Github repository[5]. Detailed information on the number of parameters and FLOPs of the models during training is presented in Table 2.

---

[3] https://openi.pcl.ac.cn/OpenMedIA.

[4] https://openi.pcl.ac.cn/OpenMedIA/Transformer3DSeg.

[5] https://github.com/Project-MONAI/research-contributions/tree/main/Swin UNETR.

Table 2. Comparison of FLOPs and numbers of parameters for various networks in full learning experiments on ImageCHD. The encoders of UNETR and CardiacSeg follow the setting of ViT-B.

Networks	FLOPs (G)	#(Params) (M)		
		Encoder	Decoder	Total
CoTr	31.01	29.82	12.05	41.87
UNETR	29.46	90.14	2.67	92.81
Swin UNETR	47.91	42.58	19.62	62.20
CardiacSeg	28.76	90.14	7.51	97.65

Table 3. Quantitative comparisons among different networks on ImageCHD with full and few-shot training samples. The training data ratio (Trn %) are listed in the second column. The metrics shown in the table are the means of Dice score. LV: left ventricle, RV: right ventricle, LA: left atrium, RA: right atrium, Myo: myocardium, AO: aorta, PA: pulmonary artery.

Networks	Trn %	Overall	LV	RV	LA	RA	Myo	AO	PA
CoTr*	100	0.822	0.838	0.809	0.867	0.866	0.837	0.808	0.733
CoTr†		0.855	0.882	0.833	0.889	0.895	0.865	0.831	0.795
CoTr		0.863	0.895	0.842	0.885	0.897	0.869	0.872	0.782
UNETR		0.863	0.901	0.843	0.890	0.898	0.877	0.848	0.790
Swin UNETR		0.870	0.895	0.848	0.898	0.896	0.875	0.862	0.822
CardiacSeg		**0.875**	0.918	0.861	0.899	0.899	0.878	0.861	0.813
CoTr*	10	0.583	0.628	0.572	0.625	0.700	0.678	0.444	0.451
CoTr†		0.572	0.666	0.568	0.640	0.702	0.667	0.407	0.364
CoTr		0.581	0.704	0.565	0.598	0.642	0.654	0.462	0.450
UNETR		0.493	0.277	0.471	0.534	0.672	0.777	0.374	0.372
Swin UNETR		0.701	0.744	0.638	0.785	0.766	0.702	0.607	0.664
CardiacSeg		**0.711**	0.737	0.627	0.782	0.786	0.756	0.651	0.641

Results on ImageCHD. As shown in Fig. 2, the CardiacSeg results have distinct improvements in boundaries of right ventricle and right atrium areas. The quantitative comparisons are shown in Table 3, which statistically proved the superior performance of the proposed CardiacSeg. In the 10% few-shot learning test, a subset containing 8 samples of the original training set is used for training, while the test set remains unchanged. CardiacSeg exhibits its strong learning ability in small-sample learning tasks, with an overall Dice score that is 1%, 21.8%, and 13% higher than that of Swin UNETR, UNETR, and CoTr, respectively. Compared to the strongest baseline Swin UNETR, our overall Dice sore shows 0.5% and 1% improvement in the full and few-shot learning experiments, respectively.

Results on MM-WHS. The Dice results with standard deviations on MM-WHS dataset are shown in Table 4, where CardiacSeg achieved the best performance. The qualitative comparisons of the segmentation results are illustrated in Fig. 3.

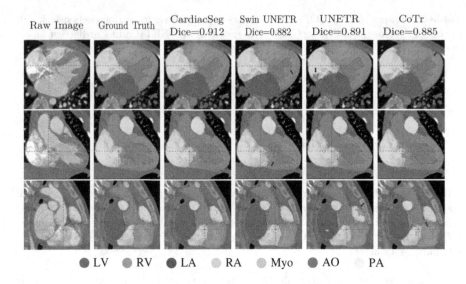

Fig. 2. Comparisons among the ground truth and results of different segmentation models toward the same test image from ImageCHD in axial, coronal, and sagittal views. The intersection of dashed lines in different views is the same pixel. The parts indicated by arrows are where the segmentation differ greatly from the ground truth.

Table 4. Segmentation comparisons on MM-WHS. The metrics shown in the table are the means of Dice score.

Networks	Overall	LV	RV	LA	RA	Myo	AO	PA
CoTr*	0.849	0.855	0.904	0.890	0.901	0.858	0.779	0.759
CoTr†	0.851	0.849	0.898	0.904	0.887	0.895	0.764	0.760
CoTr	0.870	0.877	0.918	0.901	0.895	0.88.1	0.832	0.784
UNETR	0.879	0.861	0.902	0.906	0.910	0.89.4	0.878	0.804
Swin UNETR	0.816	0.823	0.902	0.909	0.741	0.719	0.856	0.759
CardiacSeg	**0.901**	0.897	0.913	0.923	0.921	0.921	0.922	0.811

Ablation Studies. To verify the effectiveness of the scaling feature pyramid and the pre-training, we conducted ablation experiments on ImageCHD. According to Table 5, the network settings of the first two and last two rows are actually UNETR and CardiacSeg network settings, therefore the results of models with and without pre-training are compared on the basis of these two networks. The Dice of pre-trained UNETR and CardiacSeg increased by 6% and 4.3% respectively compared to those of training from scratch. However, CardiacSeg is superior to UNETR in both circumstances because of its designed scaling feature pyramid. CardiacSeg only selects the output of the last transformer block (*i.e.* z_{12}), which contains more accurate feature representations. Additionally, it chooses to convert single-scale feature maps into more competent scales.

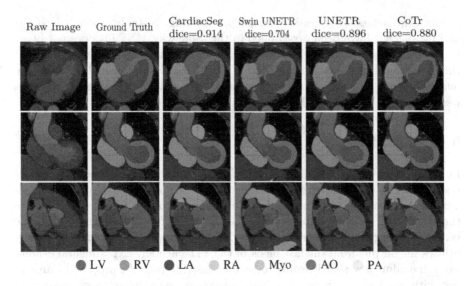

| Raw Image | Ground Truth | CardiacSeg dice=0.914 | Swin UNETR dice=0.704 | UNETR dice=0.896 | CoTr dice=0.880 |

● LV ● RV ● LA ● RA ● Myo ● AO ● PA

Fig. 3. Comparisons among the ground truth and results of different segmentation models toward the same test image from MM-WHS in axial, coronal, and sagittal views. The intersection of dashed lines in different views is the same pixel. In the results of SwinUNETR, UNETR and CoTr, there are obvious missing segmentation or incorrect segmentations.

Table 5. Ablation studies towards distinct settings of pre-training and scaling feature pyramid for CardiacSeg on ImageCHD.

Networks	Pre-training	Features	Pyramid scale	Dice
UNETR	✗	(z_3, z_6, z_9, z_{12})	$\left(\frac{1}{16}, \frac{1}{8}, \frac{1}{4}, \frac{1}{2}\right)$	0.803
	✓			0.863
/	✓	z_{12}	$\left(\frac{1}{16}, \frac{1}{8}, \frac{1}{4}, \frac{1}{2}\right)$	0.867
/	✓	(z_3, z_6, z_9, z_{12})	$\left(\frac{1}{32}, \frac{1}{16}, \frac{1}{8}, \frac{1}{4}\right)$	0.864
CardiacSeg	✗	z_{12}	$\left(\frac{1}{32}, \frac{1}{16}, \frac{1}{8}, \frac{1}{4}\right)$	0.832
	✓			**0.875**

Discussion. It is worth mentioning that UNETR, Swin UNETR, and CardiacSeg all utilise transformer-based encoders and have been pre-trained in this study. Intuitively, our method and other transformer-based segmentation networks rely heavily on pre-training. Swin UNETR which was trained with a much larger dataset of 5050 CT images, achieved the second-best performance in the full and few-shot learning experiments on ImageCHD (Table 3), which demonstrates the effectiveness of the self-supervised pre-training with a large dataset. As for the statistics shown in Table 4, the performance of Swin UNETR on MM-WHS was not that satisfactory, which further demonstrates the consistent power of our CardiacSeg with masked image pre-training. Notably, though UNETR and CardiacSeg have been pre-trained in the same way, UNETR does not benefit from pre-training as much as CardiacSeg. This confirms our hypothesis that the hierarchical skip-connections used in UNETR are sub-optimal. In addition, our model can be easily adapted to other "large scale" ViT-based backbones, *i.e.* volumetric ViT-L, ViT-H [4], and ViT-G [21].

5   Conclusion

In this paper, we proposed CardiacSeg, a pre-training based scaling pyramid transformer for 3D cardiac image segmentation. CardiacSeg is composed of a volumetric ViT backbone, a scaling feature pyramid, and a CNN decoder. Unlike the classic hierarchical skip connected fashion, CardiacSeg designed a simple yet effective segmentation architecture with a scaling feature pyramid that only utilises the output from the last block of the ViT encoder. Experimental evaluations, including few-shot learning tests, demonstrate that the proposed Cardiac-Seg with masked image pre-training achieved state-of-the-art performance on both the ImageCHD and MM-WHS datasets. Extensive ablation studies further confirm the effectiveness of the scaling feature pyramid proposed in CardiacSeg.

Acknowledgements. This work is supported in part by the Major Key Project of PCL (grant No. PCL2023AS7-1) and the National Natural Science Foundation of China (grant No. U21A20523). The computing resources of Pengcheng Cloudbrain are used in this research. We acknowledge the support provided by OpenI Community (https:// openi.pcl.ac.cn/).

References

1. Brown, T., et al.: Language models are few-shot learners. Adv. Neural. Inf. Process. Syst. **33**, 1877–1901 (2020)
2. Chen, C., et al.: Deep learning for cardiac image segmentation: a review. Front. Cardiovasc. Med **7**, 25 (2020)
3. Çiçek, Ö., Abdulkadir, A., Lienkamp, S.S., Brox, T., Ronneberger, O.: 3D U-Net: learning dense volumetric segmentation from sparse annotation. In: Ourselin, S., Joskowicz, L., Sabuncu, M.R., Unal, G., Wells, W. (eds.) MICCAI 2016. LNCS, vol. 9901, pp. 424–432. Springer, Cham (2016). https://doi.org/10.1007/978-3-319-46723-8_49
4. Dosovitskiy, A., et al.: An image is worth 16×16 words: transformers for image recognition at scale. In: International Conference on Learning Representations (2020)
5. Fang, Y., et al.: EVA: exploring the limits of masked visual representation learning at scale. In: Proceedings of the IEEE/CVF Conference on Computer Vision and Pattern Recognition, pp. 19358–19369 (2023)
6. Hatamizadeh, A., Nath, V., Tang, Y., Yang, D., Roth, H.R., Xu, D.: Swin UNETR: swin transformers for semantic segmentation of brain tumors in MRI images. In: International MICCAI Brainlesion Workshop, pp. 272–284. Springer, Cham (2022). https://doi.org/10.1007/978-3-031-08999-2_22

7. Hatamizadeh, A., et al.: UNETR: transformers for 3D medical image segmentation. In: Proceedings of the IEEE/CVF Winter Conference on Applications of Computer Vision, pp. 574–584 (2022)

8. He, K., Chen, X., Xie, S., Li, Y., Dollár, P., Girshick, R.: Masked autoencoders are scalable vision learners. In: Proceedings of the IEEE/CVF Conference on Computer Vision and Pattern Recognition, pp. 16000–16009 (2022)

9. Kenton, J.D.M.W.C., Toutanova, L.K.: BERT: pre-training of deep bidirectional transformers for language understanding. In: Proceedings of NAACL-HLT, pp. 4171–4186 (2019)

10. Li, Y., Mao, H., Girshick, R., He, K.: Exploring plain vision transformer backbones for object detection. In: Computer Vision-ECCV 2022: 17th European Conference, Tel Aviv, Israel, October 23–27, 2022, Proceedings, Part IX, pp. 280–296. Springer (2022). https://doi.org/10.1007/978-3-031-20077-9_17

11. Lin, T.Y., Dollár, P., Girshick, R., He, K., Hariharan, B., Belongie, S.: Feature pyramid networks for object detection. In: Proceedings of the IEEE Conference on Computer Vision and Pattern Recognition, pp. 2117–2125 (2017)

12. Liu, Z., Lin, Y., Cao, Y., Hu, H., Wei, Y., Zhang, Z., Lin, S., Guo, B.: Swin transformer: Hierarchical vision transformer using shifted windows. In: Proceedings of the IEEE/CVF International Conference on Computer Vision, pp. 10012–10022 (2021)

13. Milletari, F., Navab, N., Ahmadi, S.A.: V-Net: fully convolutional neural networks for volumetric medical image segmentation. In: 2016 Fourth International Conference on 3D Vision (3DV), pp. 565–571. IEEE (2016)

14. Ronneberger, O., Fischer, P., Brox, T.: U-Net: convolutional networks for biomedical image segmentation. In: Navab, N., Hornegger, J., Wells, W.M., Frangi, A.F. (eds.) MICCAI 2015. LNCS, vol. 9351, pp. 234–241. Springer, Cham (2015). https://doi.org/10.1007/978-3-319-24574-4_28

15. Roth, H., et al.: A new 2.5 D representation for lymph node detection in CT [dataset]. The Cancer Imaging Archive. https://wiki.cancerimagingarchive.net/display/Public/CT+Lymph+Nodes (Accessed on 8 Apr 2021) (2015)

16. Tang, Y., et al.: Self-supervised pre-training of swin transformers for 3D medical image analysis. In: Proceedings of the IEEE/CVF Conference on Computer Vision and Pattern Recognition, pp. 20730–20740 (2022)

17. Van Der Linde, D., et al.: Birth prevalence of congenital heart disease worldwide: a systematic review and meta-analysis (2011)

18. Xie, Y., Zhang, J., Shen, C., Xia, Y.: CoTr: efficiently bridging CNN and transformer for 3D medical image segmentation. In: International Conference on Medical Image Computing and Computer-Assisted Intervention, pp. 171–180. Springer (2021). https://doi.org/10.1007/978-3-030-87199-4_16

19. Xu, X., et al.: ImageCHD: a 3D computed tomography image dataset for classification of congenital heart disease. In: Martel, A.L., et al. (eds.) MICCAI 2020. LNCS, vol. 12264, pp. 77–87. Springer, Cham (2020). https://doi.org/10.1007/978-3-030-59719-1_8

20. Yang, L., Jiang, H., Song, Q., Guo, J.: A survey on long-tailed visual recognition. Int. J. Comput. Vis. **130**(7), 1837–1872 (2022)

21. Zhai, X., Kolesnikov, A., Houlsby, N., Beyer, L.: Scaling vision transformers. In: Proceedings of the IEEE/CVF Conference on Computer Vision and Pattern Recognition, pp. 12104–12113 (2022)

22. Zhuang, J.X., et al.: OpenMedIA: open-source medical image analysis toolbox and benchmark under heterogeneous AI computing platforms. In: Pattern Recognition and Computer Vision: 5th Chinese Conference, PRCV 2022, Shenzhen, China, November 4–7, 2022, Proceedings, Part I, pp. 356–367. Springer (2022). https://doi.org/10.1007/978-3-031-18907-4_28

23. Zhuang, X.: Multivariate mixture model for myocardial segmentation combining multi-source images. IEEE Trans. Pattern Anal. Mach. Intell. **41**(12), 2933–2946 (2018)

Voxel2Hemodynamics: An End-to-End Deep Learning Method for Predicting Coronary Artery Hemodynamics

Ziyu Ni[1], Linda Wei[2], Lijian Xu[1,3(✉)], Qing Xia[4], Hongsheng Li[5], Shaoting Zhang[1], and Dimitris Metaxas[6]

[1] Shanghai Artificial Intelligence Laboratory, Shanghai, China
bmexlj@gmail.com
[2] Shanghai Jiao Tong University, Shanghai, China
[3] Centre for Perceptual and Interactive Intelligence, The Chinese University of Hong Kong, Hong Kong, China
[4] SenseTime Research, Shanghai, China
[5] Department of Electronic Engineering, The Chinese University of Hong Kong, Hong Kong, China
[6] Rutgers University, Camden, USA

Abstract. Local hemodynamic forces play an essential role in determining the functional significance of coronary arterial stenosis and understanding the mechanism of coronary disease progression. Computational fluid dynamics (CFD) has been widely performed to simulate hemodynamics non-invasively from coronary computed tomography angiography (CCTA) images. However, fast and accurate computational analysis is still limited by the complex construction of patient-specific modeling and time-consuming computation. In this work, we proposed an end-to-end deep learning framework, which could predict the coronary artery hemodynamics from CCTA images. The model was trained and evaluated on the hemodynamic data obtained from 3D simulations of ideal synthetic and real datasets. Extensive experiments demonstrated that the predicted hemodynamic distributions by our method agreed well with the CFD-derived results. Quantitatively, the proposed method has the capability of predicting the fractional flow reserve with an average error of 0.5% and 2.5% for the ideal synthetic dataset and real dataset, respectively. This study demonstrates the feasibility and great potential of our end-to-end deep learning method as a fast and accurate approach for hemodynamic analysis. The code can be reached through https://github.com/lullcant/Voxel2Hemodynamic/tree/main.

Keywords: Deep Learning · Computational Fluid Dynamics · Hemodynamic Analysis

---

Supplementary Information The online version contains supplementary material available at https://doi.org/10.1007/978-3-031-52448-6_2.

O. Camara et al. (Eds.): STACOM 2023, LNCS 14507, pp. 15–24, 2024.
https://doi.org/10.1007/978-3-031-52448-6_2

1 Introduction

Coronary artery disease (CAD) is one of the most common types of cardiovascular disease in the world, which is mainly caused by plaque buildup in the arterial wall [5]. In clinical procedures, revascularization is routinely performed in the treatment of severe myocardial ischemia, where the degree of stenosis is usually regarded as the criterion for surgical intervention [1,12]. Nevertheless, biomechanical and hemodynamic alterations have been speculated to play an essential role in the pathogenesis of CAD and long-term outcomes of treatments [3,15]. For instance, fractional flow reserve (FFR) has been established as the golden standard for diagnosis of intermediate stenosis in patients with chronic CAD [18]. Besides, wall shear stress (WSS) is a measure of the shear force exerted on the arterial inner surface and the abnormal WSS has been found to exert a negative influence on endothelial function [7]. In this context, the quantitative evaluation of hemodynamic characteristics would contribute to the early diagnosis of coronary diseases. Computational fluid dynamics (CFD) based methods have been extensively examined to obtain hemodynamic parameters non-invasively [2,16,19]. Using the patient-specific geometry and boundary conditions obtained from medical imaging data, invivo hemodynamics could be reproduced accurately and non-invasively. The fidelity of a CFD model in reproducing hemodynamics relies on the accurate patient-specific modeling and various assumptions involved in model setup [16]. Another important limitation of CFD modeling is associated with the vast computational resources and long computing times required. In order to promote the clinical application of hemodynamics, it is necessary to develop a novel method in balance of the accuracy and computational cost.

Deep learning has achieved state-of-the-art results in automated segmentation of coronary computed tomography angiography (CCTA) images [10,22] and emerged as a potential approach to improve the efficiency of traditional physical modeling methods [4,8,11,14]. Advanced deep learning algorithms and high-performance GPUs could greatly reduce computing times while ensuring high accuracy. Several studies have already been introduced concerning predictions of coronary artery hemodynamics from point cloud [11] or geometrical features simply [4,8,14], which however are limited due to the ignorance of patient-specific geometrical and physiological features. Itu et al. put forward a DNN-based model to predict FFR of synthetically generated coronary anatomies with extracted feature input [8]. However, due to the complex geometric structure of real coronary artery data, previous works struggled to accurately predict hemodynamic parameters for 3D real coronary artery datasets. Considering that high-density 3D point clouds can provide a more detailed and high-resolution description of such complex geometry, Li et al. first employed PointNet [13], to do 3D hemodynamic prediction. They simply fed point cloud coordinates to the model to give prediction on hemodynamics of non-ideal cardiovascular model [11], but the patient-specified information has not been taken into consideration. Numerical studies have demonstrated that such simplified models could lead to marked deviations of model outputs from real in vivo hemodynamic conditions, which highlights the importance of patient-specifically modeling [6,20].

The hemodynamics of the coronary artery tree is complicated by its intricate geometry with massive coronary branches, upstream hemodynamics as well as downstream resistance ratio dominated by the microcirculation. Therefore, accurate prediction of coronary hemodynamics requires patient-specific modeling, which should be carefully considered in the deep learning method.

In this work, we developed an end-to-end deep learning method to predict hemodynamic parameters, such as velocity, pressure, WSS, and FFR. Traditional CFD pipelines were utilized to produce training data for both synthetic and real datasets with different geometries and physiological conditions. With the generated training data, the model was trained in an end-to-end manner to directly predict hemodynamic parameters from CCTA inputs. We further compared the hemodynamic distributions predicted by our method with those obtained by CFD simulations, demonstrating the effectiveness of our proposed method.

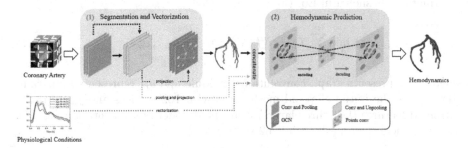

Fig. 1. Overview of the proposed end-to-end framework consisting of two stages. The first stage segments and vectorizes the inner diameter of coronary arteries to generate a high-resolution point cloud. The second stage comprises a point cloud module that predicts the characteristics of blood flow, using the point cloud data generated in the first stage.

2   Proposed Method

To achieve end-to-end hemodynamic prediction, our model consists of two stages (see Fig. 1). In the first stage, we use a geometry-based method [21] proposed by the authors to generate meshes from CCTA. The meshes will be converted to point cloud for the second stage. In the second stage, we feed the point cloud data to a point cloud network to predict hemodynamics. Li et al. [11] have demonstrated the ability of point cloud to represent the geometric structure and flow field distribution of a model, even for complex anatomies such as coronary arteries.

Stage One (Segmentation and Vectorization): Firstly, a classic U-shaped segmentation module was utilized to segment the inner diameter of coronary arteries and extract the geometric features from inputs. High-level perceptual features extracted from decoder were projected and concatenated as auxiliary input for graph convolutional module and points cloud module separately. To guide the segmentation process, we used cross-entropy loss (L_{ce}) and dice loss (L_{dice}), as shown in Eq. (1):

$$\min_{\theta_u} \sum_{i=1}^{n} (L_{ce}\left(f_u\left(x_{u_i};\theta_u\right),y_{u_i}\right) + L_{dice}(f_u\left(x_{u_i};\theta_u\right),y_{u_i})) \tag{1}$$

where f_u represents segmentation module and θ_u represents its parameters. n is the number of batch size and x_{ui} is the input patch with $32 \times 32 \times 32$ cropped size. Secondly, the intermediate point cloud was constructed in a coarse-to-fine way by graph convolutional module, which was composed of three graph convolutional network (GCN) [9] blocks with graph unpooling layer. Following Pixel2Mesh [17], mesh loss was applied to constrain the shape during mesh deformation, as shown in Eq. (2):

$$\min_{\theta_m} \sum_{i=1}^{n} L_{mesh}\left(f_m\left(x_m, f_{u-d3};\theta_m\right),y_{m_i}\right) \tag{2}$$

where f_m represents graph convolutional module and θ_m represents its parameters. x_m is the initial ellipsoid mesh input with 162 vertices and f_{u-d3} is the extracted features (last three layers) from the decoder of segmentation module. y_{mi} is the deformed mesh with 2562 vertices.

Stage Two (Hemodynamic Prediction): The whole point cloud of coronary tree was merged from the patch data and fed into point cloud module. During training and testing phase, input points were sampled to a fixed size (e.g. 20000 points) for our private dataset. To enhance the capability of our model at stenosis region, features in the last decoder layer of segmentation module were extracted and projected to jointly guide the hemodynamic prediction process. More specifically, we use global average pooling to extract global features and then project them to features with a size of $B \times N \times M$, where B is the batch size, N is the number of points generated by GCN, and M is the dimension of the feature space. In addition, we also included physiological conditions (pressure and cardiac output) as additional features. As a result, the dimension of input point cloud will be expanded to $B \times N \times (3 + M + 2)$. Mean absolute error (mae) was employed as the loss function (L_{mae}) for training point cloud module, as shown in Eq. (3):

$$\min_{\theta_p} \sum_{i=1}^{n} L_{mae}\left(f_p\left(x_{p_i}, f_{u-d1}, h_t;\theta_p\right),y_{p_i}\right) \tag{3}$$

where f_p represents point cloud module and θ_p represents its parameters. f_{u-d1} is the extracted features (last layer) from segmentation module decoder. h_t is the extra physiological input and y_{p_i} is hemodynamic parameters calculated by CFD.

Loss and Training Strategy: The overall loss function for our proposed framework is shown in Eq. (4):

$$\lambda_1 L_{ce} + \lambda_2 L_{dice} + \lambda_3 L_{mesh} + \lambda_4 L_{mae} \tag{4}$$

where $\lambda_1, \lambda_2, \lambda_3, \lambda_4$ are hyper parameters to adjust the weight between losses. In our experiments, $\lambda_1, \lambda_2, \lambda_3$ are set to 0.5 and λ_4 is set to 1.0.

To optimize the framework from end to end, the initialized whole point cloud for each sample was constructed by the pretrained first-stage modules. In training phase, whole point cloud was updated by patch-result at each iteration, and all three modules were jointly optimized. In testing phase, normalized mean absolute error (NMAE) was used as evaluation metric, as shown in Eq. (5):

$$\text{NMAE} = \frac{1}{n} \frac{\sum_{i=1}^{n} |p_i - \hat{p}_i|}{\max |p| - \min |p|} \tag{5}$$

where p_i and $\hat{p}_i$ represent CFD and our method result respectively.

For model training, we first trained the first-stage modules independently for 100 epochs to obtain initialized point cloud representations for each sample. Subsequently, we conducted end-to-end joint training of the two modules for another 100 epochs. We used the Adam optimizer with an initial learning rate of 10^{-3} and a batch size of 16. Our approach was developed using the PyTorch framework and trained across 4 NVIDIA GeForce GTX 1080 Ti GPUs. The total training time required was approximately 26 h to complete. When evaluating the trained model on real-world data, end-to-end inference time took less than 20 s using a single GPU.

3 Experiments

3.1 Datasets

To validate the effectiveness of our method, we conducted training and validation on two different datasets, an ideal synthetic dataset and a real private dataset.

Ideal Synthetic Dataset: The idealized LAD vessel models were generated with a unified length of 2 cm, a reference diameter of 0.3 cm and a single stenosis. The location of stenosis beginning varied from 0.5 cm to 1.5 cm, with the length of the stenosis region ranging from 0.2 cm to 0.4 cm. The degree of stenosis ($DS = (1 - r_{sten}/r) \times 100\%$) varied from 50% to 70%, where r_{sten} is the minimum radius in the stenosis region, and r is the unified radius of the idealized vessel.

Real Dataset: A total of 150 CCTA images were further collected from one clinical institution retrospectively. The average in-plane resolution and slice thickness are 0.038 cm and 0.047 cm, respectively. To address the imbalance between positive and negative samples, we introduced synthetic stenosis randomly at the

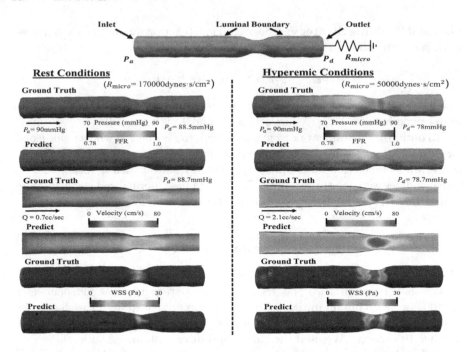

Fig. 2. The ground truth (CFD) and model predicted hemodynamics of blood flow through an idealized LAD model with a reference diameter of 0.3 cm and a 50% diameter stenosis. Rest and hyperemic myocardial blood flow are respectively simulated by adjusting the downstream microcirculatory resistance. The velocity is set at zero along the luminal boundary.

three main branches: the left anterior descending artery (LAD), left circumflex artery (LCx), and right coronary artery (RCA). The severity of stenosis was sampled between zero and two. This approach aimed to create a more balanced dataset by generating stenosis in cases where it was originally absent, considering that the majority of cases did not initially have stenosis. A total number of 900 CCTA images and masks were produced and named STENOSIS-900, where 720 cases generated from the original 120 data were used for training, and 180 cases generated from the other original 30 data were used for testing.

Hemodynamic Dataset: The hemodynamic dataset was produced by the traditional automated pipeline of image-based CFD simulation consisting of the following steps: 1) CCTA images were firstly segmented to reconstruct geometrical models of coronary arteries. Subsequently, the fluid domain of the geometrical model was divided using tetrahedral elements, followed by a mesh refinement with prism layers. Mesh sensitivity studies were further conducted and verified that the adopted mesh density was sufficient to yield numerically acceptable results; 2) All simulations were herein conducted with the OpenFOAM package where the Navier-Stokes (NS) equations were discretized and solved with the

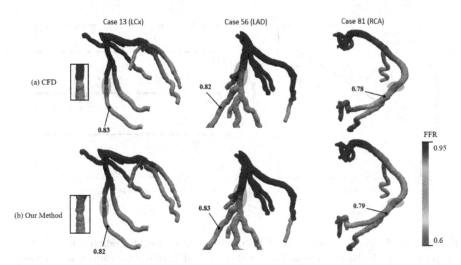

Fig. 3. FFR distributions of coronary artery tree in three representative examples utilizing CFD (a) and our deep learning method (b). The specific FFR values are recorded at the 2 cm downstream of stenosis locations marked by red circles. (Color figure online)

finite volume method. Gauss upwind was employed for the spatial discretization. Coronary artery flow was assumed to be an incompressible fluid with a density of 1060 kg/m^3 governed by the unsteady three-dimensional NS equations. The typical Carreau model was employed to calculate the blood viscosity. Additionally, arterial walls were assumed rigid where the non-slip boundary conditions were imposed. Resistance outflow boundary was applied by placing a resistance distal to each outlet to simulate the physiological flow division [19].

3.2   Results

An idealized LAD model with a 50% diameter stenosis is herein employed to evaluate the model performance at rest and hyperemic conditions (see Fig. 2). Hyperemic conditions are assumed to assess model predictions under myocardial stress similar to the environment of clinical measurement. The constant pressure ($P_a = 90$ mmHg) is applied at the inlet boundary, while the downstream microcirculatory resistance ($R_{micro} = 170000$ or 50000 dynes·s/cm^2) is prescribed at the outlet boundary to simulate the rest and hyperemic conditions, respectively. Compared with the CFD results, our model predicts consistent distributions of all hemodynamic metrics (e.g. FFR, pressure drop, velocity and WSS) at rest conditions. Besides, the hemodynamic alterations (e.g. larger pressure drop/velocity/WSS across the stenosis region and smaller downstream FFRs) are accurately predicted when switching rest conditions to hyperemic myocardial blood flow conditions. FFR is a pressure-based metric used in cardiology, which is calculated as a ratio of mean blood pressure at the corresponding

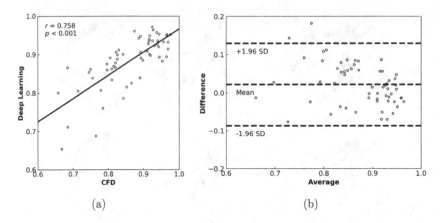

Fig. 4. Correlation (a) and Bland-Altman (b) plots of FFRs predicted by CFD vs. Deep Learning. FFRs are recorded at locations about 2 cm downstream from stenoses. All results are from STENOSIS-900 test set.

location and the mean aortic pressure under hyperemic conditions. Our model quantitatively achieves downstream FFR of 0.874 at hyperemia conditions, while CFD derives a value of 0.867.

For the real dataset, Fig. 3 shows the CFD and model-predicted FFR results of three patient-specific cases, where the stenosis is located at the three branches (i.e. LAD, LCx, RCA) respectively. Our predicted FFR distributions agree well with the CFD results throughout the coronary tree for all three cases. Besides, the specific model-derived FFR value downstream of the stenosis is quite close to the CFD-derived result. Clinicians usually adopt a fixed cut-off value (e.g. FFR < 0.8) to identify patients with myocardial ischemia. With the proposed method, our model successfully identifies the non-ischemic patients (case 13 and case 56) and ischemic patients (case 81) although all three patients suffered moderate coronary stenosis. RCA stenosis is responsible for myocardial ischemia based on the predicted FFRs for case 81, which agrees with the clinical descriptions collected. Quantitatively, NMAE is used as the evaluation metric. For the idealized dataset, the NMAE is found to be 0.005 ± 0.004, and for the real dataset, it is 0.025 ± 0.022. Specifically, the NMAE for the LAD, LCx, and RCA branches is 0.027 ± 0.019, 0.025 ± 0.018, and 0.021 ± 0.020, respectively.

Furthermore, we find a significant correlation between the estimations of CFD-derived FFRs and their counterparts predicted with our method ($r = 0.758$, $p < 0.001$). The FFR varies in a small range around a mean value of 0.021 as shown in the Bland-Altman plot (see Fig. 4), which shows good consistency between our Deep Learning approach and CFD results.

4 Conclusion

In this work, we presented an end-to-end deep learning approach to predict coronary artery hemodynamic indicators (e.g. velocity, pressure, WSS and FFR). By incorporating image features and physiological parameters, our method demonstrated the capability to predict hemodynamics under various physiological conditions and yielded fast and accurate results. In the future, our method could be generalized to estimate the hemodynamic metrics of other vascular-related diseases, which may contribute to identifying patients at high risk of cardiovascular disease and patient-specific treatment. We will introduce more datasets and do more extensive experiments and parameter adjustments to verify the robustness of our method.

References

1. Bakhshi, H., et al.: Comparative effectiveness of CT-derived atherosclerotic plaque metrics for predicting myocardial ischemia. JACC: Cardiovasc. Imaging **12**(7 Part 2), 1367–1376 (2019)
2. Colebank, M.J., et al.: Influence of image segmentation on one-dimensional fluid dynamics predictions in the mouse pulmonary arteries. J. R. Soc. Interface **16**(159), 20190284 (2019)
3. Evans, P.C., Kwak, B.R.: Biomechanical factors in cardiovascular disease. Cardiovasc. Res. **99**(2), 229–231 (2013)
4. Gharleghi, R., Samarasinghe, G., Sowmya, A., Beier, S.: Deep learning for time averaged wall shear stress prediction in left main coronary bifurcations. In: 2020 IEEE 17th International Symposium on Biomedical Imaging (ISBI), pp. 1–4. IEEE (2020)
5. Gheorghiade, M., Bonow, R.O.: Chronic heart failure in the United States: a manifestation of coronary artery disease. Circulation **97**(3), 282–289 (1998)
6. Goubergrits, L., et al.: The impact of MRI-based inflow for the hemodynamic evaluation of aortic coarctation. Ann. Biomed. Eng. **41**, 2575–2587 (2013)
7. Himburg, H.A., Dowd, S.E., Friedman, M.H.: Frequency-dependent response of the vascular endothelium to pulsatile shear stress. Am. J. Physiol.-Heart Circulatory Physiol. **293**(1), H645–H653 (2007)
8. Itu, L., et al.: A machine-learning approach for computation of fractional flow reserve from coronary computed tomography. J. Appl. Physiol. **121**(1), 42–52 (2016)
9. Kipf, T.N., Welling, M.: Semi-supervised classification with graph convolutional networks. arXiv preprint arXiv:1609.02907 (2016)
10. Lee, M.C.H., Petersen, K., Pawlowski, N., Glocker, B., Schaap, M.: Template transformer networks for image segmentation (2019)
11. Li, G., et al.: Prediction of 3D cardiovascular hemodynamics before and after coronary artery bypass surgery via deep learning. Commun. Biol. **4**(1), 99 (2021)
12. Morice, M.C., et al.: A randomized comparison of a sirolimus-eluting stent with a standard stent for coronary revascularization. N. Engl. J. Med. **346**(23), 1773–1780 (2002)
13. Qi, C.R., Su, H., Mo, K., Guibas, L.J.: PointNet: deep learning on point sets for 3D classification and segmentation. In: Proceedings of the IEEE Conference on Computer Vision and Pattern Recognition, pp. 652–660 (2017)

14. Sklet, V.: Exploring the capabilities of machine learning (ML) for 1D blood flow: application to coronary flow. Master's thesis, NTNU (2018)
15. Thondapu, V., Bourantas, C.V., Foin, N., Jang, I.K., Serruys, P.W., Barlis, P.: Biomechanical stress in coronary atherosclerosis: emerging insights from computational modelling. Eur. Heart J. **38**(2), 81–92 (2017)
16. Valen-Sendstad, K., et al.: Real-world variability in the prediction of intracranial aneurysm wall shear stress: the 2015 international aneurysm CFD challenge. Cardiovasc. Eng. Technol. **9**, 544–564 (2018)
17. Wang, N., Zhang, Y., Li, Z., Fu, Y., Liu, W., Jiang, Y.G.: Pixel2Mesh: generating 3D mesh models from single RGB images. In: Proceedings of the European Conference on Computer Vision (ECCV), pp. 52–67 (2018)
18. Windecker, S., et al.: 2014 ESC/EACTS guidelines on myocardial revascularization. Kardiologia Polska (Polish Heart J.) **72**(12), 1253–1379 (2014)
19. Xu, L., Liang, F., Gu, L., Liu, H.: Flow instability detected in ruptured versus unruptured cerebral aneurysms at the internal carotid artery. J. Biomech. **72**, 187–199 (2018)
20. Xu, L., Liang, F., Zhao, B., Wan, J., Liu, H.: Influence of aging-induced flow waveform variation on hemodynamics in aneurysms present at the internal carotid artery: a computational model-based study. Comput. Biol. Med. **101**, 51–60 (2018)
21. Yang, X., Xu, L., Yu, S., Xia, Q., Li, H., Zhang, S.: Segmentation and vascular vectorization for coronary artery by geometry-based cascaded neural network. arXiv preprint arXiv:2305.04208 (2023)
22. Zhang, X., et al.: Progressive deep segmentation of coronary artery via hierarchical topology learning. In: Wang, L., Dou, Q., Fletcher, P.T., Speidel, S., Li, S. (eds.) Medical Image Computing and Computer Assisted Intervention-MICCAI 2022: 25th International Conference, Singapore, 18–22 September 2022, Proceedings, Part V, pp. 391–400. Springer, Cham (2022). https://doi.org/10.1007/978-3-031-16443-9_38

Deep Learning for Automatic Strain Quantification in Arrhythmogenic Right Ventricular Cardiomyopathy

Laura Alvarez-Florez[1,2,3]($\boxtimes$), Jörg Sander[1,2], Mimount Bourfiss[4],
Fleur V. Y. Tjong[3,7], Birgitta K. Velthuis[5], and Ivana Išgum[1,2,6,7]

[1] Department of Biomedical Engineering and Physics, Amsterdam University
Medical Center, University of Amsterdam, Amsterdam, The Netherlands
l.alvarezflorez@amsterdamumc.nl
[2] Informatics Institute, University of Amsterdam, Amsterdam,
The Netherlands
[3] Heart Center, Department of Clinical and Experimental Cardiology,
Amsterdam University Medical Center, University of Amsterdam, Amsterdam,
The Netherlands
[4] Department of Cardiology, University Medical Center Utrecht, Utrecht,
The Netherlands
[5] Department of Radiology and Nuclear Medicine, University Medical Center
Utrecht, Utrecht, The Netherlands
[6] Department of Radiology and Nuclear Medicine, Amsterdam University Medical
Center, University of Amsterdam, Amsterdam, The Netherlands
[7] Amsterdam Cardiovascular Sciences, Amsterdam, The Netherlands

Abstract. Quantification of cardiac motion with cine Cardiac Magnetic
Resonance Imaging (CMRI) is an integral part of arrhythmogenic right
ventricular cardiomyopathy (ARVC) diagnosis. Yet, the expert evaluation of motion abnormalities with CMRI is a challenging task. To automatically assess cardiac motion, we register CMRIs from different time
points of the cardiac cycle using Implicit Neural Representations (INRs)
and perform a biomechanically informed regularization inspired by the
myocardial incompressibility assumption. To enhance the registration
performance, our method first rectifies the inter-slice misalignment inherent to CMRI by performing a rigid registration guided by the long-axis
views, and then increases the through-plane resolution using an unsupervised deep learning super-resolution approach. Finally, we propose to
synergically combine information from short-axis and 4-chamber long-
axis views, along with an initialization to incorporate information from
multiple cardiac time points. Thereafter, to quantify cardiac motion, we
calculate global and segmental strain over a cardiac cycle and compute
the peak strain. The evaluation of the method is performed on a dataset
of cine CMRI scans from 47 ARVC patients and 67 controls. Our results
show that inter-slice alignment and generation of super-resolved volumes
combined with joint analysis of the two cardiac views, notably improves
registration performance. Furthermore, the proposed initialization yields

---

L. Alvarez-Florez and J. Sander—These authors contributed equally to this work.

© The Author(s), under exclusive license to Springer Nature Switzerland AG 2024
O. Camara et al. (Eds.): STACOM 2023, LNCS 14507, pp. 25–34, 2024.
https://doi.org/10.1007/978-3-031-52448-6_3

more physiologically plausible registrations. The significant differences in the peak strain, discerned between the ARVC patients and healthy controls suggest that automated motion quantification methods may assist in diagnosis and provide further understanding of disease-specific alterations of cardiac motion.

Keywords: Implicit Neural Representations · Image Registration · Strain · Cardiac Motion · Arrhythmogenic Right Ventricular Cardiomyopathy

1   Introduction

Heart motion abnormalities serve as indicator of cardiac disease and its severity. For arrhythmogenic right ventricular cardiomyopathy (ARVC) patients, characterization of wall motion abnormalities is an integral part of diagnosis. In clinical practice, cardiac magnetic resonance imaging (CMRI) is considered the reference standard for the assessment of motion abnormalities [13]. The assessment typically relies on visual inspection by radiologists. This process is challenging, and therefore subjective and lacks a quantitative description [8]. Automated deep learning methods may offer accurate and reproducible quantification, and allow subsequent interpretation of cardiac motion.

Unlike other methods quantifying cardiac motion and strain such as CMRI tagging, motion quantification with cine CMRI is derived as a post-processing and does not require additional imaging or long acquisition and processing time [1]. Classic computation methods for motion quantification in cine CMRI include feature tracking and image registration [8]. More recent approaches used deep learning [5–7,9,14]. These methods compute the displacement fields by mapping image intensities or anatomical landmarks across different time points within the cardiac cycle. These methods have shown the ability to perform registration fast, making them appealing for clinical use [16]. Nevertheless, their performance does not necessarily outperform classical registration methods [17]. Registration using Implicit Neural Representations (INRs), a distinctive recently proposed method that employs a neural network to represent the implicit transformation function between two images, outperforms previous CNN-based methods and offers fast analysis [17]. Lopez et al. recently applied INR based networks to deformable image registration of CMR images [4]. However, CMRI registration is challenged by image acquisition throughout a cardiac cycle that leads to misalignment between consecutive slices and low through-plane resolution. In response to these limitations, recent studies incorporated preliminary alignment of images before computing anatomical displacements [11,14].

To quantify cardiac motion addressing these challenges, we propose a method that performs CMRI registration between different time points in the cardiac cycle and then computes the strain. First, we employ inter-slice alignment and unsupervised deep learning super-resolution to enhance the performance of implicit neural representations. Second, we include different cardiac views during registration. Third, our method utilizes transfer learning across different time points to incorporate temporal information into the initialization of INRs. Additionally, in line with earlier works [9], our registration method constrains the network to adhere to biomechanically informed rules by incorporating a regularization technique. Lastly, we compute the strain in 3D from the displacements leveraged from the proposed registration method. We evaluate the method with a set of ARVC patients and control subjects. The novelty of our work lies on the addition of super resolution along with preliminary slice alignment and the incorporation of multiple views into INRs. Moreover, we propose a unique way to transfer the registration transformation learned by INRs from one point in the cardiac cycle to the next.

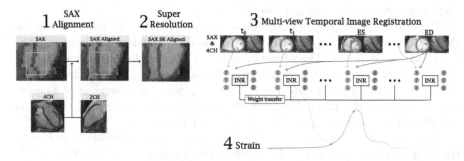

Fig. 1. Overview of the proposed method. First alignment and then super-resolution. The SAX and 4CH views are fed into the INR. The weights from each network are transferred to initialize the succeeding one. Lastly, the strain is computed.

2   Method

During cine CMRI acquisitions, each short axis (SAX) slice is typically captured during breath-hold. Variability in a patient's breath-hold relative to the imaging volume can result in misalignment of the SAX slices (Fig. 1). To address this and enhance registration performance, we first rigidly align slices in the through-plane direction of the SAX view. To guide the alignment, the 4-chamber (4CH) and 2-chamber (2CH) views are used as reference. In addition, we use segmentations of the left ventricle (LV) blood pool from the SAX, 2CH and 4CH views derived by a deep learning segmentation method [10]. Initially, the SAX, 2CH and 4CH views are transformed to the same coordinate space. To determine the rigid registration parameters, we optimize a translation matrix with learnable weights through an adaptive moment estimation (Adam) optimizer. To calculate the loss, first the SAX image is resampled using bilinear interpolation based on the alignment parameters. Then, we warp the aligned SAX image into each of the long-axis views (i.e. 4CH and 2CH). The computation of the loss that guides the learning process is composed of two components: the image loss and the segmentation loss. The image loss is calculated over the masked area of the myocardium. For this, the Normalized Cross Correlation loss (NCC) is combined with a differentiable version of Normalized Mutual Information (NMI) [15] loss. The NMI is then scaled by a factor to ensure both losses are of the same magnitude. The Dice loss is used to compute the segmentation loss. Both components of the loss (i.e. image and segmentation) are calculated for the 2CH view and for the 4CH view and combined into one component. The method is optimized for a total of 2,000 iterations with a learning rate of 0.01.

Given the highly anisotropic resolution of CMRI, increasing the through-plane resolution prior to registration of the cardiac volumes from the cine acquisition might lead to smoother displacement fields. Therefore, we employ an unsupervised deep learning super-resolution method [12] that generates intermediate slices as a linear combination of preceding and succeeding slices interpolated in the latent space. We up-sampled the volumes using a factor of 6, increasing the average number of slices from 15 to 85.

To perform registration of the different volumes from the cardiac cycle, we use INRs. INRs are trained using a subset of randomly selected coordinates from the fixed and moved images. The network employs a loss function that minimizes the discrepancy between the intensity of the warped moving and the fixed images. We use the architecture proposed by [17], the implementation consists of 3 layers, each with 256 hidden units. The network employs a periodic activation function and the Adam optimizer with a learning rate of 10^{-4}.

To improve the performance of INRs, we incorporate information from both the 4CH and SAX image coordinates into the implicit image registration. To combine the 4CH and SAX views during registration, the coordinates are transformed into a unified canonical space. In every iteration, we forward a batch of 10,000 coordinates from the SAX and another batch of 10,000 from the 4CH view. The loss is the combined NCC from both views, enabling the network to optimize both transformations simultaneously.

We introduce an adaptation to the Jacobian regularization of INRs [17]. Rather than applying a uniform regularization across the entire image, we incorporate a weighting parameter that assigns greater importance to the myocardium (MYO). The Jacobian regularization encourages the neural network to learn transformations that preserve local volumes, promoting smoother displacements that are more physiologically plausible and preventing folding. Our weighting strategy ensures that the regularization is focused on the foreground (i.e. MYO), while applying a more relaxed penalty to the background of the image. By enforcing this greater penalty on the foreground, this regularization endorses the myocardial incompressibility assumption. The overall loss $\mathcal{L}$ for our method is presented in Eq. 1.

$$\mathcal{L} = \mathcal{NCC}^{sax} + \mathcal{NCC}^{4ch} + \alpha_{fg} \cdot J_{fg}^{sax} + \alpha_{bg} \cdot J_{bg}^{sax} + \alpha_{fg} \cdot J_{fg}^{4ch} + \alpha_{bg} \cdot J_{bg}^{4ch} \quad (1)$$

where $\mathcal{NCC}^{sax}$ and $\mathcal{NCC}^{4ch}$ are the NCC losses for the SAX and 4CH. The J_{fg}^{sax}, J_{bg}^{sax}, J_{fg}^{4ch}, and J_{bg}^{4ch} components represent the foreground and background regularization terms for the SAX and 4CH views, and α_{fg} and α_{bg} are the weights applied to the regularization terms for the foreground and background.

INRs require optimizing a new network for each registration. However, we can exploit the information from the registration of cardiac volumes between earlier time points to the later time points of a cine acquisition. Hence, while requiring to train n separate networks for registration of n time points in the cardiac cycle, we make this process more efficient and leverage the information learned from one time point to the next. Specifically, given an initial time point t_0 and a final time point t_n, representing the start and end of the cardiac cycle at end-diastole (ED), we train a network for each image volume pair starting from $\{t_i, t_n\}$ where $i=0,...,n-1$. The optimized weights from each iteration are carried forward to initialize the succeeding network $\{t_{i+1}, t_n\}$. This approach ensures that the network starts by mapping small transformations. Hence, when the network is faced with the more intricate task of mapping from contraction to relaxation, it does not start anew but leverages a pre-established optimized setting. By employing an optimized setting as a foundation for each successive time point computation, our approach also aims to diminish the variability in initial network states and constrains the exploratory solution space.

Lastly, to analytically calculate the deformation gradient and subsequently the strain tensor, we leverage the automatic differentiation ability of deep learning frameworks. We compute the circumferential and longitudinal components of the strain. These components represent changes in wall thickness (radial strain), and circumferential length (circumferential strain). To facilitate an interpretation of the strain that is aligned with the intrinsic physiological shape of the LV, the resulting Cartesian-based strain tensor is transformed into a polar coordinate system. In contrast to the LV, which can be approximated as a circular or ellipsoid structure, the right ventricle (RV) exhibits a more complex and irregular shape. To account for this, we compute the outward-pointing normals of the RV contour, providing a unique local direction for each voxel. These normals define the radial direction, essentially pointing outwards from the center of the ventricle. The circumferential direction is determined as orthogonal to the radial direction within the contour plane. Figure 1 illustrates the proposed method.

Moving	Fixed	Warped	Displacement	Error

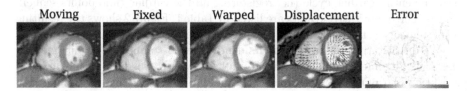

Fig. 2. Evaluation of the registration results. This results were achieved by our proposed method. The error represents the difference between the warped and fixed images.

3 Evaluation

3.1 Dataset

We evaluate our method on a dataset of conventional steady-state free precession sequences. The dataset was compiled from 47 ARVC patients and 67 control subjects. The data consists of SAX, 2CH and 4CH views captured during breath hold. Each sequence included 25 to 40 phases spanning one cardiac cycle, with a repetition time ranging from 2.6 to 3.4 ms. The end-systole (ES) and ED time points were identified by the expert radiology technicians as a part of clinical workup. Following this, segmentations for the LV blood pool, LV MYO and RV blood pool of the SAX, 2CH and 4CH slices were conducted automatically [10].

3.2   Evaluation Metrics

To evaluate the proposed registration method, we compare the segmentations between the fixed and warped images using overlap and distance metrics. The overlap between segmentations is measured with the Dice Similairty Coefficient (DSC), and the distance between the segmentations' boundaries with the Hausdorff Distance (HD). Furthermore, to evaluate the quality of the displacement fields, we calculate the determinant of the Jacobian matrix, which provides a measure of local volume changes. If the Jaccobian determinant is equal to 1, the transformation is volume-preserving. To discern the differences in strain between the ARVC patients and controls, we calculate the peak strain. This is defined as the maximum point of strain reached, representing the largest deformation experienced by the cardiac tissue. We employ a Kruskal-Wallis test to determine whether there was a statistically significant difference between the two groups. Significance was defined with a p-value lower than 0.05.

4   Experiments and Results

4.1   Evaluation of the Registration

We performed a quantitative and qualitative evaluation of the method. We limited the quantitative evaluation to a subset of 32 patients from the dataset, and registered ES to ED. The qualitative evaluation was performed on one subject over the entire cardiac cycle (i.e. registering all the cardiac time points sequentially). The quantitative results are listed in Table 1. Figure 2 shows the resulting displacement vector fields of the proposed method.

We quantitatively assessed the influence of each component in our method (i.e. slice alignment, super-resolution, multiple views, weighted Jacobian regularization and our proposed transfer learning initialization) by performing iterations with and without this steps. The results are listed in Table 1. The overall best method was achieved when including all the proposed components. The largest contribution to the DSC, for all cardiac structures (LV, MYO and RV), came from performing preliminary alignment and super-resolution along with our proposed initialization of weights. The initialization showed to have a significant positive impact in the DSC, particularly for the RV, in both the SAX and 4CH views. Including super-resolved volumes showed the largest reduction in the HD, which was further improved with the proposed initialization. When adding multiple views, there was a considerable increase in the DSC for the 4CH view. Incorporating the weighted strategy (with $\alpha_{fg} = 0.05$ and $\alpha_{bg} = 0.0001$) into the Jacobian

Table 1. Results showing the benefit of including different components from the method. INR represents the baseline registration method, weighted Jacobian (WJ) regularization, preliminary inter-slice alignment (AL), super-resolution (SR), transfer learning initialization (INIT), and multiple views (MV). The proposed method includes (INR+WJ+AL+SR+MV+INIT).

Experiment	HD	DICE SAX			|JACOBIAN| -1			DICE 4CH		
	AVG	LV	MYO	RV	LV	MYO	RV	LV	MYO	RV
Proposed	**10.33**	**0.92**	**0.79**	**0.87**	**0.04**	**0.04**	**0.05**	**0.92**	**0.85**	**0.90**
INR + WJ + AL + SR + MV	15.74	0.90	0.78	0.84	0.05	**0.04**	**0.05**	0.91	0.83	0.81
INR + WJ + AL + SR	14.01	0.91	0.78	0.84	0.05	**0.04**	**0.05**	0.90	0.76	0.79
INR + WJ + AL	23.16	0.88	0.71	0.80	**0.04**	**0.04**	**0.05**	0.88	0.72	0.77
INR + WJ	19.49	0.87	0.71	0.80	**0.04**	**0.04**	**0.05**	0.88	0.71	0.76
INR	19.46	0.81	0.70	0.76	0.17	0.18	0.22	0.79	0.71	0.71

regularization, compared to applying the regularization uniformly (with $\alpha = 0.05$) across the image, had the largest reduction in the Jacobian determinant.

For the qualitative evaluation, we registered each time point of the cine sequence to the ED, set as the fixed image. We compare our proposed initialization method, that transfers the optimization state from one time point to the next one, with the initialization from [17], which is a modified version of the Xavier initialization [2]. Figure 3 shows the impact of including the transfer learning strategy to initialize the proposed method: the proposed initialization leads to a more physiologically plausible registration (left); to visually compare, we computed the DSC scores for both strategies (right). The results show that the proposed method, compared to using a standard initialization, results in a better registration performance for larger deformations (i.e. ES to ED). This is most prominent for the RV, specially in the 4CH view.

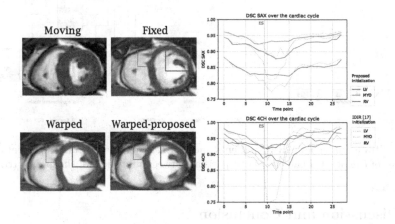

Fig. 3. Warped image obtained using the initialization from [17], and the one obtained with our proposed initialization (left). The boxes highlight regions from the image that are more challenging to register. Comparison of the DSC between the two initialization strategies for the same subject over all the cardiac time points (right).

4.2   Evaluation of the Strain

For the evaluation of the strain, we compute the radial and circumferential strain for the basal, mid and apical slices of each volume. The derived strain curves for each segment are presented in Fig. 4, left. When comparing the average strain curves between the two groups, a slight reduction in the RV radial and circumferential strain is present, specially on the basal and mid slices. This decrease is not so evident for the LV strain. The normalized peak strain for each group is detailed in Fig. 4, right. The strain results are analogous across the board, however, a significant difference ($p = 0.01$) in the radial peak strain of the RV was identified between the ARVC patients and the controls. Another weak significance was found for the radial LV strain.

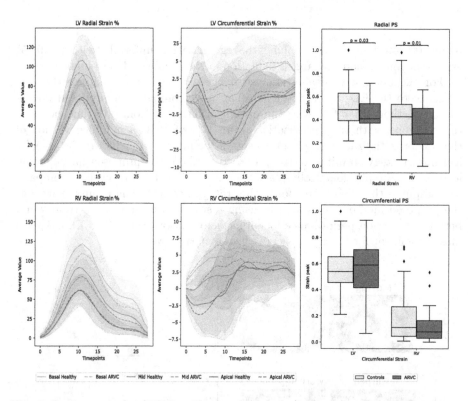

Fig. 4. Strain curves for ARVC patients and controls (left), each color represents one segment of the heart. Peak strain comparison between mid slices of the groups (right).

5   Discussion and Conclusion

We presented a method for temporal registration of cine CMRI that includes aligned super-resolved volumes, a biomechanically informed regularization and incorporates different cardiac views and time points into implicit neural networks. The method was evaluated on a dataset including ARVC patients and controls. The results showed that the method addressed the inherent limitations of cardiac imaging through preliminary slice alignment, super-resolution and inclusion of multiple views, leading to a substantial improvement in the performance of the registration method. Our proposed initialization based on transfer learning from previous time points yielded more accurate and physiologically plausible registrations. Furthermore, adopting a weighted Jacobian regularization resulted in more realistic volume-preserving transformations. For ARVC patients, a reduction in the strain, especially in the RV, is expected. This anticipated difference in strain was confirmed upon inspection of the radial and circumferential strain values for the basal, middle, and apical sections of the heart, which revealed differences in the peak radial strain between the ARVC patients and controls. Compared to the values reported in literature, our computed baseline radial strain was on average higher. This can be due to the computation of the strain in 3D instead of in 2D, as commonly done by clinical software [5]. This discrepancy is more pronounced for our method possibly due to the computation of these values in high-resolution super-resolved volumes. Additionally, our model encountered challenges when calculating the circumferential strain for the RV, particularly for the more basal slices. This difficulty might explain the lack of observable but expected difference in the circumferential strain between ARVC patients and controls [3]. A limitation of this study is the absence of comparison to established strain values obtained from clinically validated methods like tagged MRI or feature tracking. Another limitation, intrinsic to INRs, is the necessity to train a separate network for each patient and each of their respective image time pairs. However, we alleviate the process by introducing a transfer learning strategy that has the potential to speed up this process. As accuracy is enhanced by sharing weights from one time point registration to the next, fewer epochs may be sufficient to achieve accurate registration. This will be further explored and evaluated in our future research. While registration methods show promise for providing new insights in the quantification of motion abnormalities, further research evaluating clinical value of the approach is warranted.

References

1. Bucius, P., et al.: Comparison of feature tracking, fast-SENC, and myocardial tagging for global and segmental left ventricular strain. ESC Heart Failure **7**(2), 523–532 (2020)
2. Glorot, X., Bengio, Y.: Understanding the difficulty of training deep feedforward neural networks. In: Proceedings of the Thirteenth International Conference on Artificial Intelligence and Statistics, pp. 249–256. JMLR Workshop and Conference Proceedings (2010)
3. Heermann, P., et al.: Biventricular myocardial strain analysis in patients with arrhythmogenic right ventricular cardiomyopathy (ARVC) using cardiovascular magnetic resonance feature tracking. J. Cardiovasc. Magn. Reson. **16**(1), 1–13 (2014). https://doi.org/10.1186/s12968-014-0075-z

4. López, P.A., Mella, H., Uribe, S., Hurtado, D.E., Costabal, F.S.: WarpPINN: cine-MR image registration with physics-informed neural networks. Med. Image Anal., 102925 (2023)
5. Meng, Q., et al.: MulViMotion: shape-aware 3D myocardial motion tracking from multi-view cardiac MRI. IEEE Trans. Med. Imaging **41**(8), 1961–1974 (2022). https://doi.org/10.1109/tmi.2022.3154599
6. Morales, M.A., et al.: DeepStrain: a deep learning workflow for the automated characterization of cardiac mechanics. Front. Cardiovasc. Med. **8**, 730316 (2021)
7. Puyol-Antón, E., et al.: Fully automated myocardial strain estimation from cine MRI using convolutional neural networks. In: 2018 IEEE 15th International Symposium on Biomedical Imaging (ISBI 2018), pp. 1139–1143 (2018). https://doi.org/10.1109/ISBI.2018.8363772
8. Qiao, M., Wang, Y., Guo, Y., Huang, L., Xia, L., Tao, Q.: Temporally coherent cardiac motion tracking from cine MRI: traditional registration method and modern CNN method. Med. Phys. **47**(9), 4189–4198 (2020)
9. Qin, C., Wang, S., Chen, C., Bai, W., Rueckert, D.: Generative myocardial motion tracking via latent space exploration with biomechanics-informed prior. Med. Image Anal. **83**, 102682 (2023). https://doi.org/10.1016/j.media.2022.102682
10. Sander, J., de Vos, B.D., Išgum, I.: Automatic segmentation with detection of local segmentation failures in cardiac MRI. Sci. Rep. **10**(1), 21769 (2020)
11. Sander, J., de Vos, B.D., Bruns, S., Planken, N., Viergever, M.A., Leiner, T., Išgum, I.: Reconstruction and completion of high-resolution 3D cardiac shapes using anisotropic CMRI segmentations and continuous implicit neural representations. Comput. Biol. Med., 107266 (2023). https://doi.org/10.1016/j.compbiomed.2023.107266
12. Sander, J., Vos, B.D.D., Išgum, I.: Autoencoding low-resolution MRI for semantically smooth interpolation of anisotropic MRI. Med. Image Anal. **78**, 102393 (2022). https://doi.org/10.1016/j.media.2022.102393
13. Scatteia, A., Baritussio, A., Bucciarelli-Ducci, C.: Strain imaging using cardiac magnetic resonance. Heart Fail. Rev. **22**, 465–476 (2017)
14. Upendra, R.R., et al.: Motion extraction of the right ventricle from 4D cardiac cine MRI using a deep learning-based deformable registration framework. In: 2021 43rd Annual International Conference of the IEEE Engineering in Medicine & Biology Society (EMBC), pp. 3795–3799 (2021). https://doi.org/10.1109/embc46164.2021.9630586
15. de Vos, B.D., van der Velden, B.H., Sander, J., Gilhuijs, K.G., Staring, M., Išgum, I.: Mutual information for unsupervised deep learning image registration. Med. Imaging 2020: Image Process. **11313**, 155–161. SPIE (2020)
16. Wang, J., Zhang, M.: DeepFLASH: an efficient network for learning-based medical image registration. In: Proceedings of the IEEE/CVF Conference on Computer Vision and Pattern Recognition, pp. 4444–4452 (2020)
17. Wolterink, J.M., Zwienenberg, J.C., Brune, C.: Implicit neural representations for deformable image registration. In: Konukoglu, E., Menze, B., Venkataraman, A., Baumgartner, C., Dou, Q., Albarqouni, S. (eds.) Proceedings of The 5th International Conference on Medical Imaging with Deep Learning. Proceedings of Machine Learning Research, vol. 172, pp. 1349–1359. PMLR (2022)

Patient Stratification Based on Fast Simulation of Cardiac Electrophysiology on Digital Twins

Dolors Serra[1], Pau Romero[1], Miguel Lozano[1], Ignacio Garcia-Fernandez[1],
Diego Penela[3], Antonio Berruezo[3], Oscar Camara[2], Miguel Rodrigo[1],
Miriam Gil[1], and Rafael Sebastian[1(✉)]

[1] Computational Multiscale Simulation Lab (COMMLAB), Department of Computer
Science, Universitat de Valencia, Valencia, Spain
rafael.sebastian@uv.es
[2] Physense, BCN Medtech, Department of Information and Communication
Technologies, Universitat Pompeu Fabra, Barcelona, Spain
[3] Cardiology Department, Heart Institute, Teknon Medical Center, Barcelona, Spain

Abstract. After a myocardial infarction, electrophysiologists must assess the risk of the patient to develop a lethal arrhythmia. A cardiac magnetic resonance can help to evaluate the infarcted region, and provide insights into the electrically remodeled regions. However, it is not possible to evaluate the heart function or predict the behavior of slow conduction channels without an electrophysiology study. In this paper, we present a fully automatic screening approach based on fast simulation of cardiac electrophysiology, to determine arrhythmogeneity of myocardial infarction on patients. We show the accuracy and potential of the computer based approach, by comparing CathLab and simulation protocols to induce VT in 16 patients, obtaining a match between them.

Keywords: Slow conduction channel · cardiac electrophysiology simulation · therapy planning · ventricular tachycardia

1 Introduction

Patients who have experienced a heart attack are at an increased risk of developing ventricular tachycardia (VT). The scarred area of the heart creates an abnormal electrical environment that contributes to the initiation and persistence of VT episodes. Of particular importance are the slow conduction channels (SCC) that develop within the scarred tissue, further exacerbating the complexity of the electrical propagation on that region [5,11].

Catheter ablation is a frequently employed treatment for patients who have experienced VT episodes following a heart attack [4]. This procedure involves selectively ablating certain tissues to prevent them from triggering VT or maintaining the electrical circuits responsible for the arrhythmia. Electrophysiologists

© The Author(s), under exclusive license to Springer Nature Switzerland AG 2024
O. Camara et al. (Eds.): STACOM 2023, LNCS 14507, pp. 35–43, 2024.
https://doi.org/10.1007/978-3-031-52448-6_4

typically rely on electro-anatomical maps (EAMs) and contrast-enhanced magnetic resonance imaging (MRI) to evaluate the morphology and characteristics of the scarred region before performing ablation, as well as to assess VT inducibility.

However, it is important to note that catheter ablation is typically not performed on patients who have not exhibited any type of arrhythmia, even if they have experienced a myocardial infarction. However, there is growing recognition of the potential benefits of preventive ablation in high-risk patients, particularly those at risk of sudden cardiac death. Currently, non-invasive imaging studies utilizing delay enhancement magnetic resonance are the only clinical technique available to assess the risk of tachycardia [4].

Computer simulations offer an alternative approach to evaluate arrhythmic susceptibility and plan catheter ablation procedures. Typically, these simulations are based on detailed multiscale biophysical models, requiring expertise in computing and access to high-performance computing facilities [13, 14]. In this study, we demonstrate the potential of simplified electrophysiological simulations in assessing the risk of arrhythmia in a cohort of 16 patients. Our approach, implemented in the arrhythmic3D solver software [10], enables us to conduct thousands of simulations, considering numerous potential scenarios for each patient's digital twin. We outline the simulation procedure and present the results of arrhythmic susceptibility, which are then compared to the clinical data.

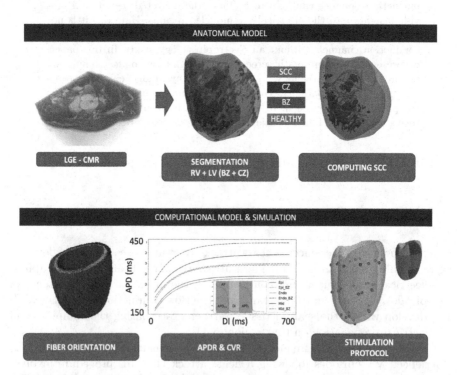

Fig. 1. Fully automatic pipeline to build the patient digital twin from de-MRI scans, and perform large simulation studies of arrhythmia inducibility.

2   Material and Methods

2.1   Imaging Data

We conducted a study involving 16 consecutive patients (10 male and 6 female) who had experienced a myocardial infarction, and were treated at the Teknon Clinical Center in Barcelona. In order to assess the extent of the infarcted region, all patients underwent late gadolinium enhancement LGE-MRI with a resolution between 0.7 mm and 1 mm in-plane, and 1.4 through. Among these patients, catheter studies were prescribed for those with varying degrees of scar weight. To analyze the inducibility of VT a detrimental pace-mapping protocol S1-S2 was used, with up to three S2 stimuli. If VT was confirmed, a catheter-based ablation procedure was performed.

There were important differences in several clinical indices obtained from the patients. Left ventricular mass ranged from 60 g to 206 g, while tissue at border zone (BZ) and core zone (CZ) mass ranged from 0.27 g to 65 g, and from 0.07 g to 40 g, respectively. For all those patients, we have a subset that have already presented clinical VTs (i.e. admitted for emergency), another group that was prescribed a catheter study (positive VTs, i.e., induced at cathlab) and a final group that have not shown any VT, or was negative at cathlab induction.

2.2   Construction of Cardiac Digital Twins

Figure 1 illustrates the automatic pipeline developed to build each digital twin. LGE-MRI datasets are segmented using Adas3D software. The software allowed us to extract the endocardial and epicardial surfaces of the left ventricle (LV), as well as the surfaces corresponding to the CZ and BZ. Following the volumentric hexahedral mesh is constructed and labeled with the tissue and cell properties. Finally, the stimulation protocol is configured for the case and simulations launched.

In addition to the surface information, we also obtained the potential slow conduction channels (SCCs) within the scar region. SCCs were defined as continuous pathways of viable tissue within the remodeled BZ that traverse the scar region, connecting healthy tissue from remote areas. Figure 2 presents two examples displaying the CZ (blue) and the extracted SCCs (depicted as magenta tubes). Both patients exhibited an extensive network of SCCs with numerous interconnections, indicating a complex structure.

Subsequently, we constructed a volumetric model using regular hexahedral elements based on the MRI voxelization, which had an isotropic resolution of approximately 1 mm. Each element was assigned the local myofibre orientation and a label to indicate healthy or scar tissue, as well as the specific regions such as endocardium, mid-myocardium, epicardium, BZ, and CZ. Additionally, the myocardial fiber orientation for each element was determined using a rule-based model [6]. Automatic labels were assigned to identify the 17 AHA (American Heart Association) regions, as well as the endocardium, epicardium, and mid-myocardium.

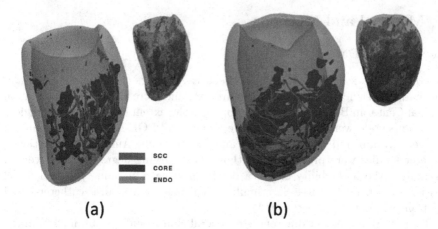

Fig. 2. Two digital twins, including the extraction of SCCs (purple tubes) that extend within the BZ surrounded by CZ (dark blue surface). (Color figure online)

2.3 Simulation Study

As we did not have data to personalize the electrophysiological properties of the digital twins, we opted to define a wide range of scenarios that encompassed the physiological properties of the most important variables. The base cell model used in the solver was the ionic ten Tusscher cell model, with six variations accounting for the endocardial, mid-myocardial, and epicardial layers in both healthy tissue and the border zone as in [8]. From this base model, we further introduced variations in the action potential duration (APD) by ±10% for each cell type. The conduction velocity (CV) properties of healthy tissue and the border zone were also varied by ±10%. Additionally, the degree of coupling between neighboring cells was adjusted within the range of 70% to 85%.

Furthermore, we replicated the clinical protocol known as pace-mapping in the simulator to elicit arrhythmia. Taking advantage of the simulation capabilities, virtual catheter stimulation was performed at 34 locations, corresponding to the 17 American Heart Association (AHA) segments at the endocardium and epicardium. The stimulation followed an S1-S2 protocol, with pacing starting from 305 ms and gradually decreasing to 270 ms in intervals of 5 ms for the S2 stimulus. The procedure involved delivering a train of six S1 stimuli with a constant basic cycle length (BCL) of 600 ms, followed by S2 stimuli (ranging from two to four) with a shorter BCL, similar to the protocol used in the EP-Lab (Electrophysiology Laboratory). The combination of all the considered variables resulted in 3000–4000 simulations per case. It is important to note that many configurations could not initiate electrical activation. For example, if the virtual catheter is placed on a segment primarily composed of CZ tissue or if the S2 frequency is too high and the tissue was still in the refractory period.

In summary, the simulation encompassed various scenarios with diverse electrophysiological properties, allowing us to explore different pacing protocols and analyze the responses to stimulation, as performed in the EP-Lab setting.

3 Results

After constructing the 16 digital twins, it was observed that most patients exhibited potential SCCs (see Fig. 2) that could sustain VT if triggered by an ectopic focus. However, confirming arrhythmogeneity based solely on imaging data was challenging due to the complexity of the scar and SCCs. Although patients with large scar regions tend to be more proarrhythmic, we had several patient with small BZ mass that also show VT episodes. We performed an statistical analysis to determine if there was correlation between patients that show large BZ or CZ, with the appearance of VT, but the results was negative.

In the simulation study, for each successful activation, the system checked whether a sustained VT persisted for at least 15 cycles. Reentry patterns occurred due to the heterogeneity of action potential duration (APD) in scar regions, leading to conduction blocks when the S2 frequency reached a certain threshold. It is worth noting that in many cases, the difference between tested configurations was the frequency of S2 stimuli, producing the same reentry pattern. Therefore, the focus was placed on studying the location and number of different VT onset sites within the scar region. Figure 3 displays the location of the onset sites (depicted as green squares) for two cases. It can be observed that these onset sites are clustered around a few regions (represented by red ellipsoids), row labeled 'E', for the number of different American Heart Association (AHA) segment onset sites found in each case). While the depolarization wavefront could reach the scar region from different sides depending on the location of the virtual catheter (indicated by blue squares in Fig. 3), the exit sites were observed only in a few hot-spots (red ellipsoids). Therefore, the exit sites were not uniformly distributed around the scar region and depended on the internal arrangement of the tissue.

The comparison between the exit sites obtained in the simulations and the proximity of segmented SCCs revealed a visual correlation, as expected. The simulations provided the advantage of predicting which SCCs were likely supporting the VT and therefore should be targeted for ablation.

It is worth noting that for all patients in whom the electrophysiologist successfully triggered a VT, different types of reentry patterns were consistently observed in the solver. For cases where clinical catheter stimulation was not successful, the patients were divided into two groups: i) no VT induced by either catheter or simulations and ii) (no VT induced by catheter, but sustained reentries obtained by simulations). It is important to consider that the electrophysiologists only stimulated the ventricles from the right ventricular (RV) septal wall, which limited their ability to cover a larger area, whereas the solver utilized 34 source locations.

In the majority of cases, VT was successfully triggered from the peri-infarct zone rather than remote regions. Cases such as P8 and P10 exhibited a limited number of reentries with two potential exit sites. In such cases, it is crucial to review the electrophysiology parameters used in the simulations to assess whether they fall within the physiological ranges, as a lower risk of reentry may be expected. On the other hand, cases P8 to P14, where VT was induced in the EP-Lab, demonstrated a large number of potential configurations that resulted in reentry in Arrhythmic3D.

The findings highlight the value of simulations in predicting and understanding the mechanisms underlying VT, particularly in cases where clinical catheter stimulation alone may not capture the full range of possibilities due to spatial limitations.

Table 1. Results on arrhythmic susceptibility per patient. See Table 1 caption; N: number of reentries by simulation; E: Number of different AHA Segment Exit sites found; VT C: clinical VT; VT I: catheter based induced VT

	P1	P2	P3	P4	P5	P6	P7	P8	P9	P10	P11	P12	P13	P14	P15	P16
N	0	22	23	0	7	13	26	5	0	6	25	19	0	29	0	36
E	0	3	4	0	2	5	3	2	0	2	5	4	0	1	0	5
VT C	0	1	0	0	1	0	0	0	0	0	0	0	0	0	0	0
VT I	0	1	1	0	0	0	0	0	0	0	1	0	0	0	0	1

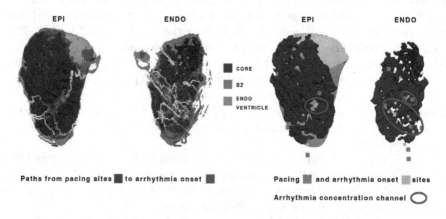

Fig. 3. Exemplary results on arrhythmic susceptibility on a patient. The blue squares indicate the pacing locations from which we could induce a VT. Green squares are the exit sites from the scar that initiated the VT. Colorful lines show the paths followed by the VT circuits. Red ellipsoids remark that exit sites are usually clustered on a few hot-spots. (Color figure online)

4   Discussion

In this study, we have shown the feasibility of using fast simplified models to simulate on infarcted digital twins a wide range of physiological scenarios to assess the risk of developing a VT. Each simulation (one scenario) of up to 5 s, could be solved in 5 min. We can analyze more than 3000 scenarios for each digital twin thanks to the efficiency of the solver arrhtythmic3D (described in [10]), considering differences in patient electrophysiology, and with a minimal user interaction, as opposed to detailed biophysical solvers. Then, as a result, we provide the scenarios that could trigger a reentry on the patient, together with a summary of VT exit sites. An alternative is to personalize the model, but it is necessary to collect patient-specific electrical data, such as EAMs [3] or body surface potential maps, and estimate specific variables, which are not commonly performed in the clinics [12].

The development of patient-specific ventricular models to analyze arrhythmia inducibility have shown great promise for patient stratification and therapy planning [1,13]. One of the differences of detailed patient-specific models compared to our approach is that they are very demanding in terms of computational resources and require the construction of complex digital twins. Those models take into account the underlying subcellular processes that are not necessary in many cases to assess the risk of reentry formation, where population based parameter are sufficient. As a consequence, only a few scenarios per patient are considered on those detailed biophysical based studies, not exploring all the scenarios that could trigger an arrhythmia in the patient. Recently, precision medicine has evolved to combine simulations and AI technology to help in patient's stratification [2,9]. This approach reduces the times required to obtain the patient's risk, but depends on the data used to train the model, and could be difficult to generalize to a broader population, where each patient shows its particular scar morphology and electrophysiological properties. However, they are still a great tool to augment the data population to improve models, since they only have to be run once [7].

In our study involving 16 patients, we did not observe a correlation between the mass of the scar border zone (BZ), core zone (CZ), or segmented continuous paths of viable tissue (SCCs) and the probability of triggering a VT in the computational model. Despite the presence of SCCs crossing the CZ in all patients, their complexity made it difficult to visually determine which SCCs would sustain a VT. Instead, it appears that the morphology of the BZ/CZ and the location of the ectopic trigger may be more crucial factors to consider. By utilizing our solver, we were able to test all critical reentry paths while considering a range of properties for healthy and remodeled tissue, accounting for the unknown specific variables of each patient.

However, it is important to acknowledge some limitations of the model. One limitation is the absence of the Purkinje system, which could potentially be the source of arrhythmia or play a role in sustaining it. Therefore, our model cannot simulate Purkinje-mediated VTs. Additionally, the model does not take into account specific mutations or the effects of drugs on myocyte behavior, which could be relevant for certain pathologies. For these cases, it would be necessary to simulate the corresponding pathological action potentials and train

Arrhythmic3D with the resulting solutions to incorporate the effects of specific mutations or drugs.

Despite these limitations, our study provides valuable insights into the use of computational modeling to assess the risk of VT in patients with myocardial infarction. By considering a wide range of scenarios and incorporating patient-specific factors, such as scar morphology and ectopic trigger location, our approach offers a comprehensive evaluation of arrhythmia susceptibility.

5   Conclusions

The combination of fast electrophysiology solvers developed and optimized for analyzing complex infarct regions, along with advanced imaging, simulation, and AI tools, holds significant promise for the analysis and stratification of patients at risk of developing arrhythmias. These tools play a crucial role in accurately assessing the risk of arrhythmias, preventing sudden cardiac death, and optimizing the planning of catheter ablation interventions. By utilizing precision medicine approaches, clinicians can adopt a more personalized and targeted approach to patient care, leveraging advanced imaging techniques like LGE-MRI, computational modeling, and AI technology to gain a comprehensive understanding of cardiac structure, electrical properties, and arrhythmia mechanisms. This integrated approach enables the identification of critical factors contributing to arrhythmia susceptibility and facilitates the development of patient-specific treatment strategies. As these tools continue to evolve, they have the potential to revolutionize the field of arrhythmia management, allowing for optimized and personalized therapeutic interventions based on individual patient characteristics and pathology.

References

1. Arevalo, H.J., et al.: Arrhythmia risk stratification of patients after myocardial infarction using personalized heart models. Nat. Commun. **7**, 11437 (2016). https://doi.org/10.1038/ncomms11437
2. Aronis, K.N., et al.: Characterization of the electrophysiologic remodeling of patients with ischemic cardiomyopathy by clinical measurements and computer simulations coupled with machine learning. Front. Physiol. **12**, 684149 (2021). https://doi.org/10.3389/fphys.2021.684149
3. Barber, F., et al.: Estimation of personalized minimal Purkinje systems from human electro-anatomical maps. IEEE Trans. Med. Imaging **40**(8), 2182–2194 (2021)
4. Cronin, E.M., et al.: 2019 hrs/ehra/aphrs/lahrs expert consensus statement on catheter ablation of ventricular arrhythmias: executive summary. J. Arrhythm. **36**(1), 1–58 (2020). https://doi.org/10.1002/joa3.12264
5. Deng, D., Prakosa, A., Shade, J., Nikolov, P., Trayanova, N.A.: Characterizing conduction channels in postinfarction patients using a personalized virtual heart. Biophys. J . **117**(12), 2287–2294 (2019). https://doi.org/10.1016/j.bpj.2019.07.024

6. Doste, R., et al.: A rule-based method to model myocardial fiber orientation in cardiac biventricular geometries with outflow tracts. Int. J. Numer. Method Biomed. Eng. **35**(4), e3185 (2019). https://doi.org/10.1002/cnm.3185

7. Godoy, E.J., et al.: Atrial fibrosis hampers non-invasive localization of atrial ectopic foci from multi-electrode signals: a 3D simulation study. Front. Physiol. **9**, 404 (2018). https://doi.org/10.3389/fphys.2018.00404

8. Lopez-Perez, A., Sebastian, R., Izquierdo, M., Ruiz, R., Bishop, M., Ferrero, J.M.: Personalized cardiac computational models: from clinical data to simulation of infarct-related ventricular tachycardia. Front. Physiol. **10**, 580 (2019). https://doi.org/10.3389/fphys.2019.00580

9. Maleckar, M.M., et al.: Combined in-silico and machine learning approaches toward predicting arrhythmic risk in post-infarction patients. Front. Physiol. **12**, 745349 (2021). https://doi.org/10.3389/fphys.2021.745349

10. Serra, D., et al.: An automata-based cardiac electrophysiology simulator to assess arrhythmia inducibility. Mathematics **10**(8), 1293 (2022)

11. Soto-Iglesias, D., et al.: Cardiac magnetic resonance-guided ventricular tachycardia substrate ablation. JACC Clin. Electrophysiol. **6**(4), 436–447 (2020). https://doi.org/10.1016/j.jacep.2019.11.004

12. Sung, E., Etoz, S., Zhang, Y., Trayanova, N.A.: Whole-heart ventricular arrhythmia modeling moving forward: mechanistic insights and translational applications. Biophys. Rev. (Melville) **2**(3) (2021). https://doi.org/10.1063/5.0058050

13. Trayanova, N.A., Doshi, A.N., Prakosa, A.: How personalized heart modeling can help treatment of lethal arrhythmias: a focus on ventricular tachycardia ablation strategies in post-infarction patients. Wiley Interdiscip. Rev. Syst. Biol. Med. **12**(3), e1477 (2020). https://doi.org/10.1002/wsbm.1477

14. Zhou, S., et al.: Feasibility study shows concordance between image-based virtual-heart ablation targets and predicted ECG-based arrhythmia exit-sites. Pacing Clin. Electrophysiol. **44**(3), 432–441 (2021). https://doi.org/10.1111/pace.14181

Deep Conditional Shape Models for 3D Cardiac Image Segmentation

Athira J. Jacob[1,2]($\boxtimes$), Puneet Sharma[1], and Daniel Ruckert[2]

[1] Digital Technology and Innovation, Siemens Healthineers, Princeton, NJ, USA
athira.jacob@siemens-healthineers.com
[2] AI in Healthcare and Medicine, Klinikum rechts der Isar, Technical University of Munich, Munich, Germany

Abstract. Delineation of anatomical structures is often the first step of many medical image analysis workflows. While convolutional neural networks achieve high performance, these do not incorporate anatomical shape information. We introduce a novel segmentation algorithm that uses Deep Conditional Shape models (DCSMs) as a core component. Using deep implicit shape representations, the algorithm learns a modality-agnostic shape model that can generate the signed distance functions for any anatomy of interest. To fit the generated shape to the image, the shape model is conditioned on anatomic landmarks that can be automatically detected or provided by the user. Finally, we add a modality dependent, lightweight refinement network to capture any fine details not represented by the implicit function. The proposed DCSM framework is evaluated on the problem of cardiac left ventricle (LV) segmentation from multiple 3D modalities (contrast-enhanced CT, non-contrasted CT, 3D echocardiography-3DE). We demonstrate that the automatic DCSM outperforms the baseline for non-contrasted CT without the local refinement, and with the refinement for contrasted CT and 3DE, especially with significant improvement in the Hausdorff distance. The semi-automatic DCSM with user-input landmarks, while only trained on contrasted CT, achieves greater than 92% Dice for all modalities. Both automatic DCSM with refinement, and semi-automatic DCSM achieve equivalent or better performance compared to inter-user variability for these modalities.

Keywords: Implicit shape · neural shape representations · cardiac segmentation · multi-modality

1 Introduction

Delineation of anatomical structures is a fundamental task in medical image analyses and often forms the first step in many clinical quantification and diagnoses workflows. Deep learning (DL) has become the most widely used approach for cardiac image segmentation over the recent years [1]. While techniques such as fully convolutional networks demonstrate state-of-the-art segmentation with accuracy [2], they have some important weaknesses: First, unlike natural images, the domain of medical images face the issue of data and/or label scarcity [3]. Annotating medical images can be expensive and tedious,

O. Camara et al. (Eds.): STACOM 2023, LNCS 14507, pp. 44–54, 2024.
https://doi.org/10.1007/978-3-031-52448-6_5

often requiring trained clinical experts. In addition, it is difficult to generalize domain knowledge across different modalities, even though the underlying anatomy might be the same. Utilizing information across various modalities becomes even more critical when some modalities inherently have more or higher fidelity data.

To address these challenges, we propose the Deep Conditional Shape models (DCSMs), an intuitive approach to anatomical segmentation using shape priors (Fig. 1). Using deep implicit shape representations [4–7], a modality agnostic shape model is learnt that can represent the anatomy of interest via signed distance functions. To fit the generated shape to the image, the shape model is conditioned on anatomic landmarks that can be automatically generated or provided by the user. Finally, we add a modality dependent, lightweight refinement network to capture any fine details not represented by the implicit function. As a proof of concept, we evaluate DCSM on the problem of cardiac left ventricle (LV) segmentation from multiple 3D modalities (contrast-enhanced CT, non-contrast CT, 3D echocardiography – 3DE). The DCSM is trained on a large dataset of contrasted CT from 1474 patients and evaluated on all modalities. Compared to the baseline, the automatic DCSM obtains 3.6 to 28.9 mm decrease in average Hausdorff distances, and 0.2 to 7.4 point decrease in average relative volume errors, depending on the modality. In the case of non-contrast CT and 3DE, the semi-automatic DCSM obtains an increase of 13 and 6 points in average Dice scores and 35 and 9 mm decrease in average Hausdorff distances.

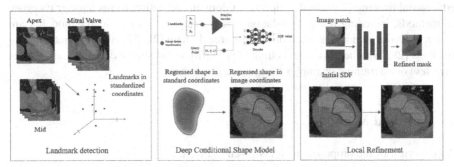

Fig. 1. Proposed method. 1) Landmark detection: 9 anatomically relevant reference points are detected from the image and converted to standard coordinate system. 2) Shape regression: A deep conditional shape model, trained on a dataset of left ventricle (LV) shapes, outputs a LV segmentation conditioned on the provided landmarks. 3) Local refinement: For modalities with image information, a lightweight image to image model is used to refine the initial output of the shape distance function (SDF) using local information

2   Related Work

Shape Priors in Medical Image Segmentation. Shape priors offer a way to integrate anatomical information into segmentation tasks. Shape priors have been traditionally used in segmentation through statistical shape models [8]. Recent methods in deep learning have tried to learn joint image and mask representations from data to encode the shape implicitly [9–10] through additional neural networks, cost functions, variational learning [11] or adversarial training [12]. These methods require training modality-specific neural networks, or only modelling the shapes implicitly. Graph based methods [13] have also been explored for medical image segmentation. While these model shapes explicitly, training such models often require dense meshes with point-to-point correspondences, and can only model fixed topologies.

Neural Shape Representations. Recently, neural implicit functions have been used extensively as shape representations [4, 5, 14, 15]. A neural network is fed with a latent code and a query point coordinate to predict a signed distance function or the binary occupancy at that location. Querying continuous points used in implicit function learning allows predicting in continuous space, creating robustness to a wide range of resolutions, and being more computationally efficient than voxel-based methods. It also allows for learning from partial data. Implicit shape functions have shown state of the art results in 3D shape generation [14], reconstruction [16, 17], hyper-resolution [18] etc. Raju et al. [19] use implicit shape functions in medical image delineation, in a two step manner, with shape positioning followed by shape regression, learnt in an end to end manner for a single modality.

Multi-modality Segmentation. Transfer learning has been used to learn from modalities with large, annotated datasets and improve performance in a target domain with more scarce data/annotations through fine-tuning [20, 21], multi-task learning [22, 23] and adversarial training [24, 25]. Unsupervised domain adaptation forms another category of methods, including image translation where images of a more scarce modality are generated from data and annotations in a source domain [26, 27] and few shot learning methods [28, 29] where a model is trained primarily in the source domain, and requires very few or no training examples from the target domain.

3   Proposed Method

An overview of the proposed model is given in Fig. 1.

3.1   Shape Regression: Conditional Implicit Shape Model

We learn a signed distance function (SDF), which is a continuous function that represents the distance of each spatial point to the closest surface. The surface of the shape is represented by the zero level-set of the neural net. To learn the positioning of the anatomy in 3D image space, we condition the model on input landmarks. Given a set of standardized landmark points $P = \{p_i\}_{i=1}^{N} \in \mathbb{R}^3$ with N points, and a query point $x \in \mathbb{R}^3$, we learn a conditional SDF ϕ such that, $SDF(x, P) = \phi(\gamma(x), \gamma(P))$, which gives the signed distance of point x from the nearest LV surface (+ for outside and − for inside the structure). Here, $\gamma(P)$ is a Fourier feature-based mapping [30], defined as $\gamma(P) = [\cos(2\pi BP), \sin(2\pi BP)]^T$, where each entry in $B \in \mathbb{R}^{m \times 3}$ is sampled from gaussian $\mathcal{N}(0, \sigma^2)$. The use of Fourier feature mapping facilitates enhanced learning of the high frequency information, as compared to coordinate-based approach. We perform this mapping for both x and P. The effect of number of input landmarks was studied and given in Appendix 1.1. The effect of the Fourier Mapping is shown in Appendix 1.2.

We train the continuous SDF model on a database of parametric LV meshes. We first preprocess the data to extract SDF samples for each training mesh. All training samples are standardized by affine alignment to a mean mesh Q calculated from the training data and centered in a unit cube. We sample 100,000 spatial points from the unit cube along with their SDF values, with around 80% of the points lying near the surface.

For our experiments, we use a modified PointNet with residual connections [31] as the encoder architecture. We use 5 down-sampling blocks, with both hidden dimension and latent dimension of 128. The decoder consists of 5 blocks, with each block consisting of 1D convolutions, batch norm and ReLU layers, with a hidden size of 256. We use L_1 loss function and ADAM optimizer, with a learning rate of $1e^{-4}$. We set $m = 32$ and $\sigma = 1$ for the Fourier feature mapping. We also train the shape model with random, normally distributed perturbations in the locations of the input landmarks to introduce robustness to landmark uncertainty.

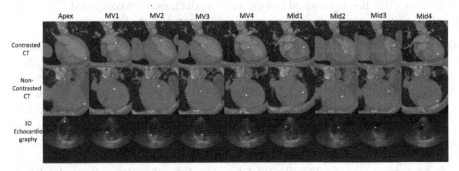

Fig. 2. Input cardiac landmarks. Each landmark is shown on one case from each modality. MV: Mitral Valve.

3.2 Cardiac Landmarks

The shape model is conditioned on nine anatomic landmarks: the apex, four on Mitral valve annulus (MV) and four at mid-level (Mid) of the LV (Fig. 2). For training, the input landmarks were extracted automatically from parametric ground truth (GT) LV meshes, where each point corresponded roughly to the same anatomic location. During inference, these landmarks can either be provided externally (such as by a user) for a semi-automatic workflow, or automatically detected on the modality. For the automatic pipeline, we train deep reinforcement learning (DRL) based landmark detectors [32] on all three modalities. Any missed landmarks are estimated from the mean mesh (Appendix 1.3).

3.3 Local Refinement

Using landmarks as the input to the shape model can result in overlooking of fine-grained details available in the image. For the contrast CT and echocardiography data, we thus add a local refinement step that takes the input image and initial output of the shape model as two separate channels, and outputs a final segmentation. This was trained with GT segmentations of the respective datasets. For both modalities, the refinement is done using $64 \times 64 \times 64$ patches, and uses a lightweight 3D UNet model with 3 downsampling blocks, each with a Conv3D, BatchNorm and Leaky RelU layers. The patches are selected to lie on the edges of the initial segmentation. The refinement is not performed on the non-contrast CT due to the lack of local image information.

3.4 Study Design

We train the shape model on LV landmarks and ground truth segmentations from contrasted CT scans, and evaluate on contrasted CT, non-contrasted CT and 3D TTE echocardiograms (3DE). It is to be noted that the same model trained on contrasted CT is used for all use cases without any further retraining. We compare the following:

a) Baseline: A 3D UNet is trained on each dataset as the baseline. The 3D UNet has 5 downsampling blocks, with increasing number of channels: 32, 64, 128, 256, 256). Each downsampling block has Conv3D, BatchNorm and Leaky RelU layers. Upsampling is done through nearest neighbor interpolation. Each network is trained for 300 epochs and best epoch is chosen using the validation dataset. We use Jaccard loss function and ADAM optimizer, with a learning rate of 0.001. All data is resampled to 1 mm resolution before training. Training is done patch-wise, due to GPU memory limitations. A patch size of $128 \times 128 \times 32$ is used for the CT modalities, with 128 in-plane patch size and 32 in the axial direction. Since the echocardiography data is isotropic in all direction, a patch size of $128 \times 128 \times 128$ is used. Random rotations and translations in 3D are used as data augmentation.

And DCSM based experiments:

b) Fully Automatic DCSM: The input landmarks are automatically found. Patients with missing landmarks are excluded.
c) Fully Automatic DCSM w. Estimation: Same as (b), but missing landmarks are estimated from the mean model and included.
d) Fully Automatic DCSM w. Estimation + Refinement: Same as (c), and with the refinement step (all except non-contrast CT).
e) Semi-automatic DCSM: We assume minimal error in the input landmarks, representing the upper limit of achievable accuracies. The input landmarks are extracted from the parametric GT meshes.

Reported metrics are Dice score as percentage, average surface distance, Hausdrorff distance and absolute and relative volume errors.

4 Data

Contrast-Enhanced CT. The data consists of gated coronary CT angiography (CCTA) scans from 1474 patients. The data was randomly split on the patient level, into training (1197), validation (149) and testing (128) sets, which was maintained for all models trained on the data. The data was anonymized according to HIPAA standards. The datasets were acquired using Siemens SOMATOM CT scanners (Force, Definition Flash, Definition AS+). GT was obtained by running a previously validated segmentation algorithm [33] and visual review of every case to ensure accuracy of contours. The GT meshes are parametrized by 545 vertices, and including mitral annulus and outflow track.

Non-contrast CT. The data consists of gated, non-contrast scans acquired for calcium scoring for the same cohort as before. Since annotating chambers in non-contrast scans is a difficult task with high inter user variability, the GT contours were obtained from the corresponding contrasted scans for each patient. GT contours registered to the non-contrast scans using header information from the scans, followed by visual review to ensure accuracy. Additionally, we only include patients with contrasted and non-contrasted data acquired within 20% of the cardiac phase from each other, to minimize inaccuracies due to phase differences. Following these criteria, we obtain a cohort of 715 patients with non-contrast data and GT, which were divided into 595, 58 and 62 patients for training, validation and testing respectively. The training and validation datasets were only used for developing the landmark detection model. The patient splits were chosen to align with the splits in contrasted CT data.

3D TTE Echocardiograms. The data consists of TTE 3DE images from an independent cohort of 1287 patients from multiple centers. The data was randomly split patient wise into training (1005), validation (106) and testing (176) sets. The data was collected using the Siemens Healthineers SC2000, and ED and ES frames were selected for annotation. Each frame is a grayscale 8-bit 3D image, with isotropic spatial resolution of 1mm, of size $256 \times 256 \times 256$. GT segmentations for the left ventricle were created by trained annotators. The annotator aligned the multiplanar view to obtain the true long axis of the LV in the A3C and A2C view. Landmarks were placed on the mitral annulus and apex. A mean LV mesh was positioned and adjusted manually where needed.

5 Results

Input Landmarks. We obtain an average error of 4 mm, 7 mm and 8 mm for the contrasted CT, non-contrast CT and 3DE datasets respectively. The errors for each landmark and number of missed landmarks in each dataset are given in Appendix 1.4. Apex and mitral valve landmarks have lower errors in general compared to the mid-level landmarks, as they are more anatomically distinct.

Segmentation. The results of the evaluation for each dataset are given in Table 1, 2 and 3. Inter-user variability for each modality was evaluated on a sub-set of the data by having two expert annotators manually delineate the LV. In case of non-contrast CT, we observe the best results using the fully automatic DCSM method. For the contrasted CT and 3DE dataset, the automatic DCSM followed by refinement outperforms the baseline in all metrics. As seen in Fig. 3, while the shape model gives reasonable outputs, the refinement step allows better attention to detail. In all modalities, semi-automatic DCSM shows good performance, despite being developed on only contrasted CT data.

Table 1. Results on testing set for contrasted CT. All metrics are expressed as mean ± standard deviation (SD)

Model	#Patients	Dice (%)	Average Surface Distance (mm)	Hausdorff Distance (mm)	Volume errors (mL)	Relative Volume Errors (%)
Inter-user variability	10	89.56 ± 4.0	1.98 ± 0.6	17.61 ± 3.6	13.26 ± 12.9	8.37 ± 8.1
Baseline	128	94.97 ± 9.6	0.86 ± 0.9	14.29 ± 16.4	4.75 ± 12.2	3.53 ± 9.7
DCSM - Automatic	125	88.12 ± 3.3	2.19 ± 0.5	17.73 ± 2.8	9.63 ± 13.0	5.57 ± 11.0
DCSM - Automatic – w. Estimation	128	87.90 ± 3.6	2.22 ± 0.5	17.83 ± 2.9	9.74 ± 13.3	5.62 ± 11.1
DCSM - Automatic – w. Estimation + Refinement	128	**96.25 ± 1.7**	**0.69 ± 0.2**	**10.67 ± 6.5**	**1.18 ± 4.2**	**1.84 ± 3.9**
DCSM - Semi-automatic	128	93.28 ± 2.2	1.28 ± 0.2	16.43 ± 3.1	2.53 ± 3.3	2.14 ± 3.3

Table 2. Results on testing set for non-contrasted CT. All metrics are expressed as mean ± SD

Model	#Patients	Dice (%)	Average Surface Distance (mm)	Hausdorff Distance (mm)	Volume Errors (mL)	Relative Volume Errors (%)
Inter-user Variability	12	77.26 ± 7.6	3.45 ± 1.4	12.75 ± 4.3	−22.58 ± 23.9	−29.8 ± 28.9
Baseline	62	80.76 ± 7.8	3.98 ± 2.2	39.49 ± 39.5	−7.30 ± 27.7	−9.43 ± 27.6
DCSM - Automatic	61	**81.02 ± 7.5**	**3.28 ± 1.2**	**10.56 ± 3.3**	2.23 ± 27.1	−2.10 ± 28.2
DCSM - Automatic – w. Estimation	62	80.86 ± 7.6	3.31 ± 1.2	10.60 ± 3.2	**2.20 ± 26.9**	**−2.06 ± 27.9**
DCSM - Semi-automatic	62	94.56 ± 2.1	0.92 ± 0.2	4.18 ± 1.0	−0.50 ± 3.3	−0.44 ± 3.7

Table 3. Results on testing set for 3DE. All metrics are expressed as mean ± SD

Model	#Patients	Dice (%)	Average Surface Distance (mm)	Hausdorff Distance (mm)	Volume Errors (mL)	Relative Volume Errors (%)
Inter-user variability	150	87.76 ± 5.7	2.44 ± 1.1	10.10 ± 4.3	−4.20 ± 26.7	−3.63 ± 16.3
Baseline	176	86.02 ± 5.4	2.49 ± 1.0	14.61 ± 11.1	2.51 ± 22.3	1.13 ± 20.4
DCSM - Automatic	161	83.46 ± 6.0	2.81 ± 1.1	9.81 ± 3.0	14.67 ± 20.8	11.10 ± 16.7
DCSM - Automatic – w. Estimation	176	82.80 ± 6.5	2.93 ± 1.2	10.12 ± 3.2	14.18 ± 21.6	10.82 ± 17.9
DCSM - Automatic – w. Estimation + Refinement	176	**87.42 ± 5.9**	**2.20 ± 1.0**	**9.62 ± 6.5**	**2.09 ± 19.8**	**0.91 ± 17.2**
DCSM - Semi-automatic	176	92.04 ± 1.4	1.39 ± 0.3	5.60 ± 1.1	−6.5 ± 6.5	−5.59 ± 4.8

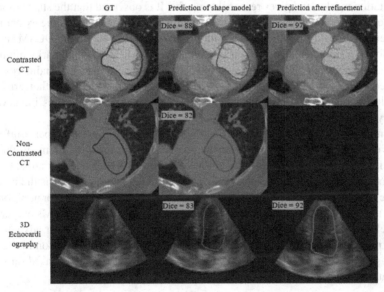

Fig. 3. GT and detections for some representative cases. Note: Refinement is not done for non-contrasted C

6 Discussion

In this work, we have presented Deep Conditional Shape models (DCSMs) for medical image segmentation. The DCSM uses continuous, implicit neural representations at its core to represent the anatomical shape. The shape model is modality agnostic and can be trained with any available dataset and re-used for other modalities without any further re-training. In this work we train the shape model only on contrasted CT data, due to its large availability and ease of annotation, and apply it on non-contrast CT and 3DE data. Using the shape model improves the largest segmentation errors, as observed from significant improvements in the Hausdorff distances.

The shape model uses anatomical landmarks to constrain the position of the anatomy in 3D space. Thus, the trained shape model can be applied to a new modality given the landmarks, which can be obtained from a user (semi-automatic), or fully automatically using trained landmark detectors. The semi-automatic method obtains high performance with average dice greater than 92% and average distance error of less than 1.5 mm for all modalities. In this scenario, the DCSM is only trained on contrasted CT data, yet shows very high accuracy in the other unseen modalities. This is representative of scenarios where a user gives the landmarks or corrects the automatically detected landmarks with high confidence. This is also useful in multi-modality scenarios where landmarks from a high-fidelity modality (e.g.: contrasted CT) can be transformed to a lower fidelity one (e.g.: non-contrasted CT) using scanner information.

We also train DRL based networks to detect the landmarks automatically since manually locating the landmarks might not be feasible or convenient in all scenarios. This only requires the annotation of landmarks on the target modality, as opposed to full segmentation masks, and gives reasonable masks. It is observed that the shape model is sensitive to errors in the input landmarks, which can be mitigated to some extent using input perturbations during the model training. However, the automatic DCSM method shows lower accuracies than the semi-automatic one. The DRL landmarks might not produce landmarks in cases with low confidence, and these missing landmarks were estimated using the mean shape model. This estimation introduces further errors, as shown in the results. More sophisticated estimation methods could thus improve the accuracy in cases with missing landmarks.

A limitation of the shape model is that the input landmarks provide sparse information, as opposed to the fine-grained details provided by an image. To remedy this, we add a refinement step for modalities with image information such as the contrast CT and 3D echocardiography. The automatic DCSM with refinement outperforms the baseline in all metrics in contrasted CT and 3DE. The refinement step is not done for non-contrast CT due to the lack of image information. In that case, the automatic DCSMs outperform the baseline without the refinement. Using the shape model gives a clear advantage in case of modalities with very little image information, such as the non-contrast CT. In other cases, the refinement step allows for local refinement of details based on image information.

Disclaimer: The concepts and information presented in this paper/presentation are based on research results that are not commercially available. Future commercial availability cannot be guaranteed.

References

1. Chen, C., et al.: Deep learning for cardiac image segmentation: a review. Front. Cardiovasc. Med. **7**, 25 (2020)

2. Yousef, R., Gupta, G., Yousef, N., Khari, M.: A holistic overview of deep learning approach in medical imaging. Multimed. Syst. **28**, 881–914 (2022)
3. Zhang, P., Zhong, Y., Deng, Y., Tang, X., Li, X.: A survey on deep learning of small sample in biomedical image analysis (2019)
4. Park, J.J., Florence, P., Straub, J., Newcombe, R., Lovegrove, S.: DeepSDF: learning continuous signed distance functions for shape representation (2019). ArXiv190105103 Cs
5. Mescheder, L., Oechsle, M., Niemeyer, M., Nowozin, S., Geiger, A.: Occupancy networks: learning 3D reconstruction in function space (2019). ArXiv181203828 Cs
6. Peng, S., Niemeyer, M., Mescheder, L., Pollefeys, M., Geiger, A.: Convolutional occupancy networks (2020). ArXiv200304618 Cs
7. Chou, G., Chugunov, I., Heide, F.: GenSDF: two-stage learning of generalizable signed distance functions (2022)
8. Heimann, T., Meinzer, H.-P.: Statistical shape models for 3D medical image segmentation: a review. Med. Image Anal. **13**, 543–563 (2009)
9. Oktay, O., Ferrante, E., Kamnitsas, K., et al.: Anatomically constrained neural networks (ACNNs): application to cardiac image enhancement and segmentation. IEEE Trans. Med. Imaging **37**, 384–395 (2018)
10. Li, S., Zhang, C., He, X.: Shape-aware semi-supervised 3D semantic segmentation for medical images. In: Martel, A.L., et al. (eds.) Medical Image Computing and Computer Assisted Intervention – MICCAI 2020. MICCAI 2020. LNCS, vol. 12261, pp 552–561. Springer, Cham (2020). https://doi.org/10.1007/978-3-030-59710-8_54
11. Painchaud, N., Skandarani, Y., Judge, T., Bernard, O., Lalande, A., Jodoin, P.-M.: Cardiac segmentation with strong anatomical guarantees. IEEE Trans. Med. Imaging **39**, 3703–3713 (2020)
12. Yang, D., et al.: Automatic liver segmentation using an adversarial image-to-image network. In: Descoteaux, M., Maier-Hein, L., Franz, A., Jannin, P., Collins, D., Duchesne, S. (eds.) Medical Image Computing and Computer Assisted Intervention – MICCAI 2017. MICCAI 2017. LNCS, vol. 10435, pp. 507–515. Springer, Cham (2017). https://doi.org/10.1007/978-3-319-66179-7_58
13. Ding, K., et al.: Graph convolutional networks for multi-modality medical imaging: methods, architectures, and clinical applications (2022)
14. Chen, Z., Zhang, H.: Learning Implicit Fields for Generative Shape Modeling (2019). ArXiv181202822 Cs
15. Michalkiewicz, M., Pontes, J.K., Jack, D., Baktashmotlagh, M., Eriksson, A.: Deep level sets: implicit surface representations for 3d shape inference (2019)
16. Xu, Q., Wang, W., Ceylan, D., Mech, R., Neumann, U.: DISN: deep implicit surface network for high-quality single-view 3D reconstruction (2021)
17. Amiranashvili, T., Menze, B., Zachow, S.: Learning shape reconstruction from sparse measurements with neural implicit functions. In: Proceedings of the 5th International Conference on Medical Imaging with Deep Learning
18. Chen, Y., Liu, S., Wang, X.: Learning continuous image representation with local implicit image function. In: 2021 IEEECVF Conference on Computer Vision and Pattern Recognition (CVPR), pp. 8624–8634. IEEE, Nashville, TN, USA (2021)
19. Raju, A., Miao, S., Jin, D., Lu, L., Huang, J., Harrison, A.P.: Deep implicit statistical shape models for 3D medical image delineation (2022). ArXiv210402847 Cs
20. Zhou, Z., Shin, J., Zhang, L., Gurudu, S., Gotway, M., Liang, J.: Fine-tuning convolutional neural networks for biomedical image analysis: actively and incrementally. In: 2017 IEEE Conference on Computer Vision and Pattern Recognition (CVPR), pp. 4761–4772. IEEE, Honolulu, HI (2017)
21. Tajbakhsh, N., et al.: Convolutional neural networks for medical image analysis: full training or fine tuning? IEEE Trans. Med. Imaging **35**, 1299–1312 (2016)

22. Samala, R.K., Chan, H.-P., Hadjiiski, L.M., Helvie, M.A., Cha, K.H., Richter, C.D.: Multi-task transfer learning deep convolutional neural network: application to computer-aided diagnosis of breast cancer on mammograms. Phys. Med. Biol. **62**, 8894–8908 (2017)

23. Moeskops, P., et al.: Deep learning for multi-task medical image segmentation in multiple modalities, pp 478–486 (2016)

24. Moeskops, P., Veta, M., Lafarge, M.W., Eppenhof, K.A.J., Pluim, J.P.W.: Adversarial training and dilated convolutions for brain MRI segmentation (2017)

25. Javanmardi, M., Tasdizen, T.: Domain adaptation for biomedical image segmentation using adversarial training. In: 2018 IEEE 15th International Symposium on Biomedical Imaging (ISBI 2018), pp. 554–558. IEEE, Washington, DC (2018)

26. Yang, Q., Li, N., Zhao, Z., Fan, X., Chang, E.I.-C., Xu, Y.: MRI cross-modality image-to-image translation. Sci. Rep. **10**, 3753 (2020)

27. Hoffman, J., et al.: CyCADA: cycle-consistent adversarial domain adaptation (2017). https:// doi.org/10.48550/ARXIV.1711.03213

28. Snell, J., Swersky, K., Zemel, R.S.: Prototypical networks for few-shot learning (2017)

29. Bian, C., Yuan, C., Ma, K., Yu, S., Wei, D., Zheng, Y.: Domain adaptation meets zero-shot learning: an annotation-efficient approach to multi-modality medical image segmentation. IEEE Trans. Med. Imaging **41**, 1043–1056 (2022)

30. Tancik, M., et al.: Fourier features let networks learn high frequency functions in low dimensional domains (2020)

31. Desai, A., Parikh, S., Kumari, S., Raman, S.: PointResNet: residual network for 3d point cloud segmentation and classification (2022)

32. Ghesu, F.-C., et al.: Multi-scale deep reinforcement learning for real-time 3D-landmark detection in CT scans. IEEE Trans. Pattern Anal. Mach. Intell. **41**, 176–189 (2019)

33. Zheng, Y., Barbu, A., Georgescu, B., Scheuering, M., Comaniciu, D.: Four-chamber heart modeling and automatic segmentation for 3-D cardiac CT volumes using marginal space learning and steerable features. IEEE Trans. Med. Imaging **27**, 1668–1681 (2008)

Global Sensitivity Analysis of Thrombus Formation in the Left Atrial Appendage of Atrial Fibrillation Patients

Zineb Smine[✉], Paolo Melidoro, Ahmed Qureshi, Stefano Longobardi, Steven E. Williams, Oleg Aslanidi, and Adelaide De Vecchi

School of Biomedical Engineering and Imaging Sciences, King's College London, London, UK
zineb.smine@outlook.com

Abstract. Atrial Fibrillation (AF) is the most common type of cardiac arrhythmia. Most AF-related thrombi originate within the left atrial appendage (LAA). This study investigated the key factors influencing thrombus formation in the LAA using global sensitivity analysis (GSA) based on computational fluid dynamics (CFD) simulations. GSA was conducted to assess the effects of four physiological input parameters: initial thrombin location within the LAA, fibrinogen (Fg) concentration in the blood, sensitivity to activated protein C (K3 constant), and inlet velocity. A total of 160 CFD simulations were performed using a 2D idealized left atrial geometry with the most common LAA morphologies: Cactus (CA), Chickenwing (CW), Windsock (WS), and Broccoli (BR). The area under the curve (AUC) of fibrin, which is a precursor of thrombus formation, was computed in the LAA to quantify net fibrin formation over time. Gaussian Process Emulators (GPE) were trained using the simulations' results to predict the Sobol indices from the input parameters. Fg concentration, initial thrombin location, and their interaction exhibited the largest Sobol indices in all LAA morphologies, impacting both average and maximum AUC. Inlet velocity affected the average AUC in BR, and its interaction with the initial thrombus location was significant for this morphology. Additionally, K3 contributed to the output variance in CW and BR. These findings emphasize the overall significance of Fg concentration and initial thrombin location, along with their interaction, in thrombus formation. The impacts of inlet velocity and K3 concentration appear to be morphology-specific. The distinct values obtained from maximum and average fibrin AUC provide complementary insights into thrombus formation.

Keywords: Atrial Fibrillation · Left atrial appendage · Global sensitivity analysis · Computational fluid dynamics · Ischaemic stroke

1 Introduction

Atrial Fibrillation (AF) is the most common form of cardiac arrhythmia [24]. It is characterized by rapid and irregular heartbeats resulting from abnormal

O. Camara et al. (Eds.): STACOM 2023, LNCS 14507, pp. 55–65, 2024.
https://doi.org/10.1007/978-3-031-52448-6_6

electrical impulses typically originating in the pulmonary vein inlets of the left atrium (LA). This condition poses significant morbidity and mortality risks as it is the primary cause of thromboembolic events, accounting for nearly one-third of all strokes [11]. It is estimated that over 90% of thrombi associated with AF are located in the left atrial appendage (LAA) [7], a muscular extension of the LA which is a remnant of the original embryonic LA [10,25]. Consequently, there has been a growing focus on investigating the quantitative characteristics of the LAA in recent years. Earlier studies have classified the LAA into four distinct anatomical morphologies based on their appearance: chicken wing (CW), windsock (WS), cactus (CA), and broccoli (BR) [5]. Among these morphologies, the BR is generally the most prone to the formation of life-threatening thrombi [19]. However, the mechanistic relationship between the LAA geometries and the risk of stroke remains poorly understood.

The higher likelihood of stroke in patients with AF has sparked the need for accurate prediction of stroke risks, and thus stratification of risk levels in individuals. Currently, the most used criteria adopted for AF patients' strat-ification is the CHA2DS2-VASc score - a point-based system assigning a risk score based on the patient's comorbidities, age and gender. However, this scor-ing system has limitations since it does not consider the effects of the blood coagulation dynamics and its underlying mechanisms on thrombus formation, which are described by the so-called Virchow's triad: blood stasis, hypercoagula-bility and endothelial damage [9]. The CHA2DS2-VASc subsequently disregards important physiology-based risk factors for thromboembolism [12].

Therefore, computational fluid dynamics (CFD) serve as a valuable tool to enhance such scoring systems, as it allows for quantification identifying of blood flow metrics within complex fluid dynamic models [21]. In addition to blood velocity and pressure, our CFD models are able to quantify the spatio-temporal dynamics of concentrations of key proteins and enzymes driving coagulation in response to macroscopic changes in blood flow [20]. To identify the most rele-vant metrics associated with thrombus formation in each LAA morphology, we performed a global sensitivity analysis (GSA) on our CFD simulations spanning a range of input parameters for each morphology, with the goal to assess the individual impact of each parameter on thrombus formation inside the LAA.

2   Methods

The complete workflow for performing a GSA on thrombus formation in 2D LAA models is illustrated in Fig. 1. First, Latin Hypercube Sampling (LHS) was used to generate a set of points within physiological bounds for four input parameters, namely (i) inlet velocity through the pulmonary veins, (ii) fibrinogen concentra-tion, (iii) sensitivity to activated protein C, and (iv) initial thrombin location. Second, CFD simulations were performed for each LHS points producing a total of 160 simulations. Third, the output features related to the concentration of fib-rin (a precursor of thrombus) obtained from the simulations were used to train a Gaussian Process Emulator (GPE). The accuracy of the GPE predictions was then assessed on a separate testing LHS. Finally, the GPE was used to generate additional results and ultimately estimate Sobol sensitivity indices.

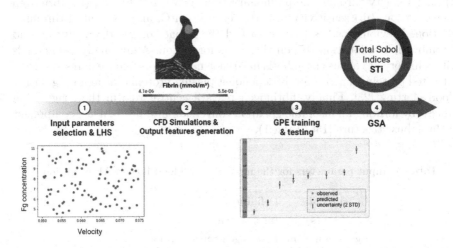

Fig. 1. Workflow for generating input points for CFD simulations to train a GPE and perform a GSA on thrombus formation

2.1 2D Mesh Models and CFD Simulations

Simplified 2D finite-element mesh models were generated for the four LAA shapes. The outlines of the BR, CW, WS and CA morphologies were designed on the CAD tool *Shapr3D* [19]. The 2D triangular meshes were created in the Simmodeler software package (Simmetrix Inc., USA) with approximately 50,000 linear triangular elements with a boundary layer, for each morphology. Seven boundaries were defined: 4 inlets for the pulmonary veins inflow (PV1-4), 1 outlet for the mitral valve (MV), 1 boundary for the LAA, and 1 boundary for the LA wall Fig. 2. All

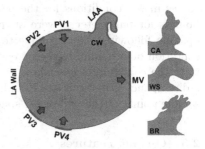

Fig. 2. Definition of the boundaries in semi-realistic 2D simulations of LA flow and coagulation [19]

simulations were performed in CHeart, a Finite Element software for biomedical research developed at King's College London [14]. Using a time step of 0.01 s and simulating for 6000 steps, the total simulation time amounted to 60 s of real-time cardiac flow. A simulation takes one hour to run on the 640 core SGI Altix-UV high performance computing (HPC) cluster with Nehalem-EX architecture at King's College London.

2.2   Input Parameters

Four input parameters were selected to study thrombus formation in the LAA: (i) inlet velocity through the pulmonary veins (PV1-4 in Fig. 2), (ii) fibrinogen concentration, (iii) sensitivity to activated protein C, and (iv) initial thrombin location. Previous studies have identified the impact of blood velocity [2] and thrombus initial location [17] on thrombus formation. Additionally, individuals with a history of venous thrombosis have shown a high prevalence of resistance to activated protein C, a vitamin-K dependent plasma protein crucial for regulating blood clotting [23]. Finally, fibrinogen plays a central role in the coagulation process by forming an insoluble gel upon conversion into fibrin, a key component of thrombus structure [13] (Table 1).

Table 1. Input parameters for the model of the blood flow in AF conditions

Label	Unit	Definition
v	m/s	Inlet velocity
FgC	mmol/m	Fibrinogen concentration
K3	Nm^{-1}	Sensitivity to activated protein C
p	cm	Initial location of thrombin

The lower and upper bounds for the LHS were determined from literature values in AF conditions for the parameters **v** [6], **FgC** [15] and **K3** [23]. The bounds for parameter **p** were defined as the geodesic distance from the LAA opening following the centreline to the tip of the LAA [1]. We used an initial condition of thrombin to mimic endothelial damage in our simulations and initiate the coagulation cascade [17]. The quality of the LHS is characterised by a discrepancy criteria of 0.0173 indicating an evenly distributed arrangement of sample points in the parameter space [3,8], as shown in Fig. 1.

2.3   Output Features

Since thrombus formation is driven by the conversion of fibrinogen into fibrin by the action of serine protease thrombin, the output features were chosen based on the concentration of fibrin [4]. Therefore, the average and maximum fibrin concentrations inside the LAA, as obtained from each CFD simulation, were integrated over time and the corresponding area under the curve was computed to assess thrombus formation (AUC_{Max} and AUC_{Avg} respectively) [18].

On one hand, AUC_{Avg} provides insight into the overall dispersion of Fn concentration within the entire LAA. On the other hand, AUC_{Max} indicates the degree of fibrin concentration at a particular location within the LAA.

2.4   Modelling AF-Induced Thrombogenicity

Three clotting proteins define the final stages of coagulation of blood - thrombin (Th), fibrinogen (Fg) and fibrin (Fn). When thrombin interacts with fibrinogen, it triggers the formation of fibrin. The concentration rate of each of these factors in the blood can be modeled by the corresponding reaction-diffusion-convection (RDC) equations (Eqs. (1) to (4)) [19]. These RDC equations are coupled to the system of 2D Navier-Stokes Equation (NSE) governing the fluid motion via the blood velocity u in the convective term on the right-hand side:

$$\frac{\partial Th}{\partial t} = D_{Th}\Delta Th - u.\nabla Th + R_{Th} \tag{1}$$

$$R_{Th} = K_1(1 + K_2Th)[K_3Th(1 + K_4Th)] \times \left(1 - \frac{Th}{u_0}\right) - K_5Th \tag{2}$$

$$\frac{\partial Fg}{\partial t} = D_{Fg}\Delta Fg - u.\nabla Th - K_{eff}.Fg.Th \tag{3}$$

$$\frac{\partial Fn}{\partial t} = D_{Fn}\Delta Fn - u.\nabla Th + K_{eff}.Fg.Th \tag{4}$$

Diffusion coefficients are set to $D_{Th} = 4.6e{-}11$ m^2/s and $D_{Fg} = D_{Fn} = 2.0e{-}11$ m^2/s, and the effective reaction coefficient is set to Keff $= 7180$ m^{3}mol s [19]. The velocity u of blood flow in the deforming left atrium was determined by solving the incompressible NSE equations within the arbitrary Eulerian-Lagrangian reference frame [19].

2.5   Modelling Thrombus Viscosity

The model captures the transition from thrombogenic proteins in the blood to a fibrin gel thrombus at the injury site by employing the Hill equation to adjust the viscosity μ throughout the entire domain, with n $= 3$, K $= 0.5$ and μmax $= 2$ Pa s similarly to the viscosity of fibrin gel.

$$\mu = \frac{\mu_{max}\left(\frac{Fn}{Fg}\right)^n}{K^n + \left(\frac{Fn}{Fg}\right)^n} \tag{5}$$

Initial concentration for Fibrin was set to 0 while the initial concentration of thrombin was set to 0.09 mmol/m^3 at the site of injury. The RDC equations were coupled with the viscosity which was updated at each time step based on the ratio of fibrin to fibrinogen, as shown in Eqs. (1) to (5).

2.6   Variance-Based Global Sensitivity Analysis

A GSA was conducted to assess the relative contributions of individual parameters to the model output variance. More specifically, the Sobol variance-based method breaks down the overall output variance into the sum of the variances of the various inputs and the interactions between them [26].

The GSA was performed using the GPE implementation [16] with GPy-Torch through the Saltelli method. The total number of model evaluations required by the Saltelli method to estimate the sensitivity indices was determined using the formula $n*(2*D+2)$ [22], where n represents the desired number of Monte Carlo samples, and D corresponds to the problem's dimension. A value of $n = 1024$ was selected for the Monte Carlo samples, and the problem had a dimension of $D = 4$, totalling for 10240 model evaluations per GPE. The emulator's uncertainty was addressed by predicting the indices 1000 times, yielding an index distribution rather than individual values. The resulting indices were determined as the mean value of these distributions.

3   Results

3.1   Model Emulation Accuracy

A total of eight GPEs were trained (one for each of the four LAA morphology and for each of the two output features AUC_{Max} and AUC_{Avg}). A separate testing set consisting of eight additional points generated from an independent LHS was used to evaluate the performance of each emulator. The quality of the LHS for the testing set was evaluated with a discrepancy criterion of 0.054.

Table 2. Emulator accuracy metrics for each LAA morphology. The table presents the averaged R2 score and mean squared error between the AUC_{Max} and AUC_{Avg} emulators of each LAA morphology.

LAA morphology	Mean R2 Score	Mean Squared error
CA	0.98	0.0003
BR	0.89	0.0071
WS	0.85	0.0078
CW	0.79	0.0049

Table 2 shows the mean R2 score and mean squared error (MSE) for each emulator. The overall accuracy assessment of the model through all eight configurations exhibited an R2 score of 0.88 and a MSE of 0.0051.

The emulator for the CA morphology achieved the highest performance, particularly for the AUC_{Max} output feature (R2 Score = 0.9935, MSE = 0.0006). On the other hand, the least accurate emulator was the CW emulator, especially for the AUC_{Avg} output feature (R2 Score = 0.7849, MSE = 0.0005).

3.2   Model Global Sensitivity Analysis

The contribution of the input parameters to each output's total variance (ST_i) in the LAA morphologies is illustrated in Fig. 3. The most significant parameters in thrombus formation were **p**, **FgC** and the interaction between those two parameters for all morphologies and for both AUC_{Avg} and AUC_{Max}. For WS, BR and CA, the sensitivity indices for **p** ($ST_i = 0.70$, 0.82 and 0.78 respectively, averaged between the AUC_{Avg} and AUC_{Max} indices) have a greater contribution in explaining the outputs' variance than **FgC** ($ST_i = 0.32$, 0.21, 0.27 similarly). However, **FgC** and **p** were found to have comparable impacts on the outputs' total variance for the CW morphology - with a slightly higher impact for **FgC** ($ST_i = 0.56$ for **FgC** against 0.48 for **p**, averaged between the AUC_{Avg} and AUC_{Max} indices). The parameter **v** exhibited an impact on the variance of AUC_{Avg} for the BR morphology ($S_i = 0.02$). In this morphology, the interaction between the **v** and **p** parameters was also significant ($S_{ij} = 0.05$). Finally, the **K3** parameter contributed to the variance of the output for the AUC_{Avg} in the CW morphology ($S_i = 0.04$), and for the AUC_{Max} for the BR morphology ($S_i = 0.02$).

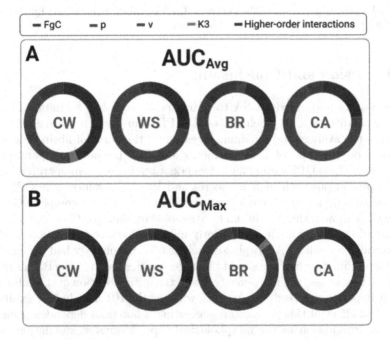

Fig. 3. Impact of input parameters v, FgC, K3 and p in explaining the output variance in (a) AUC_{Avg} and (b) AUC_{Max} models, for each LAA morphology. Each input parameter's contribution is represented by adding its first-order effect S_i and, if applicable, its second-order effect S_{ij}.

Figure 4 shows the influence of **p** on fibrin concentrations in three CFD simulations for CA LAAs, with similar values for **v** (6.7 cm/s) and for **FgC** (7.6 mmol/m^3). The spatial isocontours of fibrin concentration showed higher fibrin levels in the simulations where the initial thrombin location was closer to the tip of the LAA.

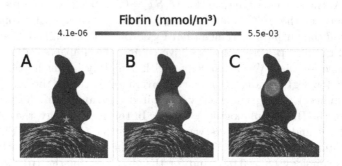

Fig. 4. Simulation snapshots of CA LAAs captured at t = 3 s. Yellow arrows show the blood velocity field. The fibrin concentration at the initial thrombin location p (measured as the distance from the centre of LAA opening) is identified by the green stars at A) p = 0.45 cm, B) 1.4 cm and C) 2.3 cm.

4   Discussion and Conclusion

We successfully performed a GSA to characterize thrombus formation in LAA under AF conditions. This pipeline uses 2D CFD simulations to train eight GPEs to emulate the average and maximum area under the curve of fibrin concentration based on the type of LAA geometry and four physiological parameters. Evaluation of these GPEs using an independent testing set demonstrated highly accurate predictions with high R2 scores and low MSE values. The GPEs are then used to produce 81920 model evaluations, allowing for a complete GSA.

The GSA showed that the initial location of thrombin, the fibrinogen concentration, and their interaction significantly influenced the concentration of fibrin in the blood across all LAA morphologies. The CW morphology had the smallest orifice, approximately half the size of the largest orifice found in the BR morphology. The CW was also the morphology where the initial location of thrombin had the least impact on the output variance, whereas the BR morphology exhibited the greatest effect of this parameter, suggesting a potential dependency on the orifice size in addition to the morphological type. Moreover, simulations indicated that initial thrombin locations closer to the tip of the LAA favored fibrin accumulation and thrombus formation, aligning with previous clinical studies [20, 23]. Finally, introducing the two output features, i.e. average and maximum fibrin concentrations inside the LAA, allowed for characterization of different morphology-specific behaviors.

Limitations of this work include the use of simplified 2D models, and the limited number of geometrical parameters in the GSA. Future simulations could integrate additional thrombotic proteins and factors which influence thrombus formation. However, simulating more complex coagulation processes would increase the computational costs, which wasn't warranted for this study.

Overall, our study introduces a novel pipeline for GSA based on CFD simulations that integrate all aspects of Virchow's triad. This tool provides insight into the physiological and anatomical factors that impact thrombus formation in AF patients. This proof of concept shows that our approach can provide physiologically plausible results in a 2D LA model. The next step would be to test it and validate it in 3D patient-specific models, to assess its potential to improve stroke risk stratification scores and anti-coagulation therapy selection in AF patients.

References

1. Aguado, A.M., et al.: In silico optimization of Left Atrial Appendage Occluder implantation using interactive and modeling tools. Front. Physiol. **10**, 237 (2019). https://doi.org/10.3389/fphys.2019.00237
2. Al-Saady, N.M., Obel, O.A., Camm, A.J.: Left atrial appendage: structure, function, and role in thromboembolism. Heart **82**(5), 547–554 (1999). https://doi.org/10.1136/hrt.82.5.547
3. Androulakis, E., Drosou, K., Koukouvinos, C., Zhou, Y.: Measures of uniformity in experimental designs: a selective overview. Commun. Stat. Theor. Meth. **45**, 3782–3806 (2016). https://doi.org/10.1080/03610926.2014.966843
4. Ataullakhanov, F., Zarnitsyna, V., Kondratovich, A., Lobanova, E., Sarbash, V.: A new class of stopping self-sustained waves: a factor determining the spatial dynamics of blood coagulation. Phys. Usp. **45**, 619 (2002). https://doi.org/10.1070/PU2002v045n06ABEH001090
5. Beigel, R., Wunderlich, N.C., Ho, S.Y., Arsanjani, R., Siegel, R.J.: The left atrial appendage: anatomy, function, and noninvasive evaluation. JACC Cardiovasc. Imaging **7**(12), 1251–1265 (2014). https://doi.org/10.1016/j.jcmg.2014.08.009
6. Chao, T.H., Tsai, L.M., Tsai, W.C., Li, Y.H., Lin, J.L., Chen, J.W.: Effect of atrial fibrillation on pulmonary venous flow patterns assessed by doppler transesophageal echocardiography. Chest **117**(6), 1546–1550 (2000). https://doi.org/10.1378/chest.117.6.1546
7. Cresti, A., et al.: Prevalence of extra-appendage thrombosis in non-valvular atrial fibrillation and atrial flutter in patients undergoing cardioversion: a large transoesophageal echo study. EuroIntervention **15**, e225–e230 (2019). https://doi.org/10.4244/EIJ-D-19-00128
8. Damblin, G., Couplet, M., Iooss, B.: Numerical studies of space-filling designs: optimization of Latin Hypercube samples and subprojection properties. J. Simul. **7**, 276–289 (2013). https://doi.org/10.1057/jos.2013.16
9. Ding, W.Y., Gupta, D., Lip, G.Y.: Atrial fibrillation and the prothrombotic state: revisiting Virchow's triad in 2020. Heart **106**, 1463–1468 (2020). https://doi.org/10.1136/heartjnl-2020-316977
10. Faletra, F.F., Narula, J.: Imaging of cardiac anatomy, 5th edn. In: Ellenbogen, K.A., Wilkoff, B.L., Kay, G.N., Lau, C.P., Auricchio, A. (eds.) Clinical Cardiac Pacing, Defibrillation and Resynchronization Therapy, pp. 15–60. Elsevier (2017). https://doi.org/10.1016/B978-0-323-37804-8.00002-X

11. Freedman, B., Potpara, T.S., Lip, G.Y.: Stroke prevention in atrial fibrillation. Lancet **388**(10046), 806–817 (2016). https://doi.org/10.1016/S0140-6736(16)31257-0
12. Hindricks, G., et al.: 2020 ESC guidelines for the diagnosis and management of atrial fibrillation developed in collaboration with the European Association for Cardio-Thoracic Surgery (EACTS): the task force for the diagnosis and management of atrial fibrillation of the European Society of Cardiology (ESC) developed with the special contribution of the European Heart Rhythm Association (EHRA) of the (ESC). Eur. Heart J. (2020). https://doi.org/10.1093/eurheartj/ehaa612
13. Kattula, S., Byrnes, J.R., Wolberg, A.S.: Fibrinogen and fibrin in hemostasis and thrombosis. Arterioscler. Thromb. Vasc. Biol. **37**(3), e13–e21 (2017). https://doi.org/10.1161/ATVBAHA.117.308564
14. Lee, J., et al.: Multi-physics computational modeling in CHeart. SIAM J. Sci. Comput. **38**(3), C150–C178 (2016). https://doi.org/10.1137/15M1014097
15. Lip, G.Y., Lowe, G.D., Rumley, A., Dunn, F.G.: Fibrinogen and fibrin d-dimer levels in paroxysmal atrial fibrillation: evidence for intermediate elevated levels of intravascular thrombogenesis. Am. Heart J. **131**(4), 724–730 (1996). https://doi.org/10.1016/s0002-8703(96)90278-1
16. Longobardi, S., et al.: Predicting left ventricular contractile function via gaussian process emulation in aortic-banded rats. Philos. Trans. R. Soc. A Math. Phys. Eng. Sci. **378**, 20190334 (2020). https://doi.org/10.1098/rsta.2019.0334
17. Marroquin, L., Tirado-Conte, G., Pracoń, R., et al.: Management and outcomes of patients with left atrial appendage thrombus prior to percutaneous closure. Heart **108**, 1098–1106 (2022). https://doi.org/10.1136/heartjnl-2021-319811
18. Prakhya, K.S., Luo, Y., Adkins, J., Hu, X., Wang, Q.J., Whiteheart, S.W.: A sensitive and adaptable method to measure platelet-fibrin clot contraction kinetics. Res. Pract. Thromb. Haemost. **6**(5), e12755 (2022). https://doi.org/10.1002/rth2.12755
19. Qureshi, A., et al.: Left atrial appendage morphology impacts thrombus formation risks in multi-physics atrial models. In: 2021 Computing in Cardiology (CinC), Brno, Czech Republic, pp. 1–4 (2021). https://doi.org/10.23919/CinC53138.2021.9662901
20. Qureshi, A., et al.: Modelling Virchow's triad to improve stroke risk assessment in atrial fibrillation patients. In: Computing in Cardiology (CinC), vol. 498, pp. 1–4. IEEE (2022). https://doi.org/10.22489/CinC.2022.378
21. Qureshi, A., Lip, G.Y., Nordsletten, D.A., Williams, S.E., Aslanidi, O., de Vecchi, A.: Imaging and biophysical modelling of thrombogenic mechanisms in atrial fibrillation and stroke. Front. Cardiovasc. Med. **9**, 1074562 (2023). https://doi.org/10.3389/fcvm.2022.1074562
22. Saltelli, A., Annoni, P., Azzini, I., Campolongo, F., Ratto, M., Tarantola, S.: Variance based sensitivity analysis of model output. Design and estimator for the total sensitivity index. Comput. Phys. Commun. **181**, 259–270 (2010). https://doi.org/10.1016/j.cpc.2009.09.018
23. Svensson, P.J., Dahlbäck, B.: Resistance to activated protein c as a basis for venous thrombosis. N. Engl. J. Med. **330**(8), 517–522 (1994). https://doi.org/10.1056/NEJM199402243300801
24. Wong, G., Singh, G.: Transcatheter left atrial appendage closure. Methodist Debakey Cardiovasc. J. **19**, 67–77 (2023). https://doi.org/10.14797/mdcvj.1215

25. Yaghi, S., Song, C., Gray, W., Furie, K., Elkind, M., Kamel, H.: Left atrial appendage function and stroke risk. Stroke **46** (2015). https://doi.org/10.1161/STROKEAHA.115.011273
26. Zhang, X.Y., Trame, M., Lesko, L., Schmidt, S.: Sobol sensitivity analysis: a tool to guide the development and evaluation of systems pharmacology models. CPT Pharmacomet. Syst. Pharmacol. **4** (2015). https://doi.org/10.1002/psp4.6

Sparse Annotation Strategies for Segmentation of Short Axis Cardiac MRI

Josh Stein[1,2], Maxime Di Folco[1(✉)], and Julia A. Schnabel[1,2,3]

[1] Institute of Machine Learning in Biomedical Imaging, Helmholtz Munich, Neuherberg, Germany
maxime.difolco@helmholtz-munich.de
[2] Technical University of Munich, Munich, Germany
[3] King's College London, London, UK

Abstract. Short axis cardiac MRI segmentation is a well-researched topic, with excellent results achieved by state-of-the-art models in a supervised setting. However, annotating MRI volumes is time-consuming and expensive. Many different approaches (e.g. transfer learning, data augmentation, few-shot learning, etc.) have emerged in an effort to use fewer annotated data and still achieve similar performance as a fully supervised model. Nevertheless, to the best of our knowledge, none of these works focus on *which* slices of MRI volumes are most important to annotate for yielding the best segmentation results. In this paper, we investigate the effects of training with sparse volumes, i.e. reducing the number of cases annotated, and sparse annotations, i.e. reducing the number of slices annotated per case. We evaluate the segmentation performance using the state-of-the-art nnU-Net model on two public datasets to identify which slices are the most important to annotate. We have shown that training on a significantly reduced dataset (48 annotated volumes) can give a Dice score greater than 0.85 and results comparable to using the full dataset (160 and 240 volumes for each dataset respectively). In general, training on more slice annotations provides more valuable information compared to training on more volumes. Further, annotating slices from the middle of volumes yields the most beneficial results in terms of segmentation performance, and the apical region the worst. When evaluating the trade-off between annotating volumes against slices, annotating more slices than volumes is a better strategy.

Keywords: Cardiac MRI · Segmentation · Sparse annotations

1 Introduction

Cardiac image segmentation constitutes a fundamental initial phase in various applications. Segmentation is often the first step in evaluating cardiac functionality in order to diagnose disease. By partitioning the image into distinct semantically meaningful regions, typically aligned with anatomical structures, it facilitates the extraction of quantitative measures critical for further analysis

O. Camara et al. (Eds.): STACOM 2023, LNCS 14507, pp. 66–76, 2024.
https://doi.org/10.1007/978-3-031-52448-6_7

and interpretation [1]. Deep learning methods have become the state-of-the-art approach for this task, but they require the collection and annotation of data, which are time-consuming and laborious processes.

Much effort has been spent improving methods that require fewer ground truth annotations. Some popular approaches include data augmentation [2], transfer learning [3–5], semi-supervised learning [6,7], and self-supervised learning [8,9] - many others exist [10].

In general, the data used for these methods are *sparse* or *limited*. The exact definition varies from context to context and usually falls into one of three broad categories [10]. First, the annotation of data volumes could be sparse, where only particular patients may be annotated. Second, the annotation of slices within volumes could be sparse. Instead of having a fully annotated 3D volume, only particular slices within the volume may be annotated. Third, there could be sparsity in the slice annotations themselves. Instead of having a pixel-accurate ground truth annotation, there may be bounding boxes, scribbles, or particular labelled points. These categories are not mutually exclusive - for example, there may be sparse slices that are sparsely annotated.

In cardiac imaging, segmentation of the short-axis view on MRI data has been well studied, thanks to technical challenges, where teams compete to achieve the best-performing models [11,12]. Nowadays, we consider fully supervised short-axis cMRI segmentation a well-researched task, with state-of-the-art approaches surpassing human performance.

In this paper, we purposely choose not to use any approaches for sparse data to answer the following questions, which are still unclear in the literature:

1. How many sparse data does are needed for achieving reasonable results with state-of-the-art nnU-Net?
2. Which cardiac regions (basal, mid or apical) contribute the most to segmentation performance?
3. Is there a particular annotation strategy which one should prefer between annotating volumes or annotating slices?

In this work, we investigate the effects on segmentation performance on two public datasets [11,12], when removing volumes (reducing the number of annotated case), removing slice annotations (both randomly and from particular cardiac regions) and the balance between these two.

2   Related Work

Recent works on cardiac imaging segmentation have focused on using sparse annotations while still achieving results that compare to using a fully annotated dataset. Bitarafan et al. [13] use a single annotated 2D slice with registration and self-training to propagate and train on label propagations. They achieve approximately a 10% reduction in Dice score using a single annotation compared to using a fully annotated volume. Bai et al. [14] also use label propagation in combination with a recurrent neural network (RNN) to incorporate both spatial and temporal information. Using two annotated frames, they are able to out-compete a baseline U-Net model (trained on all available annotated frames). Contrastive learning strategies have also been explored. Zeng et al. [9] use a contrastive loss between slice position in a self-supervised pre-training stage that achieves a similar Dice score to a fully annotated dataset using only 15 annotated volume (the achieved Dice score is 3% lower compared to using the fully annotated set). You et al. [15] present a contrastive semi-supervised 2D medical segmentation framework for very limited annotations and accomplish a Dice score of 0.82 using only 1% of the labels and similar performance to fully annotated set using only 10% of the labels.

3   Methods

3.1   Segmentation Network

We use nnU-Net [16], the current state-of-the-art model for cardiac segmenta-tion (achieving first place in ACDC and M&Ms challenges). We do not modify the standard nnU-Net processing pipeline, except to change the data sampling strategy. We evaluate on mean foreground Dice score, Hausdorff Distance (HD) and Mean Absolute Distance (MAD).

We evaluate both 3D and 2D nnU-Net models. For both datasets, we evaluate only the 3D high resolution model (i.e. not the low resolution or cascaded model). The reader is referred to the original nnU-Net paper [16] for clarification on the differences between these models and their processing pipelines. The 2D models are trained by randomly sampling a slice from a given volume. We train all models for 20 epochs.

3.2   Definition of Sparsity

For each dataset, we investigate the effects of removing volumes (i.e. enforcing sparsity of volumes), zeroing out slices (i.e. enforcing sparsity of slices) and the balance between these two.

Sparse Volumes: To investigate the sparsity of volumes, we randomly sample a percentage of the total patient cardiac volumes, which are then used for training. By iteratively training on smaller samples of the dataset we want to determine the number of needed annotated volumes to achieve comparable results to train-ing on the full dataset.

Sparse Slices: We investigate slice sparsity by sampling and training on several slices from within a volume. All non-sampled slices are zeroed, which allows us to maintain volume shape. This prevents the need to modify network parameters, which would otherwise need to continuously adapt based on the input volume. Slices can either be sampled randomly (from within the entire volume) or explic-itly sampled from the apical, mid-ventricular, and/or basal regions. We assume that each volume is split into equal thirds, where the first third corresponds to basal slices, the second mid-ventricular slices and the third apical slices. Sampling from various permutations of these regions allows us to investigate which (if any) regions are most important for segmentation performance.

Annotation Strategy: Finally, we investigate the balance and relative importance of sparse volumes vs sparse slices by sampling a percentage of the volumes and then randomly sampling different percentages of slices from within the sampled volumes.

4 Experiments and Results

4.1 Dataset

We use the Automatic Cardiac Diagnosis Challenge (ACDC) [11] and the Multi-Centre, Multi-Vendor and Multi-Disease Cardiac Segmentation Challenge (M&Ms) [12]. These datasets are popular in the literature, and have been used to investigate a variety of supervised and unsupervised segmentation methods. They are both inherently sparse (across volumes) as they provide annotations at only end-diastolic and end-systolic phases.

ACDC has a set of 100 training cases, each of which has two fully annotated volumes (one at end-diastole, one at end-systole). We train on all 200 volumes using 160 volumes for training and the remaining 40 volumes for validation. The test set is composed of 50 cases, each of which is again fully annotated at end-diastole and end-systole. After nnU-Net preprocessing transformations, all volume have 20 slices. The total number of available training slices is therefore $20 \times 160 = 3200$.

Similarly, M&Ms has a set of 150 training cases, each of which has end-diastole and end-systole volumes annotated (300 training volumes). We split the data into 240 training volumes and 60 validation volumes. The test set is composed of 136 cases. After nnU-Net preprocessing transformations, there are 14 slices per volume - the total number of available training slices is therefore $14 \times 240 = 3360$. The M&Ms dataset contains images from different vendors and centers. We keep the same dataset split described in [12]. Please refer to the corresponding paper for details on the acquisition protocol.

4.2 Sparse Volumes

The results of using a reduced dataset for both ACDC and M&Ms are shown in Table 1. First, we observe that the networks trained on ACDC generally outperform those trained on M&Ms. We believe this is due to the different MRI domains present in the M&Ms dataset. Second, we note that using more than 48 volumes (approximately 30% of the ACDC dataset, and 20% of the M&Ms dataset), regardless of dataset or model dimensionality, yields a Dice score greater than 0.85. However, using fewer volumes leads to increases in corresponding HD and MAD scores, and a decrease in Dice scores. Further, using fewer than 48 volumes leads to a worse performance in networks trained on ACDC data compared

Table 1. Effect of training on sparse annotated volumes. Note that the ACDC dataset only has 160 volumes.

Network	Dataset	Evaluation metric	Number of training volumes							
			8	24	32	48	80	160	192	240
2D nnU-Net	ACDC	Dice	0.62	0.71	0.74	0.85	0.89	**0.91**	-	-
		HD (mm)	36.03	22.15	18.56	14.61	7.20	**5.06**	-	-
		MAD (mm)	10.16	5.67	8.95	2.81	1.65	**1.16**	-	-
3D nnU-Net	ACDC	Dice	0.57	0.66	0.71	0.85	0.85	**0.91**	-	-
		HD (mm)	54.93	43.63	29.34	8.81	7.75	**4.40**	-	-
		MAD (mm)	18.03	13.61	8.95	2.17	1.94	**1.16**	-	-
2D nnU-Net	M&Ms	Dice	0.60	0.82	0.83	0.85	0.85	0.86	0.86	**0.87**
		HD (mm)	32.43	9.3	9.42	8.81	7.75	6.84	6.87	**6.54**
		MAD (mm)	8.9	2.39	2.38	2.17	1.94	1.74	1.74	**1.74**
3D nnU-Net	M&Ms	Dice	0.54	0.82	0.82	0.84	0.85	0.86	0.86	**0.87**
		HD (mm)	37.41	9.10	8.86	6.98	6.44	5.89	**5.80**	6.02
		MAD (mm)	11.26	2.25	2.32	1.79	1.65	1.56	**1.53**	1.60

to those trained on M&Ms. However, when the number of volumes is severely restricted (e.g. using 8 volumes) we see a similarly poor performance for both datasets. Finally, we note that the difference in performance between 2D and 3D networks is more pronounced for networks trained on ACDC than those trained on M&Ms. This is especially true when considering the difference in surface distance metrics between 2D and 3D networks. We conclude that having a variety of domains within the M&Ms dataset makes it more difficult to achieve higher Dice scores, while simultaneously allowing for better generalisation when removing annotated volumes.

4.3 Sparse Annotations

The results of training 3D nnU-Net only on particular cardiac regions are shown in Table 2. Qualitative results are shown for an illustrative sample in Fig. 1. As expected, the best performance is achieved by using all three cardiac regions (i.e. the most slices). Using only two regions, all combinations achieve a Dice score greater than 0.8, except for the network trained on the apical and basal combination trained on M&Ms data, which achieves 0.78. We also observe that training on combinations using mid-ventricular slices (i.e. apical and mid, or basal and mid combinations) yield the best-performing networks. The worst performance is achieved by networks trained on only a single cardiac region (the network trained on M&Ms data trained solely on the apical region is a very poor-performing network). In general, the best performing networks are those trained on mid-ventricular and basal regions, either in combination or individually.

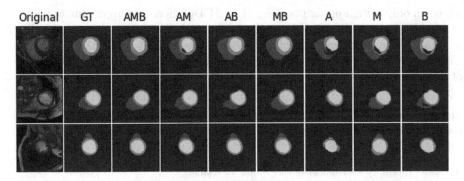

Original   GT   AMB   AM   AB   MB   A   M   B

Fig. 1. Qualitative results of segmentations produced by networks trained on different cardiac regions on the ACDC dataset. From top to bottom are slices from the basal region, the mid-ventricular region and the apical region respectively. Each slice is extracted from a different 3D segmentation volume. GT = ground truth. AMB means the network was trained on apical/middle-ventricular/basal slices (and permutations thereof).

Table 2. Effect of training on sparse annotated slices from different cardiac regions (A = apical slices, M = middle slices, B = basal slices). The 3D network is evaluated on all regions.

Dataset	Metric	Cardiac regions trained on						
		A + M + B	A + M	M + B	A + B	A	M	B
ACDC	Dice	**0.91**	0.86	0.89	0.82	0.53	0.75	0.55
	HD (mm)	**4.17**	7.10	5.46	9.52	55.99	49.12	37.56
	MAD (mm)	**1.08**	1.76	1.58	2.66	21.43	18.78	12.38
M&Ms	Dice	**0.87**	0.82	0.84	0.78	0.04	0.65	0.65
	HD (mm)	**5.52**	10.00	7.53	10.50	51.75	96.10	29.16
	MAD (mm)	**1.47**	3.60	1.94	2.82	24.10	33.62	11.65

We then train networks on randomly sampled slices from all three cardiac regions, the results of which are show in Table 3. The aforementioned cardiac regions correspond to using one-third of all slices per region. For ACDC, a single cardiac region corresponds to sampling approximately 6 slices, and two cardiac regions to approximately 13 slices, and 5 slices and 10 slices for M&Ms, respectively.

Randomly sampling a third of the available slices yields better results than sampling from any single cardiac region (although there is still a similar drop in performance when using a limited number of slices). Randomly sampling two thirds of available slices yields similar results as sampling from either the apical and middle slices or the middle and basal slices, which in turn is similar to using the full set of available slices. We also observe a Dice score greater than

0.8 when only using approximately 40% of ACDC slices and approximately 60% of M&Ms slices, with slight decreases in the surface distance metrics. 10 slice annotations are sufficient to achieve a Dice greater than 0.85 for both datasets. This corresponds to half the slices annotated for ACDC and approximately 70% of slices for M&Ms.

Table 3. Influence of training with randomly sampled and sparsely annotated slices from all three cardiac regions using 3D nnU-Net. Note that there are only 14 slices per volume for the M&Ms dataset.

Dataset	Metric	Number of slices used for training										
		1	2	4	5	6	8	10	13	14	16	20
ACDC	Dice	0.01	0.28	0.62	0.75	0.77	0.81	0.87	0.89	0.90	0.90	**0.91**
	HD (mm)	82.18	24.79	19.34	11.07	10.83	8.58	6.88	4.88	5.00	4.78	**4.63**
	MAD (mm)	33.09	8.34	6.18	3.46	2.95	2.39	1.71	1.29	1.30	1.22	**1.20**
M&Ms	Dice	0.01	0.28	0.68	0.72	0.79	0.83	0.85	0.85	**0.86**	-	-
	HD (mm)	76.06	28.2	14.10	12.48	11.11	7.58	6.33	6.54	**5.64**	-	-
	MAD (mm)	35.3	10.31	4.27	3.80	3.82	2.40	1.67	1.62	**1.49**	-	-

4.4   Sparse Volumes vs Sparse Annotations

In this section, we investigate annotation strategies using different balances of reduced volume annotations and reduced slice annotations while keeping a fixed number of total slices annotated. Table 4 shows the results for approximately 1400 slices annotated across both datasets. For both datasets, when keeping the total number of slices the same, better results are achieved when using more slices per volume. This is also shown qualitatively in Fig. 2.

Table 4. Segmentation performance of a 3D nnU-Net when changing the number of training volumes (V) vs the number of slices (S) while keeping the total number of slices approximately 1400. Note that the M&Ms dataset has a total of 14 slices per volume, and that the ACDC dataset has a total of 160 volumes.

Dataset	Metric	Proportionality constant $V \times S = {\sim}1400$				
		65 V, 20 S	100 V, 14 S	120 V, 12 S	160 V, 9 S	240 V, 6 S
ACDC	Dice	0.87 ± 0.02	**0.88 ± 0.02**	0.88 ± 0.03	0.82 ± 0.09	-
	HD (mm)	8 ± 1.47	8.52 ± 0.72	**6.70 ± 0.5**	8.27 ± 1.45	-
	MAD (mm)	1.98 ± 0.3	2.08 ± 0.18	**1.6 ± 0.08**	2.01 ± 0.33	-
M&Ms	Dice	-	**0.86 ± 0.03**	0.84 ± 0.02	0.83 ± 0.06	0.77 ± 0.08
	HD (mm)	-	**6.10 ± 0.95**	6.72 ± 1.25	6.93 ± 0.59	9.28 ± 0.78
	MAD (mm)	-	**1.58 ± 0.16**	1.86 ± 0.15	1.91 ± 0.22	2.75 ± 0.23

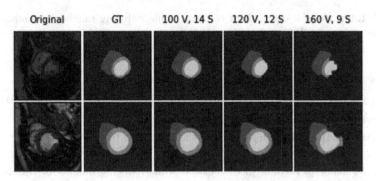

Fig. 2. Qualitative results demonstrating improved performance as the network is trained on an increased number of annotated slices/decreased number of volumes. The first row is a random slice from a 3D segmentation on the ACDC dataset. The second row is the same from the M&Ms dataset. V = volumes, S = slices and represents the total number of volumes/slices each network was trained on respectively.

This is further observed in Tables 5 and 6 where we compare 1400 to 700 annotated slices per dataset. Again, we note how better performance is achieved with more slices, even if a smaller number of volumes is annotated. We observe that in general, the networks trained on ACDC perform better than those trained on M&Ms when using the same number of slices and volumes. Note that since the overall number of slices is quite similar for both datasets (3200 for ACDC, 3360 for M&Ms) the relative proportion for a given slice/volume trial is also similar. Finally, we note that ACDC seems to be more affected by using fewer volumes - that is, the drop in performance when halving the number of volumes (and keeping the number of slices fixed) is slightly larger compared to M&Ms. Despite this, we see the same overall pattern that using fewer slices leads to worse performance.

The cost of annotating additional volumes is very expensive. However, Tables 5 and 6 show how doubling the number of annotated volumes usually only slightly improves metric scores. In contrast, the cost of annotating additional slices is usually much less expensive. Annotating only 3–4 more slices can yield significant metric gains (annotating 12 slices instead of 9 for ACDC improves Dice by 9%, while annotating 10 slices instead of 6 for M&Ms improves the score by 8%).

Table 5. Influence of keeping slices constant while reducing number of training volumes. Trained on **ACDC** with a 3D nnU-Net network.

Slices	9		10		12		14		17		20	
Volumes	80	160	77	144	60	120	50	100	40	80	32	65
Dice	0.78	0.82	0.85	0.87	0.87	0.88	0.87	0.88	0.84	**0.89**	0.70	0.87
HD (mm)	9.61	8.27	8.50	7.67	9.21	6.70	8.83	8.52	11.35	**6.15**	32.4	8.00
MAD (mm)	2.41	2.01	2.16	1.81	2.18	1.60	2.35	2.08	3.59	**1.51**	9.79	1.98

Table 6. Influence of keeping slices constant while reducing number of training volumes. Trained on **M&Ms** with a 3D nnU-Net network.

Slices	6		8		9		10		12		14	
Volumes	120	240	96	192	80	160	77	144	60	120	50	100
Dice	0.75	0.77	0.81	0.82	0.82	0.83	0.83	0.84	0.85	0.84	0.84	**0.86**
HD (mm)	12.11	9.28	7.77	8.17	7.79	6.93	6.92	7.07	6.30	6.72	7.14	**6.10**
MAD (mm)	3.76	2.75	2.15	2.34	2.22	1.91	1.83	1.93	1.67	1.86	1.79	**1.58**

5   Discussion and Conclusion

In this paper, we investigate the influence of different sparse annotation strategies for the segmentation of short-axis view in cardiac MRI. We use the state-of-the-art nnU-Net on two public datasets and evaluate using standard segmentation metrics. We show that good segmentation results can be achieved even when using a severely restricted dataset. We note that using more than 48 volumes is sufficient to achieve a Dice score greater than 0.85, and using more than 80 volumes is comparable to using the full dataset (160 volumes in ACDC, 240 volumes in M&Ms). This corresponds to using half of all available volumes for ACDC (a total of 1600 slices), and one-third of available volumes for M&Ms (a total of 1120 slices).

Further, experiments with sparse annotations demonstrate that using more than two-thirds of available slices yields results comparable to using the full set of available annotations. We observe that randomly sampling these slices from throughout all cardiac regions results in better performance than sub-sampling from particular regions. If two regions are sub-sampled, the mid-ventricular region contributes the most to segmentation performance and allows better generalisation. This is also true when sampling only a single region. As expected, the apical region generalises (and performs) the worst due to differences in ventricular sizes compared to the middle and basal regions.

Finally, we demonstrate the importance of using more slices relative to volumes. When we use the full set of available training volumes with a limited number of slices, we achieve only poor results. However, even when the number of volumes is reduced, good performance can still be achieved if there is a large number of slices to learn from. For both datasets, annotating upwards of 60%

of the slices provides the best results. We also note that the cost of annotating additional volumes is very expensive, but yields small improvements. In contrast, the cost of annotating additional slides is less expensive, but can yield very large improvements. Therefore, we recommend annotating as many slices as possible in each volume instead of annotating more volumes with fewer slices.

Future work will build on this baseline using state-of-the-art nnU-Net and compare different approaches for sparse annotations, such as transfer learning or semi-supervised learning, to evaluate the most appropriate strategy. Further, the study will be extended to include more datasets with different cardiac MRI views or modalities (e.g. ultrasound, CT).

References

1. Chen, C., Qin, C., Qiu, H., et al.: Deep learning for cardiac image segmentation: a review. Front. Cardiovasc. Med. **7**, 25 (2020)
2. Zhao, A., Balakrishnan, G., Durand, F., et al.: Data augmentation using learned transformations for one-shot medical image segmentation. In: Proceedings of the IEEE/CVF Conference on Computer Vision and Pattern Recognition, pp. 8543–8553 (2019)
3. Bai, W., Sinclair, M., Tarroni, G., et al.: Automated cardiovascular magnetic resonance image analysis with fully convolutional networks. J. Cardiovasc. Magn. Reson. **20**(1), 65 (2018)
4. Khened, M., Kollerathu, V.A., Krishnamurthi, G.: Fully convolutional multi-scale residual DenseNets for cardiac segmentation and automated cardiac diagnosis using ensemble of classifiers. Med. Image Anal. **51**, 21–45 (2019)
5. Chen, S., Ma, K., Zheng, Y.: Med3D: transfer learning for 3D medical image analysis. arXiv (2019)
6. Qin, C., et al.: Joint learning of motion estimation and segmentation for cardiac MR image sequences. In: Frangi, A.F., Schnabel, J.A., Davatzikos, C., Alberola-López, C., Fichtinger, G. (eds.) MICCAI 2018. LNCS, vol. 11071, pp. 472–480. Springer, Cham (2018). https://doi.org/10.1007/978-3-030-00934-2_53
7. Can, Y.B., Chaitanya, K., Mustafa, B., Koch, L.M., Konukoglu, E., Baumgartner, C.F.: Learning to segment medical images with scribble-supervision alone. In: Stoyanov, D., et al. (eds.) DLMIA/ML-CDS -2018. LNCS, vol. 11045, pp. 236–244. Springer, Cham (2018). https://doi.org/10.1007/978-3-030-00889-5_27
8. Bai, W., et al.: Self-supervised learning for cardiac MR image segmentation by anatomical position prediction. In: Shen, D., et al. (eds.) MICCAI 2019. LNCS, vol. 11765, pp. 541–549. Springer, Cham (2019). https://doi.org/10.1007/978-3-030-32245-8_60
9. Zeng, D., et al.: Positional contrastive learning for volumetric medical image segmentation. In: de Bruijne, M., et al. (eds.) MICCAI 2021. LNCS, vol. 12902, pp. 221–230. Springer, Cham (2021). https://doi.org/10.1007/978-3-030-87196-3_21
10. Peng, J., Wang, Y.: Medical image segmentation with limited supervision: a review of deep network models. IEEE Access **9**, 36 827–36 851 (2021)
11. Bernard, O., Lalande, A., Zotti, C., et al.: Deep learning techniques for automatic MRI cardiac multi-structures segmentation and diagnosis: is the problem solved? IEEE Trans. Med. Imaging **37**(11), 2514–2525 (2018)

12. Campello, V.M., Gkontra, P., Izquierdo, C., et al.: Multi-centre, multivendor and multi-disease cardiac segmentation: the M and Ms challenge. IEEE Trans. Med. Imaging **40**(12), 3543–3554 (2021)
13. Bitarafan, A., Nikdan, M., Baghshah, M.S.: 3D image segmentation with sparse annotation by self-training and internal registration. IEEE J. Biomed. Health Inform. **25**(7), 2665–2672 (2021). https://doi.org/10.1109/JBHI.2020.3038847
14. Bai, W., Suzuki, H., Qin, C., et al.: Recurrent neural networks for aortic image sequence segmentation with sparse annotations (2018). https://doi.org/10.48550/arXiv.1808.00273
15. You, C., Dai, W., Liu, F., et al.: Mine your own anatomy: Revisiting medical image segmentation with extremely limited labels. arXiv preprint arXiv:2209.13476 (2022)
16. Isensee, F., Jaeger, P.F., Kohl, S.A., et al.: nnU-Net: a self-configuring method for deep learning-based biomedical image segmentation. Nat. Meth. **18**(2), 203–211 (2021)

Contrast-Agnostic Groupwise Registration by Robust PCA for Quantitative Cardiac MRI

Xinqi Li[1]($\boxtimes$), Yi Zhang[1], Yidong Zhao[1], Jan van Gemert[2], and Qian Tao[1]

[1] Department of Imaging Physics, Delft University of Technology, Delft,
The Netherlands
Xinqi.Li@mdc-berlin.de
[2] Computer Vision Lab, Delft University of Technology, Delft, The Netherlands

Abstract. Quantitative cardiac magnetic resonance imaging (MRI) is an increasingly important diagnostic tool for cardiovascular diseases. Yet, co-registration of all baseline images within the quantitative MRI sequence is essential for the accuracy and precision of quantitative maps. However, co-registering all baseline images from a quantitative cardiac MRI sequence remains a nontrivial task because of the simultaneous changes in intensity and contrast, in combination with cardiac and respiratory motion. To address the challenge, we propose a novel motion correction framework based on robust principle component analysis (rPCA) that decomposes quantitative cardiac MRI into low-rank and sparse components, and we integrate the groupwise CNN-based registration backbone within the rPCA framework. The low-rank component of rPCA corresponds to the quantitative mapping (i.e. limited degree of freedom in variation), while the sparse component corresponds to the residual motion, making it easier to formulate and solve the groupwise registration problem. We evaluated our proposed method on cardiac T1 mapping by the modified Look-Locker inversion recovery (MOLLI) sequence, both before and after the Gadolinium contrast agent administration. Our experiments showed that our method effectively improved registration performance over baseline methods without introducing rPCA, and reduced quantitative mapping error in both in-domain (pre-contrast MOLLI) and out-of-domain (post-contrast MOLLI) inference. The proposed rPCA framework is generic and can be integrated with other registration backbones.

Keywords: Quantitative MRI · Groupwise registration · Robust PCA · motion correction

1 Introduction

Quantitative cardiac MRI, such as T1 and T2 mapping [22], is an increasingly important imaging modality to examine cardiovascular diseases [24]. However,

---

Supplementary Information The online version contains supplementary material available at https://doi.org/10.1007/978-3-031-52448-6_8.

the quality of quantitative mapping is negatively affected by respiratory and cardiac motion during the MR acquisition procedure [28]. Such motion leads to misalignment of tissue across baseline images, resulting in deteriorated accuracy and precision of the final quantitative mapping [18]. To improve the quality of quantitative cardiac MRI, motion correction by deformable image registration is an essential part of the post-processing pipeline [2,8,21].

Conventionally, deformable image registration is implemented in a pairwise fashion: each time two images are registered, with one designated as fixed and one moving. However, for quantitative cardiac MRI, the number of images is highly variable (ranging from 3 to >20) depending on the specific sequence. This makes pairwise registration not a natural option, as the "best" fixed image is hard to define. Moreover, registration error easily propagates across the baseline images, given all pairwise registration steps are independently performed. The alternative approach of *groupwise image registration*, which registers all baseline images simultaneously, was proposed for quantitative MRI motion correction [12,13,17,26]. Groupwise registration promises improved robustness across a sequence of images by optimizing a global metric which promotes co-registration of *all* frames, including those with extremely poor contrast and hence difficult to register in a pairwise fashion. Groupwise image registration can be divided into two paradigms: classical iterative optimization methods that are relatively slow [12,15–17,20,26] and deep-learning-based methods that promise fast inference [1,7,10,14,29].

A special challenge in motion correction for quantitative cardiac MRI is that the change in image contrast and intensity can vary drastically across baseline images, completely agnostic to the image registration pipeline [28]. The pattern of variation, which is determined by the underlying MR signal model, differs per quantitative sequences; even with the same signal model, the contrast is still dependent on the exact scheme of acquisition, which differs again among MRI machines. This makes it difficult to design a consistently reliable registration metric for optimization. Conventional registration metrics, such as NCC and NMI, can still be sensitive to agnostic contrast changes and fail [5,19,27]. Therefore, finding a robust registration metric in the face of agnostic contrast changes is of great interest.

Furthermore, we observed that an under-studied phenomenon is the degenerated solution of groupwise registration, in the format of ghosting artefacts or pixel collapse [27]. These degenerated solutions lead to an optimal metric, but are implausible because they violate the anatomical consistency. In this paper, we will further investigate the susceptibility of NCC and NMI to such artifacts.

In this work, we set out to tackle the agnostic contrast change in quantitative cardiac MRI by designing a novel registration framework, which integrates robust PCA (rPCA) [6] with state-of-the-art image registration backbones. Our rationale of introducing rPCA is as follows: firstly, the signal model, which is typically governed by physics principles, has a limited degree of freedom [9,22], underlying the *low-rank* component of rPCA. Secondly, the motion of quantitative cardiac MRI is *sparse* in the sense that it is often concentrated around

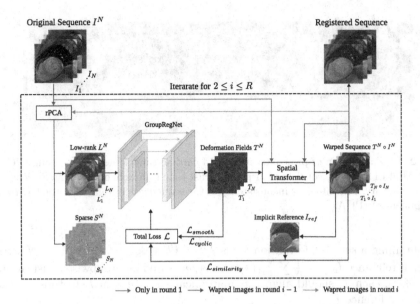

Fig. 1. Overview of the proposed framework for contrast-agnostic registration. The dotted rectangle denotes the iterative registration pipeline that progressively corrects motion from round 2 to maximal round R.

the heart, induced by non-ideal breath-hold and heart rate variability, while the background, e.g., rib cage and lung, stay largely static. Decomposition of the two components creates ease for registration algorithms. In this paper, we propose to integrate rPCA with the state-of-the-art deep-learning groupwise registration method [29] for fast, reliable motion correction of quantitative cardiac MRI. Our main contributions are:

1. We propose a novel groupwise image registration framework, which is, to the best of our knowledge, the first attempt to utilize rPCA in groupwise registration with a deep learning backbone. This generic framework can be integrated with any existing registration methods, either classical optimization or modern deep learning methods.
2. We evaluated and demonstrated the generalizability of our contrast-agnostic method on out-of-domain quantitative MRI sequences.
3. We further investigated the fitness of two popular metrics, NCC and NMI, for groupwise registration. We showed empirically that NCC could give rise to registration artefacts, leading to unwanted anatomical deformation.

2   Methods

2.1   Problem Formulation

Given a sequence of baseline images $I^N = \{I_i \in \mathbb{R}^{H \times W} | i = 1, ..., N\}$, the goal of groupwise registration is to align all I_i into one common coordinate system

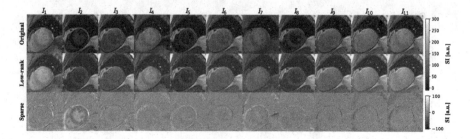

Fig. 2. Decomposition of pre-contrast MOLLI cardiac time-series images using rPCA. Each MOLLI sequence consists of 11 pre-contrast time frames in our setting. SI denotes signal intensity. The intensity inconsistency of the sequence is mitigated in the low-rank component of rPCA.

by obtaining a set of deformation fields $T^N = \{T_i \in \mathbb{R}^{2 \times H \times W}; i = 1, ..., N\}$. An implicit reference $I_{ref} = \frac{1}{N} \sum_{n=1}^{N} (T_n \circ I_n)$ is generated for groupwise registration. Therefore, each T_i should align the anatomical structures in I_i to those in the implicit reference I_{ref}.

The proposed rPCA framework is as follows: rPCA first decomposes the input baseline images I^N into the low-rank matrix L^N and sparse matrix S^N. The low-rank component is fed to the deep learning backbone to learn the deformation field T^N. Then T^N is applied to the input I^N to obtain warped images $T^N \circ I^N$, which serves as the input of the next iteration of rPCA, until the maximal iteration number is reached. The framework then progressively corrects the motion in the original input, but for each iteration, rPCA enables us to work only on the low-rank part, which is easier to register than the original input. This rationale will be revisited later in the paper. The diagram of our proposed framework is shown in Fig. 1.

2.2   Robust Principal Component Analysis

Robust principal component analysis (rPCA) [6], as its name suggests, is a robust version of PCA for matrix decomposition: For a given data matrix M, where in our case M is the matrix of vectorized grouped images I^N, the rPCA decomposes $M \in \mathbb{R}^{m \times n}$ into the sum of a low-rank matrix L and a sparse matrix S via solving the following optimization problem:

$$\text{minimize } \|L\|_* + \lambda \|S\|_1, \text{ subject to } L + S = M, \tag{1}$$

where $\| \cdot \|_*$ denotes the nuclear norm, $\| \cdot \|_1$ denotes the l_1 norm, and λ is a hyperparameter trading off the two components, which is often set by default as $\lambda = 1/\sqrt{\max(m, n)}$. Such optimization problems can be solved by well-established algorithms, such as proximal gradient descent methods [11].

An illustration of rPCA on pre-contrast cardiac MRI is shown in Fig. 2. It can be seen that the sparse matrix captures abrupt changes in baseline images, either of contrast (such as in I_2) or motion (such as in I_7), which are usually more difficult to handle by registration algorithms. The low-rank matrix, in contrast, consists of the components in baseline images which have relatively lower variations and are easier to align. Note that the amplitude of the sparse component is much lower than that of the low-rank component, the latter capturing the majority of information in the original matrix.

2.3 Loss Functions

The optimization problem for finding the deformable mapping T^N can be formulated as follows:

$$T^N = \arg\min_{T^N} \mathcal{L}_{\text{similarity}} + \lambda_0 \mathcal{L}_{\text{smooth}} + \lambda_1 \mathcal{L}_{\text{cyclic}}, \tag{2}$$

where $\mathcal{L}_{\text{similarity}}$, $\mathcal{L}_{\text{smooth}}$, and $\mathcal{L}_{\text{cyclic}}$ denote similarity function, smoothness regularization, and cyclic consistency, with weight parameters λ_0 and λ_1.

Similarity Functions: We employed the normalized mutual information (NMI) to measure the similarity between the input images I^N to the warped images $T^N \circ I^N$, which can measure alignment in the face of contrast changes [27]. The NMI between two images is defined as:

$$NMI(I_1, I_2) = \frac{2MI(I_1, I_2)}{H(I_1) + H(I_2)}, \tag{3}$$

where $MI(I_1, I_2)$ denotes the mutual information between I_1 and I_2, $H(I_1)$ is the entropy of image I_1, and $H(I_2)$ for image I_2, respectively. For groupwise registration, the similarity loss $L_{similarity}$ is then defined as:

$$\mathcal{L}_{\text{similarity}} = -\frac{1}{N} \sum_{n=1}^{N} NMI(T_n \circ I_n, I_{ref}). \tag{4}$$

Another popular similarity loss is also considered and discussed, which is the local normalized cross-correlation (NCC) [3], defined as

$$NCC(I_1, I_2) = \frac{1}{H \times W} \sum_{i,j \in H, W} \frac{\sum_{x \in \Omega}(I_1(x) - \bar{I}_1(i,j))(I_2(x) - \bar{I}_2(i,j))}{\sqrt{\hat{I}_1(i,j)\hat{I}_2(i,j)}}, \tag{5}$$

where H and W corresponds to the height and width of the image, Ω indicates the neighborhood voxels around the voxel at position (i,j) and $\bar{I}(i,j)$ and $\hat{I}(i,j)$ denote the local mean and variance.

Smoothness Regularization: The smoothness of the deformation field is regularized through B-spline registration [25]. We adopted B-spline because it can prevent the image from folding and inherently lead to smooth deformation fields:

$$\mathcal{L}_{\text{smooth}} = \frac{1}{H \times W} \sum_{n=1}^{N} \int_0^H \int_0^W \left[\left(\frac{\partial^2 \hat{T}_n}{\partial x^2} \right)^2 + \left(\frac{\partial^2 \hat{T}_n}{\partial y^2} \right)^2 + 2 \left(\frac{\partial^2 \hat{T}_n}{\partial xy} \right)^2 \right] dx dy, \tag{6}$$

where $\hat{T}_n = T_n + \sum_{l=0}^{k} \sum_{m=0}^{k} B_l(u) B_m(v) \phi_{i+l,j+m}$, and B_l is the l–th B-spline basis function, k is the order of B-spline, and $\phi_{i,j}$ denotes the control points with uniform space across the image. B-Spline control points will affect the surrounding deformation fields based on the kernel function.

Cyclic Consistency: For groupwise registration, the cyclic consistent regularization keeps the estimated implicit reference at the center of all baseline images in the manifold by minimizing the deformation field to the implicit reference [29]:

$$\mathcal{L}_{\text{cyclic}}(T^N) = \sqrt{\frac{1}{2(H \times W)} \sum_{i,j \in H,W} \left(\sum_n T_n(i,j) \right)^2}, \tag{7}$$

where $T_n(i,j)$ denotes the value of T_n at coordinate (i,j). This term prevents the degenerated solution where textures in all images collapse.

2.4 CNN-Based Neural Network Architecture

The convolution neural network architecture follows that of the VoxelMorph [4], and GroupRegNet [29], based on the UNet [23] architecture consisting of encoding and decoding layers with skip connection. Both encoder and decoder use convolutional blocks consisting of a 2D convolution and a Leaky ReLU activation function. The encoder captures the hierarchical features of the input images with multiple convolution blocks. The number of decoder layers was controlled by the B-spline kernel size k [23]. The larger kernel size indicates less decode layers, which makes the deformation field more homogeneous. This enables the coarse-to-fine representation of the two-channel deformation field. The final deformation field is computed by B-spline free form deformation (FFD) transformation model [25] based on the decoder output.

2.5 Evaluation Methods

T1 Fitting Error: In this paper, we used myocardial T1 mapping by the modified Look Locker inversion recovery sequence (MOLLI), one of the most widely used mapping modalities in clinical practice [22]. T1 mapping follows a three-parameter model, expressed by

$$y(T_I) = A - B e^{-T_I/T_1^*}, \tag{8}$$

where y denotes the signal intensity, T_I denotes the inversion time for acquisition of each baseline image, and A, B, and $T1^*$ are parameters to be estimated. Since motion correction leads to a better fitting of this MR physics model at each pixel, here we measure the performance through the T1 mapping within the ROI (myocardium and left ventricle) and the standard deviation (SD) error [18] as an indication of the fitting error. A lower SD error indicates better motion correction. We used both the native (pre-contrast) T1 mapping and post-contrast T1 mapping sequences (after Gadolinium administration). To test the generalizability of our framework, we trained our NN exclusively on pre-contrast T1 mapping, while testing it on both pre-contrast (in-domain) and post-contrast (out-of-domain) sequences.

Dissimilarity Metrics $\mathcal{D}_{PCA}$: We evaluate the warped images using $\mathcal{D}_{PCA}$, the ratio of the top-K eigenvalues to the sum of eigenvalues of the correlation matrix [17]. The higher the ratio, the better the performance of registration.

Baseline Methods: We compared our proposed framework with two methods: (1) the conventional groupwise method, Elastix-PCA [17], and (2) the groupwise registration method [29] without rPCA, denoted by *GroupRegNet**, which follows [29], but using NMI as the optimization metric. We also performed experiments on *GroupRegNet** using the NCC metric as in the original work and compared the results with NMI.

3   Experiments and Results

Dataset: We used a cardiac MRI dataset including 48 subject, with both pre-contrast and post-contrast MOLLI sequences (Philips 3.0T). Each subject had 1 to 3 slices acquired at the base, mid-ventricular, and apex levels. In total 120 pre-contrast and 120 post-contrast MOLLI sequences were included. All images were resampled to a $224 \times 224 \times 11$ grid with $1\,\text{mm}^3$ isotropic resolution and then cropped to $112 \times 112 \times 11$ at the center. The training comprised 100 random images from only the *pre-contrast* MOLLI sequences. The rest 20 pre-contrast MOLLI sequences and their corresponding post-contrast sequences, in total 40, formed the test set. We note here that the pre-contrast sequences are the *in-domain* test data, while the post-contrast sequences are the *out-of-domain* test data, given their contrast changes follow a different pattern governed by much higher relaxation rate due to the contrast agent (e.g. lower T1).

Table 1. Experiment results on T1 mapping. We compare T1 SD and $\mathcal{D}_{PCA}(K = 1)$ before and after registration. A higher $\mathcal{D}_{PCA}(K = 1)$ indicates larger power in the principle components thus better alignment. The SD measures the $T1$ fitting error within the ROI. Lower SD indicates lower fitting error thus better alignment. Our method (w/ rPCA) outperforms the GroupRegNet* on both pre-contrast and post-contrast data in terms of both SD and $\mathcal{D}_{PCA}$. The bold values demonstrates the best result and underlined values is the second highest performance for each metric.

Modality	Method	SD (ms)↓	$\mathcal{D}_{PCA}\%$ ↑	Time (s)
Pre-GD	Elastix-PCA	<u>54.5 ± 21.7</u>	93.9	≈600
	GroupRegNet*	55.4 ± 21.4	94.0	1.28
	Ours (w/ rPCA)	**53.9 ± 21.9**	**94.4**	7.11
Post-GD	Elastix-PCA	**21.5 ± 16.1**	91.7	≈600
	GroupRegNet*	24.5 ± 13.3	81.5	1.28
	Ours (w/ rPCA)	<u>21.8 ± 12.5</u>	**92.1**	7.11

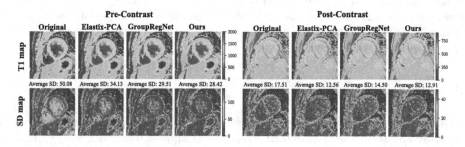

Fig. 3. Representative figures of the quantitative CMR. The figure demonstrated the T1 map (top row) and SD map (bottom row) of the same subject on both pre- and post-contrast. We compared our proposed method with conventional method (Elastix-PCA) and deep learning based GroupRegNet, and the reported average SD within the ROI indicated that our method outperformed others.

Implementation Details: Robust PCA was implemented with the GoDec algorithm [30]. In each round, the rank of L was set to be half of the sequence length, which was 5 in our case. Empirically we applied a decoder with 4 layers. In this case, the decoder included 2 convolution blocks and the output deformation field was $31 \times 31 \times 11 \times 2$. The final deformation field was transformed to $112 \times 112 \times 11 \times 2$ using B-spline FFD. The smooth regulation's weight λ_0 is set to 0.001 and cyclic regulation's weight λ_1 is set to 0.01 empirically.

Choice of Similarity Functions: Two similarity functions, NCC and NMI, are evaluated. We observed that NCC loss led to undesirable deformation as well as altered distribution of the T1 values (details in Supplementary). As suggested in [27], NCC naturally favors homogeneous distribution of pixel intensities and lead to over-smooth myocardium textures that fail the purpose of quantitative mapping, while NMI maintained the shape and texture of the ROI.

Results: The quantitative results of registration and quantitative mapping are shown in Table 1 and two representative figures are shown in Fig. 3. Note that to demonstrate the generalizability of the learned model, we train the model **only** on pre-contrast data (denoted as Pre-Gd) and tested on both pre- and post-contrast data. Our method performs best on pre-contrast datasets according to SD within the ROI and outperforms the GroupRegNet* on post-contrast data. Elastix-PCA gives slightly better performance on post-contrast data because the optimization is per datasets (no training and inference). However, it takes around 10 min for each subject, which is much slower compared to our method, with an average inference time of 7.1 s per sequence.

4   Conclusion

In conclusion, we proposed a novel rPCA framework for robust motion correction of quantitative cardiac MRI. We aim for robust performance despite agnostic image contrast changes, which are typical of quantitative MRI. We showed that the introduction of rPCA, which separates low-rank and sparse components of baseline images, led to improved registration performance and facilitated the generalization of the trained network on out-of-domain data.

In addition, our work also compared the two commonly used metrics for groupwise registration, namely, NCC and NMI, and showed that NCC might give rise to potential loss-specific artifacts in heart anatomy and quantitative mapping. Future investigations are warranted to focus not only on the performance of image registration but also on the fidelity of quantitative mapping.

References

1. Ahmad, S., Fan, J., Dong, P., Cao, X., Yap, P.T., Shen, D.: Deep learning deformation initialization for rapid groupwise registration of inhomogeneous image populations. Front. Neuroinform. **13**, 34 (2019)
2. Ashburner, J.: A fast diffeomorphic image registration algorithm. Neuroimage **38**(1), 95–113 (2007)
3. Avants, B.B., Epstein, C.L., Grossman, M., Gee, J.C.: Symmetric diffeomorphic image registration with cross-correlation: evaluating automated labeling of elderly and neurodegenerative brain. Med. Image Anal. **12**(1), 26–41 (2008)
4. Balakrishnan, G., Zhao, A., Sabuncu, M.R., Guttag, J., Dalca, A.V.: Voxelmorph: a learning framework for deformable medical image registration. IEEE Trans. Med. Imaging **38**(8), 1788–1800 (2019)
5. Brudfors, M., Balbastre, Y., Ashburner, J.: Groupwise multimodal image registration using joint total variation. In: Papież, B.W., Namburete, A.I.L., Yaqub, M., Noble, J.A. (eds.) MIUA 2020. CCIS, vol. 1248, pp. 184–194. Springer, Cham (2020). https://doi.org/10.1007/978-3-030-52791-4_15

6. Candès, E.J., Li, X., Ma, Y., Wright, J.: Robust principal component analysis? J. ACM (JACM) **58**(3), 1–37 (2011)
7. Che, T., et al.: Deep group-wise registration for multi-spectral images from fundus images. IEEE Access **7**, 27650–27661 (2019)
8. Chen, X., Diaz-Pinto, A., Ravikumar, N., Frangi, A.F.: Deep learning in medical image registration. Prog. Biomed. Eng. **3**(1), 012003 (2021)
9. Chow, K., Flewitt, J.A., Green, J.D., Pagano, J.J., Friedrich, M.G., Thompson, R.B.: Saturation recovery single-shot acquisition (SASHA) for myocardial T1 mapping. Magn. Reson. Med. **71**(6), 2082–2095 (2014)
10. Fechter, T., Baltas, D.: One-shot learning for deformable medical image registration and periodic motion tracking. IEEE Trans. Med. Imaging **39**(7), 2506–2517 (2020)
11. Feng, J., Xu, H., Yan, S.: Online robust PCA via stochastic optimization. In: Advances in Neural Information Processing Systems, vol. 26 (2013)
12. Feng, Q., et al.: Liver DCE-MRI registration in manifold space based on robust principal component analysis. Sci. Rep. **6**(1), 34461 (2016)
13. Geng, X., Christensen, G.E., Gu, H., Ross, T.J., Yang, Y.: Implicit reference-based group-wise image registration and its application to structural and functional MRI. Neuroimage **47**(4), 1341–1351 (2009)
14. Gonzales, R., et al.: MOCOnet: robust motion correction of cardiovascular magnetic resonance T1 mapping using convolutional neural networks. Front. Cardiovasc. Med. **8**, 768245 (2021)
15. Guyader, J.M., et al.: Groupwise image registration based on a total correlation dissimilarity measure for quantitative MRI and dynamic imaging data. Sci. Rep. **8**(1), 13112 (2018)
16. Hamy, V., et al.: Respiratory motion correction in dynamic MRI using robust data decomposition registration-application to DCE-MRI. Med. Image Anal. **18**(2), 301–313 (2014)
17. Huizinga, W., et al.: PCA-based groupwise image registration for quantitative MRI. Med. Image Anal. **29**, 65–78 (2016)
18. Kellman, P., Arai, A.E., Xue, H.: T1 and extracellular volume mapping in the heart: estimation of error maps and the influence of noise on precision. J. Cardiovasc. Magn. Reson. **15**(1), 1–12 (2013)
19. Klein, A., et al.: Evaluation of 14 nonlinear deformation algorithms applied to human brain MRI registration. Neuroimage **46**(3), 786–802 (2009)
20. Li, Y., Wu, C., Qi, H., Si, D., Ding, H., Chen, H.: Motion correction for native myocardial T1 mapping using self-supervised deep learning registration with contrast separation. NMR Biomed. **35**(10), e4775 (2022)
21. Makela, T., et al.: A review of cardiac image registration methods. IEEE Trans. Med. Imaging **21**(9), 1011–1021 (2002)
22. Messroghli, D.R., Radjenovic, A., Kozerke, S., Higgins, D.M., Sivananthan, M.U., Ridgway, J.P.: Modified look-locker inversion recovery (MOLLI) for high-resolution T1 mapping of the heart. Magn. Reson. Med. Official J. Int. Soc. Magn. Reson. Med. **52**(1), 141–146 (2004)
23. Ronneberger, O., Fischer, P., Brox, T.: U-Net: convolutional networks for biomedical image segmentation. In: Navab, N., Hornegger, J., Wells, W.M., Frangi, A.F. (eds.) MICCAI 2015 Part III. LNCS, vol. 9351, pp. 234–241. Springer, Cham (2015). https://doi.org/10.1007/978-3-319-24574-4_28
24. de Roos, A., Higgins, C.B.: Cardiac radiology: centenary review. Radiology **273**(2S), S142–S159 (2014)

25. Rueckert, D., Sonoda, L.I., Hayes, C., Hill, D.L., Leach, M.O., Hawkes, D.J.: Non-rigid registration using free-form deformations: application to breast MR images. IEEE Trans. Med. Imaging **18**(8), 712–721 (1999)

26. Tao, Q., van der Tol, P., Berendsen, F.F., Paiman, E.H., Lamb, H.J., van der Geest, R.J.: Robust motion correction for myocardial T1 and extracellular volume mapping by principle component analysis-based groupwise image registration. J. Magn. Reson. Imaging **47**(5), 1397–1405 (2018)

27. de Vos, B.D., van der Velden, B.H., Sander, J., Gilhuijs, K.G., Staring, M., Išgum, I.: Mutual information for unsupervised deep learning image registration. In: Medical Imaging 2020: Image Processing, vol. 11313, pp. 155–161. SPIE (2020)

28. Xue, H., et al.: Motion correction for myocardial T1 mapping using image registration with synthetic image estimation. Magn. Reson. Med. **67**(6), 1644–1655 (2012)

29. Zhang, Y., Wu, X., Gach, H.M., Li, H., Yang, D.: Groupregnet: a groupwise one-shot deep learning-based 4D image registration method. Phys. Med. Biol. **66**(4), 045030 (2021)

30. Zhou, T., Tao, D.: Godec: randomized low-rank & sparse matrix decomposition in noisy case. In: Proceedings of the 28th International Conference on Machine Learning. ICML 2011 (2011)

FM-Net: A Fully Automatic Deep Learning Pipeline for Epicardial Adipose Tissue Segmentation

Fan Feng[1]([✉]), Carl-Johan Carlhäll[2,3,4], Yongyao Tan[1], Shaleka Agrawal[1],
Peter Lundberg[2,3,5], Jieyun Bai[1,6], John Zhiyong Yang[1], Mark Trew[1],
and Jichao Zhao[1]

[1] Auckland Bioengineering Institute, University of Auckland, Auckland, New Zealand
ffen908@aucklanduni.ac.nz
[2] Division of Diagnostics and Specialist Medicine, Department of Health, Medicine and Caring Sciences, Linköping University, Linköping, Sweden
[3] Center for Medical Image Science and Visualization (CMIV), Linköping University, Linköping, Sweden
[4] Department of Clinical Physiology in Linköping, and Department of Health, Medicine and Caring Sciences, Linköping University, Linköping, Sweden
[5] Department of Medical Radiation Physics, and Department of Health, Medicine and Caring Sciences, Linköping University, Linköping, Sweden
[6] School of Information Science and Technology, Jinan University, Guangzhou, China

Abstract. Epicardial adipose tissue (EAT) has been recognized as a risk factor and independent predictor for cardiovascular diseases (CVDs), due to its intimate relationship with the myocardium and coronary arteries. Dixon MRI is widely used to depict adipose tissue by deriving fat and water signals. The purpose of this study was to automatically segment and quantify EAT from Dixon MRI data using a fully automated deep learning pipeline based on fat maps (FM-Net). Data used in this study was from a sub-study (HEALTH) of the Swedish CArdioPulmonarybiolmage Study (SCAPIS), with 6504 Dixon MRI 2D images from 90 participants (45 each for type 2 diabetes and controls). FM-Net was comprised of a double Res-UNet CNN architecture, designed to compensate for the severe class imbalance and complex geometry of EAT. The first network accurately detected the region of interest (ROI) containing fat, and the second network performed targeted regional segmentation of the ROI. Performance of fat segmentation was improved by using fat maps as input of FM-Net, to enhance fat features by combining out-of-phase, water, and fat phase images. Performance was evaluated using dice similarity coefficient (DSC) and 95% Hausdorff distance (HD95). Overall, FM-Net obtained a promising DSC of 86.3%, and a low HD95 of 3.11 mm, outperforming existing state-of-the-art methods. The proposed method enables automatic and accurate quantification of EAT from Dixon MRI data, which could enhance the understanding of the role of EAT in CVDs.

Keywords: Deep Learning · Epicardial Adipose Tissue · Cardiovascular Disease · Dixon MRI

© The Author(s), under exclusive license to Springer Nature Switzerland AG 2024
O. Camara et al. (Eds.): STACOM 2023, LNCS 14507, pp. 88–97, 2024.
https://doi.org/10.1007/978-3-031-52448-6_9

1 Introduction

Epicardial adipose tissue (EAT) also known as epicardial fat is an ectopic fat deposit surrounding the heart, distributed between the visceral serous pericardium (epicardium) and the parietal serous pericardium, and therefore is both anatomically and functionally in contact with the myocardium and coronary arteries due to the lack of a fascial boundary [1, 2]. EAT has gained great attention as it has been demonstrated to play an important role in cardiovascular diseases (CVDs) [3], such as atrial fibrillation and coronary artery disease [4–6]. EAT could be modifiable via pharmacologic treatment and may be a therapeutic target [7, 8]. Recent studies have suggested that EAT could be a potential imaging biomarker for the risk of CVDs. Hence, there is a growing interest in the segmentation and quantification of EAT.

EAT can be visualized and quantified non-invasively by echocardiography, computed tomography (CT) and magnetic resonance imaging (MRI) [9]. Among them, MRI relying on the Dixon water-fat-separated technique is currently regarded as the gold standard for assessing subcutaneous and visceral adipose tissue with high spatial resolution, consequently making it a natural choice for quantifying and analyzing the segmented 3D geometry of EAT. However, systematic quantification of EAT from medical imaging remains challenging in clinical practice. The manual process is time-consuming, prone to human error, and exhibits significant differences between observers due to varying size, shape, and distribution of EAT. To overcome these limitations, an automated method is indispensable to perform this task routinely and reliably in an efficient manner.

Deep learning has emerged as a powerful technique in medical image analysis, enabling end-to-end automatic segmentation [10, 11]. So far, many methods based on deep learning have been developed to segment EAT [12–15], such as Zhang et al. utilized two U-Nets and morphological layers to achieve better EAT segmentation results [16], and He et al. also proposed a 3D attention U-Net architecture for EAT segmentation [17]. However, all of them were developed to segment EAT on non-contrast or contrast CT images or high-resolution coronary computed tomography angiography (CCTA), with a few studies focusing on cardiac MRI [18, 19]. To the best of our knowledge, no deep learning algorithm has yet been published specifically to EAT segmentation on Dixon MRI data. The Dixon MRI sequence provides complementary information, in-phase (IP) images, out-of-phase (OP) images, fat phase (F) images and water phase (W) images, that can aid in identifying and segmenting EAT regions accurately and can also enable a deep learning algorithm to differentiate adipose tissue from other structures effectively. Fat fraction (FF) was commonly computed using fat and water phase images, FF = F/(F + W). By setting an appropriate threshold in the fat fraction, it becomes possible to segment voxels that predominantly contain fat [20], and some studies used fat fraction as input data to segment adipose tissue [21, 22].

This study aimed to automate the segmentation and quantification of EAT using Dixon MRI data through a fully automated deep learning pipeline (FM-Net), which utilized novel fat maps as input. The fat map was defined by combining OP, W and F images from the Dixon MRI sequence. By incorporating these different phase image signals, the fat map enhanced the visibility and distinctiveness of fat features, leading to improved segmentation accuracy. FM-Net was designed based on the double Res-UNet architecture to effectively capture the complex geometry of EAT. The first network in the pipeline focused on detecting the region of interest (ROI) encompassing the (fibrous) pericardium and EAT. The second network was dedicated to analyzing the segmented ROI from the first network. This study can provide valuable insights into quantifying EAT based on Dixon MRI.

2   Methods

2.1   Data and Pre-processing

Data used in this study was from a sub-study (HEALTH) of the Swedish CArdioPul-monarybiolmage Study (SCAPIS), with 90 subjects [23]. In this study, 45 control subjects without diabetes and 45 subjects with type 2 diabetes (T2D) were included. The control group was carefully matched to the diabetes group on an individual basis in terms of age, sex, and smoking habits. All participants provided written informed consent to participate. The study was performed in accordance with the Declaration of Helsinki and approved by the Linköping Ethical Review Board (Dnr 2023-02089-02). The images were reconstructed at a spatial resolution of 0.8 mm × 0.8 mm × 1.5 mm. Each 3D Dixon MRI data consisted of a varying number of slices along the Z direction, ranging from 50 to 82 slices of each phase and the dimension of each axial slice was 448 × 448 pixels (Fig. 1A).

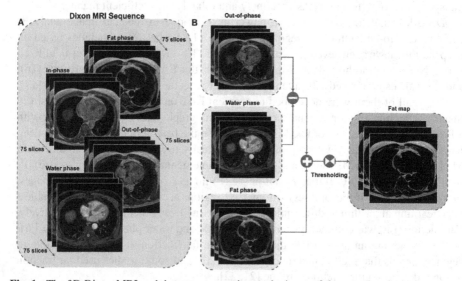

Fig. 1. The 3D Dixon MRI and data pre-processing as the input of the automated segmentation pipeline (FM-Net). **A,** The Dixon MRI sequences consisted of fat phase (F), in phase (IP), out of phase (OP) and water phase (W) images, and each phase contained the same number of slices. **B,** Fat map was obtained from out-of-phase, water phase and fat phase images.

By combining the information from OP, W and F, the fat map was proposed as the input data to enhance the accuracy and effectiveness of EAT segmentation. The fat map was derived from the three phases and specifically showed fat signal only (Fig. 1B), via removing the water-related information and highlighting the fat, reconstructed as:

$$Fat\ map = f_{threshold}(OP - W + F) \tag{1}$$

To distinguish fat and non-fat signals in each slice, an optimal threshold value for each slice was automatically determined by combining Otsu's method and the mean of grayscale value using formula (2) and pixels with intensities lower than the threshold were considered as background or non-fat signals.

$$threshold = (Otsu's\ value + mean\ value)/2 \tag{2}$$

In this study, fat fraction images were compared with fat maps, and their threshold was automatically determined using the same thresholding method described above. The (fibrous) pericardium of each subject was determined by experts slice by slice from the most caudal slice where the myocardium was visible to the first slice in the cranial direction where the lumen of the right branch of the pulmonary artery was observed continuity (Fig. 2C–E). EAT was identified by fat maps, and pixels within the pericardium were deemed EAT (Fig. 2F–H).

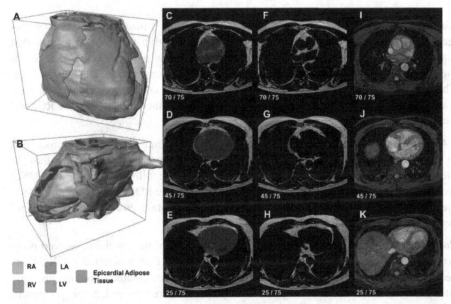

Fig. 2. Manual epicardial adipose tissue (EAT) segmentation was performed on Dixon MRI images from the control group (with EAT volume of 147.42 cm³). Reconstruction of the EAT (orange) and fibrous pericardium (light grey) surrounding the heart. **C–D,** Fat map images in transversal view, showing the segmented region of interest (blue) following the pericardium. **F–H,** EAT was identified as interior pixels in orange and water phase images (**I–K**). Top, middle, and bottom rows correspond to the cranial, middle and caudal parts of the heart. RA, right atrium; LA, left, atrium; RV, right ventricle; LV, left ventricle. (Color figure online)

2.2 FM-Net Pipeline for EAT Segmentation

The deep learning pipeline was designed as shown in Fig. 3. The FM-Net was structured into three distinct stages to facilitate the automated segmentation and quantification of EAT.

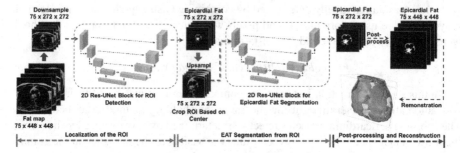

Fig. 3. The proposed deep learning based pipeline (FM-Net) for automatic segmentation of the epicardial adipose tissue (EAT) by using the fat map as input data (obtained from Dixon MRI sequence) and dual 2D Res-UNets, consisting of three stages.

Localization of the ROI. The first stage of the pipeline utilized the Res-UNet architecture, to accurately identify and delineate the area within the fat map images that encompass the target structures. The input of this stage was a down-sampled version of the entire 3D fat map with a size of $75 \times 272 \times 272$ pixels, processed slice-by-slice. The center of mass of the EAT was calculated from the coarse segmentation in each slice, and a 272×272 patch was cropped around this point, emphasizing the ROI and excluding most background pixels.

EAT Segmentation from ROI. Once the ROI was determined, the pipeline proceeded to the second stage, which involved the slice-by-slice segmentation of EAT within the identified ROIs. The network employed in the second stage followed the same Res-UNet architecture used in the first stage. For each slice within the ROI, the network processed the cropped input image of size 272×272 pixels. This stage ensured an accurate and detailed segmentation of EAT.

Post-processing and Reconstruction. The final stage of the pipeline involved post-processing and reconstruction steps on the predicted EAT segmentation results. Connected component analysis was performed within each slice obtained from the previous stage, and small holes or small objects under 50 pixels were removed, helping to ensure coherent and consistent segmentation. The individual segmentations of each slice were stacked together and zero-padded to an initial size of $75 \times 448 \times 448$ to obtain the final segmentation.

2.3 Training and Evaluation

The networks of the two stages were trained using the Dice similarity coefficient (DSC) as the loss function [24], evaluating the overlap between the predicted segmentation and the ground truth. Optimization was performed by using the Adam method. At each epoch, online augmentation was used to randomly augment each data with a probability of 50% and included translation, scaling, rotation and flipping. The 4488 2D slices derived from 62 subjects (31 with T2D and 31 matched subjects without T2D) were used for training, and 2016 2D slices from the remaining 28 subjects (14 control subjects and 14 matched subjects with T2D) served as the test dataset to evaluate the performance of the FM-Net. Four-fold cross-validation experiments were conducted. This involved partitioning the dataset from 62 3D data into four subsets, with 48 data samples used for training and 14 data for validation in each fold. After training, the CNN with the highest cross-validation accuracy was selected as the final model and was tested on the test dataset with 28 3D data, allowing for an unbiased evaluation of the model's performance on unseen data. The network training was performed on an NVIDIA Tesla V100 GPU with 5120 CUDA cores and 32 GB RAM. The network performance was compared using the fat map and fat fraction [25] as inputs. This comparison allows for an assessment of the effectiveness of each input type in accurately segmenting EAT. To assess the segmentation performance of the model, DSC and 95% Hausdorff distance (HD95) were utilized to indicate segmentation accuracy and provide insights into the sensitive boundary agreement, respectively. The proposed FM-Net was compared against state-of-the-art segmentation methods, including Res-UNet, UNet + +, and Double-VNet. To ensure a fair comparison, all methods utilized publicly available implementations based on MONAI [26].

3   Results and Discussion

Table 1 summarized the comparison between the proposed FM-Net and state-of-the-art methods using the fat map or fat fraction as input for EAT segmentation. FM-Net with fat map achieved the highest DSC of 86.3% and a low HD95 of 3.11 mm. The double 2D Res-UNet architecture was employed in the proposed pipeline to obtain the ROI and perform regional segmentation. It was noticed that 2D UNet++ also obtained promising DSC and HD95 on the fap map. Although 2D UNet++ increased the parameters and computation time, it showed potential for improving the segmentation performance. Double-VNet using 3D data achieved the lowest DSC among the compared methods. This could be attributed to the distributed nature of EAT surrounding hearts, lacking consistency in different directions. Directional connectivity-based segmentation methods and orthogonal annotation may benefit segmenting EAT from 3D data and improve border identification [27, 28]. Fat maps obtained superior segmentation performance compared to the fat fraction, suggesting that the fat map could improve the segmentation performance.

The volumes of the automatically segmented EAT structures from the 28 test data were compared to the ground truth volumes using the Pearson correlation coefficient R and Bland-Altman analysis. Figure 4 illustrated agreement between predicted and ground truth volumes, showing high arrangement ($R = 0.9614$, $P < 0.001$, Fig. 4A). No significant bias was observed in the automated quantification. The average volumetric difference was 0.52 cm^3 ([-15, 18] for ± 1.96 SD, Fig. 4B). Two examples of EAT segmentation from the control group and T2D group were presented in Fig. 5, respectively.

Table 1. Comparison of the proposed method FM-Net, and the current state-of-art approaches (Res-UNet [29], UNet++ [30] and Double-VNet [31]) and the proposed fat map versus fat fraction as the input for the testing data (N = 28).

Methods	Input	DSC ↑	HD95 (mm) ↓
Res-UNet	Fat fraction (2D)	82.8	6.73
	Fat map (2D)	84.4	7.80
UNet++	Fat fraction (2D)	84.5	8.05
	Fat map (2D)	86.1	4.68
Double-VNet	Fat fraction (3D)	80.1	6.46
	Fat map (3D)	81.5	5.95
FM-Net	Fat fraction (2D)	84.0	4.35
	Fat map (2D)	**86.3**	**3.11**

DSC, Dice similarity coefficient; HD95, Hausdorff distance (95%).

In the 3D view, false predictions outside the EAT volume from the reference standard can be visualized, despite the good DSC. Different 2D slices from superior, median, and inferior parts of the heart showed that superior parts consistently exhibited better segmenting results, potentially due to difficulty identifying structures from the raw images, which could lead to reduced accuracy and consistency. The main limitation of this study was the lack of directional connectivity analysis of the referenced EAT to maintain anatomical consistency, specifically in the Z-direction. To address this limitation and improve the accuracy of the segmentation performance, future work could involve incorporating a directional connectivity analysis during training and inference and comprehensive evaluation of the distribution of EAT across different slices. Additionally, multiple expert readers should be included in the evaluation process to mitigate potential biases and variations arising from individual interpretations. Furthermore, while the fat map demonstrated promising results, incorporating multi-channel inputs, such as IP (in phase) and W (water phase), into the deep learning network would be beneficial, allowing for a more detailed analysis and enabling whole heart segmentation (left/right atrium and left/right ventricle). This expanded segmentation capability would enhance the understanding of the role of EAT in CVDs by studying EAT distribution and fat infiltration throughout the heart.

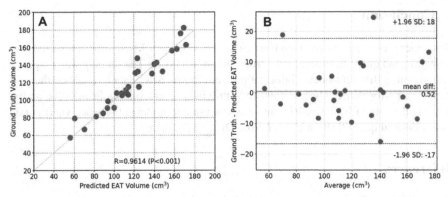

Fig. 4. Correlation and Bland-Altman plots for the EAT volume based on the test dataset (N = 28), comparing the automatic segmentation and the ground truth of epicardial adipose tissue (EAT). **A,** Correlation plot; **B,** Bland-Altman plot with 95% Limits of Agreement.

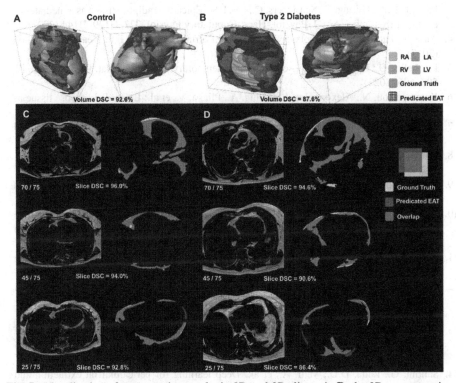

Fig. 5. Visualization of segmentation results in 3D and 2D slices. **A, B,** the 3D reconstruction of the hearts from control (left) or with (right) T2D, respectively; referenced EAT volumes were 81.42 cm³ (**A**) and 176.05 cm³ (**B**). EAT (orange), predicted EAT (blue) fibrous pericardium (light grey). **C, D,** the ground truth and prediction were visualized in three representative 2D images corresponding to **A, B**. See abbreviations in Fig. 2 (Color figure online)

4 Conclusion

This paper presented a novel deep learning based pipeline, FM-Net, for automated segmentation and quantification of EAT from Dixon MRI data. The fat map was proposed by combining different phasic information from Dixon MRI data to enhance the segmentation performance. Overall, this study provided a valuable implication for understanding the role of EAT in CVDs and risk assessment in clinical practice via accurate and automated quantification of EAT from Dixon MRI data.

References

1. Iacobellis, G., Corradi, D., Sharma, A.M.: Epicardial adipose tissue: anatomic, biomolecular and clinical relationships with the heart. Nat. Rev. Cardiol. **2**, 536–543 (2005)
2. Sacks, H.S., Fain, J.N.: Human epicardial adipose tissue: a review. Am. Heart J. **153**, 907–917 (2007)
3. Patel, K.H.K., Hwang, T., Se Liebers, C., Ng, F.S.: Epicardial adipose tissue as a mediator of cardiac arrhythmias. Am. J. Physiol.-Heart Circulatory Physiol. **322**, H129–H144 (2022)
4. Auer, J.: Fat: an emerging player in the field of atrial fibrillation. Eur. Heart J. **38**, 62–65 (2017)
5. Iacobellis, G.: Epicardial adipose tissue in contemporary cardiology. Nat. Rev. Cardiol. **19**, 593–606 (2022)
6. Karampetsou, N., Alexopoulos, L., Minia, A., Pliaka, V., et al.: Epicardial adipose tissue as an independent cardiometabolic risk factor for coronary artery disease. Cureus (2022)
7. Konwerski, M., Gąsecka, A., Opolski, G., Grabowski, M., Mazurek, T.: Role of epicardial adipose tissue in cardiovascular diseases: a review. Biology **11**, 355 (2022)
8. Hatem, S.N., Sanders, P.: Epicardial adipose tissue and atrial fibrillation. Cardiovasc. Res. **102**, 205–213 (2014)
9. Bertaso, A.G., Bertol, D., Duncan, B.B., et al.: Epicardial fat: definition, measurements and systematic review of main outcomes. Arquivos Brasileiros de Cardiologia (2013)
10. Xiong, Z., Fedorov, V.V., Fu, X., et al.: Fully automatic left atrium segmentation from late gadolinium enhanced magnetic resonance imaging using a dual fully convolutional neural network. IEEE Trans. Med. Imaging **38**, 515–524 (2019)
11. Xiong, Z., Xia, Q., Hu, Z., Huang, N.,et al.: A global benchmark of algorithms for segmenting late gadolinium-enhanced cardiac magnetic resonance imaging. 26 (2021)
12. Commandeur, F., Goeller, M., Betancur, J., et al.: Deep learning for quantification of epicardial and thoracic adipose tissue from non-contrast CT. IEEE Trans. Med. Imaging **37**, 1835–1846 (2018). https://doi.org/10.1109/TMI.2018.2804799
13. Commandeur, F., Goeller, M., Razipour, A., et al.: Fully automated CT quantification of epicardial adipose tissue by deep learning: a multicenter study. Radiol. Artif. Intell. **1**, e190045 (2019). https://doi.org/10.1148/ryai.2019190045
14. Santini, G., Latta, D.D., Vatti, A., et al.: Deep learning for pericardial fat extraction and evaluation on a population study. Radiol. Imaging (2020)
15. Li, X., Sun, Y., Xu, L., et al.: Automatic quantification of epicardial adipose tissue volume. Med. Phys. **48**, 4279–4290 (2021)
16. Zhang, Q., Zhou, J., Zhang, B., et al.: Automatic epicardial fat segmentation and quantification of CT scans using dual U-Nets with a morphological processing layer. IEEE Access **8**, 128032–128041 (2020)

17. He, X., Guo, B.J., Lei, Y., et al.: Automatic segmentation and quantification of epicardial adipose tissue from coronary computed tomography angiography. Phys. Med. Biol. **65**, 095012 (2020)
18. Daudé, P., Ancel, P., Confort Gouny, S., et al.: Deep-learning segmentation of epicardial adipose tissue using four-chamber cardiac magnetic resonance imaging. Diagnostics **12**, 126 (2022)
19. Bard, A., Raisi-Estabragh, Z., Ardissino, M., et al.: Automated quality-controlled cardiovascular magnetic resonance pericardial fat quantification using a convolutional neural network in the UK biobank. Front. Cardiovasc. Med. **8**, 677574 (2021)
20. Homsi, R., Meier-Schroers, M., Gieseke, J., et al.: 3D-Dixon MRI based volumetry of peri- and epicardial fat. Int. J. Cardiovasc. Imaging **32**, 291–299 (2016)
21. Homsi, R., Sprinkart, A.M., Gieseke, J., et al.: 3D-Dixon cardiac magnetic resonance detects an increased epicardial fat volume in hypertensive men with myocardial infarction. Eur. J. Radiol. **85**, 936–942 (2016)
22. Langner, T., Hedström, A., Mörwald, K., et al.: Fully convolutional networks for automated segmentation of abdominal adipose tissue depots in multicenter water–fat MRI. Magn. Reson. Med. **81**, 2736–2745 (2019)
23. Edin, C., Ekstedt, M., Scheffel, T., et al.: Ectopic fat is associated with cardiac remodeling—a comprehensive assessment of regional fat depots in type 2 diabetes using multi-parametric MRI. Front. Cardiovasc. Med. **9**, 813427 (2022)
24. Milletari, F., Navab, N., Ahmadi, S.-A.: V-Net: fully convolutional neural networks for volumetric medical image segmentation. In: 2016 Fourth International Conference on 3D Vision (3DV), Stanford, CA, USA, pp. 565–571. IEEE (2016)
25. Henningsson, M., Brundin, M., Scheffel, T., et al.: Quantification of epicardial fat using 3D cine Dixon MRI. BMC Med. Imaging **20**, 80 (2020)
26. Cardoso, M.J., Li, W., Brown, R., et al.: MONAI: an open-source framework for deep learning in healthcare. http://arxiv.org/abs/2211.02701 (2022)
27. Yang, Z., Farsiu, S.: Directional connectivity-based segmentation of medical images (2023)
28. Cai, H., Li, S., Qi, L., et al.: Orthogonal annotation benefits barely-supervised medical image segmentation. http://arxiv.org/abs/2303.13090 (2023)
29. Alom, Z., Taha, T.M., Asari, V.K.: Recurrent residual convolutional neural network based on U-Net (R2U-Net) for medical image segmentation. 12 (2018)
30. Zhou, Z., Siddiquee, M.M.R., Tajbakhsh, N., Liang, J.: UNet++: a nested U-Net architecture for medical image segmentation. http://arxiv.org/abs/1807.10165 (2018)
31. Xia, Q., Yao, Y., Hu, Z., Hao, A.: Automatic 3D atrial segmentation from GE-MRIs using volumetric fully convolutional networks. In: Pop, M., Sermesant, M., Zhao, J., Li, S., McLeod, K., Young, A., Rhode, K., Mansi, T. (eds.) STACOM 2018. LNCS, vol. 11395, pp. 211–220. Springer, Cham (2019). https://doi.org/10.1007/978-3-030-12029-0_23

Automated Quality-Controlled Left Heart Segmentation from 2D Echocardiography

Bram W. M. Geven[1,2], Debbie Zhao[1(✉)], Stephen A. Creamer[1],
Joshua R. Dillon[1], Gina M. Quill[1], Nicola C. Edwards[3,4], Malcolm E. Legget[4],
Robert N. Doughty[3,4], Alistair A. Young[5], Thiranja P. Babarenda Gamage[1],
and Martyn P. Nash[1,6]

[1] Auckland Bioengineering Institute, University of Auckland, Auckland, New Zealand
debbie.zhao@auckland.ac.nz
[2] Department of Biomedical Engineering, Eindhoven University of Technology,
Eindhoven, The Netherlands
[3] Green Lane Cardiovascular Service, Auckland City Hospital, Auckland,
New Zealand
[4] Department of Medicine, University of Auckland, Auckland, New Zealand
[5] School of Biomedical Engineering and Imaging Sciences, King's College London,
London, UK
[6] Department of Engineering Science and Biomedical Engineering, University of
Auckland, Auckland, New Zealand

Abstract. Segmentation of 2D echocardiography (2DE) images is an important prerequisite for quantifying cardiac function. Although deep learning can automate analysis, variability in image quality and limitations in model generalisability can result in inaccurate segmentations. We present an automated quality control (QC) methodology to identify invalid segmentations, and propose post-processing techniques to automatically correct erroneous segmentations. A workflow was developed to utilise a deep learning model, trained using the CAMUS dataset, for segmenting all frames within apical two-chamber and four-chamber 2DE images from an independent dataset containing 91 participants (28 females; 51 healthy controls and 40 patients with mixed cardiac pathologies). Single- and multi-frame QC and post-processing techniques were applied, and subsequently validated against manual QC in a sample of 50 randomly selected participants. Cardiac indices derived from the automated segmentations using 2DE were compared to reference values obtained through expert manual analysis on the same subjects. Single-frame QC improved the proportion of usable frames from 76% to 96%. Multi-frame QC indicated failures in 53% of the images, and while the resulting specificity was 96%, correction only achieved a sensitivity of 42% with respect to manual assessment. The exclusion of the rejected images resulted in improvements in the reliability between predicted and manual measurements. These results demonstrated that applying automated QC to deep learning segmentation methods can enhance the reliability of 2DE segmentations.

Keywords: Cardiac segmentation · Quality control · 2D echocardiography · Deep learning

© The Author(s), under exclusive license to Springer Nature Switzerland AG 2024
O. Camara et al. (Eds.): STACOM 2023, LNCS 14507, pp. 98–107, 2024.
https://doi.org/10.1007/978-3-031-52448-6_10

1   Introduction

2D echocardiography (2DE) is a highly accessible imaging modality that allows non-invasive and real-time examination of the geometry, motion, and deformation of the heart. Accurate segmentation of cardiac structures including the left ventricular (LV) cavity (LV_{cav}), myocardium (LV_{myo}), and left atrium (LA), is crucial for deriving clinical indices used to assess cardiac function. Recent studies have shown the feasibility of using deep learning models to automatically segment the left heart from 2DE, substantially accelerating image analysis [16].

Despite numerous efforts to create robust deep learning models for cardiac segmentation, the ability to perform effectively across different datasets remains a major challenge [5]. This issue is exemplified by ongoing concerns regarding the accuracy and presence of erroneous segmentations when employing such models in cross-dataset segmentation tasks. Incorporating post-processing can improve flawed segmentations, such that a larger pool of valuable images can be used for clinical analysis [10].

While minor segmentation issues can generally be resolved with simple post-processing operations, correcting some faults can be challenging due to ambiguity in the images. Combining automated quality control (QC) methodology with segmentation processes can help to address errors and improve the reliability and accuracy of the generated segmentations [14]. QC eliminates the need for time-consuming and subjective manual review processes and can highlight prevalent failure modes in deep learning algorithms. This can be used to apply targeted improvements to automated segmentations, which may result in more accurate and reliable predictions of standard clinical indices.

Various QC methodologies and metrics have been proposed, including abnormality detection [14], real-time Dice Similarity Coefficients (DSC) [12], reverse classification accuracy [11], Bayesian uncertainty-based methods [3,13], and convexity scores [20]. Despite these advancements, no open-source QC methodologies are currently available.

To our knowledge, this study presents the first publicly accessible automated and quality-controlled workflow for left heart segmentation from 2DE images using a state-of-the-art deep learning model. Single-frame QC criteria were used to identify appropriate post-processing steps, and multi-frame QC criteria were applied to assess the reliability of standard cardiac indices derived from the segmentations. The dependability of the workflow was assessed by comparing the generated cardiac indices with those derived from expert manual analysis.

2   Methods

The proposed QC workflow consists of automated left heart segmentation, single-frame QC assessment, followed by post-processing procedures and multi-frame QC analysis. Subsequently, this workflow is utilised for the calculation of routine clinical cardiac indices. A schematic representation of this workflow is depicted in Fig. 1.

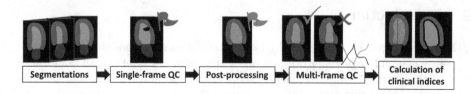

Fig. 1. Schematic representation of the proposed quality control (QC) workflow for the automated calculation of routine left heart indices from 2D echocardiography.

2.1 Datasets

CAMUS. The publicly available CAMUS dataset [7] comprised apical two-chamber (A2CH) and four-chamber (A4CH) 2DE views acquired from 500 patients at end-diastole (ED) and end-systole (ES). All images were acquired using a GE Vivid E95 ultrasound scanner (GE Vingmed Ultrasound, Horten, Norway), with a GE M5S probe (GE Healthcare, US). Expert manual labels were available for LV_{cav}, LV_{myo}, and LA on the ED and ES frames.

CARDIOHANCE. The private CARDIOHANCE dataset consists of echocardiograms for 91 participants conducted at the University of Auckland. Ethical approval for this study was granted by the Health and Disability Ethics Committee of New Zealand (17/CEN/226), and all research was performed in accordance with relevant guidelines and regulations. Written informed consent was obtained from each participant. This dataset comprised 51 healthy controls (20 females) and 40 patients (8 females) with mixed cardiac pathologies. The primary diagnoses were: cardiac amyloidosis (11), LV hypertrophy (10), aortic regurgitation (7), dilated cardiomyopathy (5), hypertrophic cardiomyopathy (4), heart transplant (2), and coronary artery disease (1). 2DE image sequences were available with both the LV_{cav} and LA clearly visible. The images were obtained with a Siemens ACUSON SC2000 ultrasound scanner, equipped with a 4Z1c matrix array transducer (Siemens Medical Solutions, Issaquah, WA, USA).

2.2 Deep Learning Model for Segmentation

A self-configuring segmentation framework (2D nnU-Net (v2) [4]), was trained on the CAMUS dataset to segment the LV_{cav}, LV_{myo}, and LA. This dataset was split into a 90/10 training/testing ratio, and the model was trained for 100 epochs using five-fold cross-validation. The nnU-Net model was used to perform cross-dataset segmentations of the left heart on all images in the CARDIOHANCE dataset for which no expert labels were available.

2.3 Single-frame Quality Control and Post-processing

Following cross-dataset segmentation, various issues in the segmentations became apparent, including missing or duplicated structures and the presence of holes or discontinuities within or between structures. To address this, a single-frame QC assessment was implemented to identify and flag erroneous segmentations. Subsequently, the flagged segmentations were post-processed.

Initially, mean centroids of LV_{cav} and LA were computed across the entire image sequence, excluding any frames that were flagged during the QC assessment. Structures lacking the mean centroid were removed, thereby eliminating redundant structures. The LV_{myo} was identified by including only structures bordering LV_{cav}. Additionally, any holes within or between structures were filled. The issue of missing structures could not be resolved by simple post-processing.

2.4 Multi-frame Quality Control

Cycle Selection. In clinical practice, image sequences often consist of multiple cardiac cycles. In this study, we defined a cardiac cycle as the duration between two consecutive ED points. The ED and ES points were taken as the locations of the maximum and minimum LV_{cav} areas in the area-time curve, respectively. The cardiac cycle was selected based on having the fewest flagged frames and the highest contrast-to-noise ratio [9] between the myocardium and blood pool.

Structural Information. After selecting the best cycle from each sequence, frames with a disconnected LV_{myo} or LA extending beyond the imaging field of view, indicative of partial cutoff, were identified and flagged. If two or more frames were flagged within the selected cardiac cycle, the cycle was excluded.

Area-Time Curve Analysis. Area-time curves were generated for the LV_{cav} and LA across the entire cardiac cycle to differentiate between well-segmented and poorly-segmented cycles. A population prior was established as a reference, based on the selected cycles from all cases in the CARDIOHANCE dataset. Using the ES timings from all cycles, an average ES point was computed to temporally align the area-time curves for each subject. Subsequently, all curves were normalised, thereby accounting for variations in heart size, and a mean curve was computed for both LV_{cav} and LA.

The similarity between the original area-time curves of each image and the reference curve was assessed using Dynamic Time Warping (DTW) [1]. DTW measures the cumulative Euclidean distance between curves after DTW alignment. Images were excluded if the DTW distance for either the LV_{cav} or LA exceeded thresholds of 1 and 2, respectively. These thresholds were empirically chosen, with the threshold for the LA being larger due to greater variation observed in the area-time curves of the LA compared to those of LV_{cav}.

Validation. To validate the multi-frame QC, an expert observer manually reviewed segmentations from one cardiac cycle for 50 randomly selected participants in order to identify any erroneous segmentations or temporal incoherencies. Sensitivity and specificity were computed based on the results of the automated QC and manual review.

2.5 Calculation of Clinical Indices

LV end-diastolic volume (EDV) and end-systolic volume (EDV) were computed from ED and ES segmentations using the biplane method of disks summation [2], allowing for LV ejection fraction (EF) determination. The contours extracted from the segmentations of the LV_{cav} at ED and ES were used to compute the endocardial LV global longitudinal strain (GLS). Furthermore, the maximum area of the LA was calculated.

Expert Manual Analysis. Alongside the calculation of clinical indices from the predicted segmentations, a sonographer carried out manual analysis of the 2DE images using TOMTEC-ARENA 2.31 2D CPA (TOMTEC Imaging Systems GmbH, Unterschleißheim, Germany). This analysis included LV EDV, ESV, EF, GLS assessment, and measurement of the maximum LA area.

2.6 Statistics

The agreement between the clinical indices derived from the predicted segmentations and expert manual analysis on the images in the CARDIOHANCE dataset was quantified using the average measure, two-way mixed effects intraclass correlation coefficient (ICC) [6]. Paired-sample t-tests were performed to identify statistically significant differences (p-values < 0.05). All tests were conducted using Python [18] v3.10, with the Pingouin [17] and SciPy [19] packages.

3 Results

3.1 Evaluation of Deep Learning Segmentation Performance

The segmentation performance on the CAMUS test set, measured by DSC, Hausdorff Distance (HD) and Mean Absolute Distance (MAD), is presented in Table 1. The LV_{cav} achieved the highest DSC compared to LV_{myo} and LA, while the HD and MAD were lowest for the LV_{cav}. The LA exhibited the highest variability across all metrics.

Table 1. Segmentation accuracy on the CAMUS test set (n=50) for left ventricular cavity (LV_{cav}), myocardium (LV_{myo}) and left atrium (LA). Metrics include Dice Similarity Coefficient (DSC), Hausdorff Distance (HD) and Mean Absolute Distance (MAD), with values given as mean $\pm$ standard deviation. Results are averaged across apical two-chamber and four-chamber views in end-diastole (ED) and end-systole (ES) phases.

	LV_{cav}			LV_{myo}			LA		
	DSC(-)	HD (mm)	MAD (mm)	DSC (-)	HD (mm)	MAD (mm)	DSC (-)	HD (mm)	MAD (mm)
ED	0.953 ± 0.020	3.9 ± 1.9	1.2 ± 0.6	0.889 ± 0.036	4.5 ± 1.6	1.4 ± 0.5	0.916 ± 0.065	4.2 ± 3.0	1.4 ± 0.9
ES	0.938 ± 0.033	3.8 ± 1.6	1.2 ± 0.6	0.896 ± 0.036	4.6 ± 1.7	1.4 ± 0.5	0.933 ± 0.036	4.2 ± 2.4	1.3 ± 0.7

3.2 Single-frame Quality Control and Post-processing

Table 2 presents the percentages of flagged frames per failure mode identified through single-frame QC. Before post-processing, 20750 frames passed, representing 76% of the 27406 frames in the CARDIOHANCE dataset. The number of passed frames increased to 26233 after post-processing, accounting for 96% of the total number of frames.

The main reason for flagging frames before post-processing was the presence of holes within the LV_{cav} (9.4%) or between LV_{cav} and LV_{myo} (10.0%). After post-processing, frames were mainly flagged due to the presence of multiple distinct LV_{myo} structures (3.9%). Figure 2 illustrates examples of post-processing procedures.

Table 2. Percentage of flagged frames in the CARDIOHANCE dataset (n = 27406) before and after post-processing, categorised by failure mode for left ventricular cavity (LV_{cav}), myocardium (LV_{myo}), and left atrium (LA) during single-frame QC.

Failure mode	before	after
No LV_{cav}	<0.1%	0.1%
No LV_{myo}	<0.1%	<0.1%
No LA	0.1%	0.2%
Duplicate LV_{cav}	<0.1%	<0.1%
Duplicate LV_{myo}	4.1%	3.9%
Duplicate LA	0.7%	<0.1%
Holes within LV_{cav}	9.4%	0.6%
Holes within LV_{myo}	2.1%	0.4%
Holes within LA	2.3%	<0.1%
Holes between LV_{cav} and LV_{myo}	10.0%	0.8%
Holes between LV_{cav} and LA	0.5%	0.1%
Holes between LV_{myo} and LA	0.6%	<0.1%

3.3 Multi-frame Quality Control

Table 3 presents a comparison between clinical indices derived from predicted segmentations and expert manual analysis using TOMTEC software, both before and after exclusion based on multi-frame QC assessment. The comparison revealed significant biases in all indices before and after exclusion, with the exception of LV EF after exclusion. After QC exclusion, all ICC values increased, indicating enhanced reliability between the segmentations and TOMTEC measurements.

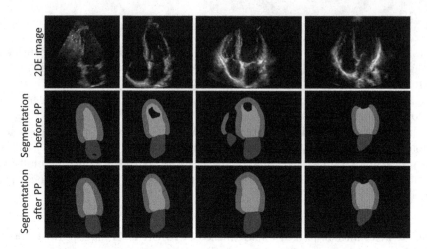

Fig. 2. Examples demonstrating the effect of post-processing (PP) on segmentations. Top row: 2D echocardiography (2DE) images; middle row: segmentations before PP; bottom row: segmentations after PP. The examples showcase different scenarios, including (**A**) filling holes in the left atrium, and (**B**) filling holes in the left ventricular cavity (LV_{cav}), (**C**) removing redundant structures and filling holes between LV_{cav} and myocardium (LV_{myo}), and (**D**) unimproved unconnected LV_{myo}.

Table 3. Comparison of clinical indices derived from segmentations and expert manual analysis, before and after multi-frame QC exclusion, including intraclass correlation coefficients (ICC) and biases (mean ± standard deviation). Statistically significant differences (p-values < 0.05) between indices derived from segmentations and expert manual analysis are indicated by asterisks (*). The indices included left ventricular (LV) end-diastolic volume (EDV), end-systolic volume (ESV), ejection fraction (EF), global longitudinal strain (GLS) and left atrium (LA) maximum area for the apical two-chamber (A2CH) and four-chamber (A4CH) views.

Indices	Before exclusion		After exclusion	
	ICC	Bias	ICC	Bias
LV EDV (ml)	0.763	*−28 ± 22	0.798	*−29 ± 22
LV ESV (ml)	0.852	*−14 ± 14	0.877	*−13 ± 15
LV EF (%)	0.866	*1.6 ± 5.4	0.874	0.5 ± 5.5
LV GLS A2CH (%)	0.287	*7.0 ± 7.0	0.445	*4.1 ± 5.3
LV GLS A4CH (%)	0.536	*3.1 ± 4.7	0.653	*2.7 ± 4.3
LA maximum area A2CH (mm^2)	0.648	*−4.8 ± 4.0	0.793	*−3.6 ± 2.4
LA maximum area A4CH (mm^2)	0.754	*−3.8 ± 3.3	0.848	*−2.7 ± 2.0

In total, 53% of all images were excluded through the multi-frame QC assessment. Manual validation further demonstrated a sensitivity of 42% and a specificity of 96% for excluding images based on both structural faults and temporal inconsistencies.

4   Discussion

The 2D nnU-Net model that was trained in this study outperforms all methods listed on the CAMUS challenge website[1] for LV_{cav}, LV_{myo} and LA segmentations. Despite its good performance on the CAMUS test set, several issues arose during cross-dataset segmentation on the CARDIOHANCE dataset. Single-frame QC analysis, presented in Table 2, revealed a substantial improvement in the availability of valid frames after post-processing. Only a small fraction of the frames still exhibited an issue with unconnected cardiac labels after post-processing, as depicted in Fig. 2D. To address this issue, it may be beneficial to incorporate a statistical shape model in the post-processing step.

After multi-frame QC, over half of all images were excluded. For the acceptable images, there was a low sensitivity of 42% with respect to manual assessment, indicating that a considerable number of flawed segmentations remained undetected, which could be problematic if used in a clinical setting. On the other hand, a high specificity of 96% was observed, indicating that only a small proportion of accurately segmented images was flagged erroneously. To enhance sensitivity in detecting faults and abnormalities, additional QC criteria can be introduced, potentially including structural properties like convexity and simplicity of anatomical structures [8, 20].

Table 3 demonstrates a significant underestimation of the LV EDV and ESV, as well as the maximum LA areas. This discrepancy could be caused by apical undersegmentation in the LV_{cav}. This can be attributed to the inherent limitations of the expert manual labels in the CAMUS dataset, which have also been associated with volume underestimation [20]. Moreover, the LV GLS values were found to be overestimated compared to the values obtained with TOMTEC in both A2CH and A4CH views, possibly due to the presence of apical undersegmentation in the LV_{cav}. This undersegmentation can result in higher GLS measurements [15].

The exclusion of flagged images resulted in an improved agreement of LV GLS and LA maximum area in A4CH views with respect to expert manual analysis. Initially categorised as poor (ICC < 0.5) and moderate (0.5 ≤ ICC < 0.75) reliability, the LV GLS and LA maximum area measurements were reassessed as exhibiting moderate and good (0.75 ≤ ICC < 0.9) reliability, respectively, after excluding the indices of flagged images. However, poor reliability persisted in the measurements of LV GLS in A2CH views (ICC = 0.445).

In this study, the manual determination of multi-frame QC criteria raises concerns regarding optimality and dataset-specific applicability. It may therefore be beneficial to explore more generalised criteria, reducing dependence on qualitative manual threshold determination. Furthermore, the utilisation of a single average population prior in area-time curve analysis overlooks potential variations in curve shapes between healthy controls and patients with cardiac pathologies. Future studies could investigate the use of population priors tailored to specific pathologies, thereby enhancing the effectiveness of the analysis.

---

[1] https://www.creatis.insa-lyon.fr/Challenge/camus/results.html

5   Conclusion

In this study, we have presented an automated QC workflow for left heart analysis from 2DE, with the objective of enhancing the accuracy and reliability of automatically segmented images when applied to cross-dataset segmentation. The implementation of QC, which identifies, flags and applies corrections to faulty segmentations, is shown to be crucial for improving the reliability of the derived cardiac indices when compared to expert manual analysis. Application of the workflow resulted in an improvement in the proportion of usable frames from 76% to 96%. By releasing the code associated with the proposed automated QC workflow, we aim to facilitate collaborative efforts and further developments by the community.

Acknowledgements. We gratefully acknowledge the participants of the CARDIO-HANCE study for volunteering their time, and the staff at the Centre for Advanced MRI at the University of Auckland for their expertise and assistance with the imaging components of this study.

Funding Information. This study was funded by the Health Research Council of New Zealand (programme grant 17/608).

Code Availability. Our code is publicly available on GitHub: https://github.com/bgeven/AQC-left-heart-segmentation.

References

1. Berndt, D.J., Clifford, J.: Using dynamic time warping to find patterns in time series. In: KDD Workshop. vol. 10, pp. 359–370. Seattle, WA, USA (1994)
2. Folland, E.D., Parisi, A.F., Moynihan, P.F., Jones, D.R., Feldman, C.L., Tow, D.E.: Assessment of left ventricular ejection fraction and volumes by real-time, two-dimensional echocardiography. A comparison of cineangiographiy and radionuclide techniques. Circulation **60**(4), 760–766 (1979). https://doi.org/10.1161/01.cir.60.4.760
3. Hann, E., Gonzales, R.A., Popescu, I.A., Zhang, Q., Ferreira, V.M., Piechnik, S.K.: Ensemble of deep convolutional neural networks with Monte Carlo dropout sampling for automated image segmentation quality control and robust deep learning using small datasets. In: Papież, B.W., Yaqub, M., Jiao, J., Namburete, A.I.L., Noble, J.A. (eds.) MIUA 2021. LNCS, vol. 12722, pp. 280–293. Springer, Cham (2021). https://doi.org/10.1007/978-3-030-80432-9_22
4. Isensee, F., Jaeger, P.F., Kohl, S.A., Petersen, J., Maier-Hein, K.H.: nnU-Net: a self-configuring method for deep learning-based biomedical image segmentation. Nat. Methods **18**(2), 203–211 (2021). https://doi.org/10.1038/s41592-020-01008-z
5. Keshavan, A., Datta, E., M. McDonough, I., Madan, C.R., Jordan, K., Henry, R.G.: Mindcontrol: a web application for brain segmentation quality control. NeuroImage **170**, 365–372 (2018). https://doi.org/10.1016/j.neuroimage.2017.03.055
6. Koo, T.K., Li, M.Y.: A guideline of selecting and reporting intraclass correlation coefficients for reliability research. J. Chiropr. Med. **15**(2), 155–163 (2016). https://doi.org/10.1016/j.jcm.2016.02.012

7. Leclerc, S., et al.: Deep learning for segmentation using an open large-scale dataset in 2D echocardiography. IEEE Trans. Med. Imaging **38**(9), 2198–2210 (2019). https://doi.org/10.1109/TMI.2019.2900516
8. Leclerc, S., et al.: Deep learning segmentation in 2D echocardiography using the CAMUS dataset: automatic assessment of the anatomical shape validity. In: International Conference on Medical Imaging with Deep Learning - Extended Abstract Track, London, United Kingdom (2019)
9. Meyers, B., Brindise, M., Kutty, S., Vlachos, P.: A method for direct estimation of left ventricular global longitudinal strain rate from echocardiograms. Sci. Rep. **12**(1), 4008 (2022). https://doi.org/10.1038/s41598-022-06878-1
10. Painchaud, N., Skandarani, Y., Judge, T., Bernard, O., Lalande, A., Jodoin, P.M.: Cardiac segmentation with strong anatomical guarantees. IEEE Trans. Med. Imaging **39**(11), 3703–3713 (2020). https://doi.org/10.1109/TMI.2020.3003240
11. Robinson, R., et al.: Automated quality control in image segmentation: application to the UK Biobank cardiovascular magnetic resonance imaging study. J. Cardiovasc Magn. Reson. **21**(1), 18 (2019). https://doi.org/10.1186/s12968-019-0523-x
12. Robinson, R., et al.: Real-time prediction of segmentation quality. In: Frangi, A.F., Schnabel, J.A., Davatzikos, C., Alberola-López, C., Fichtinger, G. (eds.) MICCAI 2018. LNCS, vol. 11073, pp. 578–585. Springer, Cham (2018). https://doi.org/10.1007/978-3-030-00937-3_66
13. Roy, A.G., Conjeti, S., Navab, N., Wachinger, C.: Inherent brain segmentation quality control from fully ConvNet Monte Carlo sampling. In: Frangi, A.F., Schnabel, J.A., Davatzikos, C., Alberola-López, C., Fichtinger, G. (eds.) MICCAI 2018. LNCS, vol. 11070, pp. 664–672. Springer, Cham (2018). https://doi.org/10.1007/978-3-030-00928-1_75
14. Ruijsink, B., et al.: Fully automated, quality-controlled cardiac analysis from CMR. JACC Cardiovasc. Imaging **13**(3), 684–695 (2020). https://doi.org/10.1016/j.jcmg.2019.05.030
15. Smiseth, O.A., Donal, E., Penicka, M., Sletten, O.J.: How to measure left ventricular myocardial work by pressure-strain loops. Eur. Heart J. Cardiovasc. Imaging **22**(3), 259–261 (2020). https://doi.org/10.1093/ehjci/jeaa301
16. Thrall, J.H., Li, X., Li, Q., Cruz, C., Do, S., Dreyer, K., Brink, J.: Artificial intelligence and machine learning in radiology: opportunities, challenges, pitfalls, and criteria for success. J. Am. Coll. Radiol. **15**(3, Part B), 504–508 (2018). https://doi.org/10.1016/j.jacr.2017.12.026
17. Vallat, R.: Pingouin: statistics in Python. J. Open Source Softw. **3**(31), 1026 (2018). https://doi.org/10.21105/joss.01026
18. Van Rossum, G., Drake, F.L.: Python 3 Reference Manual. CreateSpace, Scotts Valley, CA (2009)
19. Virtanen, P., et al.: SciPy 1.0 Contributors: SciPy 1.0: fundamental algorithms for scientific computing in Python. Nat. Methods **17**, 261–272 (2020). https://doi.org/10.1038/s41592-019-0686-2
20. Zhang, X., et al.: Generalizability and quality control of deep learning-based 2D echocardiography segmentation models in a large clinical dataset. Int. J. Cardiovasc. Imaging (2022). https://doi.org/10.1007/s10554-022-02554-7

Impact of Hypertension on Left Ventricular Pressure-Strain Loop Characteristics and Myocardial Work

Stephen A. Creamer[1]([✉]), Debbie Zhao[1], Gina M. Quill[1],
Abdallah I. Hasaballa[1], Vicky Y. Wang[1], Thiranja P. Babarenda Gamage[1],
Nicola C. Edwards[2,3], Malcolm E. Legget[3], Boris S. Lowe[2],
Robert N. Doughty[2,3], Satpal Arri[2], Peter N. Ruygrok[2], Alistair A. Young[4],
Julian F. R. Paton[5], Gonzalo D. Maso Talou[1], and Martyn P. Nash[1,6]

[1] Auckland Bioengineering Institute, University of Auckland, Auckland, New Zealand
stephen.creamer@auckland.ac.nz
[2] Green Lane Cardiovascular Service, Auckland City Hospital,
Auckland, New Zealand
[3] Department of Medicine, University of Auckland, Auckland, New Zealand
[4] School of Biomedical Engineering and Imaging Sciences,
King's College London, London, UK
[5] Department of Physiology, University of Auckland, Auckland, New Zealand
[6] Department of Engineering Science and Biomedical Engineering,
University of Auckland, Auckland, New Zealand

Abstract. Hypertension is a major risk factor for cardiovascular disease. Pressure-strain loop analysis has recently been introduced as a clinical tool to quantify the pumping efficiency of the left ventricle (LV), as an alternative to the classical pressure-volume loop analysis. The aims of this study were to: (i) combine global longitudinal strain (GLS) from 3D transthoracic echocardiography (TTE) with LV catheter pressure measurements to compute myocardial work indices—namely the global work index (GWI), wasted work index (WWI), and constructive work index (CWI); and (ii) compare work indices between normotensive and hypertensive patients, with further categorisation into medicated and non-medicated sub-groups. 3D TTE and LV pressures were measured in 143 patients (49 females), with 53, 42, 28, and 20 patients in groups of controlled hypertensives, uncontrolled hypertensives, normotensive controls, and untreated hypertensives (e.g., recent diagnoses), respectively. Statistically significant differences ($p < 0.05$) in GWI and CWI were observed between patients with high blood pressure and the normotensive controls, with largest GWI values in the untreated hypertensive group (1954 ± 322 mmHg%), and smallest values in the controlled hypertensive group (1531 ± 449 mmHg%). Examining LV energetics using pressure-strain (GLS) loop analysis may provide greater understanding into cardiac function and disease progression in hypertension, and help inform appropriate treatment strategies for cardiac patients with pathologies such as heart failure.

© The Author(s), under exclusive license to Springer Nature Switzerland AG 2024
O. Camara et al. (Eds.): STACOM 2023, LNCS 14507, pp. 108–118, 2024.
https://doi.org/10.1007/978-3-031-52448-6_11

Keywords: Hypertension · Cardiac work · Pressure-strain loops · LV energetics · Echocardiography

1   Introduction

Systemic hypertension is a widely prevalent medical condition and a leading risk factor for heart attacks, stroke, and kidney failure [9]. The underlying mechanics of hypertension and how it impacts cardiac function and contractility are not completely understood. This study investigated the correlation and impact of hypertension and pharmacological treatment on left ventricular (LV) work.

Hypertension medication is critical for blood pressure control and maintenance of pressure targets [16]. The American Society of Hypertension and the European Society of Cardiology recommend reducing blood pressure to as low as the patient can tolerate (down to 120 mmHg) [17]. However, the impact of anti-hypertensives on cardiac energetics has not been extensively explored. Important determinants of heart and cardiovascular system function relate to geometric and pressure-related factors. Geometric, kinematic, and volumetric information about the heart can be captured using medical imaging techniques, such as 3D transthoracic echocardiography (TTE). Gold-standard measurements of chamber pressures can be collected through cardiac catheterisation; however, due to the invasive nature of this procedure, brachial cuff pressure recordings are most commonly recorded, which only gives information about arterial systolic and diastolic pressure. The combination of geometric and invasive pressure information allows indices of cardiac work performed during the heart cycle to be calculated.

The use of LV global longitudinal strain (GLS) as a measure of cardiac function has gained attention due to its higher sensitivity with respect to disease progression compared to conventional volumetric measures, such as ejection fraction (EF) [10]. Increased use of GLS has been a precursor to the use of pressure-strain loops to examine cardiac work. The global work index (GWI) [13] is a surrogate index of the energy that the myocardium expends during a cardiac cycle. It has been proposed that a more detailed understanding of cardiac mechanics can be derived through the so-called constructive work index (CWI) [2] and wasted work index (WWI) [12], defined as work contributing to the flow of blood around the body, and work that does not, respectively. Relative changes in estimates of constructive and wasted work may serve as indicators of impaired cardiac function, and have been shown to be prognostically significant in the presence of renal failure [21] and myocardial infarction [8].

This study investigates cardiac work indices in hypertensive patients and normotensive controls. Furthermore, the effects of hypertensive medication on indices of constructive and wasted work are also examined in these groups. The results of our analysis provide a more detailed understanding of the effects of hypertension and its treatment on cardiac energetics.

2   Methods

2.1   Patient Recruitment and Data Collection

Patients presenting for coronary angiography and left heart catheterisation at Auckland City Hospital were recruited for subsequent imaging with 3D TTE within one hour of LV pressure measurement (example shown in Fig. 1). Ethical approval for this study was granted by the Health and Disability Ethics Committee of New Zealand (17/NTB/46), and written informed consent was obtained from each participant. Throughout each procedure, a standard 3-lead electrocardiogram (ECG) was recorded. Standard clinical and demographic (i.e., age, sex, height, weight) information were recorded.

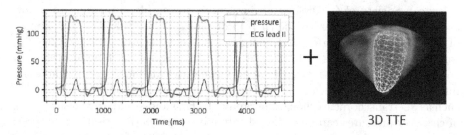

Fig. 1. Left ventricular pressure waveform with concurrent electrocardiogram (ECG) and 3D transthoracic echocardiogram (3D TTE) with corresponding geometric mesh of the left ventricle from which global longitudinal strain can be derived.

A cohort of 143 participants (49 females) was split into four sub-groups according to the European Society of Cardiology guidelines for hypertension classification (i.e., systolic blood pressure >140 mmHg, or diastolic >90 mmHg [18] based on standard brachial cuff measurements taken in clinic prior to the catheterisation procedure), and medication status (i.e., whether or not the patient had been prescribed anti-hypertensive medication).

2.2   Generation of Pressure-Strain Loops

Continuous LV pressure traces were divided into individual cardiac cycles by determining end-diastolic time points through the use of a shear-wave transformation method [1]. Strain curves over one cardiac cycle were extracted from 3D TTE recordings using a fully automated method for LV segmentation [20]. Corresponding R-R intervals for each cardiac cycle were extracted from the concurrent ECG recordings during pressure measurement and 3D TTE. For each GLS trace, the pressure cycle with the closest R-R interval was selected. Subsequently, the GLS trace was piecewise-linearly up-sampled to match the recording rate (240 Hz) of the pressure trace. The temporally aligned pressure and GLS data were plotted to form a set of pressure-strain loops for each patient, following which the median pressure-strain loop was taken for subsequent analyses.

2.3   Calculation of Work Indices

The GWI is defined as the area enclosed within a single pressure-strain loop. The WWI is defined as the sum of the area corresponding to shortening during isovolumic relaxation, plus the area corresponding to lengthening during isovolumic contraction (see Fig. 2 for examples). The CWI was defined as GWI – WWI, and was also expressed as a percentage of the GWI.

2.4   Statistics

To test for statistically significant differences in GWI, CWI, and WWI between the patient sub-groups, independent-samples (Welch's) t-tests were performed for each index between all possible sub-group pairings. A p-value < 0.05 was considered statistically significant.

3   Results

3.1   Patient Demographics and Conventional Cardiac Indices

The age range of the entire cohort was 19 to 84 years. The participant cohort was split into four sub-groups based on diagnosis (HTN: hypertensive; NT: normotensive), and whether or not anti-hypertensive medication had been prescribed. Demographic and conventional cardiac indices are shown in Table 1. On a per-group basis, the numbers of patients taking each of the specific blood pressure medications are listed in Table 2.

Compared to the normotensive controls, while there were statistically significant differences between end-diastolic volumes (EDV) for each sub-group, the end-systolic volume (ESV) was only significantly different in the controlled hypertensive group. The diastolic and systolic pressures in each of the uncontrolled and untreated hypertensive groups were each significantly different to the normotensive controls.

3.2   Cardiac Work Indices

Measured distributions of cardiac work indices are shown in Table 3. Statistically significant differences ($p < 0.05$) in GWI were found between four of the six possible combinations, with no statistical significance being found between the two normal blood pressure groups (normotensive controls, controlled hypertensives), and no statistical significance between the two high blood pressure groups (uncontrolled hypertensives, untreated hypertensives).

The CWI distributions exhibited similar statistical significance comparisons as GWI for all patient groups (i.e., statistically significant differences between the controlled hypertensive group and both high blood pressure groups, as well as between the normotensive controls and both high blood pressure groups), as shown in Fig. 3. With regard to the WWI, no statistically significant differences were observed between any pairings of patient groups.

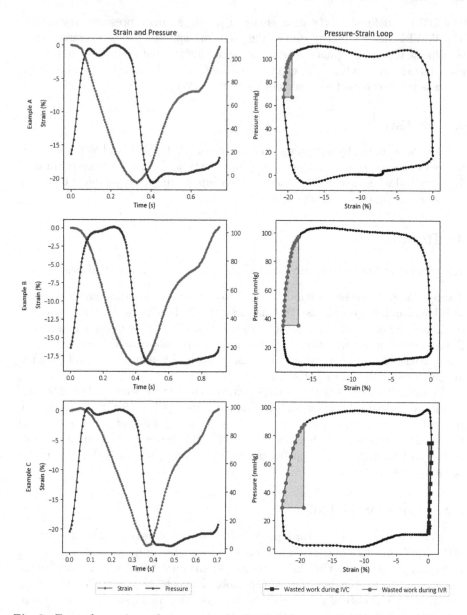

Fig. 2. Example strain and pressure recordings with corresponding pressure-strain loops showing areas of wasted work due to lengthening during isovolumic contraction (dark blue shading), and shortening during isovolumic relaxation (light blue shading). (Color figure online)

Table 1. Patient demographics, conventional cardiac indices, and mean numbers of types of anti-hypertensive medications (#anti-HTN) prescribed per patient. All values are mean ± SD. EDV: end-diastolic volume; ESV: end-systolic volume; EF: ejection fraction; GLS: global longitudinal strain; SBP/DBP: systolic/diastolic brachial blood pressure; HR: heart rate. Sub-group means were each compared to the normotensive (NT) control group, with asterisks (*) indicating significant differences ($p<0.05$). Male is denoted by "M", and female is denoted by "F".

Group	NT controls (n = 28) [15M:13F]	Controlled HTN (n = 53) [38M:15F]	Uncontrolled HTN (n = 42) [27M:15F]	Untreated HTN (n = 20) [14M:6F]	All (n = 143) [94M:49F]
Age (years)	54 ± 13	58 ± 11	59 ± 14	60 ± 7	58 ± 12
Height (cm)	171 ± 7	173 ± 8	171 ± 9	173 ± 8	172 ± 8
Weight (kg)	77 ± 14	*87 ± 20	*85 ± 18	*82 ± 15	84 ± 18
EDV (ml)	122 ± 38	*154 ± 57	*142 ± 36	*142 ± 28	143 ± 46
ESV (ml)	52 ± 26	*70 ± 37	58 ± 23	58 ± 14	61 ± 30
EF (%)	59 ± 12	56 ± 10	60 ± 5	59 ± 4	58 ± 9
Peak GLS (%)	–19 ± 3	*–18 ± 3	–19 ± 3	–19 ± 2	–19 ± 3
SBP (mmHg)	123 ± 13	125 ± 13	*148 ± 12	*147 ± 12	134 ± 17
DBP (mmHg)	74 ± 11	75 ± 11	*92 ± 13	*93 ± 18	82 ± 17
HR (bpm)	74 ± 14	72 ± 15	70 ± 15	76 ± 10	72 ± 14
HR mismatch (ms)	59 ± 79	62 ± 83	67 ± 97	48 ± 51	53 ± 60
#anti-HTN	–	*1.8 ± 0.9	*1.8 ± 0.9	–	1.2 ± 1.1

Table 2. Numbers of patients in the controlled hypertensive (Controlled HTN) and uncontrolled hypertensive (Uncontrolled HTN) groups that were prescribed each type of anti-hypertensive medication. See Table 1 for group numbers and demographics.

Group	Controlled HTN	Uncontrolled HTN	All
No. on BP lowering medication	53	42	95
No. on diuretics	15	14	29
No. on beta-blockers	24	15	39
No. on ACE inhibitors	28	17	45
No. on calcium channel blockers	18	19	37
No. on alpha-blockers	5	3	8
No. on nitrate-based vasodilator	0	1	1
No. on angiotensin II receptor blockers	6	7	13

Table 3. Cardiac work indices (mean ± SD) for each patient group. The global work index (GWI), wasted work index (WWI), constructive work index (CWI) and CWI ratio were each calculated from the personalised pressure-strain loops. Sub-group means were each compared to the normotensive (NT) control group, with asterisks (*) indicating significant differences ($p < 0.05$).

Group	NT controls (n = 28)	Controlled HTN (n = 53)	Uncontrolled HTN (n = 42)	Untreated HTN (n = 20)	All (n = 143)
GWI (mmHg%)	1708 ± 334	1531 ± 449	*1913 ± 438	*1954 ± 322	1737 ± 442
WWI (mmHg%)	129 ± 87	123 ± 89	111 ± 96	94 ± 107	117 ± 93
CWI (mmHg%)	1580 ± 339	1408 ± 427	*1802 ± 475	*1860 ± 305	1620 ± 448
CWI ratio (%)	92 ± 5	92 ± 6	93 ± 7	*95 ± 5	93 ± 6

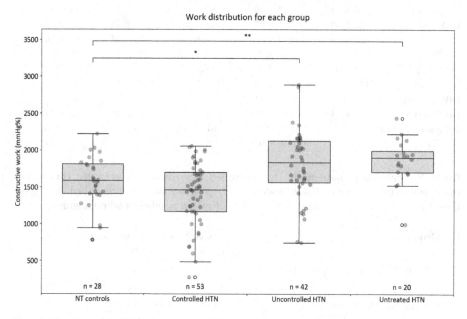

Fig. 3. Boxplots of global constructive work index for each patient sub-group, showing the medians and quartiles. Asterisks indicate statistically significant differences between groups (* $p<0.05$; ** $p<0.01$).

4   Discussion

Pressure-strain loops were generated for cardiac patients using 3D transthoracic echocardiography and LV catheter pressure recordings. The cohort was split into groups based on patients' hypertensive and medicative status. Statistically significant differences were found in the brachial blood pressures, which was expected due to the group selection being based on clinical guidelines [18]. For all hypertensive sub-groups, the EDV means were significantly larger than that of the normotensive controls, whereas only the controlled hypertension group had a significantly larger ESV. Despite the differences in volumes, there were no significant differences in systolic function in terms of EF, and only the controlled hypertension group exhibited statistically significant lower peak GLS compared to the normotensive controls, although this was not substantial (−18% versus −19%, respectively).

In this study, statistically significant differences in GWI and CWI were found between the controlled hypertensive group and the uncontrolled and untreated hypertensive sub-groups. These differences indicate that the amount of work done increases with blood pressure, and is consistent with the notion that anti-hypertensive drugs play an important role in reducing the afterload, and therefore work required, to maintain homeostatic blood flow [15]. Likewise, the similarity in work indices between uncontrolled and untreated hypertensive groups suggests that this effect is present regardless of medication. The observed difference between the untreated and controlled hypertensive groups suggests a further reduction of energy expenditure due to the medication.

The lack of statistical differences in WWI and CWI ratio (see Table 3) means that we cannot conclude that hypertension or medication causes a change in the energetic efficiency of the heart. The lack of statistical difference in WWI differs from existing literature which found an increase in wasted work for hypertensive patients [5,6]. This difference could be due to the use of invasive pressure used in this study, compared with pressure that is calculated non-invasively using the method proposed by Russel et al. [13] commonly adopted by other studies, indicating that wasted work could be effected by the pressure mapping scheme. Further investigation is therefore needed to better understand the discrepancy between these results.

While it may be that certain types of anti-hypertensive medications are more effective at reducing cardiac work than others [19], the number of patients per drug group in this study were too small to draw conclusions in this regard. Nevertheless, PSL analyses provide a new dimension with which to examine anti-hypertensive drug efficacy—a readout that may be more specific to protecting cardiac function beyond the simple lowering of blood pressure. Pressure-strain loop analysis might also help optimise personalised selection of type(s) of anti-hypertensive medication to improve patient care.

Hypertension is a precursor for myocardial hypertrophy and stiffening [14], and can lead to heart failure with preserved ejection fraction (HFpEF) [4]. There is no effective medical therapy for HFpEF, and it is difficult to reverse the damage caused by the disease [3,7,11]. Pressure-strain loop analysis has potential to provide a new, more effective means of detecting and assessing HFpEF at an earlier stage (which is important as earlier intervention leads to better outcomes), and may provide a method for informing the selection of drug therapy type(s) and dose(s) for improved prevention or amelioration of HFpEF.

Possible confounders of this study include variable levels of adherence of patients to taking their prescribed anti-hypertensive medication, and the possible cardio-depressive effects of sedation (which was administered as needed to patients during cardiac catheterisation). Furthermore, as pressure and strain measurements could not be acquired simultaneously, there may be differences in haemodynamic loading conditions which are not accounted for. Lastly, the invasive nature of catheterisation in order to obtain LV pressure traces over the full cardiac cycle limits the widespread uptake of the proposed method for hypertension stratification. The extrapolation of these findings using non-invasive surrogates (such as brachial cuff pressure measurements) is therefore needed to ensure the more general utility of pressure-strain analysis in clinical practice.

5   Conclusions

Global cardiac work, assessed using PSL analysis, is lower in normotensive controls compared to patients with systolic blood pressure greater than 140 mmHg, or diastolic blood pressure greater than 90 mmHg, regardless of medication status. We demonstrated that, although constructive work differed significantly between these groups, the same findings were not evident through analysis of wasted work. Analysis of cardiac work using pressure-strain analysis may provide a means for improved stratification of cardiac patients into functional groupings that respond more consistently to the various available therapies, which can help improve cardiac patient management and care.

Acknowledgements. This research was funded by the Health Research Council of New Zealand (grant 17/608). We gratefully thank our research nurses, Mariska Oakester Bals, Jane Hannah, Anna Taylor, Gracie Hoskin, and Nicole Prince, for their clinical expertise and invaluable assistance with patient recruitment and data collection.

References

1. Balmer, J., et al.: Pre-ejection period, the reason why the electrocardiogram Q-wave is an unreliable indicator of pulse wave initialization. Physiol. Meas. **39**(9), 95005 (2018). https://doi.org/10.1088/1361-6579/aada72
2. Galli, E., et al.: Role of myocardial constructive work in the identification of responders to CRT. Eur. Heart J. Cardiovasc. Imaging **19**(9), 1010–1018 (2018). https://doi.org/10.1093/ehjci/jex191
3. Gard, E., Nanayakkara, S., Kaye, D., Gibbs, H.: Management of heart failure with preserved ejection fraction. Aust. Prescriber **43**(1), 12 (2020). https://doi.org/10.18773/AUSTPRESCR.2020.006
4. Garg, P., et al.: Left ventricular fibrosis and hypertrophy are associated with mortality in heart failure with preserved ejection fraction. Sci. Rep. **11**(1), 617 (2021). https://doi.org/10.1038/s41598-020-79729-6
5. Huang, H., Fu, L., Ruan, Q., You, Z., Yan, L.: Segmental and global myocardial work in hypertensive patients with different left ventricular geometry. Cardiovasc. Ultrasound **21**(1), 1–11 (2023). https://doi.org/10.1186/S12947-023-00310-Y
6. Jaglan, A., Roemer, S., Perez Moreno, A.C., Khandheria, B.K.: Myocardial work in Stage 1 and 2 hypertensive patients. Eur. Heart J. Cardiovasc. Imaging **22**(7), 744–750 (2021). https://doi.org/10.1093/EHJCI/JEAB043

7. Kittleson, M.M., et al.: 2023 ACC expert consensus decision pathway on management of heart failure with preserved ejection fraction: a report of the American college of cardiology solution set oversight committee. J. Am. Coll. Cardiol. **81**(18), 1835–1878 (2023). https://doi.org/10.1016/J.JACC.2023.03.393

8. Mahdiui, M.E., et al.: Myocardial work, an echocardiographic measure of post myocardial infarct scar on contrast-enhanced cardiac magnetic resonance. Am. J. Cardiol. **151**, 1–9 (2021). https://doi.org/10.1016/j.amjcard.2021.04.009

9. Mills, K.T., Stefanescu, A., He, J.: The global epidemiology of hypertension. Nat. Rev. Nephrol. **16**(4), 223–237 (2020). https://doi.org/10.1038/s41581-019-0244-2

10. Potter, E., Marwick, T.H.: Assessment of left ventricular function by echocardiography: the case for routinely adding global longitudinal strain to ejection fraction. JACC: Cardiovas. Imaging **11**(2, Part 1), 260–274 (2018). https://doi.org/10.1016/j.jcmg.2017.11.017

11. Reddy, Y.N., Borlaug, B.A.: Heart failure with preserved ejection fraction. Curr. Probl. Cardiol. **41**(4), 145–188 (2016). https://doi.org/10.1016/J.CPCARDIOL.2015.12.002

12. Russell, K., et al.: Assessment of wasted myocardial work: a novel method to quantify energy loss due to uncoordinated left ventricular contractions. Am. J. Physiol.-Heart Circulatory Physiol. **305**(7), H996–H1003 (2013). https://doi.org/10.1152/ajpheart.00191.2013

13. Russell, K., et al.: A novel clinical method for quantification of regional left ventricular pressure-strain loop area: a non-invasive index of myocardial work. Eur. Heart J. **33**(6), 724–733 (2012). https://doi.org/10.1093/eurheartj/ehs016

14. Saheera, S., Krishnamurthy, P.: Cardiovascular changes associated with hypertensive heart disease and aging. Cell Transplant. **29**, 0963689720920830 (2020). https://doi.org/10.1177/0963689720920830

15. Takami, T., Hoshide, S., Kario, K.: Differential impact of antihypertensive drugs on cardiovascular remodeling: a review of findings and perspectives for HFpEF prevention. Hypertens. Res. **45**(1), 53–60 (2021). https://doi.org/10.1038/s41440-021-00771-6

16. The SPRINT Research Group: A Randomized Trial of Intensive versus Standard Blood-Pressure Control. New Engl. J. Med. **373**(22), 2103–2116 (2015). https://doi.org/10.1056/NEJMoa1511939

17. Whelton, P.K., Carey, R.M., Mancia, G., Kreutz, R., Bundy, J.D., Williams, B.: Harmonization of the American college of cardiology/American heart association and European Society of cardiology/European society of hypertension blood pressure/hypertension guidelines: comparisons, reflections, and recommendations. Circulation **146**(11), 868–877 (2022). https://doi.org/10.1161/CIRCULATIONAHA.121.054602

18. Williams, B., et al.: 2018 ESC/ESH Guidelines for the management of arterial hypertension: the task force for the management of arterial hypertension of the European society of cardiology (ESC) and the European Society of Hypertension (ESH). Eur. Heart J. **39**(33), 3021–3104 (2018). https://doi.org/10.1093/eurheartj/ehy339

19. You, S.C., et al.: Comprehensive comparative effectiveness and safety of first-line β-blocker monotherapy in hypertensive patients: a large-scale multicenter observational study. Hypertension **77**(5), 1528–1538 (2021). https://doi.org/10.1161/HYPERTENSIONAHA.120.16402

20. Zhao, D., et al.: Correcting bias in cardiac geometries derived from multimodal images using spatiotemporal mapping. Sci. Rep. **13**(1), 8118 (2023). https://doi.org/10.1038/s41598-023-33968-5

21. Zhu, H., et al.: Non-invasive myocardial work index contributes to early identification of impaired left ventricular myocardial function in uremic patients with preserved left ventricular ejection fraction. Biomed. Eng. Online **21**(1), 57 (2022). https://doi.org/10.1186/s12938-022-01023-5

Automated Segmentation of the Right Ventricle from 3D Echocardiography Using Labels from Cardiac Magnetic Resonance Imaging

Joshua R. Dillon[1]([✉]), Debbie Zhao[1], Thiranja P. Babarenda Gamage[1],
Gina M. Quill[1], Vicky Y. Wang[1], Nicola C. Edwards[2,3], Timothy M. Sutton[4],
Boris S. Lowe[2], Malcolm E. Legget[3], Robert N. Doughty[2,3],
Alistair A. Young[5], and Martyn P. Nash[1,6]

[1] Auckland Bioengineering Institute, University of Auckland, Auckland, New Zealand
jdil469@aucklanduni.ac.nz
[2] Green Lane Cardiovascular Service, Auckland City Hospital,
Auckland, New Zealand
[3] Department of Medicine, University of Auckland, Auckland, New Zealand
[4] Counties Manukau Health Cardiology, Middlemore Hospital,
Auckland, New Zealand
[5] School of Biomedical Engineering and Imaging Sciences,
King's College London, London, UK
[6] Department of Engineering Science and Biomedical Engineering,
University of Auckland, Auckland, New Zealand

Abstract. Segmentation of the right ventricle (RV) from 3D echocardiography (3DE) is a challenging task. In comparison to the left ventricle (LV), the complex geometry of the RV hinders accurate and reproducible volume quantification. While more accessible, 3DE falls short of gold-standard cardiac magnetic resonance (CMR) imaging for volume quantification due to its low spatial resolution and poor contrast-to-noise ratio. The use of machine learning can overcome these challenges to improve 3DE RV segmentation. This study assessed this approach by leveraging segmentations derived from CMR as ground truth labels, and including LV labels as contextual information. Forty subjects (20 females; 20 with cardiac diseases of mixed origin; 20 healthy controls) were imaged with transthoracic 3DE and cine CMR <1 h apart. Biventricular segmentations from CMR were spatially registered to corresponding end-diastolic and end-systolic 3DE images. Paired 3DE images and CMR labels from 32 subjects were used to train deep-learning models for RV segmentation from 3DE. One model was trained with RV labels only, and a second was trained with both RV and LV labels. Using the 8 test cases, the model trained with biventricular labels predicted an end-diastolic volume of 158 ± 36 ml, end-systolic volume of 105 ± 40 ml, and ejection fraction of 36 ± 11 %, which were not statistically significantly different to values measured using CMR (165 ± 30 ml, 115 ± 32 ml and 31 ± 8 %, respectively; P=NS). Inclusion of LV labels improved segmentation accuracy in cases with RV free wall signal dropout. These results indicate that leveraging CMR-derived labels for deep-learning can facilitate reliable clinical assessment of RV function from 3DE.

Keywords: Right ventricle · 3D echocardiography · Cardiac magnetic resonance · Deep learning · Cardiac segmentation

1   Introduction

Segmentation of the right ventricle (RV) from echocardiography is important for assessing the severity and progression of a variety of cardiac conditions. Despite greater clinical emphasis on the left ventricle (LV), the quantification of standard RV volume indices—end-diastolic volume (EDV), end-systolic volume (ESV), and ejection fraction (EF)—can help with prognosis, particularly in patients with heart failure [7]. Segmentation of the RV from echocardiography is more challenging than that of the LV. The retrosternal position, irregular crescent-like shape, and complex 3D geometry of the RV hinders the use of routine 2D echocardiography (2DE) with its plane-based geometric assumptions [16]. On the other hand, 3D echocardiography (3DE) is a fast and accessible modality for full-volume imaging. However, compared to gold-standard cardiac magnetic resonance (CMR) imaging, 3DE is limited by poor spatial resolution, low contrast-to-noise ratio, and is susceptible to RV free wall signal dropout due to limited reliability of transthoracic acoustic windows.

Despite integration into clinical guidelines for RV assessment [12], the use of 3DE is impeded by the ongoing difficulties of manual analysis [6]. Furthermore, manual segmentation of 3DE RV suffers from high inter- and intra-observer variabilities, which limit the accuracy and reproducibility of derived indices [13]. In addition to being subjecgtive and time intensive, manually derived volumes from 3DE tend to provide underestimates compared to those derived from CMR [16]. Such challenges have motivated the application of machine learning (ML) methods to 3DE analyses of the RV. Some studies have demonstrated that RV volume indices derived from ML-based commercial software show reasonably good agreement with CMR-derived indices [1,6,19]. However, these approaches typically rely on manual corrections, increasing analysis time and introducing observer subjectivity. The validity of such approaches remains unclear, due to opaque training methods and unpublished segmentation performance metrics. There is therefore a need for a standardised and transparent approach for fully automated, accurate, and reliable RV segmentation of 3DE.

Recently, U-Net based deep learning approaches have been shown to produce highly accurate segmentations of the RV from CMR [3]. The inclusion of contextual information from more well-defined cardiac features, such as the LV, also has the potential to improve the quality of RV segmentations [18]. In particular, nnU-Net [9], a self-configuring deep learning model for image segmentation, has been successfully applied to 3DE segmentation of the LV by leveraging CMR-derived labels as accurate and reliable ground truth labels [17]. However, extension of this approach to the RV has not yet been established.

In this study, we apply a novel methodology for automated segmentation of the RV from 3DE that leverages both CMR-derived RV segmentations, and associated LV labels in model training. We show that this approach provides reliable ground truth labels for training deep learning models for 3DE RV segmentation, and we evaluate whether inclusion of LV labels improves segmentation accuracy and agreement between 3DE- and CMR-based assessment of RV function.

2   Methods

2.1   3DE and CMR Image Acquisition

Forty prospectively recruited participants (20 women; 20 with cardiac diseases of mixed origin; 20 healthy controls) were imaged with transthoracic 3DE and cine CMR less than one hour apart. Ethical approval for this research was granted by the Health and Disability Ethics Committee of New Zealand (17/CEN/226), and written informed consent was obtained from each participant.

3D Echocardiography. Real-time 3DE images were acquired across one cardiac cycle using a Siemens ACUSON SC2000 Ultrasound System and a 4Z1c matrix array transducer (Siemens Medical Solutions, Mountain View, CA, USA), and reconstructed into 3D Cartesian images with $1\,mm^3$ isotropic voxels. RV-focused views were acquired in 20 subjects, and LV-focused views were acquired in the remaining 20 subjects.

Cardiac Magnetic Resonance Imaging. Cine CMR imaging was performed using a Siemens Magnetom 1.5T Avanto Fit (n=27) or 3T Skyra (n=13) scanner (Siemens Healthcare, Erlangen, Germany) with a retrospectively gated balanced steady-state free precession sequence under breath-holds. Acquired planes were taken over one complete cardiac cycle according to a standardised protocol [11]. The CMR images included three LV long-axis slices (standard two-, three-, and four-chamber views), an RV two-chamber view, perpendicular slices through the RV inflow and outflow tracts, and a short-axis stack of 6–10 slices spanning both ventricles.

2.2   Image Analysis and Label Generation

To generate subject-specific labels from CMR, dynamic 3D biventricular geometries were constructed semi-automatically with Cardiac Image Modeler (CIM RVLV, Version 9.1, University of Auckland, New Zealand) [8] by a single expert. Fiducial landmarks (including LV apical and basal centroids, RV insertion points on the short-axis images, as well as mitral, aortic, tricuspid, and pulmonary valve insertion points) were manually identified to create an initial coarse geometry, and to establish the position and orientation of both ventricles. This model was subsequently refined by fitting contours to the endocardial and epicardial borders on both the long- and short-axis slices, and manually correcting in-plane breath-hold mis-registrations using the intersections between images. Static 3D biventricular CMR geometries for each subject were extracted at end-diastole (ED) and end-systole (ES), corresponding to the first frame (gated to the R-wave of the electrocardiogram) and the frame associated with the smallest LV cavity volume, respectively.

Subject-specific CMR-derived biventricular labels at ED and ES were aligned to the subject's respective 3DE frames. For 3DE, ED was also assumed to correspond to the first frame (again imaging was gated to the R-wave of the electrocardiogram), while ES was identified as the frame with the lowest normalised cross-correlation score with respect to ED [5].

Spatial registration of the CMR-derived biventricular geometries to corresponding 3DE frames for each subject was carried out by applying a series of manual translations and rotations using the Paraview software (v5.11.0) [2]. All manual alignments were carried out by a single observer, yielding 80 (40 subjects × 2 frames) paired 3DE images and biventricular CMR labels. The CMR-derived labels were then converted to masks with equal dimensions to their respective 3DE image volumes. One set of masks was saved with one foreground label class representing the RV cavity (for training an RV only model), and a second set of masks was saved with two foreground label classes, representing the RV and LV cavities. A schematic of the label generation process for RV segmentation is provided in Fig. 1. In total, preparation of biventricular labels for model training took approximately 1 h per subject.

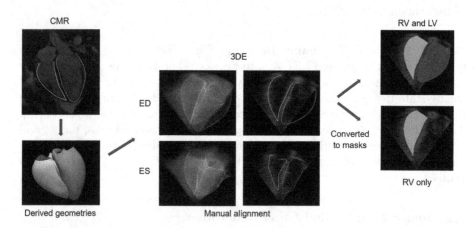

Fig. 1. CMR-derived labels registered to 3DE images (using manual alignment) to generate subject-specific ground truth labels for model training. Registration was performed separately for ED and ES images for each participant.

2.3   nnU-Net Model Training

A deep learning model based on nnU-Net [9], a self-configuring image segmentation network, was trained to predict RV segmentations using only 3DE images as inputs. Of the 40 participants, 32 were selected randomly for training (64 paired images and labels), and the remaining 8 were held out for testing (16 paired images and labels), according to an 80%/20% training/testing split. Training was carried out using 5-fold cross-validation with the 3D full-resolution nnU-Net configuration. During training, the nnU-Net algorithm applied on-the-fly data augmentations, including rotation, scaling, and downsampling, to increase model robustness. Network topology, preprocessing, and training parameters were unchanged from their fixed or self-configured values (250 iterations per epoch; batch size of 2; initial learning rate of 0.01 that was decreased according to the polyLR schedule described in [4]; Nesterov momentum of 0.99) except for epoch number, which was reduced from the default of 1000 to 80 epochs to prevent overfitting. Training loss was calculated with a combination of cross-entropy and Dice scores.

One model (nnU-Net-RV) was trained with the first set of paired labels and 3DE images, which contained only the RV cavity as a foreground label class (as described in Sect. 2.2). A second model (nnU-Net-BiV) was trained with both RV and LV cavities as separate foreground label classes. The training time was approximately 80 s per epoch (<2 h per fold) for each model on an NVIDIA Tesla V100 PCIe 32GB GPU.

2.4   Inference

Inference was carried out by each model on the 16 3DE test images (8 subjects, 2 frames each) to generate predicted RV cavity segmentations (nnU-Net-RV) or RV and LV cavity labels (nnU-Net-BiV) at ED and ES for each subject. Prediction time was approximately 15 s per frame (30 s per subject) on a NVIDIA Tesla V100 PCIe 32GB GPU.

2.5   Validation and Performance

Clinical volume indices (EDV, ESV, EF) were subsequently calculated by voxel summation and compared against CMR-derived measurements. Agreement with CMR was quantified with an intraclass correlation coefficient (ICC) using a two-way, mixed effects model for absolute agreement. Based on prevailing guidelines [10], ICC values less than 0.50, between 0.50 and 0.75, between 0.75 and 0.90, and greater than 0.90 were interpreted to indicate poor, moderate, good, and excellent reliability, respectively. The relative performance of each model was evaluated by the accuracy of predicted segmentations, characterised by Dice coefficient[1], mean surface distance (MSD)[2] and Hausdorff distance (HD)[3]. Paired-sample t-tests were used to identify statistically significant differences ($p<0.05$) in clinical indices between the ML models and CMR.

---

[1] Dice was calculated as the volume of overlap between the predicted segmentation and ground truth label, divided by the total number of voxels.

[2] MSD was calculated as the average Euclidian distance between the surface of the predicted segmentation and the ground truth label.

[3] HD was calculated as the maximum Euclidian distance between any subset of the predicted segmentation surface to the corresponding subset of the ground truth label surface.

3 Results

3.1 Participant Demographics

Table 1 defines the subjects whose 3DE images and CMR-derived labels were used in model development. The training set comprised 16 healthy controls, 9 patients with cardiac amyloidosis, 1 with hypertensive LV hypertrophy, 2 heart transplant recipients, 2 with hypertrophic cardiomyopathy, and 2 with dilated cardiomyopathy. The test set comprised 4 healthy controls, 2 patients with cardiac amyloidosis, 1 with aortic regurgitation, and 1 with dilated cardiomyopathy.

Table 1. Summary of demographics of the training and testing sets used in this study. Body surface area (BSA) was calculated using the Mosteller formula [14]. RV volume indices (EDV, ESV, and EF) were extracted from CMR-derived segmentations.

	Training (n = 32)	Testing (n = 8)	Total (n = 40)
No. diseased	16 (50 %)	4 (50 %)	20 (50 %)
No. women	16 (50 %)	4 (50 %)	20 (50 %)
Age (years)	56 ± 16	58 ± 16	56 ± 16
Height (m)	1.70 ± 0.09	1.68 ± 0.08	1.69 ± 0.09
Weight (kg)	74.7 ± 16.0	68.1 ± 7.0	73.4 ± 14.8
BSA (m^2)	1.87 ± 0.24	1.78 ± 0.11	1.85 ± 0.22
EDV (ml)	146 ± 38	165 ± 30	150 ± 37
ESV (ml)	89 ± 25	115 ± 32	95 ± 28
EF (%)	39 ± 6	31 ± 9	37 ± 8

3.2 Segmentation Accuracy

Table 2 summarises the performance of the ML models. In general, the inclusion of LV labels in model training did not improve performance, as evidenced by the distributions of segmentation accuracy metrics, which were not statistically significantly different between ML models. However, for the two test cases that

Table 2. Comparison of RV segmentation accuracy between trained *nnU-Net* models. Metrics were calculated on each frame, averaged across the test set for each ML model, and presented as mean ± SD. MSD: mean surface distance; HD Hausdorff distance.

n = 8	nnU-Net-RV	nnU-Net-BiV	p-value
Dice	0.77 ± 0.5	0.78 ± 0.4	0.38
MSD (mm)	3.3 ± 0.9	3.1 ± 0.7	0.27
HD (mm)	18 ± 4	18 ± 5	0.68

were imaged with LV-focused views, there was improvement in mean segmentation accuracy (Dice: +0.02; MSD: –0.25 mm; HD: –1.4 mm) for nnU-Net-BiV compared to nnU-Net-RV. In these subjects, there were differing levels of RV free wall signal dropout. One of these cases had a truncated RV where the ground truth label exceeded the 3DE field-of-view. In the remaining 6 test cases, inspection revealed fully intact RV labels and no signal dropout, and the differences in segmentation accuracy between ML models were negligible (Fig. 2). In the two cases with truncated RV and or free wall signal dropout, the bias in EDV compared to CMR-derived volumes for the two cases were reduced for nnU-Net-BiV compared to nnU-Net-RV (–22 ml and –10 ml compared to –32 ml and –15 ml). Such a difference was not observed with ESV. This is likely because end-diastole is normally the point in the cardiac cycle that corresponds to the largest cavity volume (and therefore presents the largest potential to be affected by label truncation and or signal dropout in the RV free wall).

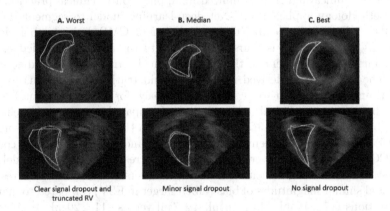

Fig. 2. RV segmentations from nnU-Net-RV (magenta contours) and nnU-Net-BiV (green) overlaid on the 3DE images. Compared with CMR-derived labels (white), the segmentation accuracy metrics (Dice, MSD, and HD) for nnU-Net-RV and nnU-Net-BiV were respectively: (**A**) (0.68, 4.9 mm, 25.3 mm) and (0.72, 4.2 mm, 21.6 mm) for the worst case (at ED); (**B**) (0.82, 2.3 mm, 15.4 mm) and (0.82, 2.2 mm, 14.8 mm) for the median case (at ED); and (**C**) (0.83, 2.3 mm, 12.2 mm) and (0.83, 2.4 mm, 11.9 mm) for the best case (at ES). (Color figure online)

3.3 Accuracy of Clinical Volume Indices

Table 3 compares clinical volume indices predicted using nnU-Net-RV or nnU-Net-BiV with those derived from CMR. Both models showed excellent agreement with CMR in EDV and ESV, and good agreement in EF. None of the predicted indices were significantly different to those derived from CMR.

Table 3. Comparison of RV volume indices between CMR and segmentations predicted for the test set using each ML model.

n = 8	CMR	nnU-Net-RV			nnU-Net-BiV		
	Mean ± SD	Bias	ICC	p-value	Bias	ICC	p-value
EDV (ml)	165 ± 30	2 ± 18	0.937	0.75	−7 ± 16	0.935	0.30
ESV (ml)	115 ± 32	−5 ± 17	0.946	0.47	−11 ± 16	0.938	0.12
EF (%)	31 ± 9	4 ± 10	0.770	0.32	4 ± 7	0.839	0.15

4 Discussion

Despite 3DE being recommended over 2DE as the preferred modality for RV volume quantification [12], the difficulty of manually extracting accurate and reproducible clinical indices has limited its application in clinical practice. Our novel methodology applies a nnU-Net deep-learning model to segment the RV from 3DE fully automatically. The application of CMR-derived label classes to 3DE RV segmentation is a first in the field to the authors' knowledge, and enables further development in the field of clinically translatable cardiac modelling. The use of CMR-derived labels as ground truth segmentations for ML model training was shown to be a viable approach for delivering accurate and reliable predictions of clinical cardiac indices. Compared to a recent benchmark for ML-based approaches to 3DE RV segmentation [1], our nnU-Net-BiV model (Table 3) showed higher agreement in EDV, ESV, and EF before manual correction (ICC of 0.94, 0.94, 0.84 versus 0.79, 0.85, 0.78, respectively). Our model also demonstrated similar agreement (0.94, 0.94, 0.84 versus 0.95, 0.96, 0.83, respectively) and similar magnitudes of bias (albeit larger in ESV) compared to manual segmentations (−7 ± 16 ml, −11 ± 16 ml, 4 ± 7 ml versus −11 ± 20 ml, 0 ± 15 ml, −3 ± 8 ml, respectively) [13], while providing advantages in non-subjectivity and shorter analysis times.

The inclusion of LV labels in ML model training provided no significant benefit compared to CMR at the group level (Table 2). However, including this contextual information was of benefit when considering LV-focussed test cases with a truncated RV and/or free wall signal dropout (Fig. 2). The incorporation of view classification algorithms could be used to select between different ML models, trained on 3DE with either LV-focused and RV-focused views, to increase the robustness of automated RV segmentation. However, performance differences between ML models trained with RV labels and biventricular labels were generally minor, and did not indicate that the performance gap in ML-based 3DE segmentation between the LV and the RV can be bridged by the inclusion of LV labels alone. Non-symmetry of the RV may require a region-focused approach to improve segmentation performance. Regional targeting approaches, such as attention U-Nets [15], may help improve segmentation performance in the worst performing regions, such as the RV free wall (Fig. 2).

This study is limited by the small sample sizes of training (n = 32) and test (n = 8) cases, which restricts the generalisability of our findings and exposes our models to potential bias. The paucity of available paired 3DE and CMR images—particularly RV-focused 3DE images—and RV labels, illustrates the challenge of establishing methodologies for automated RV segmentation from 3DE. Future extensions of this methodology should prioritise increasing the size of the dataset and the variety of cases to fully establish its benefits to clinical practice. The limitations of using CMR-derived labels is that we assume no change in cardiac function between 3DE and CMR imaging. While images could not be acquired simultaneously, the time between imaging was kept below 1 h to mitigate changes in physiological state. CMR analysis was reliant on a manual component, which may have introduced a minor observer bias in the CMR segmentations. Automated segmentations of CMR could be used to avoid this possible subjectivity, and to generate ground truth labels in larger paired datasets. The manual nature of registering CMR labels to 3DE introduced potential instabilities that may have had negative effects on the performance of the nnU-Net models. Future extensions of this methodology should investigate the feasibility of automatic registration approaches which might reduce registration inaccuracy.

5   Conclusion

We have developed a novel approach for reliable automated segmentation of the RV from 3DE by leveraging CMR-derived RV (and LV) as ground truth labels. Inclusion of LV labels in ML model training improved segmentation accuracy in test cases with a truncated RV and/or free wall signal dropout, but improvements over the entire cohort were not statistically significant. ML model performance was validated against current benchmarks in ML-based and manual volume quantification. This study represents a positive step towards accurate, reliable, fast, and fully automated RV segmentation from 3DE.

Acknowledgements. We gratefully acknowledge the study participants, and the staff at the Centre for Advanced MRI at the University of Auckland for their expertise and assistance with the imaging components of this study.

Funding Statement. This study was funded by the Health Research Council of New Zealand (programme grant 17/608).

References

1. Ahmad, A., et al.: Feasibility and accuracy of a fully automated right ventricular quantification software with three-dimensional echocardiography: comparison with cardiac magnetic resonance. Front. Cardiovasc. Med. **8**, 732893 (2021)

2. Ahrens, J., Geveci, B., Law, C.: 36 - ParaView: an end-user tool for large-data visualization. In: Hansen, C.D., Johnson, C.R. (eds.) Visualization Handbook, pp. 717–731. Butterworth-Heinemann, Burlington (2005). https://doi.org/10.1016/B978-012387582-2/50038-1, https://www.sciencedirect.com/science/article/pii/B9780123875822500381

3. Bernard, O., et al.: Deep learning techniques for automatic MRI cardiac multi-structures segmentation and diagnosis: is the problem solved? IEEE Trans. Med. Imaging **37**(11), 2514–2525 (2018)

4. Chen, L.C., Papandreou, G., Kokkinos, I., Murphy, K., Yuille, A.L.: DeepLab: semantic image segmentation with deep convolutional nets, atrous convolution, and fully connected CRFs. IEEE Trans. Pattern Anal. Mach. Intell. **40**(4), 834–848 (2018)

5. Danudibroto, A., Bersvendsen, J., Mirea, O., Gerard, O., D'hooge, J., Samset, E.: Image-based temporal alignment of echocardiographic sequences. In: Duric, N., Heyde, B. (eds.) Medical Imaging 2016: Ultrasonic Imaging and Tomography. vol. 9790, p. 97901G. International Society for Optics and Photonics, SPIE (2016). https://doi.org/10.1117/12.2216192

6. Genovese, D., et al.: Machine learning-based three-dimensional echocardiographic quantification of right ventricular size and function: Validation against cardiac magnetic resonance. J. Am. Soc. Echocardiogr. **32**(8), 969–977 (2019)

7. Ghio, S., et al.: Independent and additive prognostic value of right ventricular systolic function and pulmonary artery pressure in patients with chronic heart failure. J. Am. Coll. Cardiol. **37**(1), 183–188 (2001). https://doi.org/10.1016/S0735-1097(00)01102-5, https://www.sciencedirect.com/science/article/pii/S0735109700011025

8. Gilbert, K., et al.: An interactive tool for rapid biventricular analysis of congenital heart disease. Clin. Physiol. Funct. Imaging **37**(4), 413–420 (2017)

9. Isensee, F., Jaeger, P.F., Kohl, S.A.A., Petersen, J., Maier-Hein, K.H.: nnU-Net: a self-configuring method for deep learning-based biomedical image segmentation. Nat. Methods **18**(2), 203–211 (2021)

10. Koo, T.K., Li, M.Y.: A guideline of selecting and reporting intraclass correlation coefficients for reliability research. J. Chiropractic Med. **15**(2), 155–163 (2016). https://doi.org/10.1016/j.jcm.2016.02.012, https://www.sciencedirect.com/science/article/pii/S1556370716000158

11. Kramer, C.M., Barkhausen, J., Bucciarelli-Ducci, C., Flamm, S.D., Kim, R.J., Nagel, E.: Standardized cardiovascular magnetic resonance imaging (CMR) protocols: 2020 update. J. Cardiovasc. Magn. Reson. **22**(1), 17 (2020)

12. Lang, R.M., et al.: Recommendations for cardiac chamber quantification by echocardiography in adults: an update from the American society of echocardiography and the European association of cardiovascular imaging. J. Am. Soc. Echocardiogr. **28**(1), 1–39.e14 (2015). https://doi.org/10.1016/j.echo.2014.10.003, https://www.sciencedirect.com/science/article/pii/S0894731714007457

13. Medvedofsky, D., et al.: Novel approach to three-dimensional echocardiographic quantification of right ventricular volumes and function from focused views. J. Am. Soc. Echocardiogr. **28**(10), 1222–1231 (2015)

14. Mosteller, R.D.: Simplified calculation of body-surface area. N. Engl. J. Med. **317**(17), 1098 (1987)

15. Oktay, O., et al.: Attention U-Net: learning where to look for the pancreas. In: Medical Imaging with Deep Learning (2018). https://openreview.net/forum?id=Skft7cijM

16. Wu, V.C.C., Takeuchi, M.: Three-dimensional echocardiography: current status and real-life applications. Acta Cardiol. Sin. **33**(2), 107–118 (2017)

17. Zhao, D., et al.: MITEA: a dataset for machine learning segmentation of the left ventricle in 3D echocardiography using subject-specific labels from cardiac magnetic resonance imaging. Front. Cardiovasc. Med. **9**, 1016703 (2022)

18. Zhou, J., Du, M., Chang, S., Chen, Z.: Artificial intelligence in echocardiography: detection, functional evaluation, and disease diagnosis. Cardiovasc. Ultrasound **19**(1), 29 (2021)

19. Zhu, Y., et al.: Quantitative assessment of right ventricular size and function with multiple parameters from artificial intelligence-based three-dimensional echocardiography: a comparative study with cardiac magnetic resonance. Echocardiography **39**(2), 223–232 (2022)

Neural Implicit Functions for 3D Shape Reconstruction from Standard Cardiovascular Magnetic Resonance Views

Marica Muffoletto[1]([✉]), Hao Xu[1], Yiyang Xu[1], Steven E Williams[1,2], Michelle C Williams[2], Karl P Kunze[1,3], Radhouene Neji[1], Steven A Niederer[4], Daniel Rueckert[5], and Alistair A Young[1]

[1] School of Biomedical Engineering and Imaging Sciences, King's College London, St Thomas' Hospital, 4th Floor Lambeth Wing, Westminster Bridge Road, London SW1 7EH, UK
marica.muffoletto@kcl.ac.uk
[2] University/BHF Centre for Cardiovascular Science, University of Edinburgh, 47 Little France Crescent, Edinburgh EH16 4TJ, UK
[3] MR Research Collaborations, Siemens Healthcare Limited, Frimley, UK
[4] Cardiac Electro-Mechanics Research Group (CEMRG), Imperial College London, London, UK
[5] Institute for Artificial Intelligence and Informatics in Medicine, Klinikum Rechts der Isar, Technical University of Munich, Munich, Germany

Abstract. In cardiovascular magnetic resonance (CMR), typical acquisitions often involve a limited number of short and long axis slices. However, reconstructing the 3D chambers is crucial for accurately quantifying heart geometry and assessing cardiac function. Neural Implicit Representations (NIR) learn implicit functions for anatomical shapes from sparse measurements by leveraging a learned continuous shape prior, without the need for high-resolution ground truth data. In this study, we utilized coronary computed tomography (CCTA) images to simulate CMR sparse label maps of two types: standard (10 mm spaced short axis and 2 long axis slices) and 3-slice (single short and 2 long axis slices). Whole heart NIR reconstructions were compared to a Label Completion U-Net (LC-U-Net) network trained on the dense segmentations. The findings indicate that the LC-U-Net is not robust when tested with fewer slices than those used during training. In contrast, the NIR consistently achieved Dice scores above 0.9 for the left ventricle, left ventricle myocardium, and right ventricle labels, irrespective of changes in the training or test set. Predictions from standard views achieved average Dice scores across all labels of 0.84±0.03 and 0.88±0.03, when training on 3-slice and standard data respectively. In conclusion, this study presents promising results for 3D shape reconstruction invariant to slice position and orientation without requiring full resolution training data, offering a robust and accurate method for cardiac chamber reconstruction in CMR.

Keywords: CMR · Neural Implicit Functions · 3D Reconstruction

ⓒ The Author(s), under exclusive license to Springer Nature Switzerland AG 2024
O. Camara et al. (Eds.): STACOM 2023, LNCS 14507, pp. 130–139, 2024.
https://doi.org/10.1007/978-3-031-52448-6_13

1   Introduction

The reconstruction of anatomy from sparse data is a crucial task in medical imaging, with significant implications for time and resource efficiency. Deep learning methods have advanced this field. These normally rely on discrete representations like voxels, point clouds, or meshes [3,8]. Emerging techniques, such as super-resolution and inpainting methods, have further enriched this domain [9,12]. In particular, Neural Implicit Functions (NIFs), offer a novel approach by representing surfaces as implicit decision boundaries. NIFs provide versatile capabilities, enabling shape sampling at arbitrary resolutions and spatial points. Consequently, they prove especially well-suited for handling sparse, partial, or non-uniform data, which are common in medical imaging [5]. Furthermore, the potential for robust generalisation across diverse test datasets, which poses a significant challenge in the field, seems to be well handled by NIFs [2]. Another noteworthy advantage of NIFs lies in their ability to process each point independently, eliminating the need for complete ground truth training data. This feature opens up possibilities for various applications, including image reconstruction directly from imaging scanners [6].

Contributions and Aim

The aim of this paper is to use NIFs acting on segmentations (SEG-NIF) to reconstruct whole-heart shapes from sparse labelled slices available from Cardiovascular Magnetic Resonance (CMR) standard acquisitions. We compared this approach to Label Completion U-Net (LC-U-Net), a fully-supervised baseline method developed by [10], which uses CNNs trained on dense segmentations. Since CMR often acquires a limited number of slices, training of fully supervised networks is often challenging, hence the application of NIFs to this task is particularly beneficial. In this study, we simulated two types of MR acquisition: a standard CMR acquisition with two long axis (LAX) slices and several short axis (SAX) slices with 10mm gap, and a 3-slice with only one SAX and two LAX slices. We used coordinates sampled from these slices to learn a Neural Implicit Representation, which can be used to output a multi-label occupancy probability value for coordinates on any other slice (Fig. 1). By querying the function from points on the whole volume, we obtained a final dense output, which was compared to predictions from the LC-U-Net method described in [10]. Meaningful contributions of this study are the following: 1) new application of NIFs to a whole-heart CT dataset which simulates CMR standard acquisition and extension of method in [2] to multi-label prediction; 2) analysis of the impact of reducing the amount of labelled data during training and testing (we trained and tested on both sets of simulated slices); 3) comparison of an unsupervised method, which relies solely on sparse data, to a fully supervised method.

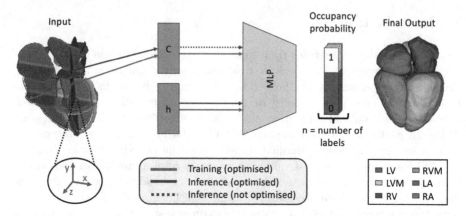

Fig. 1. Overview of the SEG-NIF method. Coordinates (c: x,y,x) from the sparse labelled slices are used for training an MLP, together with a randomly initialised latent vector (h) which is optimised at both training and inference time for each cardiac shape. The MLP predicts occupancy values per each coordinate. At inference time, after latent vector optimisation, any coordinate from an unlabelled slice can be queried to obtain the dense label map.

2 Method

2.1 Data

In this study, data from the Scottish COmputed Tomography of the HEART (SCOT-HEART) trial [1] was utilized, involving patients with suspected coronary artery disease. To generate sparse label maps, the whole-heart segmentations from the full dataset (1700 CT cases) were employed. These include left ventricle (LV), left ventricle myocardium (LVM), right ventricle (RV), right ventricle myocardium (RVM), left atrium (LA) and right atrium (RA) labels. The segmentations were previously aligned to the cardiac coordinate space according to the method described in [10]. Subsequently, a bounding box centered at the heart's centroid and 20% larger than its largest dimension was applied for cropping. To achieve data sparsification, key cardiac slices were identified. The 4-chamber (4CH) and 2-chamber (2CH) long axis slices were determined by locating slices parallel to the x-y and y-z planes, respectively, that passed through the centroid of the left ventricle (LV). Additionally, one mid short-axis slice was defined, parallel to the x-z plane and intersecting the LV centroid. These three slices constituted the 3-slice sparse label maps. Furthermore, standard label maps were generated by adding parallel slices with a 10mm gap in the short axis plane. Figure 2 provides an illustrative representation of the final label maps.

2.2 Network

The SEG-NIF method was based on the implementation described in [2]. A multilayer perceptron (MLP) with 8 linear layers, 128 dimensions each, is employed

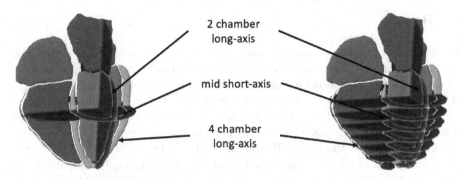

Fig. 2. Examples of sparse input data. On the left, the 3-slice sparse label map, which only includes the 3 main slices (2CH, 4CH, mid short-axis). On the right, the standard label map, which contains extra short-axis slices with 10mm distance from the mid short-axis.

as a classifier which takes as input the 3D coordinates from the training slices in the sparse label map and a latent vector. During training, latent vectors are randomly initialised for each shape, and they are optimised jointly with the parameters of the MLP (Fig. 1). The original network output probability for only one label, but we extended it to predict occupancy for the aforementioned six labels. The difference to the ground-truth occupancy is minimised through a voxel-based loss function, combining Dice score and Cross Entropy. At inference time, only the latent vector is optimised. The model is trained with a learning rate of 0.001, Adam optimiser and a batch size of 4. Since standard input data can have a variable number of labelled slices, we randomly sub-sampled 20.000 coordinates per volume, to ensure that the number of coordinates within batches are always consistent. At inference, prediction of a single case took 20 s to 60 s, depending on the amount of steps necessary for latent optimisation.

The LC-U-Net consisted in a 3D U-Net built with Pythorch, with six labels as both input and output. The network had four spatial resolutions and convolutional kernels of size $3 \times 3 \times 3$, with the number of (16, 32), (32, 64), (64, 128), and (128, 256) in the encoder and bottle neck, and (128, 128), (64, 64), (32, 6) in the decoder. Max-pooling and deconvolution with a stride of $2 \times 2 \times 2$ were used for contraction and expansion [11]. Adam optimization with a learning rate of 10^{-3} was used, and the inference time for a single case was approximately 2 s.

2.3 Experiments

We split the dataset into 1360 cases for training (80%) and 340 for testing (20%). We performed four total experiments: 1) training and testing on 3-slice label maps; 2) training on standard label maps and testing on 3-slice; 3) training on 3-slice and testing on standard; 4) training and testing on standard. We repeated the same experiments on the LC-U-Net method, keeping the same training and test set, to obtain a fully-supervised baseline comparison [10].

3   Results

We evaluated the reconstruction from sparse to dense label maps using the 3D Dice similarity (Fig. 3,4), Hausdorff Distance and Average Surface Distance (Tables 1-3). Figure 3 shows a comparison between the performance of the SEG-NIF method and the fully supervised baseline LC-U-Net. The highest performance across labels is found when training and testing LC-U-Net on standard data (LV: 0.98, LVM: 0.96, RV: 0.97). The same experiment (n°4) is also the most successful for the SEG-NIF approach (LV: 0.96, LVM: 0.94, RV: 0.96). With exception of the atria, performance of the two methods is comparable; in experiment 2, SEG-NIF outperforms the baseline (LVM: 0.92 vs 0.88, RV: 0.93 vs 0.82, RVM: 0.55 vs 0.37). Paired t-test reveals significance ($p<0.05$) between the two methods for every label, except the LV. The drop between predictions from standard (experiment 4) and 3-slices (experiment 2) is significant for all labels in LC-U-Net ($p<0.001$) and for all, except LA and RA ($p>0.05$), in SEG-NIF, but absolute difference between metrics (Table 2) is much smaller in the latter method (HD: 0.23; ASD: 0.15) than in the former one (HD: 10.93; ASD: 0.85).

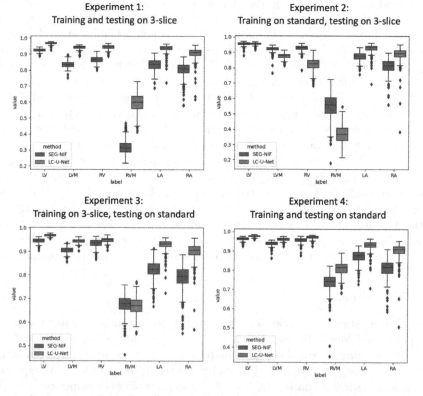

Fig. 3. Comparison of accuracy (Dice coefficient) of test-set predictions among experiments and methods (SEG-NIF vs LC-U-Net).

Table 1. CCTA test set accuracy (mean ± standard deviation). SEG-NIF (ours) vs fully-supervised baseline LC-U-Net [10]. Both trained on 3-slice sparse label maps. HD: Hausdorff distance; ASD: average surface distance. Best results for each category are highlighted in bold.

	TEST SET	METHOD	LV	LVM	RV	RVM	LA	RA
HD (mm)	3-slice	SEG-NIF	7.88±1.81	8.86±3.43	12.42±3.19	13.82±3.59	11.99±3.31	14.86±4.38
		LC-U-Net	**3.66±1.15**	**4.18±2.67**	**6.35±2.13**	**10.04±5.06**	**6.05±3.57**	**7.85±4.09**
	standard	SEG-NIF	8.92±2.24	10.52±5.38	11.45±4.11	13.80±4.08	12.32±3.25	16.26±4.72
		LC-U-Net	**3.83±2.19**	**6.00±5.77**	**7.22±2.49**	**8.84±3.23**	10.73±17.54	14.94±18.32
ASD (mm)	3-slice	SEG-NIF	1.19±0.14	1.28±0.15	1.99±0.27	2.20±0.33	2.16±0.42	2.53±0.63
		LC-U-Net	**0.54±0.08**	**0.48±0.07**	**0.90±0.20**	**0.65±0.14**	**0.92±0.47**	**1.26±0.43**
	standard	SEG-NIF	0.86±0.12	0.80±0.11	0.91±0.17	0.77±0.15	2.27±0.44	2.65±0.66
		LC-U-Net	**0.52±0.06**	**0.48±0.07**	**0.86±0.18**	0.66±0.10	1.09±0.66	1.47±0.79

Table 2. CCTA test set accuracy (mean ± standard deviation). SEG-NIF (ours) vs fully-supervised baseline LC-U-Net [10]. Both trained on standard sparse label maps. HD: Hausdorff distance; ASD: average surface distance. Best results for each category are highlighted in bold.

	TEST SET	METHOD	LV	LVM	RV	RVM	LA	RA
HD (mm)	3-slice	SEG-NIF	5.37±1.54	5.27±1.73	8.31±2.65	10.43±2.95	9.41±4.31	16.69±5.09
		LC-U-Net	4.88±1.22	9.62±3.38	20.73±5.30	24.73±5.97	30.77±30.58	11.61±10.28
	standard	SEG-NIF	5.30±1.45	5.20±1.62	7.47±2.38	9.74±2.52	9.56±4.00	16.84±5.18
		LC-U-Net	**3.55±1.08**	**3.87±2.25**	**4.79±2.75**	**8.39±3.25**	**6.63±4.42**	**9.49±8.76**
ASD (mm)	3-slice	SEG-NIF	0.72±0.13	0.62±0.11	1.08±0.27	1.05±0.37	1.63±0.32	2.21±0.60
		LC-U-Net	0.71±0.09	0.89±0.10	2.26±0.58	1.58±0.56	2.03±2.40	1.51±0.80
	standard	SEG-NIF	0.63±0.10	0.51±0.08	0.70±0.17	0.65±0.18	1.67±0.37	2.24±0.61
		LC-U-Net	**0.39±0.05**	**0.32±0.05**	**0.48±0.39**	**0.39±0.10**	**0.97±0.49**	**1.33±0.58**

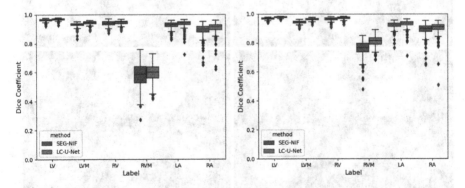

Fig. 4. Comparison of accuracy (Dice coefficient) of test-set predictions in upper-bound experiment. SEG-NIF was trained with dense label maps and LC-U-Net was trained with 3-slice label maps, both were tested on 3-slice (left) and standard (right).

We finally trained SEG-NIF on denser points, sampled from a uniform grid (approach used to sparsify the data in [2]), and hence covering most chambers. This enabled us to obtain equivalent models trained on dense data for both the Neural Implicit Representation method and the LC-U-Net. The results of this upper-bound experiment can be found in Table 3 and Fig. 4.

Table 3. CCTA test set accuracy (mean ± standard deviation). SEG-NIF (ours) results after training on dense label maps. HD: Hausdorff distance; ASD: average surface distance.

	TEST SET	LV	LVM	RV	RVM	LA	RA
HD (mm)	3-slice	4.08±1.22	4.20±1.33	6.95±2.39	7.69±2.88	**6.37±3.03**	8.25±3.10
	standard	**3.79±1.1.07**	**4.04±1.34**	**4.93±2.16**	**6.29±2.75**	6.44±2.28	**8.21±3.11**
ASD (mm)	3-slice	0.64±0.10	0.57±0.09	0.95±0.24	0.90±0.25	**1.03±0.28**	**1.50±0.50**
	standard	**0.56±0.08**	**0.48±0.07**	**0.57±0.16**	**0.53±0.15**	1.08±0.28	1.50±0.51

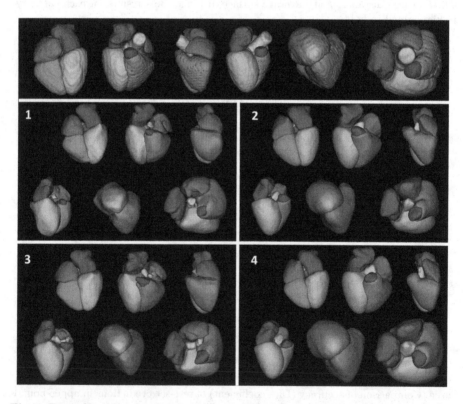

Fig. 5. Reconstruction visualisations of our method for an average case. Top row is the reference dense label map. Each quadrant contains results from a different experiment.

Figure 5 shows the prediction of an average case by the SEG-NIF method from six different views. The main chambers are mostly faithfully reconstructed, although the reference label map shows coarser surfaces than the predictions and a very thin RVM, which leads to poor performance of the metrics in Tables 1 and 2 for this label.

4   Conclusion

This paper shows that NIFs can be successfully employed to perform a full segmentation of the heart, relying solely on sparse data. The four experiments provide proof-of-concept that, even without full-resolution information, NIFs can implicitly retrieve shape information from optimisation of latent code for each shape. The result is a fully segmented prediction with Dice scores and Hausdorff distances which closely resemble the ones obtained with the LC-U-Net. It's worth considering that, although we use the same training data for SEG-NIF and LC-U-Net, the difference between the two methods makes it so that the former only learns from the few MR-simulated slices, while the LC-U-Net sees the dense maps during training. This explains the difference in performance for the atria (Fig. 3), for which the NIF only relies on one slice (4CH LAX). Although specific metrics for the Aorta (AO) and Pulmonary Artery (PA) are not included in this paper, observations from Fig. 5 show that SEG-NIF attempts to reconstruct those labels as well, despite the very limited information available from the 4CH slice. We also found that the RVM label was particularly challenging to evaluate, because the ground truth employed (Fig. 5) is missing labels for some of the voxels, prohibiting a correct analysis of the prediction. The primary contribution of this paper is showing that self-supervised methods, such as SEG-NIF, can effectively adapt and generalise to a test set with substantially less information than the training set (experiment 2). This finding gives confidence in the potential of exploring Neural Implicit Representation for heart chamber segmentation and encourages further investigation in this direction. An additional benefit of using NIFs is the reduced memory requirement. While CNNs tend to exponentially increase their dimension with spatial resolution, NIFs offer a viable and memory-efficient alternative for encoding large 3D medical data [7]. In this study, the LC-U-Net had around 50 times more parameters than the MLP approach but it was faster at inference time, which leaves the question of adaptability to clinical practice open to further debate.

Future work

Given the initial findings and the observed decrease in performance when reducing information during testing for the CNN-based approach, meaningful insights can be also provided by evaluating both methods on extremely limited test sets, containing only one LAX or SAX slice. Although the study demonstrates the effectiveness of using NIFs to complete unlabelled volumes from standard MR acquisition slices, it overlooks two crucial factors: slice position errors and breath-hold misalignments (motion artefacts) [10]. Drawing inspiration from the work of Wang et al. [9], these factors could be addressed in NIFs by optimising the latent code. In the future, segmentation of all cardiac chambers through NIFs can also lead to evaluation of volumes or surface areas [11], which can be directly correlated to diseases. Such an approach could be particularly advantageous in studies like the UK Biobank, where multiple correlations between diseases and cardiac shapes can be analysed [4], and the data is often incomplete and only partially labelled.

Acknowledgments. Research supported by ESPRC and Siemens Healthineers.

References

1. Coronary CT angiography and 5-year risk of myocardial infarction. New England Journal of Medicine 379(10), pp .924–933 (2018). https://doi.org/10.1056/NEJMoa1805971./ https://doi.org/10.1056/NEJMoa1805971, pMID: 30145934
2. Amiranashvili, T., Lüdke, D., Li, H., Menze, B., Zachow, S.: Learning shape reconstruction from sparse measurements with neural implicit functions. In: Medical Imaging with Deep Learning (2022)
3. Beetz, M., Banerjee, A., Grau, V.: Biventricular surface reconstruction from cine MRI contours using point completion networks. In: 2021 IEEE 18th International Symposium on Biomedical Imaging (ISBI), pp. 105–109 (2021). https://doi.org/10.1109/ISBI48211.2021.9434040
4. Gilbert, K., et al.: End-diastolic and end-systolic lv morphology in the presence of cardiovascular risk factors: a UK biobank study. In: Coudière, Y., Ozenne, V., Vigmond, E., Zemzemi, N. (eds.) Functional Imaging and Modeling of the Heart, pp. 304–312. Springer International Publishing, Cham (2019)
5. Hu, H., Chen, Y., Xu, J., Borse, S., Cai, H., Porikli, F., Wang, X.: Learning Implicit feature alignment function for semantic segmentation (2022)
6. Huang, W., Li, H., Pan, J., Cruz, G., Rueckert, D., Hammernik, K.: Neural implicit k-space for binning-free non-cartesian cardiac MR imaging (2023)
7. Khan, M.O., Fang, Y.: Implicit neural representations for medical imaging segmentation. In: Wang, L., Dou, Q., Fletcher, P.T., Speidel, S., Li, S. (eds.) Medical Image Computing and Computer Assisted Intervention - MICCAI 2022, pp. 433–443. Springer Nature Switzerland, Cham (2022)
8. Mescheder, L., Oechsle, M., Niemeyer, M., Nowozin, S., Geiger, A.: Occupancy networks: Learning 3D reconstruction in function space (2019)
9. Wang, S., et al.: Joint motion correction and super resolution for cardiac segmentation via latent optimisation (2021)
10. Xu, H., Muffoletto, M., Niederer, S.A., Williams, S.E., Williams, M.C., Young, A.A.: Whole heart 3D shape reconstruction from sparse views: leveraging cardiac computed tomography for cardiovascular magnetic resonance. In: Bernard, O., Clarysse, P., Duchateau, N., Ohayon, J., Viallon, M. (eds.) Functional Imaging and Modeling of the Heart, pp. 255–264. Springer Nature Switzerland, Cham (2023)

11. Xu, H., et al.: Deep learning estimation of three-dimensional left atrial shape from two-chamber and four-chamber cardiac long axis views. European Heart Journal - Cardiovascular Imaging 24(5), pp. 607–615 (02 2023). https://doi.org/10.1093/ehjci/jead010. https://doi.org/10.1093/ehjci/jead010
12. Xu, H., Zacur, E., Schneider, J.E., Grau, V.: Ventricle surface reconstruction from cardiac MR slices using deep learning. In: Functional Imaging and Modeling of the Heart (2019)

Deep Learning-Based Pulmonary Artery Surface Mesh Generation

Nina Krüger[1,2,3](✉) , Jan Brüning[1,2] , Leonid Goubergrits[1,2] ,
Matthias Ivantsits[1,2] , Lars Walczak[1,2,3] , Volkmar Falk[2,4,5] ,
Henryk Dreger[1,2,4] , Titus Kühne[1,2,4] , and Anja Hennemuth[1,2,3,4,6]

[1] Deutsches Herzzentrum der Charité (DHZC), Institute of Computer-assisted
Cardiovascular Medicine, Augustenburger Platz 1, 13353 Berlin, Germany
nina.krueger@dhzc-charite.de

[2] Charité - Universitätsmedizin Berlin, corporate member of Freie Universität Berlin
and Humboldt-Universität zu Berlin, Charitéplatz 1, 10117 Berlin, Germany

[3] Fraunhofer MEVIS, Bremen, Germany

[4] DZHK (German Centre for Cardiovascular Research), Partner Site Berlin,
Berlin, Germany

[5] Deutsches Herzzentrum der Charité (DHZC), Department of Cardiothoracic
and Vascular Surgery, Berlin, Germany

[6] University Medical Center Hamburg-Eppendorf, Department of Diagnostic
and Interventional Radiology and Nuclear Medicine, Hamburg, Germany

Abstract. Properties of the pulmonary artery play an essential role
in the diagnosis and treatment planning of diseases such as pulmonary
hypertension. Patient-specific simulation of hemodynamics can support
the planning of interventions. However, the variable complex branching structure of the pulmonary artery poses a challenge for image-based
generation of suitable geometries. State-of-the-art segmentation-based
approaches require an interactive 3D surface reconstruction to prepare
the simulation geometry. We propose a deep learning approach to generate a 3D surface mesh of the pulmonary artery from CT images suitable
for simulation. The proposed method is based on the Voxel2Mesh algorithm and includes a voxel encoder and decoder as well as a mesh decoder
to deform a prototype mesh. An additional centerline coverage loss facilitates the reconstruction of the branching structure. Furthermore, vertex
classification allows for the definition of in- and outlets. Our model was
trained with 48 human cases and tested on 10 human cases annotated by
two observers. The differences in the anatomical parameters inferred from
the automatic surface generation correspond to the differences between
the observers' annotations. The suitability of the generated mesh geometries for numerical flow simulations is demonstrated.

Keywords: Graph Neural Network · Voxel2Mesh · Surface
Reconstruction · Mesh Generation · Pulmonary Artery

© The Author(s) 2024
O. Camara et al. (Eds.): STACOM 2023, LNCS 14507, pp. 140–151, 2024.
https://doi.org/10.1007/978-3-031-52448-6_14

1 Introduction

The pulmonary artery (PA) originates from the right ventricle and delivers de-oxygenated blood towards the lungs, where it is oxygenated before reaching the left heart and thus the systemic circulation. The blood pressure within the PA is an important biomarker for assessing the severity of heart failure. Modeling and simulation are important tools for investigation of the complex hemodynamics [5,16]. Numerical simulation of PA hemodynamics requires well-defined and high-quality surface meshes enclosing the PA vessel lumen [11]. Pure voxel-based segmentation of the underlying 3D image data [4,10,13,18] is not sufficient, as these suffer from step artifacts depending ón the voxel size [3]. When transformed into a surface representation, the resulting meshes often show defects such as bad-quality triangles [2,15]. Using image-based anatomical information for flow simulation typically requires adjustments such as correcting segmentation masks, expanding inflow and outflow regions, or surface smoothing [2]. This can result in time- and resource-intensive interaction.

Data-driven models show promise to unify this entire processing pipeline, allowing to reduce or fully omit manual interaction, but also unnecessary steps, which in turn might reduce the propagation of errors. Kong et al. [8] proposed such a unified approach for mesh generation of the whole heart. However, the level of detail of the mesh representation is limited and considered insufficient for complex vascular structures.

In this study, we demonstrate an approach for the automatic generation of pulmonary artery surface meshes, which can be used for numerical assessment of intra-arterial hemodynamics. This requires the main (MPA), left (LPA) and right (RPA) pulmonary artery up to two degrees of branching vessel. The inlet as well as all outlets (for simplicity referred to as openings) are truncated for subsequent specification of boundary conditions in the numerical setup. We combine different techniques for accurate segmentation, mesh generation, and opening, compare relevant anatomical parameters of automatically and manually obtained surface meshes, and lastly, apply computational fluid dynamics to test the suitability of our surface geometries for hemodynamic assessment.

2 Methods

We propose a deep learning approach based on the Voxel2Mesh algorithm [17] for automatic pulmonary artery mesh generation from contrast-enhanced CT images of the heart. Voxel2Mesh is a neural network architecture, that directly produces 3-dimensional triangular surface meshes from volumetric image data. The Voxel2Mesh algorithm combines information from a simultaneously trained voxel-wise segmentation with a mesh decoder, that deforms a prototype mesh to approximate the target mesh (see Fig. 1). Consequently, the prototype mesh and the target mesh need to have the same topology. As the number of branches of the pulmonary artery differs this cannot be achieved with open-ended surface meshes. Hence, we train the Voxel2Mesh algorithm with closed meshes, using

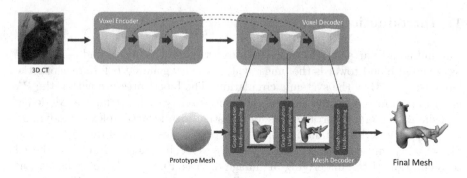

Fig. 1. An overview of the proposed architecture, including a voxel-encoder, a voxel-decoder, and a mesh-decoder: Using a 3D CT image as input, the model successively deforms a prototype mesh at increasing resolution to approximate the target mesh. For this, the mesh-decoder samples features from the latent voxel-space.

a sphere as the prototype. Additionally, our algorithm contains a vertex classification to differentiate between the vessel wall and opening, allowing for the removal of respective faces.

2.1 Data

We used retrospective CT image data of 58 aortic stenosis patients, weighing 76.8 ± 18.2 kg, with an age of 81 ± 7.6 years and a female percentage of 60% for the construction of human pulmonary artery surface meshes. CT data sets of the entire heart were acquired as part of transcatheter aortic valve implantation planning at our clinical center, with wide area-detector volume CT scanners: Aquilion One Vision (Canon Medical Systems, Tochigi, Japan) or Revolution CT (GE Healthcare, Chicago, IL, USA) with 100 kV tube voltage, $(0.390$–0.648 mm$) \times (0.390$–0.648 mm$)$ in-plane resolution and $(0.5$–1.0 mm$)$ slice thickness. The institutional review board (Ethikkommission Charité - Universitätsmedizin Berlin, approval number EA2/004/21) approved the use of retrospective data for this study and waived individual informed consent. Figure 2 shows a sample of the reconstruction pipeline from voxel-wise CT segmentation to open-ended surface mesh. We used 40 cases for training the neural network and 8 for validation. A thorough evaluation of anatomical parameters and simulation properties was done on the remaining 10 cases. Two observers generated meshes for each of these 10 cases and we compare deviations from our automatically generated meshes to inter-observer variability. We resampled all CT images to an isotropic resolution of 1.7 mm for neural network training, due to memory constraints. We close all mesh openings, ensuring that the normals are consistently pointing outwards.

Fig. 2. Illustration of the pipeline for manual reconstruction of the PA surface mesh from CT image data: The region of interest is segmented manually within the image data and a rough initial surface geometry is reconstructed from the 3D voxel mask. This surface is subsequently smoothed and all vessel endings are truncated. SVC = superior vena cava; PV = pulmonary vein.

2.2 Voxel2Mesh-Based Surface Mesh Generation

Voxel2Mesh uses several loss functions and regularization terms [17]. The cross-entropy loss

$$\mathcal{L}_{CE} = -\frac{1}{N}\sum_{n=1}^{N}\sum_{c=0}^{1} log \frac{exp(x_{n,c})}{\sum_{i=0}^{1} exp(x_{n,i})} y_{n,c} \tag{1}$$

for training the voxel encoder, where x is the prediction, y is the ground-truth, c is the class (0 for background, 1 for the pulmonary artery) and N is the number of samples within the batch. For training the mesh decoder, the Chamfer distance

$$\mathcal{L}_{CD} = \sum_{v_p \in S_p} \min_{v_g \in S_g} ||\vec{x}_p - \vec{x}_g|| + \sum_{v_g \in S_g} \min_{v_p \in S_p} ||\vec{x}_p - \vec{x}_g|| \tag{2}$$

is applied, where S_p and S_g denote the predicted and ground-truth surfaces, with v_p and v_g being the respective vertices sampled from those surfaces, and $\vec{x}_p$ and $\vec{x}_g$ the corresponding 3D coordinates. This notation also applies hereafter. The regularization terms are: the normal consistency loss

$$\mathcal{L}_{NC} = \frac{1}{|N_F|} \sum_{(f_0,f_1) \in N_F} 1 - \frac{\vec{n}_0 \cdot \vec{n}_1}{||\vec{n}_0|| \cdot ||\vec{n}_1||}, \tag{3}$$

where f_0 and f_1 denote the neighbouring faces in N_F, the set of neighbouring faces of the mesh and $\vec{n}_0$ and $\vec{n}_1$ the normal vectors of the respective neighboring faces; the Laplacian smoothing constraint

$$\gamma_L = \frac{1}{|S_p|} \sum_{v \in S_p} ||\frac{1}{|N(v)|} \sum_{v' \in N(v)} (\vec{x} - \vec{x'})||, \tag{4}$$

where $N(v)$ is the set of neighboring vertices of vertex v; and the edge length loss

$$\mathcal{L}_E = \frac{1}{|E|} \sum_{(v_0,v_1) \in E} (||\vec{x}_0 - \vec{x}_1||_2 - L_t)^2, \tag{5}$$

where E denotes the set of edges of a mesh. v_0 and v_1 are the vertices composing the edge, and L_t is a constant target length, in our case 0 to ensure edges of similar length.

We introduced an additional *centerline coverage loss* $\mathcal{L}_{CC}$ to enforce the desired branching based on the ground-truth centerlines computed with morphMan [7]:

$$\mathcal{L}_{CC} = \sum_{e \in E_o} d_e, \tag{6}$$

We penalize centerline endpoints located outside of the generated mesh. E_o is the set of centerline endpoints outside of the mesh and d_e is the distance between the endpoint e and the corresponding closest point on the mesh [12].

Thus, our Voxel2Mesh implementation balances between a smooth mesh and the formation of branches. Furthermore, we use a cross-entropy loss $\mathcal{L}_{VC}$ (analog to equation (1)) for the vertex class prediction. The final loss function L is a linear combination of all loss terms $\mathcal{L}$ weighted with their respective hyperparameters λ at different resolutions.

$$L = \lambda_{CE}\mathcal{L}_{CE} + \lambda_{CD}\mathcal{L}_{CD} + \lambda_{NC}\mathcal{L}_{NC} + \lambda_L\mathcal{L}_L + \lambda_E\mathcal{L}_E + \lambda_{CC}\mathcal{L}_{CC} + \lambda_{VC}\mathcal{L}_{VC} \tag{7}$$

Based on a grid search with 5-fold cross-validation, the loss hyperparameters were uniformly weighted across all resolutions with: $\lambda_{CE} = 5$, $\lambda_{CD} = 15$, $\lambda_{NC} = 0.1$, $\lambda_L = 0.1$, $\lambda_E = 0.1$, $\lambda_{CC} = 100$ and $\lambda_{VC} = 5$.

We rely on a uniform upsampling strategy instead of the adaptive upsampling suggested in the original Voxel2Mesh architecture [17], as too many vertices were discarded due to the close adjacency of the branches.

We performed random augmentation during training: affine (rotation -10 to $10°C$, scaling 0.9 to 1.1, translation up to 5 mm), blur (standard deviation of the Gaussian Kernel up to 2 mm), and Gaussian noise (mean of 0 and standard deviation up to 0.25 mm).

2.3   Postprocessing

Due to the topology restrictions of Voxel2Mesh, we obtain closed surface meshes, which we need to open for simulation. We approach this by additionally predicting a class for each vertex, defining whether it belongs to an opening or not. We then remove all faces containing opening vertices. The resulting boundary vertices for each opening are then adjusted to lie on the same average plane by least square approximation. At this point, few openings are not detected. Hence, we manually open the remaining closed ends for the subsequent steps.

2.4   Centerline-Based Analysis

For comparison of the anatomical features, we automatically obtained length and diameters for different segments of the PA, as well as the tortuosity of LPA

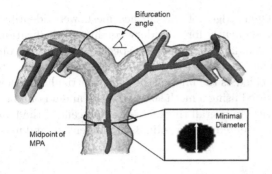
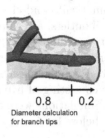

Fig. 3. Anatomical measurements from centerline-based analysis: The centerline (red) is subdivided into several edges, connected by nodes (blue), marking a bifurcation. Diameters for each edge are obtained at the midpoint (blue star) between start and end node of an edge or at start node + 0.8 * (end node - start node) for branch tips (blue triangle). The bifurcation angle between LPA and RPA is the angle between the vectors connecting the endpoint of the MPA to the endpoint of the first edge of LPA and RPA, respectively. (Color figure online)

and RPA and their bifurcation angle from a centerline-based analysis of the open-ended surfaces. The analysis was done using MeVisLab software (MeVis Medical Solutions AG, Germany).

Edges until the first bifurcation were considered to be the MPA. The LPA and RPA were defined depending on the orientation within the scanner, respectively as the longest possible path. All other edges were considered side branches.

For each edge, we evaluate the length as well as the minimal diameter of the surrounding surface. For the MPA and internal edges, the diameters were obtained at the midpoint of each edge, for branch tips closer to the endpoints (see Fig. 3). We accumulated the measurements to calculate the overall length and length-weighted diameters of the MPA, LPA and RPA, respectively. Edges with a length below 10 mm were disregarded on the length-weighted diameter calculation, as side branches will likely distort the diameter measurements. The tortuosity was calculated from the distance of the start and end nodes divided by the overall length of the considered segment. Lastly, the bifurcation angle between LPA and RPA was determined, by calculating the angle between the vectors connecting the endpoint of the MPA to the endpoint of the first edges of LPA and RPA, respectively, with an overall length of at least 10 mm.

2.5 Hemodynamic Simulation

To assess whether the automatically generated surface geometries can be used for the calculation of the pulmonary artery hemodynamics, a steady-state simulation of the peak-systolic state was conducted, as previously described by Brüning et al. [1], using a finite volume solver (STAR-CCM+, v. 17.06, Siemens PLM). Initially, the surface repair functionality of STAR-CCM+ was used to repair any persisting topology errors in the surface geometries, such as non-manifold edges or piercing surface elements. Open edges of the surface mesh were identified automatically and closed off subsequently for boundary condition specification. For all outlets, a zero-pressure boundary condition was specified, whereas a constant mass flow of 350 ml/s, which equaled the average peak-systolic flow rate observed in our sample, was specified as the inlet at the right ventricular outflow tract. A volume mesh was generated using a mesh size of 0.5 mm and a boundary layer consisting of five layers, with an overall thickness of 50 percent of the base size. Simulations were deemed converged when residuals of momentum and mass were below 1e−5.

3   Results

We validated our neural network on 8 cases and evaluated the resulting surface meshes on 10 further cases, which were each processed by two observers.

3.1   Voxel2Mesh - Comparison to Baselines

Training and inference was performed on Dell PowerEdge R7525 compute nodes with 64 AMD Epyc cores, 512 GB RAM and NVIDIA A100 40G or 80G GPU. We used Python 3.9.12 with PyTorch 1.11.0+cu113 and PyTorch3D 0.6.1. As a baseline, we use the state-of-the-art nnU-Net [6], followed by marching cubes (MC) [9] for surface generation. As the vertex class is not directly transferable, we disregard it. Furthermore, we conduct an ablation study, by evaluating the effect of our centerline coverage loss in comparison to a standard Voxel2Mesh.

Figure 4 demonstrates the effect of the centerline coverage loss, which facilitates the formation of the proper branching structure and alleviates the effect of smoothing.

Table 1. Comparative results against baselines on the validation data set of 8 patients using the Intersection over Union (IoU) and mean distance (md), as well as the F1 score for the vertex class predicted by Voxel2Mesh. Our method, Voxel2Mesh + $\mathcal{L}_{CC}$, combines the standard Voxel2Mesh loss functions with an additional centerline coverage loss.

	IoU	md [mm]	vertex class F1
nnU-Net + MC	0.93 ± 0.01	2.9 ± 1.1	N/A
Voxel2Mesh	0.64 ± 0.08	7.3 ± 0.9	0.45 ± 0.28
Voxel2Mesh + $\mathcal{L}_{CC}$	0.87 ± 0.06	2.3 ± 0.6	0.58 ± 0.11

Fig. 4. Example of a surface mesh before (left) and after (middle) introduction of the centerline coverage loss. The target mesh (right) is shown for reference. The centerline coverage loss facilitates the proper branching structure.

3.2 Centerline-Based Analysis of Anatomical Measurements

We performed an inter-observer variability analysis for the 10 CT images annotated by two observers. For this, we manually revised the openings resulting from our Voxel2Mesh algorithm. In the 10 cases, we had to open on average three (23%) additional outlets, that were not detected automatically.

For several measurements, we calculate the mean differences from our reconstructed surface mesh to the average measurements from the two observer reconstructed meshes. We contrast this with the inter-observer difference in Table 2. With a two-tailed paired t-test with a significance level of 0.05, we don't have any significant differences in the shape of the LPA and RPA and the length of the MPA. Differences in the diameter of the MPA ($p \approx 0.023$) and the bifurcation are significant. ($p \approx 0.037$).

Table 2. Centerline-based analysis of anatomical features: We report the mean difference for our automatically generated mesh to the target meshes by the two observers, as well as the inter-observer difference, averaged over all 10 cases.

Measurement		Mean Difference to Target	Inter-observer Difference
MPA	Length [mm]	2.66 ± 1.56	1.89 ± 2.83
	Diameter [mm]	1.17 ± 0.58	0.62 ± 0.56
LPA	Length [mm]	6.69 ± 4.02	8.14 ± 7.18
	Diameter [mm]	1.26 ± 0.66	1.19 ± 0.93
	Tortuosity	0.04 ± 0.03	0.06 ± 0.06
RPA	Length [mm]	5.25 ± 3.49	6.12 ± 4.52
	Diameter [mm]	1.06 ± 0.83	1.23 ± 1.14
	Tortuosity	0.06 ± 0.05	0.07 ± 0.08
Bifurcation [°]		2.38 ± 0.98	1.24 ± 0.65

3.3 Simulation

Simulation of the patient-specific hemodynamics using the automatically generated surface meshes was feasible in an automated manner. Isolated topological errors could be fixed with the automated surface remeshing algorithm used in the numerical pipeline in all instances. No subsequent manual interaction was necessary to process the surface meshes for the numerical simulation. The visualization of resulting intravascular hemodynamics showed a good agreement between flow fields calculated using manually and Voxel2Mesh generated meshes (Fig. 5).

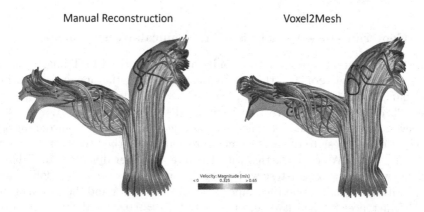

Fig. 5. Exemplary comparison of the simulation of the patient-specific hemodynamics, using the manually and the automatically generated surface mesh, illustrated with velocity-colored streamlines.

Table 3 shows a summary of simulated velocities for the meshes generated by Voxel2Mesh and our two observers. The correlations between our generated meshes and the two observers are comparable.

4 Discussion

We extended the Voxel2Mesh algorithm with a centerline coverage loss to promote pulmonary artery branching and a vertex class prediction for subsequent opening. Derived anatomical measurements from our resulting surface meshes are in accordance with results from two observers. We only see significant differences for the main pulmonary artery and the bifurcation angle. A minor change in the cut-off length of the MPA leads to a change in the measurement position for the diameter and likely also the angle, hence resulting in a significantly different diameter, while the overall shape remains the same. The differences between the bifurcation angles are in the range of 1 to 3°C, so even minor deviations lead to statistically significant differences. However, we do not expect any impact on the hemodynamic simulations from the detected differences in the

Table 3. Velocity measurements from simulations using the mesh from Voxel2Mesh (V2M) and two observers (O1 & O2): We report the mean and standard deviation (Std) for the mean and maximum velocity in meters per second. The correlation between Voxel2Mesh and the observers as well as the inter-observer correlation is shown.

	Mean Velocity [m/s]			Max Velocity [m/s]		
	V2M	O1	O2	V2M	O1	O2
Mean	0.265	0.295	0.312	0.744	0.787	0.844
Std	0.046	0.058	0.070	0.143	0.171	0.232
	V2M/O1	V2M/02	O1/O2	V2M/O1	V2M/02	O1/O2
Correlation	0.981	0.966	0.955	0.882	0.875	0.846

anatomy and the comparison showed a good agreement (Table 3). We experienced self-folding surface meshes, a known issue in surface mesh generation [14], obstructing the hemodynamic simulations. For avoiding such foldings, well-defined training meshes are of utter importance - they need to be watertight, with consistently oriented normals, and also very thin branches should be truncated, to prevent topological errors. Pretraining the segmentation without mesh deformation, reducing the learning rate as well as initially increasing the smoothing factors and slowly reducing them during the training, reduced the chance of self-foldings.

4.1   Time Investment - Manual Vs Automatic Mesh Generation

The manual mesh generation requires four steps: Image-based segmentation of a voxel mask containing the pulmonary artery lumen, generation of an initial surface mesh that is subsequently smoothed, correction of errors in surface topology, and finally truncation of vessel endings. The majority of time necessary for the procedure was spent on the initial segmentation step. Depending on the quality of the image data, this step took between 30 and 180 min per case. Substantial manual effort was required to eradicate fused branches and other artifacts, leading to errors in the deduced meshes (see Table 1). Similarly, smoothing and correction of topological errors of the surface depended strongly on the image quality and thus the quality of the initial surface mesh. This step took between 1 and 10 min. Finally, the truncation of vessel endings took approximately 3 min.

The automated procedure takes approximately 2 s per case for surface reconstruction, with an additional 1 to 3 min of manual interaction for possible error correction, including truncation of remaining vessel endings.

5   Conclusion

We have presented a deep learning-based approach for automatic pulmonary artery mesh generation from CT images with subsequent in- and outlet opening for numerical flow simulation. Apart from the last step of quality assessment and possible error correction, the procedure is fully automated and does not require any human interaction. Therefore, the method enables cost reduction and better reproducibility by minimizing operator-dependent influences. For future work, we consider directly penalizing self-foldings in our loss function or including the automatic removal in our processing pipeline. The nnU-Net demonstrated superior segmentation performance compared to our Voxel2Mesh segmentation pipeline. Thus, directly incorporating the nnU-Net into the Voxel2Mesh architecture might further improve results in the future. Apart from that, an increase in training sample size likely improves the robustness and overall quality of the proposed method. Future work will also focus on enhancing fully automatic outlet opening.

Acknowledgement. This project has received funding from the European Union's Horizon 2020 research and innovation programme under grant agreement N° 101017578.

References

1. Brüning, J., et al.: In-silico enhanced animal study of pulmonary artery pressure sensors: assessing hemodynamics using computational fluid dynamics. Front. Cardiovasc. Med. **10**, 1193209 (2023). https://doi.org/10.3389/fcvm.2023.1193209
2. Frey, P., George, P.L.: Mesh Generation: Application to Finite Elements, 2nd edn. Wiley, Ltd (2008)
3. Ballester, M.A.G., Zisserman, A.P., Brady, M.: Estimation of the partial volume effect in MRI. Med. Image Anal. **6**(4), 389–405 (2002). https://doi.org/10.1016/S1361-8415(02)00061-0
4. Guo, Q., Gao, C., Liu, M., Wang, H., Yuan, H.: Pulmonary artery segmentation based on three-dimensional region growth approach. In: Zhao, Y., Barnes, N., Chen, B., Westermann, R., Kong, X., Lin, C. (eds.) Image and Graphics, pp. 548–557. Springer International Publishing, Cham (2019). https://doi.org/10.1007/978-3-030-34110-7_46
5. Hunter, K.S., Feinstein, J.A., Ivy, D.D., Shandas, R.: Computational simulation of the pulmonary arteries and its role in the study of pediatric pulmonary hypertension. Prog. Pediatr. Cardiol. **30**(1), 63–69 (2010). https://doi.org/10.1016/j.ppedcard.2010.09.008, proceedings of the 1st International Conference on Computational Simulation in Congenital Heart Disease
6. Isensee, F., et al.: nnU-Net: self-adapting framework for U-Net-based medical image segmentation. arXiv:1809.10486 (2018)
7. Kjeldsberg, H.A., Bergersen, A.W., Valen-Sendstad, K.: morphMan: automated manipulation of vascular geometries. J. Open Source Softw. **4**(35), 1065 (2019). https://doi.org/10.21105/joss.01065
8. Kong, F., Shadden, S.C.: Whole heart mesh generation for image-based computational simulations by learning free-from deformations. In: de Bruijne, M., et al. (eds.) Medical Image Computing and Computer Assisted Intervention - MICCAI 2021, pp. 550–559. Springer International Publishing, Cham (2021). https://doi.org/10.1007/978-3-030-87202-1_53
9. Lewiner, T., Lopes, H., Vieira, A.W., Tavares, G.: Efficient implementation of marching cubes' cases with topological guarantees. J. Graph. Tools **8**(2), 1–15 (2003). https://doi.org/10.1080/10867651.2003.10487582

10. López-Linares Román, K., et al.: 3D pulmonary artery segmentation from CTA scans using deep learning with realistic data augmentation. In: Stoyanov, D., et al. (eds.) RAMBO/BIA/TIA -2018. LNCS, vol. 11040, pp. 225–237. Springer, Cham (2018). https://doi.org/10.1007/978-3-030-00946-5_23

11. Nguyen, T.D., Kadri, O.E., Voronov, R.S.: An introductory overview of image-based computational modeling in personalized cardiovascular medicine. Front. Bioeng. Biotechnol. **8**, 529365 (2020). https://doi.org/10.3389/fbioe.2020.529365

12. Osher, S., Fedkiw, R.: Level Set Methods and Dynamic Implicit Surfaces. AMS, vol. 153. Springer, New York (2003). https://doi.org/10.1007/b98879

13. Ronneberger, O., Fischer, P., Brox, T.: U-Net: convolutional networks for biomedical image segmentation. arXiv:1505.04597 (2015)

14. Sfikas, K., Perakis, P., Theoharis, T.: For^{2}m: recognition and repair of foldings in mesh surfaces. Application to 3D object degradation. arXiv:2206.09699 (2022)

15. Shewchuk, J.: What is a good linear finite element? - interpolation, conditioning, anisotropy, and quality measures. In: Proceedings of the 11th International Meshing Roundtable, vol. 73 (2002)

16. Taylor, C., Figueroa, C.: Patient-specific modeling of cardiovascular mechanics. Annu. Rev. Biomed. Eng. **11**(1), 109–134 (2009). https://doi.org/10.1146/annurev.bioeng.10.061807.160521, pMID: 19400706

17. Wickramasinghe, U., Remelli, E., Knott, G., Fua, P.: Voxel2Mesh: 3D mesh model generation from volumetric data. arXiv:1912.03681 (2019)

18. Zhang, C., Sun, M., Wei, Y., Zhang, H., Xie, S., Liu, T.: Automatic segmentation of arterial tree from 3D computed tomographic pulmonary angiography (CTPA) scans. Comput. Assist. Surg. **24**(sup2), 79–86 (2019). https://doi.org/10.1080/24699322.2019.1649077, pMID: 31401886

Impact of Catheter Orientation
on Cardiac Radiofrequency Ablation

Massimiliano Leoni[1] , Argyrios Petras[1(✉)] , Zoraida Moreno Weidmann[2] ,
Jose M. Guerra[2] , and Luca Gerardo-Giorda[1,3]

[1] RICAM - Johann Radon Institute for Computational and Applied Mathematics,
4040 Linz, Austria
{massimiliano.leoni,argyrios.petras}@ricam.oeaw.ac.at,
luca.gerardo-giorda@jku.at
[2] Hospital de la Santa Creu i Sant Pau, 08025 Barcelona, Spain
{zmoreno,jguerra}@santpau.cat
[3] Johannes Kepler University, 4040 Linz, Austria

Abstract. Radiofrequency ablation is a typical treatment for severe
cases of cardiac arrhythmias. A catheter, inserted from the patient's
groin, delivers current at frequencies of 450–500 kHz to the arrhythmogenic area, inflicting thermal damage. The electrical current delivered
to the tissue depends (among others) on the tip shape and the catheter
orientation. A modified Penne's bioheat equation with an electric source
and cooling blood convection is the standard choice for RFA models.
The incompressible Navier-Stokes equation describes the interaction of
the blood flow and the irrigated saline. The cardiac tissue is a nonlinear orthotropic hyperelastic material and the Hertz-Signorini-Moreau
contact boundary conditions model the frictionless interaction of the
electrode with the cardiac tissue. In this work, we consider a spherical
electrode tip shape and different orientation angles (from perpendicular
up to 60° from the vertical position). We perform a virtual ablation for
a standard protocol of power 30 W for a duration of 30 s on a simulated
porcine cardiac slab. We compare the contact surface of the electrode
with the tissue for the different orientations and the characteristics of
the generated lesions.

Keywords: radiofrequency ablation · catheter orientation · computer
simulation

1 Introduction

Radiofrequency ablation using a catheter is a minimally invasive treatment for
cardiac arrhythmias. Electrical current at frequencies of 450–500 kHz is delivered
to the arrhythmogenic tissue via the tip electrode of the catheter inflicting thermal damage to the tissue, destroying the arrhythmogenic area and generating a
lesion. The process is repeated until a normal cardiac rhythm is restored.

© The Author(s), under exclusive license to Springer Nature Switzerland AG 2024
O. Camara et al. (Eds.): STACOM 2023, LNCS 14507, pp. 152–162, 2024.
https://doi.org/10.1007/978-3-031-52448-6_15

The electrical current delivered to the tissue is proportional to the direct contact with the tip electrode at a perpendicular or parallel placement, namely the *electrode footprint* [17]. The electrode footprint depends (among others) on the tip shape and the catheter orientation. Ex-vivo studies on porcine cardiac tissue have shown that the electrode footprint changes the characteristics of the resulting lesion [6].

In-silico studies have explored the impact of the orientation of the catheter on the generated lesion [3,12]. However, no available studies consider the mechanical deformation of the tissue in contact with the tip electrode, which we proved to be very important for the accurate estimation of the irreversible tissue damage [8,10].

In this work we explore in-silico the impact of catheter orientation on the resulting lesion. The tissue-electrode mechanical interaction is modelled using nonlinear mechanics and the Hertz-Signorini-Moreau contact boundary conditions. We perform virtual ablations for a standard protocol of power 30 W for a duration of 30 s on a simulated porcine cardiac slab, and compare the contact surface and the characteristics of the generated lesions for the different orientations.

2   Mathematical Model

2.1   Geometry

The geometry is based on an in-vitro experimental setup [10], designed as a cylinder of height 80 mm and diameter 80 mm, which consists of a porcine cardiac tissue of height 20 mm, a blood chamber of 40 mm and a polymethyl methacrylate (PMMA) board of 20 mm. We assume the tissue slab to contain fibers pointing along the x-axis. We model a 7 Fr spherical tip electrode with a 6-pore irrigation system and a thermistor, as described in [10]. The total domain Ω consists of the tissue Ω^{tissue}, the fluid (including the blood and the saline irrigation pipes Ω^{fluid}), the board, the electrode and the thermistor, as in Fig. 1.

2.2   Electrode-Tissue Contact

The porcine cardiac tissue is considered a nearly incompressible orthotropic hyperelastic material modelled using the strain energy density [15]

$$\Psi(\overline{\mathbf{E}}) = C(e^Q - 1) + \kappa(\ln J)^2,$$

$$Q = b_{ff}\overline{E}_{ff}^2 + b_{ss}\overline{E}_{ss}^2 + b_{nn}\overline{E}_{nn}^2 + 2(b_{fs}\overline{E}_{fs}^2 + b_{sn}\overline{E}_{sn}^2 + b_{fn}\overline{E}_{fn}^2),$$

where C is the material stiffness constant, b_{ij} $(i, j \in \{f, s, n\})$ are the stiffness anisotropy coefficients in the axial fiber (ff), sheet (ss) and normal (nn) direction, and the shear fiber-sheet (fs), sheet-normal (sn) and fiber-normal (fn)

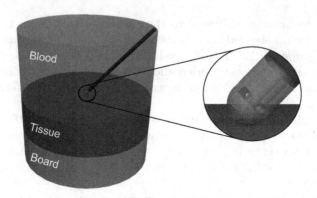

Fig. 1. The computational domain consisting of the blood, tissue, board, electrode and thermistor subdomains.

direction. The volumetric constraint part of Ψ is enforced using the bulk modulus κ and the volume ratio J, defined as

$$J = \det(\mathbf{F}),$$

where $\mathbf{F} = d\mathbf{x}/d\mathbf{X}$ is the deformation gradient tensor of the deformed body $\mathbf{x}(\mathbf{X})$ (current configuration Ω_t for some time $t > 0$) in the reference configuration $\mathbf{X}$ (undeformed configuration Ω_0).

The isochoric Green-Lagrange strain tensor $\overline{\mathbf{E}}$ is defined using the isochoric right Cauchy-Green tensor $\overline{\mathbf{C}}$ as

$$\overline{\mathbf{E}} = \frac{1}{2}(\overline{\mathbf{C}} - \mathbf{I}), \qquad \overline{\mathbf{C}} = J^{-2/3}\mathbf{C} = J^{-2/3}\mathbf{F}^T\mathbf{F},$$

where $\mathbf{I}$ is the identity tensor. The terms $\overline{E}_{ij}$ are the projections of the isochoric Green-Lagrange strain tensor $\overline{\mathbf{E}}$ in the axial and shear directions as

$$\overline{E}_{ij} = \mathbf{v}_1{}^T \overline{\mathbf{E}} \mathbf{v}_2, \qquad (i, \mathbf{v}_1), (j, \mathbf{v}_2) \in \{(f, \mathbf{f}_0), (s, \mathbf{s}_0), (n, \mathbf{n}_0)\},$$

where $\mathbf{f}_0$, $\mathbf{s}_0$ and $\mathbf{n}_0$ are the vectors pointing in the fiber, sheet and normal direction respectively.

The equilibrium equation describes the mechanical interaction of the tip electrode and the tissue, given as

$$-\nabla_{\mathbf{X}} \cdot \mathbf{P} = 0, \quad \text{in } \Omega^{tissue}$$

where the $\nabla_{\mathbf{X}}$ is the divergence operator in the reference configuration, $\mathbf{P}$ is the first Piola-Kirchhoff stress tensor and Ω^{tissue} is the tissue subdomain.

We assume frictionless contact between the tip electrode and the tissue and model it using the Hertz-Signorini-Moreau boundary conditions

$$\mathbf{n}^T \boldsymbol{\sigma} \mathbf{n} \leq 0, \quad \text{and} \quad \boldsymbol{\sigma}\mathbf{n} - (\mathbf{n}^T \boldsymbol{\sigma} \mathbf{n})\mathbf{n} = 0,$$

where σ is the Cauchy stress tensor and $\mathbf{n}$ the unit normal vector on the contact boundary in the current configuration. We consider the standard augmented Lagrangian approach for the imposition of the contact boundary conditions [18].

We set a no-displacement boundary condition at the bottom of the tissue and zero normal stress on the remaining boundaries of the geometry.

2.3   Blood-Saline Interaction

The incompressible Navier-Stokes equation models the blood interaction with the irrigated saline. Assuming a perfect mix of the two, the equations are written

$$\frac{\partial \mathbf{v}}{\partial t} + \mathbf{v} \cdot \nabla \mathbf{v} - \operatorname{div} \boldsymbol{\tau}(\mathbf{v}, p) = \mathbf{0}, \quad \text{in } \Omega^{fluid},$$

$$\operatorname{div} \mathbf{v} = 0, \quad \text{in } \Omega^{fluid},$$

where $\mathbf{v}$ is the flow velocity, p is the pressure scaled by the density ρ, $\boldsymbol{\tau}$ is the stress tensor defined as

$$\boldsymbol{\tau}(\mathbf{v}, p) = \frac{\mu}{\rho}(\nabla \mathbf{v} + \nabla \mathbf{v}^T) - p\mathbf{I},$$

and μ is the dynamic viscosity of the blood.

We set a constant blood inflow of $\mathbf{v}_b = (0.5, 0, 0)\,\text{m s}^{-1}$ based on common in-vitro experimental protocols [4] that mimic high blood flow areas within the ventricle (such as near the mitral valve). This boundary condition is applied on the part of the fluid domain's boundary satisfying $\mathbf{v}_b \cdot \mathbf{n} \leq 0$, where $\mathbf{n}$ is the outward unit normal vector. We further impose a zero-pressure outflow boundary condition opposite to the inflow. The remaining sides of the blood domain are equipped with no-slip boundary conditions. The saline inlet is imposed as an inflow boundary condition at the top of the irrigation pipe with a parabolic profile tuned to correspond to the volume rate of $17\,\text{mL/min}$. No-slip boundary conditions are imposed on all the remaining boundaries.

2.4   Electrical Source

The electrical source is modelled using the quasi-static equation augmented with a power constraint equation for constant-power ablation as

$$\operatorname{div}(\sigma(T)\nabla\Phi) = 0, \quad \text{in } \Omega$$

$$\int_\Omega \sigma(T)\nabla\Phi \cdot \nabla\Phi \, \mathrm{d}x + R_{out} \int_{\Gamma_{out}} (\sigma(T)\nabla\Phi \cdot \mathbf{n})^2 \, \mathrm{d}s = P,$$

where Φ is the electrical potential, $\sigma(T)$ is the temperature-dependent electrical conductivity, R_{out} is the resistance outside the computational domain Ω, Γ_{out} is the portion of the domain boundary that separates the blood and the tissue from the outside and P is the power of the ablation protocol.

We set a potential V_0 at the boundary between the electrode and the shaft – initially chosen to match the power and the resistance of the system $R_{sys} = 121\,\Omega$ using Ohm's law – and Robin-type boundary conditions on Γ_{out} of the form

$$-\sigma(T)\nabla\Phi \cdot \mathbf{n} = \frac{\Phi}{R_{out}}.$$

The resistance outside the computational domain is tuned to match the total power and resistance of the system. The return electrode is at the bottom of the computational domain $\{z = 0\}$, where $\Phi = 0$, while zero current flow is imposed in all the remaining boundaries, including the catheter body.

2.5   Temperature Monitoring

A modified Penne's bioheat equation monitors the changes in the temperature, with a cooling convection term due to the fluid flow and an electrical heat source. This reads as

$$\rho c(T)\left(\frac{\partial T}{\partial t} - \mathbf{v} \cdot \nabla T\right) - \text{div}(k(T)\nabla T) = \sigma(T)|\nabla\Phi|^2 \qquad \text{in } \Omega,$$

where c is the specific heat, T is the temperature and k is the thermal conductivity.

The saline temperature is set to $28\,^\circ\mathrm{C}$ at the saline inflow [14], while homogenous Neumann boundary conditions are imposed on the catheter body. We set a body temperature of $37\,^\circ\mathrm{C}$ at the remaining boundaries.

2.6   Lesion Assessment

A three-state hyperthermic cell-death model is considered for the assessment of the lesion size [9], following the procedure therein to identify the irreversibly damaged area. In brief, the three-states considered for the cardiomyocytes are the native (N), the unfolded (U) and the denaturated (D), which account for the irreversible damage as

$$N \underset{k_3(T)}{\overset{k_1(T)}{\rightleftharpoons}} U \xrightarrow{k_2(T)} D, \tag{1}$$

where $k_i(T)$, $i = 1, 2, 3$ are the temperature-dependent transition rates among the different states. The system of ordinary differential equations describing the interactions of the states is

$$\frac{dN}{dt} = -k_1(T)N + k_3(T)U,$$

$$\frac{dU}{dt} = k_1(T)N - (k_2(T) + k_3(T))U,$$

$$\frac{dD}{dt} = k_2(T)U,$$

where $k_i(T) = A_i e^{\Delta E_i/(RT)}$, $i = 1, 2, 3$, and R is the universal gas constant. To account for the slow cell death dynamics, a threshold N_{thr} at the native state cells identifies the irreversible thermal damage after the tissue cools down, as described in [9]. This threshold is chosen as $N_{thr} = 0.8$, following experimental results on several types of cells [7]. The measured dimensions are lesion volume (V), width (W) and depth (D), as measured from the undeformed endocardial surface (see Fig. 2).

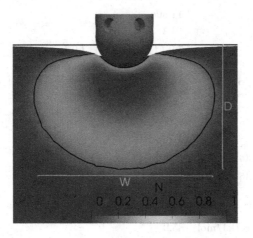

Fig. 2. The lesion size measurement as identified by the N_{thr}.

2.7   Model Parameters

We tune the material stiffness C to match the tissue-electrode contact area $2.77\,\text{mm}^2$ for $10\,\text{g}$ contact force given by the model in [11]. The remaining model parameters are drawn from the literature [1, 2, 5, 7, 9, 10, 13, 16] and summarized in Table 1.

The density, specific heat, thermal and electrical conductivities of the tissue as well as the electrical conductivity of the blood are considered temperature dependent as

$$\rho(T)c(T) = \left(3.643 - 0.003533 e^{0.06263T}\right) \times 10^6,$$
$$k(T) = 0.5655 + 5.034 \times 10^{-12} e^{0.263T},$$
$$\sigma_t(T) = 0.54 \left(1 + 0.015(T - 37)\right),$$
$$\sigma_b(T) = 1.2 \left(1 + 0.011(T - 37)\right),$$

Table 1. Summary of the model parameters

Physics	Parameters	Tissue	Blood	Board	Electrode	Thermistor
RFA	ρ (kg/m^3)	$\rho(T)$	1020	1185	21500	32
	c (J/kg/K)	$c(T)$	3454	1466	132	835
	k (W/m/K)	$k(T)$	0.3	0.209	71	0.038
	σ (S/m)	$\sigma_t(T)$	$\sigma_b(T)$	10^{-10}	4.6×10^6	10^{-5}
	$\mu\rho^{-1}$ (S/m)	–	2.52×10^{-6}	–	–	–

		Tissue		
Mechanics	C (kPa)	10.3		
	b_{ij} (–)	5.0 (ff), 6.0 (ss), 3.0 (nn),		
		10.0 (fs), 2.0 (fn), 2.0 (sn)		
	κ (kPa)	650		

		Cardiomyocytes	
Cell damage	A_1 (s^{-1})	8.87×10^{73}	$T \leq 55°C$
		3.56×10^{22}	$T > 55°C$
	A_2 (s^{-1})	5.35×10^{11}	
	A_3 (s^{-1})	1.6×10^{15}	
	ΔE_1 (kJ/mol)	467.6	$T \leq 55°C$
		144.7	$T \leq 55°C$
	ΔE_2 (kJ/mol)	85.9	
	ΔE_3 (kJ/mol)	105.1	
	N_{thr} (–)	0.8	

3   Results

We simulated virtual ablations for different orientations of the catheter using a standard RFA protocol of 30 W for 30 s, commonly used in clinical practice. The saline irrigation flow was set 17 mL/min [10]. The computational model is approximated with the Finite Elements Method using a self-developed code based on FEniCS-X (fenicsproject.org). The algorithm is the same as the one used in our previous work ([10]) and it has been converted from FEniCS-hpc to FEniCS-X. For the simulations in this work, we used $\mathbb{P}_2$ elements for the electrical potential and $\mathbb{P}_1$ elements for the temperature; at each time step we solve the coupled nonlinear system with a standard fixed-point iteration method.

The assessment of the lesion dimensions is performed by solving the three-state equations with the python interface of ParaView (https://www.paraview.org). The numerical simulations ran on the Vienna Scientific Cluster (VSC4); a 60 s simulation (30 s ablation time followed by 30 s thermal relaxation, as in [9]) takes about 3.5 h using 144 cores across 3 nodes.

First, we evaluate the variation in the electrode footprint as a function of the catheter position. We consider a fixed vertical contact force of 10 g, we vary the catheter inclination with respect to the tissue and solve for the corresponding deformation of the tissue. In Table 2 we report the contact surface percentage with the tissue of the electrode for different orientations of the catheter. It can be observed that for inclinations up to 45° the contact surface between the electrode and the tissue remains practically unchanged. A minimal increase appears at 60°.

For each configuration, we simulated the standard (30 W, 30 s) RFA ablation protocol. Figure 3 shows the final lesions, the dimensions of which appear in Table 2. It can be observed that, despite the electrode footprint being similar, the tilted configurations result in more asymmetric lesions in the tilting direction. In terms of lesion size, the depth is comparable for all orientations, with a maximum increase of 5% observed at 60°. On the other hand, the orientation has a much larger impact on the lesion width and volume, which increase up to 12% and 29.4% respectively at 60° when compared to the perpendicular case.

Table 2. Comparison of the tissue-electrode contact percentage and lesion sizes for different inclinations of the electrode. RFA protocol: contact force 10 g, power 30 W, duration 30 s, saline irrigation 17 mL/min.

Angle	0°	15°	30°	45°	60°
Electrode contact (%)	9.8	9.8	9.8	9.9	10.3
Depth (mm)	4.86	4.86	4.90	4.96	5.11
Width (mm)	7.26	7.25	7.33	7.63	8.14
Volume (mm^3)	130.6	130.4	131.8	145.0	169.0

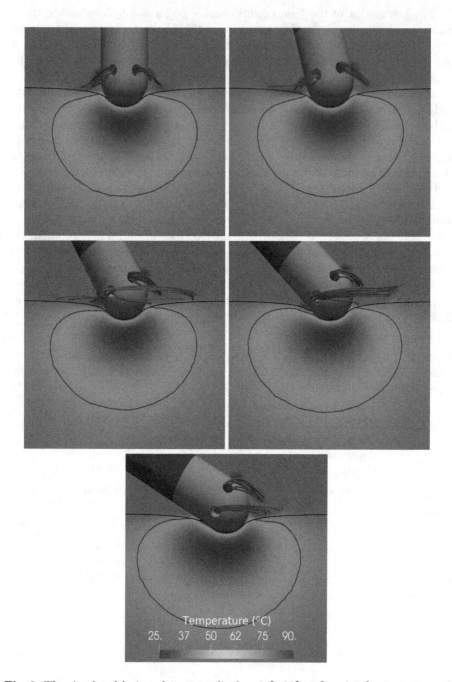

Fig. 3. The simulated lesions for perpendicular, 15°, 30°, 45° and 60° orientations of the electrode tip.

4 Discussion

This study explores the impact of catheter orientation on the electrode footprint and the resulting lesion after the RFA application. An accurate representation of the tissue-electrode deformation for angular catheter placements is considered using nonlinear contact mechanics, surpassing the limitations of the existing literature.

For a spherical tip electrode, the electrode footprint is comparable for 10 g contact force and all inclination angles up to 60°. The effect on the lesion depth is marginal, however a significant change is observed for the lesion width and volume as the catheter inclination increases. The angular placements of the catheter creates non-symmetric lesions, in the direction of the tilting.

Another relevant aspect is the tip electrode shape, which has a significant impact on the deformation of the tissue and the electrode footprint on it [11]. This aspect is part of our ongoing work.

5 Limitations

It is well known that the cardiac wall consists of fibers that rotate by 120° from the endocardial to the myocardial surface. In this work, we use fibers that are aligned with the x-axis. The incorporation of the fiber structure in the RFA model is part of our ongoing work. Additionally, no trabeculae are considered for the porcine tissue, which is a reasonable approach considering that in-vitro experiments are conducted at the smoother epicardial surface.

Acknowledgements. This work has been partially supported by the State of Upper Austria. This research was funded in part by the Austrian Science Fund (FWF) P35374N. For the purpose of Open Access, the author has applied a CC BY public copyright license to any Author Accepted Manuscript (AAM) version arising from this submission.

References

1. Augustin, C.M., et al.: A computationally efficient physiologically comprehensive 3D–0D closed-loop model of the heart and circulation. Comput. Methods Appl. Mech. Eng. **386**, 114092 (2021). https://doi.org/10.1016/j.cma.2021.114092
2. Bianchi, L., Bontempi, M., De Simone, S., Franceschet, M., Saccomandi, P.: Temperature dependence of thermal properties of Ex vivo porcine heart and lung in hyperthermia and ablative temperature ranges. Ann. Biomed. Eng. **51**(6), 1181–1198 (2023). https://doi.org/10.1007/s10439-022-03122-9
3. Gallagher, N., Fear, E.C., Byrd, I.A., Vigmond, E.J.: Contact geometry affects lesion formation in radio-frequency cardiac catheter ablation. PLoS ONE **8**, e73242 (2013). https://doi.org/10.1371/journal.pone.0073242
4. Guerra, J.M., et al.: Effects of open-irrigated radiofrequency ablation catheter design on lesion formation and complications: in vitro comparison of 6 different devices. J. Cardiovasc. Electrophysiol. **24**(10), 1157–1162 (2013). https://doi.org/10.1111/jce.12175

5. Jaspard, F., Nadi, M.: Dielectric properties of blood: an investigation of temperature dependence. Physiol. Meas. **23**(3), 547 (2002). https://doi.org/10.1088/0967-3334/23/3/306

6. Masnok, K., Watanabe, N.: Catheter contact area strongly correlates with lesion area in radiofrequency cardiac ablation: an ex vivo porcine heart study. J. Interv. Card. Electrophysiol. **63**, 561–572 (2021). https://doi.org/10.1007/s10840-021-01054-3

7. O'Neill, D.P., et al.: A three-state mathematical model of hyperthermic cell death. Ann. Biomed. Eng. **39**, 570–579 (2011). https://doi.org/10.1007/s10439-010-0177-1

8. Petras, A., Leoni, M., Guerra, J., Jansson, J., Gerardo-Giorda, L.: Effect of tissue elasticity in cardiac radiofrequency catheter ablation models. In: 2018 Computing in Cardiology Conference (CinC), pp. 3–6 (2018). https://doi.org/10.22489/CinC.2018.035

9. Petras, A., Leoni, M., Guerra, J.M., Gerardo-Giorda, L.: Calibration of a three-state cell death model for cardiomyocytes and its application in radiofrequency ablation. Physiol. Meas. **44**(6), 065003 (2023). https://doi.org/10.1088/1361-6579/acdcdd

10. Petras, A., Leoni, M., Guerra, J.M., Jansson, J., Gerardo-Giorda, L.: A computational model of open-irrigated radiofrequency catheter ablation accounting for mechanical properties of the cardiac tissue. Int. J. Numer. Methods Biomed. Eng. **35**, e3232 (2019). https://doi.org/10.1002/cnm.3232

11. Petras, A., Weidmann, Z.M., Ferrero, M.E., Leoni, M., Guerra, J.M., Gerardo-Giorda, L.: Impact of electrode tip shape on catheter performance in cardiac radiofrequency ablation. Heart Rhythm O2 **3**(6), 699–705 (2022). https://doi.org/10.1016/j.hroo.2022.07.014

12. Pérez, J.J., D'Angelo, R., González-Suárez, A., Nakagawa, H., Berjano, E., d'Avila, A.: Low-energy (360 J) radiofrequency catheter ablation using moderate power - short duration: proof of concept based on in silico modeling. J. Interv. Card. Electrophysiol. **66**, 1085–1093 (2022). https://doi.org/10.1007/s10840-022-01292-z

13. Rosentrater, K.A., Flores, R.A.: Physical and rheological properties of slaughterhouse swine blood and blood components. Trans. ASAE **40**(3), 683–689 (1997). https://doi.org/10.13031/2013.21287 https://doi.org/10.13031/2013.21287 https://doi.org/10.13031/2013.21287 https://doi.org/10.13031/2013.21287

14. Squara, F., et al.: In vitro evaluation of ice-cold saline irrigation during catheter radiofrequency ablation. J. Cardiovasc. Electrophysiol. **25**, 1125–1132 (2014). https://doi.org/10.1111/jce.12479

15. Usyk, T.P., Mazhari, R., McCulloch, A.D.: Effect of laminar orthotropic myofiber architecture on regional stress and strain in the canine left ventricle. J. Elast. **61**, 143–164 (2000). https://doi.org/10.1023/A:1010883920374

16. Wen, J., Wan, N., Bao, H., Li, J.: Quantitative measurement and evaluation of red blood cell aggregation in normal blood based on a modified Hanai equation. Sensors **19**(5), 1095 (2019). https://doi.org/10.3390/s19051095

17. Wittkampf, F.H., Nakagawa, H.: RF catheter ablation: lessons on lesions. Pacing Clin. Electrophysiol. **29**, 1285–1297 (2006). https://doi.org/10.1111/j.1540-8159.2006.00533.x

18. Yastrebov, V.A.: Numerical Methods in Contact Mechanics. Wiley (2013). https://doi.org/10.1002/9781118647974

Generating Virtual Populations of 3D Cardiac Anatomies with Snowflake-Net

Jiachuan Peng[1]([✉]), Marcel Beetz[1], Abhirup Banerjee[1,2][iD], Min Chen[1], and Vicente Grau[1][iD]

[1] Department of Engineering Science, University of Oxford, Oxford, UK
jiachuan.peng@seh.ox.ac.uk
[2] Division of Cardiovascular Medicine, Radcliffe Department of Medicine, University of Oxford, Oxford, UK
abhirup.banerjee@eng.ox.ac.uk

Abstract. High-quality virtual populations of human hearts are of significant importance for a variety of applications, such as *in silico* simulations of cardiac physiology, data augmentation, and medical device development. However, their creation is a challenging endeavor since the synthesized hearts not only need to exhibit plausible shapes on an individual level but also accurately capture the considerable variability across the true underlying population. In this work, we present Snowflake-Net as a novel approach to automatically generate arbitrarily-sized and realistic populations of 3D heart anatomies in the form of high-resolution point clouds. Our proposed method combines transformer components with point cloud-based deep learning to effectively and directly process 3D heart anatomies reconstructed from cine magnetic resonance images. We develop our approach on a large UK Biobank dataset of about 1000 subjects. We find that the Snowflake-Net achieves average reconstruction errors of 0.90 mm in terms of mean Chamfer Distances, which is considerably below the pixel resolution of the underlying MRI acquisition, and outperforms a prior state-of-the-art approach by ∼20%. Furthermore, we show the Snowflake-Net's ability to create new 3D cardiac anatomies with a high degree of realism on both an individual and population level and observe the generated virtual and the true underlying populations to be highly similar in terms of multiple generation quality metrics. Finally, we investigate how the captured 3D shape variability is encoded in the low-dimensional latent space and its effect on model interpretability.

Keywords: Virtual Heart Population · 3D Cardiac Anatomy Modeling · Snowflake-Net · Point Clouds · Geometric Deep Learning · Cardiac MRI · Generative Models

---

J. Peng and M. Beetz—Authors contributed equally.

Supplementary Information The online version contains supplementary material available at https://doi.org/10.1007/978-3-031-52448-6_16.

O. Camara et al. (Eds.): STACOM 2023, LNCS 14507, pp. 163–173, 2024.
https://doi.org/10.1007/978-3-031-52448-6_16

1   Introduction

Cardiac *in silico* trials employ computational simulations to model a variety of physiological processes of the heart on both an individual and subpopulation level [28]. The resulting virtual replicas of real cardiac functions have the potential to considerably improve and accelerate the development of new medical treatments and devices and provide clinicians with unparalleled insights into the workings of the cardiovascular system [18]. A crucial requirement for *in silico* trials is the existence of high-quality 3D representations of cardiac anatomies. While the large-scale acquisition of personalized cardiac images is difficult and costly, virtual populations of realistic synthesized anatomies can provide a more comprehensive representation of the considerable inter-person diversity and heterogeneity of the real population in a targeted and more efficient manner. Furthermore, such virtual populations are also highly useful as augmented data instances for cardiac disease classification tasks.

Consequently, many endeavors have been made to accurately analyze population-wide cardiac shape variability using traditional statistical shape models such as principal component analysis (PCA) [1,21,22,25]. More recently, deep learning approaches have also been explored to model heart anatomy represented as both voxel grids [15,16] and meshes [6,11,13,19,20]. Point clouds represent an alternative way to store cardiac surface data, and pertinent deep learning techniques have been developed for multiple 3D cardiac modeling tasks, such as deformation modeling [7,9,33], pathology classification [8,12,17], 3D surface reconstruction [3,6,30,32], and multi-modal modeling [5,10,23]. Compared to mesh representations, point clouds require no connectivity information and also no point-to-point correspondence due to their invariance properties. In addition, they are more memory efficient at storing surface-level information, particularly compared to voxel grid deep learning approaches with cubic memory requirements.

In this work, we adopt Snowflake-Net [29] as a novel point cloud deep learning approach to generate 3D biventricular point cloud anatomies on a population level. Incorporating transformer-like architecture [27] and normalizing flow [31], our model uses a coarse-to-fine point generation strategy, which ensures accurate surface reconstruction and flexible prior data distribution modeling. Compared to previous pertinent deep learning approaches [4,10], our model does not require complicated update schedules of loss weighting terms during training, which enables better stability and faster convergence of training. We first evaluate our network in the tasks of point cloud anatomy reconstruction and generation. In the reconstruction task, we show the network's ability to accurately encode and recover 3D cardiac shape information (both locally and globally) and its variation across the population. In the generation task, we demonstrate the network's ability to generate realistic virtual populations of high-resolution 3D cardiac anatomies that are suitable for *in silico* trials. For both tasks, we successfully apply our approach to a large real-world dataset with 3D point cloud anatomies at both the end-diastolic (ED) and end-systolic (ES) phases of the cardiac cycle to validate its robustness and showcase its suitability to capture shape patterns

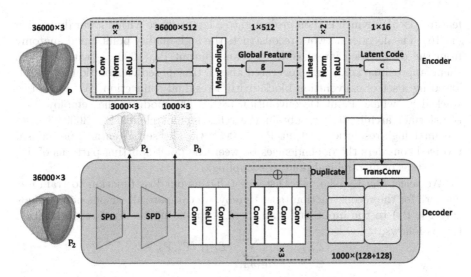

Fig. 1. The overall architecture of the proposed Snowflake-Net for capturing fine-grained spatial information of 3D biventricular anatomies. It comprises an encoder network for point feature extraction and a decoder network for point generation in a coarse-to-fine manner. Snowflake Point Deconvolution (SPD) modules are utilized to progressively generate high-resolution predictions.

at different time points. Finally, we investigate the method's interpretability by studying associations between 3D shape and variations in its individual latent space components.

2   Methods

2.1   Dataset and Preprocessing

In this work, we first create a dataset of 3D cardiac anatomies based on cine MRI acquisitions from 994 subjects of the UK Biobank study [24]. For each subject, we follow the pipeline proposed in [2,3,14] to reconstruct high-resolution point cloud representations of the biventricular anatomy at both the ED and ES phases of the cardiac cycle. The dataset was randomly split into 696, 50, and 248 cases for training, validation, and testing, respectively. We normalized each point cloud anatomy by shifting it to the 3D coordinate origin (i.e., zero mean) and scaling it by a constant value (i.e., the standard deviation of the whole dataset).

2.2   Network

As shown in Fig. 1, we use the PointNet [26] encoder to capture features of the input point clouds P (36000×3) progressively. It consists of a series of convolution blocks, interspersed max pooling operation to produce a global point feature (1×512) which is then projected to a latent space with a lower dimension (1×16). The decoder network reconstructs the high-quality point cloud anatomy from the compressed latent code. A coarse output P_0 (1000×3) with a global shape approximation is first generated after a transposed convolution layer, followed by a set of convolution blocks with a residual connection. We adopt two stacked Snowflake Point Deconvolution (SPD) [29] modules that employ skip-transformer architecture to obtain the refined point clouds P_1 (3000×3) and the final high-resolution output P_2 (36000×3). The self-attention mechanism involved considers the dependencies between the point splitting patterns of different steps of hierarchical point upsampling.

We select the L_1 Chamfer Distance (CD) as our loss function to train our network for the reconstruction task. Hereby, we compare all output point clouds (P_0, P_1, P_2) to the input and sum them, formulating the final reconstruction loss as follows:

$$\mathcal{L}_{recon} = \sum_{i\in\{0,1,2\}} \mathcal{L}_{CD_1}(P_i, P). \qquad (1)$$

Following the setting used in [29], we replace the L_1 CD term in the reconstruction loss with the L_2 CD to train our method for the generation task. Only P_i at steps 0 and 2 are used to compare with the downsampled gold standard point cloud P'_i that has the same point number of P_i, which gives the new reconstruction loss:

$$\mathcal{L}_{recon} = \sum_{i\in\{0,2\}} \mathcal{L}_{CD_2}(P_i, P'_i). \qquad (2)$$

To enable a good generative ability of the network, we employ normalizing flow [31], which uses a bijective mapping $\mathcal{F}_\alpha$ between the latent space and the data space for complex distribution modeling. It consists of a series of invertible transformations to map a simple prior distribution (e.g., isotropic Gaussian) to a complex and exact distribution. Together with the reconstruction loss, it is trained by optimizing the evidence lower bound (ELBO). We provide further implementation details in the Supplementary Material.

3   Experiments and Results

3.1   Evaluation Metrics

For the anatomy reconstruction task, the L1 CD is employed to evaluate the reconstruction quality of the point cloud. For the anatomy generation task, we follow prior work [31] and select three evaluation metrics: 1) Minimum matching distance (MMD), 2) Coverage score (COV), and 3) 1-NN classifier accuracy (1-NNA).

3.2   Anatomy Reconstruction

In this task, the network is expected to recover the cardiac shape from the latent representation, which is obtained by feeding the gold standard point clouds in the test set into the encoder. We train two separate networks, one on ED and the other on ES data. To evaluate the reconstruction ability of our method, we compare it to the Point Cloud VAE [4] by examining the point cloud anatomies reconstructed by both methods quantitatively. We also visualize the reconstruction results of our method on several examples.

We record the Chamfer Distance scored by both methods in Table 1. Our method achieves a mean of 0.98 mm on ED and 0.82 mm on ES, which is ~20% lower than that of the Point Cloud VAE. Also, as shown in Fig. 2, we randomly select two ED cases and two ES cases and visualize the input point clouds as well as our network's output reconstructions. We observe a smooth surface reconstruction across all cardiac substructures, namely the left ventricular endocardium (LV), LV epicardium, and right ventricular endocardium, as well as an agreement of the shape variations between the input anatomies and the reconstructed ones.

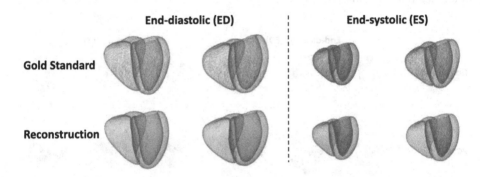

Fig. 2. Four random samples of reconstructed point clouds generated by the proposed method. Columns 1 and 2 are ED samples, and columns 3 and 4 show ES samples. Note that, ED and ES illustrated are not paired.

Table 1. Anatomy reconstruction results in terms of L1 Chamfer distance.

Metric	Phase	Snowflake-Net	Point Cloud VAE
Chamfer distance (mm)	ED	**0.98** (±0.44)	1.23 (±0.25)
	ES	**0.82** (±0.38)	1.03 (±0.22)

Values represent mean (± standard deviation).

3.3 Anatomy Generation

In this task, we assess the network's generative capability on a population level. To this end, we randomly sample latent code values from a multivariate standard normal distribution and pass them through the pre-trained decoder to generate new 3D anatomies. We repeat this procedure separately with the networks trained on ED and ES data to synthesize both an ES and an ED virtual population of 1000 cases each. We then compare the performance with the Point Cloud VAE under the same setting using the metrics described in Sect. 3.1. The results are presented in Table 2.

We observe that the Snowflake-Net achieves better results than the Point Cloud VAE in terms of all three evaluation metrics. Especially, for 1-NNA which takes both generation quality and diversity into consideration, we observe that our model outperforms the Point Cloud VAE by a large margin. Overall, we find our model achieves better scores for ES anatomy generation than ED. For a qualitative study, we visually compare multiple sample cardiac shapes generated by the proposed approach and the Point VAE with samples from the true population as the reference in Fig. 3. Diverse variations in myocardial thickness, overall heart size, and orientation can be observed, which align well with the reference point clouds.

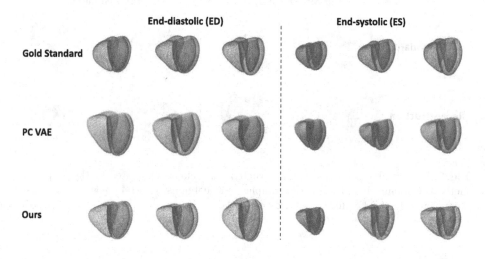

Fig. 3. Three virtual sample point clouds of the true population (row 1) and the synthetic populations generated by the Point Cloud VAE (row 2) and our proposed method (row 3). Columns 1,2,3 are generated ED anatomy samples, and columns 4,5,6 are ES anatomy samples.

3.4   Latent Space Component Analysis

To analyze the contributions of different latent space components to plausible and realistic heart generation, we conduct a component analysis. To this end, we vary the value of one dimension of the latent space while keeping the remaining dimensions unchanged, pass the resulting latent space vector through the

Table 2. Comparison of anatomy generation performance of the Snowflake-Net and the Point Cloud VAE. ↓ means the lower the better, and ↑ means the higher the better.

Metric	Phase	Snowflake-Net	Point Cloud VAE
Minimum matching distance (mm, ↓)	ED	**4.687**	5.973
	ES	**2.923**	4.228
Coverage score (%, ↑)	ED	64.9	**65.7**
	ES	**75.4**	62.5
1-NN classifier accuracy (%, ↓)	ED	**82.7**	94.2
	ES	**79.8**	96.6

decoder, and observe the associated changes in the 3D shape. We select variations of 3 and 5 times the SD in both the positive and negative directions. Figure 4 depicts the shape changes after varying three example components. For brevity, we only show the ES results here and provide the corresponding ED results in Fig. 1 of the Supplementary Material as we find similar patterns for both of them. By varying component 1, we can observe a gradual basal plane tilt of the heart. Component 2 modulates the cavity size and orientation variation, while component 3 changes the thickness of the heart. Overall, we observe a smooth and gradual cardiac attribute morphing after the variation of the latent space components.

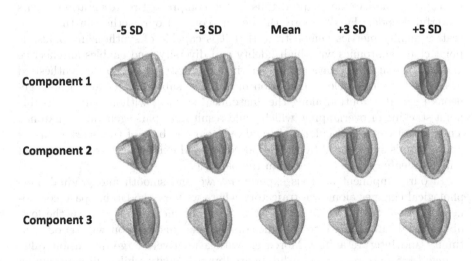

Fig. 4. Latent space components variation and associated 3D shape changes. SD stands for standard deviation. ES anatomy is used here for illustration.

4   Discussion

In this study, we present an efficacious method to model human biventricular anatomies on a population level with 3D point clouds. The PointNet encoder, together with the transformer architecture in the decoder, allow the network to better capture and reconstruct complex 3D shape information on both a local and global level as well as generate realistic virtual populations which exhibit highly similar characteristics to the true ones. We conjecture that the attention mechanism enables the network to better focus on more localized shape patterns and that the coarse-to-fine strategy allows for a gradual upsampling that smoothly connects lower and higher level topology information. Furthermore, the normalizing flow component transforms a simple probability distribution into a more complex one through a series of invertible transformations. This results in a flexible distribution parameterization which is particularly suitable to model the complex underlying distribution of cardiac anatomy variability.

We perform comprehensive experiments on real data collected from the UK Biobank study, showing the superior performance of our method by comparing it to the previous state-of-the-art method on two point cloud tasks. Our method not only obtains a mean Chamfer Distance between reconstructed and gold standard point clouds below the underlying image pixel resolution, but also a pronounced reconstruction error reduction compared to the Point Cloud VAE benchmark. We note that our method achieves this outperformance despite having a shallower encoder architecture, which indicates that it is able to better capture the cardiac shape details in the compressed representation via its powerful decoder. Furthermore, the better scores of our method in three generative quality metrics demonstrate that it is capable of synthesizing plausible point cloud anatomies with high fidelity and diversity, and enables an effective and flexible parameterization of underlying data distribution. In the synthesized hearts, we observe a clear separation of different substructures and a homogeneous point distribution along the anatomical surfaces without any noticeable point stacking or overlapping, which could result in a sparse generated anatomy. This advantage is particularly pronounced near the base of the heart with the points clouds synthesized by the proposed method exhibiting less bluriness and a higher degree of smoothness than the baseline.

In our component analysis experiment, we find smooth and gradual morphological changes along the trajectory where we vary the latent space component from −5 SD to 5 SD, highlighting that our method maps from the low-dimensional latent space to a higher-dimensional distribution with good continuity and interpretability. Moreover, with a relatively large dimension value change (± 5 SD), our model exhibits attribute diversity while still maintaining the proper cardiac shape, which showcases its good robustness and disentanglement of its latent space. We observe similarly good results for both the ED and ES datasets, demonstrating that our network can successfully handle 3D shape variability at multiple phases of the cardiac cycle.

5   Conclusion

We have proposed a novel approach for reconstructing and synthesizing populations of 3D cardiac anatomies with plausible individual shapes and realistic levels of population-wide diversity. The good results of our comparative experiments demonstrate that our model is able to capture global and local geometry details of human hearts as well as learn a flexible and interpretable transformation from a simple latent space distribution into a more complex one. In our future work, we plan to incorporate data from other modalities, for example electrocardiogram signals, to further improve model expressiveness and investigate its effectiveness in a multi-modal setting.

Acknowledgments. This research has been conducted using the UK Biobank Resource under Application Number '40161'. The authors express no conflict of interest. The work of M. Beetz was supported by the Stiftung der Deutschen Wirtschaft (Foundation of German Business). A. Banerjee is a Royal Society University Research Fellow and is supported by the Royal Society Grant No. URF\R1\221314. The work of A. Banerjee was partially supported by the British Heart Foundation (BHF) Project under Grant PG/20/21/35082. The work of V. Grau was supported by the Comp-BioMed 2 Centre of Excellence in Computational Biomedicine (European Commission Horizon 2020 research and innovation programme, grant agreement No. 823712).

References

1. Bai, W., et al.: A bi-ventricular cardiac atlas built from 1000+ high resolution MR images of healthy subjects and an analysis of shape and motion. Med. Image Anal. **26**(1), 133–145 (2015)
2. Banerjee, A., et al.: A completely automated pipeline for 3D reconstruction of human heart from 2D cine magnetic resonance slices. Philos. Trans. A Math. Phys. Eng. Sci. **379**(2212), 20200257 (2021)
3. Beetz, M., Banerjee, A., Grau, V.: Biventricular surface reconstruction from cine MRI contours using point completion networks. In: 2021 IEEE 18th International Symposium on Biomedical Imaging (ISBI), pp. 105–109. IEEE (2021)
4. Beetz, M., Banerjee, A., Grau, V.: Generating subpopulation-specific biventricular anatomy models using conditional point cloud variational autoencoders. In: Puyol Antón, E., et al. (eds.) STACOM 2021. LNCS, vol. 13131, pp. 75–83. Springer, Cham (2022). https://doi.org/10.1007/978-3-030-93722-5_9
5. Beetz, M., Banerjee, A., Grau, V.: Multi-domain variational autoencoders for combined modeling of MRI-based biventricular anatomy and ECG-based cardiac electrophysiology. Front. Physiol. **13**, 991 (2022)
6. Beetz, M., Banerjee, A., Grau, V.: Point2Mesh-net: combining point cloud and mesh-based deep learning for cardiac shape reconstruction. In: Camara, O., et al. Statistical Atlases and Computational Models of the Heart. Regular and CMRx-Motion Challenge Papers. STACOM 2022. Lecture Notes in Computer Science, vol. 13593, pp. 280–290. Springer, Cham (2022). https://doi.org/10.1007/978-3-031-23443-9_26
7. Beetz, M., Banerjee, A., Grau, V.: Modeling 3D cardiac contraction and relaxation with point cloud deformation networks. arXiv preprint arXiv:2307.10927 (2023)

8. Beetz, M., Banerjee, A., Grau, V.: Multi-objective point cloud autoencoders for explainable myocardial infarction prediction. arXiv preprint arXiv:2307.11017 (2023)

9. Beetz, M., Ossenberg-Engels, J., Banerjee, A., Grau, V.: Predicting 3D cardiac deformations with point cloud autoencoders. In: Puyol Antón, E., et al. (eds.) STACOM 2021. LNCS, vol. 13131, pp. 219–228. Springer, Cham (2022). https://doi.org/10.1007/978-3-030-93722-5_24

10. Beetz, M., et al.: Combined generation of electrocardiogram and cardiac anatomy models using multi-modal variational autoencoders. In: 2022 IEEE 19th International Symposium on Biomedical Imaging (ISBI), pp. 1–4 (2022)

11. Beetz, M., et al.: Interpretable cardiac anatomy modeling using variational mesh autoencoders. Front. Cardiovasc. Med. 9, 3258 (2022)

12. Beetz, M., et al.: 3D shape-based myocardial infarction prediction using point cloud classification networks. arXiv preprint arXiv:2307.07298 (2023)

13. Beetz, M., et al.: Mesh U-Nets for 3D cardiac deformation modeling. In: Camara, O., et al. Statistical Atlases and Computational Models of the Heart. Regular and CMRxMotion Challenge Papers. STACOM 2022. Lecture Notes in Computer Science, vol. 13593, pp. 245–257. Springer, Cham (2023). https://doi.org/10.1007/978-3-031-23443-9_23

14. Beetz, M., et al.: Multi-class point cloud completion networks for 3D cardiac anatomy reconstruction from cine magnetic resonance images. arXiv preprint arXiv:2307.08535 (2023)

15. Bertrand, A., et al.: Deep learning-based emulation of human cardiac activation sequences. In: Bernard, O., Clarysse, P., Duchateau, N., Ohayon, J., Viallon, M. (eds.) Functional Imaging and Modeling of the Heart. FIMH 2023. Lecture Notes in Computer Science, vol. 13958, pp. 213–222. Springer, Cham (2023). https://doi.org/10.1007/978-3-031-35302-4_22

16. Biffi, C., et al.: Explainable anatomical shape analysis through deep hierarchical generative models. IEEE Trans. Med. Imaging 39(6), 2088–2099 (2020)

17. Chang, Y., Jung, C.: Automatic cardiac MRI segmentation and permutation-invariant pathology classification using deep neural networks and point clouds. Neurocomputing 418, 270–279 (2020)

18. Corral-Acero, J., et al.: The 'digital twin' to enable the vision of precision cardiology. Eur. Heart J. 41(48), 4556–4564 (2020)

19. Dou, H., Ravikumar, N., Frangi, A.F.: A conditional flow variational autoencoder for controllable synthesis of virtual populations of anatomy. arXiv preprint arXiv:2306.14680 (2023)

20. Dou, H., et al.: A generative shape compositional framework: towards representative populations of virtual heart chimaeras. arXiv preprint arXiv:2210.01607 (2022)

21. Gilbert, K., et al.: Artificial intelligence in cardiac imaging with statistical atlases of cardiac anatomy. Front. Cardiovasc. Med. 7, 102 (2020)

22. Gooya, A., Davatzikos, C., Frangi, A.F.: A bayesian approach to sparse model selection in statistical shape models. SIAM J. Imag. Sci. 8(2), 858–887 (2015)

23. Li, L., et al.: Deep computational model for the inference of ventricular activation properties. In: Camara, O., et al. Statistical Atlases and Computational Models of the Heart. Regular and CMRxMotion Challenge Papers. STACOM 2022. Lecture Notes in Computer Science, vol. 13593, pp. 369–380. Springer, Cham (2023). https://doi.org/10.1007/978-3-031-23443-9_34

24. Petersen, S.E., et al.: UK Biobank's cardiovascular magnetic resonance protocol. J. Cardiovasc. Magn. Reson. 18(1), 1–7 (2015)

25. Piazzese, C., et al.: Statistical shape models of the heart: applications to cardiac imaging. In: Statistical Shape and Deformation Analysis, pp. 445–480. Elsevier (2017)
26. Qi, C.R., et al.: PointNet: deep learning on point sets for 3D classification and segmentation. In: Proceedings of the IEEE conference on computer vision and pattern recognition, pp. 652–660 (2017)
27. Vaswani, A., et al.: Attention is all you need. In: Advances in Neural Information Processing Systems, vol. 30 (2017)
28. Viceconti, M., et al.: In silico trials: verification, validation and uncertainty quantification of predictive models used in the regulatory evaluation of biomedical products. Methods **185**, 120–127 (2021)
29. Xiang, P., et al.: Snowflake point deconvolution for point cloud completion and generation with skip-transformer. IEEE Trans. Pattern Anal. Mach. Intell. **45**(5), 6320–6338 (2022)
30. Xiong, Z., et al.: Automatic 3D surface reconstruction of the left atrium from clinically mapped point clouds using convolutional neural networks. Front. Physiol. **13**, 880260–880260 (2022)
31. Yang, G., et al.: PointFlow: 3D point cloud generation with continuous normalizing flows. In: Proceedings of the IEEE/CVF International Conference on Computer Vision, pp. 4541–4550 (2019)
32. Ye, M., et al.: PC-U Net: learning to jointly reconstruct and segment the cardiac walls in 3D from CT data. In: Puyol Anton, E., et al. (eds.) STACOM 2020. LNCS, vol. 12592, pp. 117–126. Springer, Cham (2021). https://doi.org/10.1007/978-3-030-68107-4_12
33. Zakeri, A., et al.: A probabilistic deep motion model for unsupervised cardiac shape anomaly assessment. Med. Image Anal. **75**, 102276 (2022)

Effects of Fibrotic Border Zone on Drivers for Atrial Fibrillation: An In-Silico Mechanistic Investigation

Shaheim Ogbomo-Harmitt[1]([✉]) [iD], George Obada[1], Nele Vandersickel[2] [iD], Andrew P. King[1] [iD], and Oleg Aslanidi[1] [iD]

[1] School of Biomedical Engineering and Imaging Sciences, King's College London, London, UK
shaheim.ogbomo-harmitt@kcl.ac.uk
[2] Department of Physics and Astronomy, Ghent University, Ghent, Belgium

Abstract. Atrial fibrillation (AF) is a prevalent cardiac arrhythmia that also carries a high risk of stroke. Catheter ablation has emerged as an effective treatment option for paroxysmal AF, but it results in high recurrence rates in persistent AF cases. Recent studies have highlighted the potential of low voltage- and fibrosis-substrate ablation as effective strategies for reducing AF recurrence. However, there is no mechanistic explanation for the success of substrate-based ablation. We use patient imaging and computational modelling to investigate such mechanisms. Left atrial (LA) models were constructed based on late gadolinium enhanced cardiac magnetic resonance (LGE-CMR) imaging data from a cohort of nine AF patients. Tissue conductivity was inversely proportional to the LGE-CMR image intensity. The impact of low-conductive fibrotic border zone (FBZ) was investigated by comparing conduction velocity (CV) maps and phase singularity (PS) distributions associated with re-entrant drivers for AF with and without its inclusion in each patient LA model. The model simulations revealed that PSs associated with AF drivers were predominantly located within regions of low CV (high LGE-CMR image intensity), either with or without of FBZ. The presence of FBZ facilitated the formation of PSs in regions deeper inside the dense fibrotic tissue, which were characterised by the lowest CV values. This contributed to the stabilisation of AF drivers in the fibrotic areas. Therefore, fibrotic areas characterised by the lowest CV provide the most likely substrate for AF drivers and can be targeted by ablation. Such patient-specific areas can include both the FBZ and deeper-lying regions of dense fibrosis.

Keywords: Atrial Fibrillation · Medical Imaging · Cardiac Modelling · Re-entrant Drivers · Catheter Ablation

1 Introduction

Atrial fibrillation (AF) is the most prevalent form of cardiac arrhythmia worldwide, and it also increases the risk of thromboembolic stroke five-fold [1, 2]. Catheter ablation (CA) is a first-line treatment for AF and the only option with a proven curative effect. While

O. Camara et al. (Eds.): STACOM 2023, LNCS 14507, pp. 174–185, 2024.
https://doi.org/10.1007/978-3-031-52448-6_17

CA has relatively high success rate for paroxysmal AF, persistent AF patients experience AF recurrence rates of >50% [3, 4]. Low effectiveness of the current gold standard CA strategy, the pulmonary vein isolation (PVI), for persistent AF can be explained by the presence of non-pulmonary AF drivers. The latter have been linked to fibrotic tissue formation in the left atrium (LA) during AF [5]. Recently, research has focused on investigating the role of fibrotic tissue in AF arrhythmogenesis, and the development of new CA strategies such as substrate-based ablation targeting low-voltage areas (LVA) and fibrosis. Notably, LVA-substrate-based ablation has demonstrated better outcomes than PVI in randomised clinical studies [6]. However, a clinical trial comparing fibrosis-substrate based CA strategies with PVI have shown contradicting results [7]. The lack of mechanistic understanding of the effectiveness of substrate-based ablation hinders its wider application. Computational modelling studies have leveraged medical imaging data to investigate the relationship between fibrosis and AF drivers, highlighting the fibrotic border zone (FBZ) as a crucial area for the generation and maintenance of AF drivers [8–10]. This study aims to use image-based models of the LA incorporating FBZ tissue to perform patient-specific AF simulations, to understand the impact of FBZ on the dynamics and distribution of AF drivers, and thus to shed light on the mechanisms of substrate-based ablation.

2 Methods

2.1 Patient Data and Fibrosis Characterisation

The data used in this study was obtained from an earlier investigation by Roney et al. The dataset consisted of 9 patient-specific LA meshes derived from late gadolinium enhanced cardiac magnetic resonance (LGE-CMR) images, along with corresponding image intensity ratio (IIR) and fibre orientation data mapped onto the meshes (340 μm resolution) using an LA fibre atlas [11]. To evaluate the degree of dense fibrotic tissue in each patient, we employed the Utah LA fibrosis classification system, consisting of four stages (Utah stages I to IV) [12]. These stages are determined by the percentage of fibrosis in the LA tissue: Stage I ($<10\%$), Stage II ($\geq 10\%$ and $<20\%$), Stage III ($\geq 20\%$ and $<30\%$), and Stage IV ($\geq 30\%$) [12]. Within our dataset, we identified one patient in Utah Stage I and another in Utah Stage IV, as well as four patients in Utah Stage II and three patients in Utah Stage III. Each mesh was thresholded in two versions: one including FBZ and another without it. The LA tissue was categorized into six levels based on the IIR values, where level 0 represented healthy tissue, levels 1–4 corresponded to the FBZ, and level 5 represented dense fibrosis. Specifically, LA mesh elements with IIR values equal ≤ 1.2 were classified as healthy tissue (level 0), while those with IIR values of ≥ 1.32 were classified as dense fibrosis (level 5) [13]. The intermediate IIR range between the healthy tissue and dense fibrosis were sub-divided into four discrete tissue levels (levels 1–4) to represent the FBZ [9]. In LA meshes modelled without FBZ, the tissue was only thresholded into levels 0 and 5, with levels 1–4 assigned same properties as level 0 (Fig. 1).

2.2 Atrial Fibrillation Simulation and Analysis

The OpenCARP software package was used for AF simulations [14]. Specific tissue conductivities were assigned to accurately represent conduction velocities (CV) in different tissue types. Healthy tissue (level 0) had CV of 0.7 m/s, while dense fibrosis had CV of 0.2 m/s, consistent with clinical observations [15]. FBZ (levels 1–4) were assigned different velocities: level 1 - 0.6 m/s, level 2 - 0.5 m/s, level 3 - 0.4 m/s, and level 4 - 0.3 m/s. Conductivities for each velocity were determined iteratively using a function within OpenCARP based on a study by Costa et al. [16]. Conductivities were then adjusted to achieve a longitudinal-transverse anisotropy ratio of 1:4 [17].

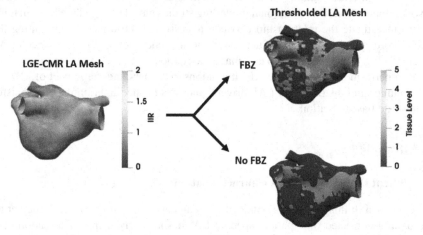

Fig. 1. Thresholding LGE-CMR data into different tissue levels on the LA mesh with and without FBZ, where red colour is dense fibrosis (level 5), blue is healthy tissue (level 0) and the in between colours are FBZ (levels 1–4). (Color figure online)

Two cell models were employed: the AF variant of Courtemanche model for tissue levels 0–4, and a cytokine-related and TGF-β1 remodelling model for tissue level 5, corresponding to dense fibrosis [18, 19]. The PEERP (pacing at end of the effective refractory period) protocol was used to initiate AF, with a pre-pacing phase of five beats at a fixed cycle length of 270 ms [20]. After achieving sustained re-entry, the simulation was run for an additional 5 s, which was used for analysis (Fig. 2A).

CV maps in the LA were assessed by calculating the spatial gradient of the local activation time (*LAT*) obtained over 1 cycle of AF simulation (Eq. 1, Fig. 2B) [21].

$$\|CV\| = \frac{1}{\|\nabla LAT\|} \tag{1}$$

To analyse the spatial distribution of phase singularities (PSs) associated with re-entrant drivers (RDs) in each AF simulation, we employed an automated pipeline for calculating PS frequency maps. By applying the Hilbert transform at 5 ms intervals, we determined the instantaneous phase (θ) of the transmembrane voltage [19]. The instantaneous phase was calculated throughout the entire 5s duration of sustained AF for each subject. PSs were detected in each frame using the Iyer-Gray method, which verified if the phase for each node satisfied the criterion in Eq. 2 [22].

$$\oint \nabla\theta \cdot dr = \pm 2\pi \tag{2}$$

We then identified and marked nodes that exhibited PS presence over a 5 s AF simulation. Finally, the PS frequency map was computed by summing the number of PS occurrences at each node throughout the entire duration of sustained AF (Fig. 2C). The PS frequency maps obtained were cross-validated using a novel directed graph mapping (DGM) method, which employs a directed graph breadth-first search algorithm to identify sustained re-entrant circuits from LAT data and cardiac meshes [23]. PSs corresponding to RDs that made at least two rotations lasting for at least 200ms are referred below as PS-RD [10, 24, 25].

To evaluate the proximity of PS locations to dense fibrotic tissue, we computed fibrotic density (FD) maps using the methods by Zahid et al. This method calculates FD on a per-element basis on the LA mesh. It involves assessing the surrounding elements within a 2.5 mm radius (corresponding to the maximum distance between two adjacent voxels in LGE-CMR images) of a given element and calculating the proportion of dense fibrotic elements (tissue level 5) within the radius [10]. A paired one-tailed hypothesis testing, with a 95% confidence level (significance level $p < 0.05$), was employed to statistically validate the findings. A Shapiro-Wilk test was conducted first to assess the data normality, followed by a t-test (for parametric hypothesis testing) or a Wilcoxon signed-rank test (for non-parametric hypothesis testing).

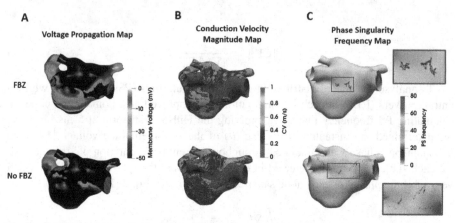

Fig. 2. Comparison of the LA voltage, CV magnitude, and PS frequency maps for simulations with and without FBZ. The top row displays examples of AF simulations with FBZ, while the bottom row represents the respective simulations without FBZ. **A.** LA voltage maps showing re-entrant wave propagation. **B.** Respective CV magnitude maps. **C.** PS frequency maps.

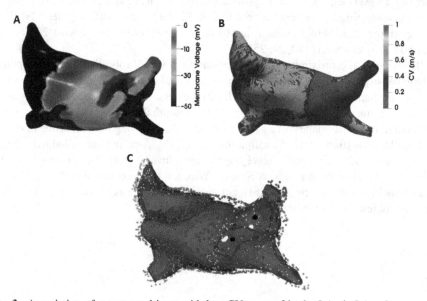

Fig. 3. Association of re-entrant drivers with low CV areas of in the LA. **A.** LA voltage maps showing propagation of two re-entrant waves. **B.** Respective CV magnitude maps, with a large area of low CV on the right. **C.** DGM map of two re-entrant circuits (blue arrows), their cores (black), high PS frequency areas (white) and LATs (coloured nodes; red is early activation). (Color figure online)

3   Results

3.1   Phase Singularity and Conduction Velocity Analysis

In AF simulations, either with or without FBZ in the LA, the predominant localisation of PSs was observed in regions characterised by low CV, as illustrated in Fig. 4. The association of PSs with re-entrant drivers was further confirmed by DGM analysis, which showed (i) co-localisation of high PS frequency areas and cores of re-entrant circuits identified by DGM, and thus (ii) localisation of the RDs in areas with low CVs (see Fig. 3). Specifically, Fig. 3 shows the correspondence of two RDs in the LA (Fig. 3A) to the region of low CV (Fig. 3B) and the locations of two high PS frequency areas and cores of two re-entrant circuits (Fig. 3C). Statistical analysis supported these observations, as the mean CV in PS areas was significantly lower than the mean CV in the LA tissue for simulations with FBZ (p = 1.76e−10) and without FBZ (p = 0.00195) (Fig. 4). Furthermore, as shown in Fig. 4, the mean CV in PS areas was consistently lower than the mean CV in the dense fibrotic tissue in both simulation scenarios (with FBZ: p = 6.89e−05; without FBZ: p = 0.00977). In areas with PS-RDs, the mean CV was significantly lower than that in dense fibrotic tissue with FBZ (p = 4.966e−05). However, when FBZ was not included, there was no significant difference (p = 0.0767). Generally, the cases with FBZ resulted in a significantly lower mean CV in PS areas compared to the cases without FBZ (p = 0.0137). This result stands when only considering PS-RDs, as simulations with FBZ had a significantly lower mean CV in PS areas compared those without FBZ (p = 4.59e−03).

3.2   Phase Singularity Stability and Distribution

The inclusion of FBZ in the LA models significantly increased the mean PS frequency (p = 0.013), as depicted in Fig. 5A. However, the mean total number of PSs was significantly higher when FBZ was not included (p = 0.028) (Fig. 5B). Figure 6 illustrates that PSs were concentrated in small regions when FBZ was present, whereas without FBZ, PSs were dispersed across the LA with a relatively low PS frequency. The mean percentage of PSs classified as PS-RDs was significantly lower in simulations with FBZ than without it (p = 0.0359) (Fig. 5D). This emphasises the enhanced stability of PSs when FBZ is included, as they become localised in a specific region more frequently, and hence there is a greater percentage of PSs classified as PS-RDs. Mean CV in PS areas was significantly lower than that in dense fibrotic areas, but there was no significant difference in mean CV between PS-RD areas and dense fibrosis when FBZ was not included in the models. This can be explained by the majority of PSs being non-RD in the latter case, corresponding to transient wavebreaks and conduction blocks located deeper in the dense fibrotic tissue. Additionally, the PSs were localised in areas with significantly higher FD in simulations with FBZ compared to those without FBZ (p = 3.48e−05) (Fig. 5C). This suggests that the FBZ may facilitate drift of re-entrant drivers resulting in localisation of PSs deeper within the dense fibrotic tissue, as illustrated in Fig. 6.

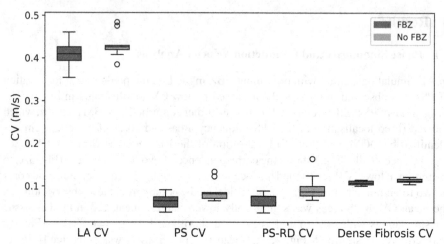

Fig. 4. Boxplots comparing mean CV in the LA tissue, PS areas, PS-RDs areas and dense fibrotic tissue across all subjects in AF simulations with (red) and without FBZ (blue). (Color figure online)

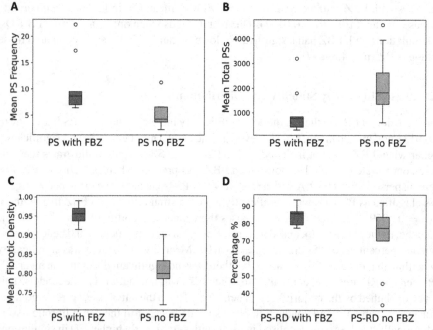

Fig. 5. Boxplots comparing mean and total PS frequencies, and the percentage of PSs in FD and PS-RD areas in simulations with (blue) and without (red) FBZ. Mean values calculated across all subjects **A**. Mean PS frequencies. **B**. Mean total number of PS. **C**. Mean number of PS in FD areas. **D**. Mean percentage of PSs classified as PS-RDs. (Color figure online)

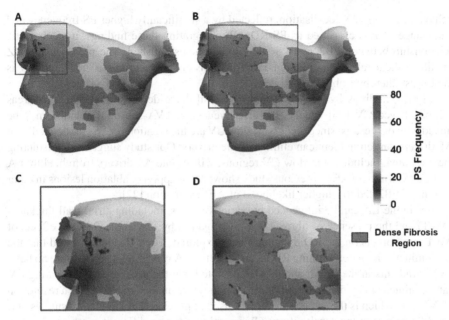

Fig. 6. PS frequency maps overlaid with regions of dense fibrosis (opaque red) in the LA (grey). **A. C.** Simulations with FBZ. **B, D.** Simulations without FBZ. (Color figure online)

4 Discussion and Conclusion

This study aimed to investigate the impact of FBZ on AF drivers, and thus to provide mechanistic insights into the efficacy of substrate-based ablation. This was achieved by conducting a comparative analysis of CV maps and PS distributions in patient-specific AF simulations, both with and without FBZ tissue. Further DGM analysis was performed for cross-validation of the locations of re-entrant drivers and PS areas. The simulations revealed multiple PSs in all simulations, which typically corresponded to between 1–3 stable re-entrant drivers (corresponding to high PS frequency areas) and several transient re-entries. DGM identifies all stable, but not transient re-entries.

All PSs (including high PS frequency areas) tended to co-localise with regions of low CV, exhibiting significantly lower mean CV in PS areas compared to the mean CV in the LA and dense fibrotic tissue. Notably, when FBZ was included in the AF simulations, CVs in PS and PS-RD areas were significantly lower. These observations can be attributed to drift of RDs through low-conducting FBZ, leading to a deeper localisation of PSs within dense fibrotic tissue compared to simulations without FBZ.

These results are consistent with a study by Roney et al., which demonstrated that PSs were present in regions exhibiting high LGE-CMR intensity [19]. Furthermore, our findings agree with the conclusions of Zahid et al., who reported a mean FD of 0.63 ± 0.17 for PS-RD areas; note that FBZ was not modelled in this study [10]. In our study, the respective mean FD for PS-RD areas was 0.754 ± 0.172, calculated across all subjects using the same approach as Zahid et al. Our study also demonstrated an increase in PS stability when FBZ was included in the simulation, as PSs exhibited reduced transient behaviour and greater localisation, reflected by a significantly higher PS frequency and percentage of PSs classified as PS-RDs. By integrating these findings, a mechanistic relationship between FBZ and PS stability can be established. The presence of FBZ facilitates the localisation of PSs in deeper regions of dense fibrotic tissue, where CV is the lowest, thereby stabilising the PSs.

Clinical findings by Roberts-Thompson et al. have demonstrated that LVA areas exhibit slowed CV [26]. Hence, the effectiveness of LVA-substrate ablation may be linked with our results: since areas with slow CV are the locations of PSs associated with AF drivers, targeting them can eliminates the drivers. Our study suggests that isolating the LVA area, including the slow CV regions, will isolate AF drivers from healthy LA tissue and terminate AF. A previous study showed that applying ablation lesions in outer regions of FBZ led to a higher likelihood of AF termination [27].

CV in the LA can be influenced by various factors, including atrial wall thickness (AWT) and the presence of fibrosis. An investigation by Roy et al. [28], the effects of AWT and fibrosis on the RD dynamics were explored. Their findings showed that the predominant factor determining RD locations in the LA was fibrosis, as there are no large AWT gradients in the LA. It should also be noted that the accuracy of calculating CV can be influenced by mesh resolution. In our study, we utilised meshes with a resolution of 340 μm, which is the same resolution as used in previous LA simulation studies [11, 29, 30]. To assess the sensitivity of CV magnitude across different mesh resolutions, we simulated a square 2D LA tissue. It featured a central region of dense fibrotic tissue encircled by healthy tissue. The same cell models and CV assignment as in the 3D LA meshes were applied to the healthy and dense fibrotic tissue, while the conductivities were adjusted for each mesh resolution using the method by Costa et al. [16]. Mean CV for a single plane wave propagating through the 2D tissue was calculated. A small 1.47% error in mean CV between mesh resolutions of 300 μm and 400 μm was observed, and a slightly increased 1.71% error between mesh resolutions of 400 μm and 500 μm. These findings align with a previous study that reported a similar <3% error when decreasing spatial resolution by ~100 μm, from 330 μm to 250 μm in finite difference simulations of the 3D atria [31]. Another study by Clayton et al. also reported minimal differences in RD tip trajectory when using mesh resolutions below 500 μm in ventricular simulations [32].

The current study suggests new possible approaches to improve ablation, based on either LGE-CMR intensity corresponding to the FBZ, or on DGM analysis of LATs measured from AF patients. Our analysis also shows the importance of using CV as a feature in deep learning models for guiding AF catheter ablation therapy. While our previous studies focused on structural features like fibrosis distribution [33, 34], this study emphasises the value of functional features like CV in enhancing the accuracy of these models due to its strong association with PSs and PS-RDs. Lastly, a more comprehensive, larger-cohort investigation is warranted to focus on the direct implementation of such ablation approaches in patient-specific AF models, before their further validation and application in the clinic.

In conclusion, this study investigated the mechanistic relationship between FBZ in the LA with the distribution and stability of PSs during AF. Our results, show that the presence of FBZ facilitates the localisation of PSs in deeper regions of dense fibrotic tissue, where CV is the lowest, thereby stabilising the PSs. Therefore, fibrotic areas characterised by the lowest CV provide the most likely substrate for AF drivers and could be targeted by ablation in persistent AF patients.

Acknowledgements. This work was supported by funding from the Medical Research Council [MR/N013700/1], the British Heart Foundation [PG/15/8/31130], and the Wellcome/EPSRC Centre for Medical Engineering [WT 203148/Z/16/Z].

References

1. Brundel, B.J.J.M., Ai, X., Hills, M.T., Kuipers, M.F., Lip, G.Y.H., de Groot, N.M.S.: Atrial fibrillation. Nat. Rev. Dis. Primers. **8**, 21 (2022). https://doi.org/10.1038/s41572-022-00347-9
2. Wolf, P.A., Abbott, R.D., Kannel, W.B.: Atrial fibrillation as an independent risk factor for stroke: the Framingham Study. Stroke **22**, 983–988 (1991). https://doi.org/10.1161/01.str.22.8.983
3. Darby, A.E.: Recurrent atrial fibrillation after catheter ablation: considerations for repeat ablation and strategies to optimize success. J. Atr. Fibrillation **9**, 1427 (2016). https://doi.org/10.4022/JAFIB.1427
4. Sultan, A., et al.: Predictors of atrial fibrillation recurrence after catheter ablation: data from the german ablation registry. Sci. Rep. **7** (2017). https://doi.org/10.1038/S41598-017-16938-6
5. Lin, W.S., et al.: Catheter ablation of paroxysmal atrial fibrillation initiated by non-pulmonary vein ectopy. Circulation **107**, 3176–3183 (2003). https://doi.org/10.1161/01.CIR.0000074206.52056.2D
6. Junarta, J., Siddiqui, M.U., Riley, J.M., Dikdan, S.J., Patel, A., Frisch, D.R.: Low-voltage area substrate modification for atrial fibrillation ablation: a systematic review and meta-analysis of clinical trials. Europace **24**, 1585–1598 (2022). https://doi.org/10.1093/EUROPACE/EUAC089
7. Marrouche, N.F., Wazni, O., McGann, C., et al.: Effect of MRI-guided fibrosis ablation vs conventional catheter ablation on atrial arrhythmia recurrence in patients with persistent atrial fibrillation: the DECAAF II randomized clinical trial. JAMA **327**, 2296–2305 (2022). https://doi.org/10.1001/JAMA.2022.8831
8. Roy, A., et al.: Identifying locations of re-entrant drivers from patient-specific distribution of fibrosis in the left atrium. PLoS Comput. Biol. **16**, e1008086 (2020). https://doi.org/10.1371/journal.pcbi.1008086
9. Morgan, R., Colman, M.A., Chubb, H., Seemann, G., Aslanidi, O.V.: Slow conduction in the border zones of patchy fibrosis stabilizes the drivers for atrial fibrillation: insights from multi-scale human atrial modeling. Front. Physiol. **7**, 474 (2016). https://doi.org/10.3389/fphys.2016.00474
10. Zahid, S., et al.: Patient-derived models link re-entrant driver localization in atrial fibrillation to fibrosis spatial pattern. Cardiovasc. Res. **110**, 443–454 (2016). https://doi.org/10.1093/CVR/CVW073
11. Roney, C.H., et al.: Predicting atrial fibrillation recurrence by combining population data and virtual cohorts of patient-specific left atrial models. Circ. Arrhythm. Electrophysiol. **15**, e010253 (2022). https://doi.org/10.1161/CIRCEP.121.010253
12. Marrouche, N.F., et al.: Association of atrial tissue fibrosis identified by delayed enhancement MRI and atrial fibrillation catheter ablation: the DECAAF study. JAMA **311**, 498–506 (2014). https://doi.org/10.1001/JAMA.2014.3

13. Benito, E.M., et al.: Left atrial fibrosis quantification by late gadolinium-enhanced magnetic resonance: a new method to standardize the thresholds for reproducibility. Europace **19**, 1272–1279 (2017). https://doi.org/10.1093/EUROPACE/EUW219

14. Plank, G., Loewe, A., Neic, A., et al.: The openCARP simulation environment for cardiac electrophysiology. Comput. Methods Programs Biomed. **208**, 106223 (2021). https://doi.org/10.1016/J.CMPB.2021.106223

15. Ohguchi, S., et al.: Regional left atrial conduction velocity in the anterior wall is associated with clinical recurrence of atrial fibrillation after catheter ablation: efficacy in combination with the ipsilateral low voltage area. BMC Cardiovasc. Disord. **22**, 1–11 (2022). https://doi.org/10.1186/S12872-022-02881-6/FIGURES/5

16. Mendonca Costa, C., Hoetzl, E., Martins Rocha, B., Prassl, A.J., Plank, G.: Automatic parameterization strategy for cardiac electrophysiology simulations. Comput. Cardiol. **40**, 373 (2013)

17. Azzolin, L., et al.: Personalized ablation vs. conventional ablation strategies to terminate atrial fibrillation and prevent recurrence. Europace. euac116–euac116 (2022). https://doi.org/10.1093/EUROPACE/EUAC116

18. Loewe, A., Wilhelms, M., Dössel, O., Seemann, G.: Influence of chronic atrial fibrillation induced remodeling in a computational electrophysiological model. Biomed. Eng./Biomedizinische Technik **59**, S929–S932 (2014). https://doi.org/10.1515/bmt-2014-5012

19. Roney, C.H., et al.: Modelling methodology of atrial fibrosis affects rotor dynamics and electrograms. EP Europace **18**, iv146–iv155 (2016). https://doi.org/10.1093/EUROPACE/EUW365

20. Azzolin, L., Schuler, S., Dössel, O., Loewe, A.: A Reproducible protocol to assess arrhythmia vulnerability in silico: pacing at the end of the effective refractory period. Front. Physiol. **12**, 420 (2021). https://doi.org/10.3389/FPHYS.2021.656411/

21. Varela, M., Colman, M.A., Hancox, J.C., Aslanidi, O.V.: Atrial heterogeneity generates re-entrant substrate during atrial fibrillation and anti-arrhythmic drug action: mechanistic insights from canine atrial models. PLoS Comput. Biol. **12**, e1005245 (2016). https://doi.org/10.1371/JOURNAL.PCBI.1005245

22. Iyer, A.N., Gray, R.A.: An experimentalist's approach to accurate localization of phase singularities during reentry. Ann. Biomed. Eng. **29**, 47–59 (2001). https://doi.org/10.1114/1.1335538

23. Van Nieuwenhuyse, E., et al.: DG-Mapping: a novel software package for the analysis of any type of reentry and focal activation of simulated, experimental or clinical data of cardiac arrhythmia. Med. Biol. Eng. Comput. **60**, 1929–1945 (2022). https://doi.org/10.1007/S11517-022-02550-Y/FIGURES/5

24. Narayan, S.M., Krummen, D.E., Rappel, W.J.: Clinical mapping approach to diagnose electrical rotors and focal impulse sources for human atrial fibrillation. J. Cardiovasc. Electrophysiol. **23**, 447–454 (2012). https://doi.org/10.1111/J.1540-8167.2012.02332.X

25. Haissaguerre, M., et al.: Driver domains in persistent atrial fibrillation. Circulation **130**, 530–538 (2014). https://doi.org/10.1161/CIRCULATIONAHA.113.005421

26. Roberts-Thomson, K.C., et al.: Fractionated atrial electrograms during sinus rhythm: relationship to age, voltage, and conduction velocity. Heart Rhythm **6**, 587–591 (2009). https://doi.org/10.1016/J.HRTHM.2009.02.023

27. Ogbomo-Harmitt, S., Qureshi, A., King, A., Aslanidi, O.: Impact of fibrosis border zone characterisation on fibrosis-substrate isolation ablation outcome for atrial fibrillation. In: 2022 Computing in Cardiology Conference (CinC), vol. 49 (2022). https://doi.org/10.22489/CINC.2022.218

28. Roy, A., Varela, M., Aslanidi, O.: Image-based computational evaluation of the effects of atrial wall thickness and fibrosis on re-entrant drivers for atrial fibrillation. Front Physiol. **9** (2018). https://doi.org/10.3389/FPHYS.2018.01352
29. Roney, C.H., et al.: Universal atrial coordinates applied to visualisation, registration and construction of patient specific meshes. Med. Image Anal. **55**, 65–75 (2019). https://doi.org/10.1016/J.MEDIA.2019.04.004
30. Roney, C.H., et al.: In silico comparison of left atrial ablation techniques that target the anatomical, structural, and electrical substrates of atrial fibrillation. Front. Physiol. **11**, 1145 (2020). https://doi.org/10.3389/fphys.2020.572874
31. Aslanidi, O.V., et al.: 3D virtual human atria: a computational platform for studying clinical atrial fibrillation. Prog. Biophys. Mol. Biol. **107**, 156–168 (2011). https://doi.org/10.1016/J.PBIOMOLBIO.2011.06.011
32. Clayton, R.H., et al.: Models of cardiac tissue electrophysiology: progress, challenges and open questions. Prog. Biophys. Mol. Biol. **104**, 22–48 (2011). https://doi.org/10.1016/J.PBIOMOLBIO.2010.05.008
33. Muffoletto, M., et al.: Toward patient-specific prediction of ablation strategies for atrial fibrillation using deep learning. Front. Physiol. **12**, 717 (2021). https://doi.org/10.3389/FPHYS.2021.674106
34. Ogbomo-Harmitt, S., Muffoletto, M., Zeidan, A., Qureshi, A., King, A.P., Aslanidi, O.: Exploring interpretability in deep learning prediction of successful ablation therapy for atrial fibrillation. Front. Physiol. **14**, 1054401 (2023). https://doi.org/10.3389/FPHYS.2023.1054401

Exploring the Relationship Between Pulmonary Artery Shape and Pressure in Pulmonary Hypertension: A Statistical Shape Analysis Study

Malak Sabry[1,2(✉)], Uxio Hermida[1], Ahmed Hassan[2,4], Michael Nagy[2], David Stojanovski[1], Irini Samuel[2,4], John Locas[2], Magdi H. Yacoub[2,3], Adelaide De Vecchi[1], and Pablo Lamata[1]

[1] Department of Biomedical Engineering, King's College London, London, UK
malak.sabry@kcl.ac.uk
[2] Aswan Heart Research Centre, Magdi Yacoub Foundation, Aswan, Egypt
[3] National Heart and Lung Institute, Imperial College London, London, UK
[4] Cardiology Department, Cairo University, Cairo, Egypt

Abstract. Pulmonary Hypertension (PH) is a progressive condition affecting the right heart, defined by a mean pulmonary arterial pressure (mPAP) greater than 20 mmHg. Measuring mPAP with a pressure catheter is the gold standard for diagnosing PH despite its associated costs and risks. As an alternative, this work investigates the inference of mPAP from pulmonary vasculature anatomy. We thus studied the shape of the main pulmonary artery trunk along with its left and right branches (MPA, LPA, and RPA) across a population of group I PH patients and investigated the relationship between shape and mPAP. Computed Tomography images from 80 confirmed PH cases were used to create a statistical shape model: anatomy was manually segmented, represented by 3 centerlines and radii at evenly spaced control points, and reduced in dimensionality by Principal Component Analysis (PCA). The correlation between each PCA mode of variation and mPAP was then used to identify relevant shape features associated with elevated pressure, which were finally combined in a linear regression model. Results reveal that changes in MPA's diameter and bulging, as well as in the angle between RPA and LPA, related to mPAP. A linear combination of the first 12 modes, with a cumulative variance of 95%, resulted in a linear regression model explaining 36% of mPAP variability in the population, while a combination of the 6 most relevant PCA modes was able to explain 34% of mPAP variability. These results provide initial evidence of the ability to infer mPAP from pulmonary arteries anatomy in PH patients.

Keywords: Pulmonary Hypertension · Statistical Shape Analysis · Remodelling

O. Camara et al. (Eds.): STACOM 2023, LNCS 14507, pp. 186–195, 2024.
https://doi.org/10.1007/978-3-031-52448-6_18

1   Introduction

Pulmonary Hypertension (PH) is a chronic disease characterised by a mean pulmonary arterial pressure (mPAP) greater than 20 mmHg at rest. PH leads to remodelling of the pulmonary vasculature and progressive right ventricular (RV) dysfunction [13,18]. The invasive right-heart catheterisation (RHC) to measure mPAP is the gold standard for diagnosing PH. This procedure typically takes place at a late stage because the primary symptoms are not disease-specific and because RHC has increased risks and expenses. There is therefore a need for alternative non-invasive markers of the disease and its progression to support suitable treatment planning. PH is accompanied by an increased pulmonary vasculature resistance, resulting in structural maladaptations such as increased wall stiffness and dilation of the main pulmonary artery (MPA).

The shape of the pulmonary artery (PA) is hypothesised to be an indicator of PH pathology. Both the spatial orientation of the PA and its morphology were shown to vary in previous studies, across control or diseased cohorts. In [11], Lee et al. show how the orientation and geometric features of the PA, such as diameters, lengths, and angles, change across cohorts (controls vs. heart failure vs. PH). A link between PA shape and hemodynamics was also shown in transposition of great arteries [10]. The angle between the main pulmonary trunk and the right pulmonary artery was found to correlate to the flow distribution between right and left pulmonary arteries, the RPA and LPA respectively.

Statistical shape modelling (SSM) is a digital twin technology [5] that has been extensively used to derive novel shape biomarkers, explore characteristic disease phenotypes, and provide new mechanistic insights about disease pathophysiology [9,12,19]. Particularly, SSM has been used to study the relationship between arterial remodelling and disease in several conditions [3,4,8]. However, such a study has not been performed on the pulmonary artery.

A better understanding of the structural changes of the PA due to elevated pressures could provide new insights into the aetiology of the disease and influence new diagnostic and treatment approaches. Therefore, this work aims to create a statistical shape model of the pulmonary artery in a population of PH patients to (1) study the shape variability of the PA in the population and its relationship with PH, and (2) derive novel mechanistic insights.

2   Methods

2.1   Clinical Dataset

We used a cohort of 80 confirmed group I PH cases. Clinical image data were acquired from patients referred for cardiac Computed Tomography (CT) exams at the Aswan Heart Centre. All subjects participated under informed consent. Data access was granted through the Aswan Heart Centre Research Ethics Committee (REC code: 20210804MYFAHC - VAPH - 20211025). Images were acquired on a Siemens dual source 128 multidetector CT machine with 0.6 mm cut sections. 3D surfaces of the pulmonary artery were manually segmented by a single observer from CT images using 3D Slicer [1,6]. A Judkins right coronary catheter was passed from the inferior vena cava to the right atrium, then to the RV and the MPA. Pressure tracings are thus obtained and the mPAP is computed.

2.2   Statistical Shape Modelling (SSM) Pipeline

A centerline-based approach was used for the SSM, where the PA shape was encoded by the centerline points and their maximum inscribed sphere radius, as in a previous study [8]. All steps were implemented using open-source Python toolkits, such as the Visualization Toolkit (VTK) [17] or the Vascular Modeling Toolkit (VMTK) [2].

Centerline Extraction. The arterial segments of interest were the MPA and its two bifurcating branches, LPA and RPA. Figure 1 shows the different steps undertaken starting from the CT images, up to the visualisation of the anatomy captured by the centerline.

All PA anatomies were manually segmented from the CT stacks. Landmarks for centerline extraction were manually placed by a single observer for all cases as follows: 1. at the level of the pulmonary valve plane; 2. at the first bifurcation of the RPA branch; 3. at the first bifurcation of the LPA branch. Centerlines were then extracted using those landmarks and merged using the bifurcation origin that results from the geometrical decomposition of the bifurcation [15]. Then, each centerline segment was resampled to a fixed number of points: 26 points for MPA, and 25 points for LPA and RPA, respectively.

To visualise the shapes encoded, the 3D shape was reconstructed using a set of spheres at each centerline point. Endpoints of the 3D reconstruction were then clipped by Boolean subtraction with a tubular surface generated with information about centerline normal and radius at endpoints.

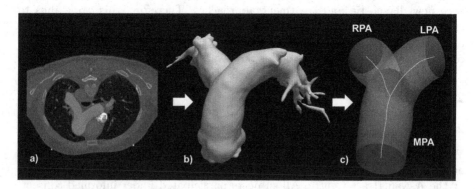

Fig. 1. Pipeline used. a) Segmentation from CT images, b) Segmented geometry, and c) Extracted centerline and illustrative reconstructed geometry.

Dimensionality Reduction. After centerline extraction, the shape of the PA is encoded with the coordinates of each point along the centerline plus their captured radius. As all the PAs were segmented using CT stacks under the same imaging protocol. Patient are all positioned in the same way in the CT machine, and therefore any variation in MPA angle would be a feature to consider in the shape analysis. Centerlines were aligned with respect to the bifurcation origin. Principal Component Analysis (PCA) was then used to reduce the high dimensionality of the problem and build a SSM of the PA anatomy. PCA finds the directions that maximise the variance in the observed shapes (i.e., PCA modes or anatomical modes of variation), and thus captures the largest and most common PA shape changes in the cohort. Note that all features were standardised before PCA to avoid biasing the SSM due to feature range differences [7].

2.3   Principal Component Linear Regression Analysis

Once the PCA modes are extracted, the task is to find their linear combination that best correlates with mPAP. As described in [19], two methods were explored to determine the 3D shape features associated with mPAP: 1. Inclusive linear regression; 2. Selective linear regression.

The inclusive linear regression model studies the gradual addition of PCA modes ranked by the variance they explain. The selective model ranks PCA modes by their ability to infer mPAP, i.e. by the absolute value of the Pearson correlation coefficient with mPAP.

The resulting linear models were tested in cross-validation by splitting the population into 90% cases for training and 10% for testing. 1000 runs were performed to remove potential biases in the population split. The root mean squared error (RMSE) between predicted and ground truth mPAP was assessed in both training and test sets, along with their respective standard deviations to assess variability across folds. The relative error between them was used as a surrogate metric for data overfitting (i.e., the more PCA modes included in the linear regression, the smaller the RMSE on the training set but the larger on the test set). The number of PCA modes to be included in the final linear regression model was selected after inspection of the RMSE for training and test sets. Both the cross-validation and the linear regression were performed using the Scikit-learn library in Python [14].

3   Results

3.1   Linear Regression Models

Inclusive Linear Regression Model. Figure 2a shows the RMSE error in the train and test sets depending on the number of PCA modes included in the linear regression model, along with the standard deviation of train and test sets across folds. As expected, the more PCA modes included, the smaller the RMSE in the training set, but the larger it is in the test set. This is reflected in an increasing relative error between train and test sets RMSE as the number of PCA modes increases. Based on the relative error, the first 12 PCA modes, capturing 95% of the shape variability in the population, were included in the final inclusive linear regression model. Table 1 shows the intercept and coefficients for the optimal inclusive linear regression model, along with its coefficient of determination (R^2). The resulting model showed good correlation (Pearson correlation = 0.60) with mPAP. The shape model explained 36% of the mPAP variability $(R^2 = 0.36)$.

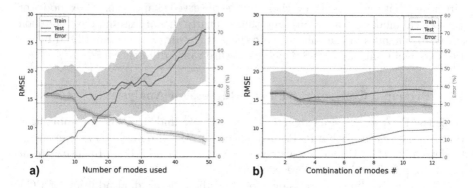

Fig. 2. RMSE between predicted value vs. actual pressure value for the test (blue) and train (orange) sets, using a) increasing number of modes, and b) different mode combinations. The standard deviation of the test set is shown in light blue. The error between test and train sets, in percentage, is shown in green. (Color figure online)

Selective Linear Regression Model. Figure 2b shows the RMSE error in the train and test sets depending on the combination of the 12 PCA modes included in the selective regression model, along with the standard deviation of train and test sets across folds. The combinations were created by progressively adding modes starting from the one with the highest Pearson correlation to mPAP (mode 11) to the one with the lowest one (mode 9). The univariate regression showed that modes 11, 1, 10, 12, 5, and 8 were the ones most correlated with mPAP, in decreasing order. Table 1 shows the intercept and coefficients for the optimal selective linear regression model. The resulting model showed good correlation (Pearson correlation 0.58) with mPAP. The shape model explained 34% of the mPAP variability $(R^2 = 0.34)$.

Table 1. Intercept (Int.), coefficients, and R^2 for inclusive (Incl.) and selective (Select.) linear regression models

Model	Int.	PCA Mode Coefficients												R^2
		1	2	3	4	5	6	7	8	9	10	11	12	
Incl.	63.26	0.48	0.11	−0.10	−0.18	−0.59	0.43	0.34	−0.68	0.08	−2.19	2.96	−2.17	0.36
Select.	63.26	0.48	–	–	–	−0.59	–	–	−0.67	–	−2.19	2.95	−2.17	0.34

3.2   3D Shape Features Associated with mPAP

Individual PCA Modes. Figure 3 shows the shape variations for the six individual PCA modes that contribute most to mPAP variations. Qualitatively, extreme shapes of modes 11 and 1 captured changes in MPA diameter and bulging. Modes 10 and 12 captured changes in the LPA-RPA angle in the side and front views, respectively. Mode 5 presented variations in the LPA and RPA lengths, while mode 8 extreme shapes displayed changes in the vessel's curvature.

An R^2 analysis was then performed between each individual mode (from 1 to 12), the mPAP, and selected geometric features, such as average segment diameters, lengths, average segment curvatures, and bifurcation angle (see Fig. 4). The analysis confirms the relevant modes to mPAP variations. Results also show that the geometric features are mostly explained by mode 1, especially the diameters which have the highest R^2 values ($R^2 > 0.75$). The bifurcation angle is mostly explained by modes 2, 3, 6, and 7. The results of the R^2 analysis between each geometric feature and the mPAP (last bin of Fig. 4) showed that the three diameters, the LPA curvature, and the bifurcation angle, were more significant to mPAP than the lengths and the remaining curvature values ($R^2 > 0.075$ vs. $R^2 < 0.02$, respectively). Checking the univariate correlations between each of the three segment diameters and mPAP, as expected, the larger the diameter, the larger the mPAP. Concerning the curvature, we should note that the mean curvature of the centerline segment calculated here is different than the wall curvature. The MPA wall curvature, related to the bulging shown in Fig. 3, is suspected to be more correlated with mPAP.

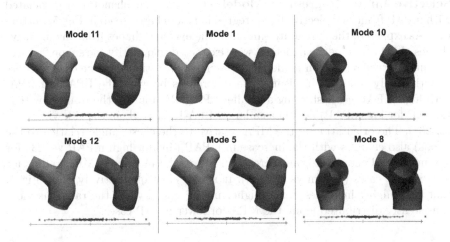

Fig. 3. Individual PA shape variations for the six modes most correlated to mPAP. Orange and purple meshes correspond, respectively, to the plus and minus 3 standard deviations (3SD) from the average mesh along the shape mode. (Color figure online)

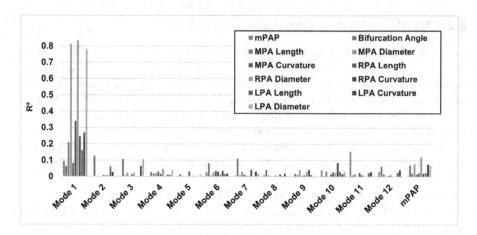

Fig. 4. Coefficient of determination (R^2) analysis between the PCA modes, mPAP, and geometric features (bifurcation angle, average segment diameters and curvatures, and segment lengths).

Inclusive Linear Regression Model. The extreme phenotypes associated with mPAP from the inclusive linear regression model are shown in Fig. 5, top row. Qualitatively, the extreme shapes captured changes in the diameters, MPA angle and bulging, and the bifurcation angle.

Selective Linear Regression Model. The extreme phenotypes associated with mPAP from the selective linear regression model are shown in Fig. 5, bottom row. As expected, they are quite similar to the extreme shapes from the inclusive shapes. A set of geometrical metrics were extracted to quantify the shape features captured by the regression model. The previously mentioned shape changes were quantitatively reflected in clear angle differences between the LPA and RPA, with high mPAP cases showing a smaller LPA-RPA angle in the front view (64° vs. 109° for low mPAP) as well as a larger LPA-RPA angle in the side view (50° vs. 26°). The orientation of the MPA relative to the chest (angle relative to the y-axis) also changes with the increase in mPAP (49° for high mPAP vs. 23° for low mPAP). There is also an increase of 6% in the maximum MPA radius for high mPAP cases (17 mm vs. 18 mm). It is also clear qualitatively, from Fig. 5, that the bulging increases in the higher mPAP cases, while the radius is more consistently distributed along the MPA for lower mPAP cases.

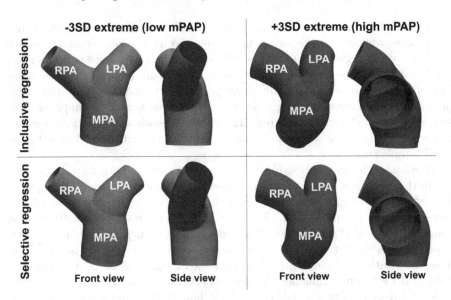

Fig. 5. Shape variation in the PA related to mPAP changes for PH cases for the inclusive and selective linear regression combinations. Orange and purple meshes correspond to the plus and minus 3 standard deviations (3SD) from the average mesh along the shape mode, respectively. (Color figure online)

4   Discussion and Conclusion

This work investigated if shape variations of the pulmonary artery can infer mPAP in PH, and found positive initial evidence. The fact that anatomy remodels in the presence of high pressure is common across several cardiovascular conditions, and this study indicates it might also be the case in the pulmonary arteries. The ability to predict 34% of the variability in mPAP with 6 morphologial descriptors is an encouraging initial quantitative result.

The main shape features found to be linked to mPAP are the angle between RPA and LPA, both in the front and side views, the MPA absolute angle, and the MPA diameter. The dilation of the vascular anatomy in the presence of PH was an expected feature, and it is mainly in the MPA where this feature seems to concentrate. PH seems to also be linked to a set of specific changes in vascular length and angulation as illustrated in Fig. 3. The R^2 analysis in Fig. 4 showed that, as expected, the diameters of all segments is the strongest signal to infer mPAP (the larger the diameter, the larger the mPAP). Diameter changes in PH seem to explain a 10% change in mPAP, by looking at the individual average diameters or at PCA mode 1. Although the centerline curvature was expected to be related to pressure changes, evidence shows that average MPA centerline curvature is not predictive of mPAP. We note, however, that the centerline curvature does not reflect the wall curvature or bulging, which should be quantified and added to the analysis in future studies. Also, bifurcation angle and the LPA curvature, which were shown to be relevant to mPAP in the R^2 analysis, do not seem to be captured by the modes in the selective regression. Nevertheless, for the bifurcation angle, the sum of the R^2 from the first 12 modes is around 0.8, which could indicate that the angle changes are captured but spread over those 12 modes. These results, the quantitative performance, and the qualitative interpretation, should however be regarded cautiously given the small sample size (80 subjects). Other dimensionality reduction methods, such as partial least squares regression for instance, can also be investigated in the future to deal with the small sample size.

A stunning linear relationship between mPAP and the duration of the vortex in the MPA has been recently reported [16]. Vortex dynamics are arguably dependent on vasculature anatomy, and thus one can hypothesise a causal link between anatomy and pressure, i.e. is there any change in vascular anatomy that causes an increase of mPAP? The 3D models reported suggest some mechanistic hypotheses, such that the MPA angle change could cause the inlet flow to differ from cases with low to high mPAP, or that the difference of angulation between RPA and LPA could influence the flow repartition between branches. Further research involving flow simulations to study how these shape changes affect the hemodynamics in the PA is needed.

References

1. 3D slicer. https://www.slicer.org/
2. Antiga, L., Piccinelli, M., Botti, L., Ene-Iordache, B., Remuzzi, A., Steinman, D.A.: An image-based modeling framework for patient-specific computational hemodynamics. Med. Biol. Eng. Comput. **46**(11), 1097–1112 (2008). https://doi.org/10.1007/s11517-008-0420-1
3. Bruse, J.L., et al.: How successful is successful? Aortic arch shape after successful aortic coarctation repair correlates with left ventricular function. J. Thorac. Cardiovasc. Surg. **153**(2), 418–427 (2017). https://doi.org/10.1016/j.jtcvs.2016.09.018
4. Bruse, J.L., et al.: A statistical shape modelling framework to extract 3D shape biomarkers from medical imaging data: assessing arch morphology of repaired coarctation of the aorta. BMC Med. Imaging **16**(1), 1–19 (2016). https://doi.org/10.1186/s12880-016-0142-z
5. Corral-Acero, J., et al.: The 'Digital Twin' to enable the vision of precision cardiology. Eur. Heart J. 1–11 (2020). https://doi.org/10.1093/eurheartj/ehaa159
6. Fedorov, A., et al.: 3D slicer as an image computing platform for the quantitative imaging network. Magn. Reson. Imaging **30**(9), 1323–1341 (2012)
7. Gewers, F.L., et al.: Principal component analysis: a natural approach to data exploration. ACM Comput. Surv. **54**(4), 1–33 (2021). https://doi.org/10.1145/3447755
8. Hermida, U., et al.: Learning the hidden signature of fetal arch anatomy: a three-dimensional shape analysis in suspected coarctation. J. Cardiovasc. Transl. Res. (2022). https://doi.org/10.1007/s12265-022-10335-9
9. Hermida, U., et al.: Left ventricular anatomy in obstructive hypertrophic cardiomyopathy: beyond basal septal hypertrophy. Eur. Heart J. Cardiovasc. Imaging **24**(6), 807–818 (2023). https://doi.org/10.1093/ehjci/jeac233

10. Kholwadwala, D., Parnell, V.A., Cooper, R.S.: Transposition of the great arteries S, D, D and absent proximal left pulmonary artery. Cardiol. Young **5**(2), 199–201 (1995). https://doi.org/10.1017/S1047951100011847

11. Lee, S.L., et al.: Spatial orientation and morphology of the pulmonary artery: relevance to optimising design and positioning of a continuous pressure monitoring device. J. Cardiovasc. Transl. Res. **9**(3), 239–248 (2016). https://doi.org/10.1007/s12265-016-9690-4

12. Lewandowski, A.J., et al.: Preterm heart in adult life: cardiovascular magnetic resonance reveals distinct differences in left ventricular mass, geometry, and function. Circulation **127**(2), 197–206 (2013). https://doi.org/10.1161/CIRCULATIONAHA.112.126920

13. Nakaya, T., et al.: Right ventriculo-pulmonary arterial uncoupling and poor outcomes in pulmonary arterial hypertension. Pulm. Circ. **10**, 2045894020957223 (2020)

14. Pedregosa, F., et al.: Scikit-learn: machine learning in Python. J. Mach. Learn. Res. **12**, 2825–2830 (2011)

15. Piccinelli, M., Veneziani, A., Steinman, D.A., Remuzzi, A., Antiga, L.: A framework for geometric analysis of vascular structures: application to cerebral aneurysms. IEEE Trans. Med. Imaging **28**(8), 1141–1155 (2009). https://doi.org/10.1109/TMI.2009.2021652

16. Reiter, G., et al.: Magnetic resonance derived 3-dimensional blood flow patterns in the main pulmonary artery as a marker of pulmonary hypertension and a measure of elevated mean pulmonary arterial pressure. Circ. Cardiovasc. Imaging **1**(1), 23–30 (2008)

17. Schroeder, W.J., Martin, K.M.: The Visualization Toolkit. No. July, Kitware, 4th edn. (2006). https://doi.org/10.1016/B978-012387582-2/50032-0

18. Vanderpool, R.R., et al.: Surfing the right ventricular pressure waveform: methods to assess global, systolic and diastolic RV function from a clinical right heart catheterization. Pulm. Circ. **10**, 1–11 (2020)

19. Varela, M., et al.: Novel computational analysis of left atrial anatomy improves prediction of atrial fibrillation recurrence after ablation. Front. Physiol. **8**(FEB), 68 (2017). https://doi.org/10.3389/fphys.2017.00068

Type and Shape Disentangled Generative Modeling for Congenital Heart Defects

Fanwei Kong[1,2,3](✉) and Alison L. Marsden[1,2,3,4]

[1] Department of Pediatrics, Stanford University, Stanford, CA, USA
amarsden@stanford.edu
[2] Institute for Computational and Mathematical Engineering, Stanford University, Stanford, CA, USA
[3] Department of Bioengineering, Stanford University, Stanford, CA, USA
fwkong@stanford.edu
[4] Department of Mechanical Engineering, Stanford University, Stanford, CA, USA

Abstract. Congenital heart diseases (CHDs) encompass a wide range of cardiovascular structural abnormalities, and CHD patients exhibit complex and unique malformations. Analysis of these unique cardiac anatomies can greatly improve diagnosis and treatment planning. However, CHDs are often rare, making it extremely challenging to acquire patient cohorts of sufficient size. Although generative modeling of cardiac anatomies capturing patient variations can generate virtual cohorts to facilitate in-silico clinical trials, prior approaches were largely designed for normal anatomies and cannot readily model the vast topological variations seen in CHD patients. Therefore, we propose a generative approach that models cardiac anatomies for different CHD types and synthesizes CHD-type specific shapes that preserve the unique topology. Our deep learning (DL) approach represents whole heart shapes implicitly using signed distance fields (SDF) based on CHD-type diagnosis, which conveniently captures divergent anatomical variations across different types. We then learn invertible deformations to deform the learned type-specific anatomies and reconstruct patient-specific geometries. Our approach has potential applications in augmenting the image-segmentation pairs for rarer CHD types for cardiac segmentation and generating CHD cardiac meshes for computational simulations. Our source code is available at https://github.com/fkong7/SDF4CHD.

Keywords: Congenital cardiac defects · Generative Shape Modeling · Implicit Neural Representations

This work was supported by NSF 1663671, NIH R01EB029362, and NIH R01LM013120.

Supplementary Information The online version contains supplementary material available at https://doi.org/10.1007/978-3-031-52448-6_19.

1   Introduction

Congenital heart defects (CHDs) are the most common birth defects and result in abnormal cardiac anatomies that can dramatically impact clinical outcomes [12,13]. CHD patients often have unique but complex cardiac malformations that warrant individualized diagnosis and treatment planning. Personalized modeling of the heart, including 3D printing [10], shape analysis [9], and computational simulations [1] can enable patient-specific surgical and treatment planning, ultimately leading to improved patient outcomes. However, validation of this personalized paradigm requires a large and diverse dataset. CHDs, particularly the more complex types (e.g. transposition of great arteries, hypoplastic left heart syndrome), are rare, making it extremely challenging to acquire sufficiently sized patient cohorts [8,21]. Generative modeling of cardiac anatomies capturing patient variations can generate virtual cohorts that facilitate in-silico clinical trials [7]. While deep learning (DL) has shown potential in cardiac shape generation [3,7], prior approaches primarily focused on cardiac structures with normal topology and do not readily generalize to CHDs with topologically unusual cardiac abnormalities such as holes in the atrial and ventricular septum, missing ventricles or pulmonary outflow tracks, and switched aortic and pulmonary positions.

Therefore, we propose an approach for type-controlled generative modeling of cardiac shapes to enable the generation of synthetic cardiac shapes specific to desired CHD types. Our key insight is to learn a disentangled type and shape representation of CHD hearts. Using a set of ground truth segmentations and corresponding CHD types as inputs, our method automatically learns both the CHD type-specific cardiac anatomical arrangements and the cardiac shape variations shared across different types. Namely, we represent the heart shapes implicitly using signed distance fields (SDF) learned by neural networks to capture the unique arrangements and topologies of cardiac shapes in different types of CHDs. We then deform the SDFs to reconstruct personalized cardiac models that preserve the structural abnormalities. Our approach can be used to augment type-specific image and segmentation pairs for training segmentation methods or to create synthetic cardiac meshes for 3D printing and computational simulations.

Related Works. Cardiac shape synthesis Prior approaches that generate cardiac shapes have largely focused on normal cardiac anatomies [2,3,15,17,18]. Statistical shape models and variational autoencoders have been common for such applications. These methods often assume a specific cardiac anatomy represented by a 3D template with a fixed topology and thus cannot be generalized to CHD patients with diverse and peculiar topologies which differ from the template.

Implicit shape modeling for medical applications DeepSDF was proposed as an effective DL method to approximate a continuous SDF to model shapes with arbitrary topology. A few studies have explored using this approach to model shapes of cells [20], and organs [16,22]. However, these approaches use mixed representations of cardiac shapes and topologies, which makes it difficult to synthesize shapes corresponding to a particular CHD.

Disentangled implicit shape modeling A few approaches outside the medical domain have proposed to separately model shape variations and structure changes by employing one network to learn implicit templates that model the topology and another network to deform the learned template [6,11]. Our approach took inspiration from this idea, yet was designed to model the complex and wide-ranging multi-structural geometries found in CHD. We use CHD diagnosis as input and thus reconstruct cardiac geometries that represent the unique cardiac abnormalities for each CHD type.

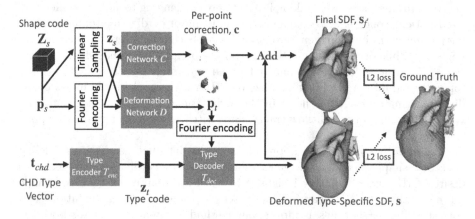

Fig. 1. Diagram of the proposed type- and shape-disentangled shape modeling method.

2   Methods

Figure 1 shows our proposed type- and shape-disentangled modeling method. It consists of a type representation network that leverages CHD-type information to predict type-specific cardiac geometries, a deformation network that changes the shape of the type-specific geometries, and a correction network that captures additional topological variations of patient-specific anatomical features.

Implicit Representations for Cardiac Structures. Implicit representations use SDFs to represent 3D shapes of arbitrary topology. Prior studies of implicit shape modeling often model only one structure [14,20,22]. Namely, an SDF assigns any point $\mathbf{p} \in \mathbb{R}^3$ a scalar value $s \in \mathbb{R}$, such that $s = \mathcal{SDF}(\mathbf{p})$, where the magnitude of s represents the distance from $\mathbf{p}$ to surfaces of cardiac structures and the sign represents whether $\mathbf{p}$ is inside or outside. However, since a heart has multiple structures, we model signed distance vector field, $\mathbf{s} = \mathcal{SDF}(p)$, where $\mathbf{s} \in \mathbb{R}^7$ is a vector representing the signed distance from the surfaces of every cardiac structure. We modeled a total of 7 cardiac structures including the myocardium (Myo), blood pools of the right atrium (RA), left atrium (LA), right ventricle (RV) left ventricle (LV), aorta (Ao), and pulmonary arteries (PA).

Learning Type-Specific CHD Geometries. Multilayer perceptrons (MLPs) have been used to model the SDFs for different shapes [14]. Namely for a shape S, a neural network approximates the mapping, $s = \mathcal{F}_\theta(\mathbf{p}, \mathbf{z}) : \mathbb{R}^3 \times \mathbb{R}^c \to \mathbb{R}$, where $\mathbf{z}$ is the latent vector that encodes shape S and c is the dimension of $\mathbf{z}$. To model type-specific CHD geometries, we first use a small MLP ($\mathcal{T}_{enc}$) to encode the CHD diagnostic information to a latent type vector $\mathbf{z}_t$, namely $\mathbf{z}_t = \mathcal{T}_{enc}(\mathbf{t_{chd}})$, where $\mathbf{z_t}$ is latent type code. The input diagnostic information $\mathbf{t}_{chd}$is a vector containing true or false values, corresponding to whether the patient modeled had the CHD types covered in the dataset. We then used the latent type code $\mathbf{z_t}$ and point coordinates $\mathbf{p}_t$ as inputs to another MLP ($\mathcal{T}_{dec} : \mathbb{R}^3 \times \mathbb{R}^c \to \mathbb{R}^7$) to model type-specific SDFs for whole heart geometries, namely,

$$\mathbf{s} = \mathcal{T}_{dec}(\mathbf{p_t}, \mathbf{z_t}) = \mathcal{T}_{dec}(\mathbf{p_t}, \mathcal{T}_{enc}(\mathbf{t}_{chd})). \tag{1}$$

Since the ReLU activation functions commonly used in MLPs are biased towards low-frequency features, we used Fourier positional encoding to augment the point coordinates $\mathbf{p}_t$ before using them as inputs [19]. The dimension of $\mathbf{t}_{chd}$ is the number of types modeled.

Learning Invertible Deformations for Cardiac Shape Modeling. To preserve the cardiac anatomies learned by the type-specific network, we leveraged the neural ordinary differential equation (NODE) approach [5,11] to learn the invertible deformation between the learned type-specific geometries (type space) and patient-specific cardiac geometries (shape space). Namely, we computed the trajectories of points from the shape to the type space at time interval $[0, K]$ by solving an ODE parameterized by an MLP ($\mathcal{D}$),

$$\mathbf{p}_t = \mathbf{p}_s + \int_0^K \mathcal{D}(\mathbf{z}_s, \mathbf{p}_k)dk, \tag{2}$$

where $\mathbf{p}_s$ and $\mathbf{p}_t$ are point locations in the shape and the type spaces, respectively, and $\mathbf{z}_s$ is a latent shape code that encodes the shape changes from the type to the patient-specific geometries. Furthermore, we adopted the idea of position-aware shape encoding [4] where the latent shape code is a latent grid $\mathbf{Z}_s$. We performed trilinear sampling to this latent grid to obtain the point-specific shape code, thus $\mathbf{z}_s = \mathbf{Z}_s(\mathbf{p}_s)$.

Correction Fields for Additional Topological Changes. CHD cardiac geometries are primarily described by their CHD type, but minor anatomical variations may exist among patients with the same type, mostly related to their vessel connections. We thus use an additional MLP ($\mathcal{C}$) to output a correction vector **c** for the predicted SDFs from the position-aware shape encodings sampled from the latent shape grid $\mathbf{Z}_s$, namely, $\mathbf{c} = \mathcal{C}(\mathbf{Z}_s(\mathbf{p}_s), \mathbf{p}_s)$. This correction describes the minor variations of cardiac anatomies at the point locations that cannot be modeled by simply deforming type-specific cardiac shapes. In summary, the final SDF prediction $\mathbf{s}_f$, for a point $\mathbf{p}_s$ in the shape space is,

$$\mathbf{s}_f = \mathbf{s} + \mathcal{C}(\mathbf{Z}_s(\mathbf{p_s}), \mathbf{p_s}). \tag{3}$$

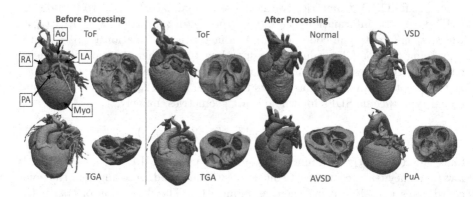

Fig. 2. Ground truth segmentations of different CHD types after manual processing and comparison with the raw segmentations provided by the "ImageCHD" dataset. We manually removed noise on the myocardium so that defects can be easily identified and are consistent with the provided diagnosis and image data, removed small vessel branches that are beyond our scope of modeling cardiac defects, added the missing myocardium segmentation for some samples, and fixed the incorrect segmentation spacing that caused some hearts in the original dataset to be un-physiologically flat along a certain axis (as illustrated for the TGA patient before processing).

Optimization. We converted ground truth segmentations in the training dataset into ground truth SDFs. We sample points near the surfaces of the heart and use the ground truth SDF values ($\mathbf{s}_{gt}$) to supervise the reconstruction accuracy for point SDF values produced after the deformation module (**s**) and the correction module ($\mathbf{s}_f$) respectively $\mathcal{L}_{recons} = \lambda_1 ||\mathbf{s} - \mathbf{s}_{gt}||_2 + ||\mathbf{s}_f - \mathbf{s}_{gt}||_2$. Furthermore, we optimized the latent shape grid $\mathbf{Z}_s$ and applied L2 regularization to $\mathbf{Z}_s$ and $\mathbf{z}_t$ to ensure compact latent spaces [14]. Namely the total objective function is $\mathcal{L}_{total} = \mathcal{L}_{recons} + \lambda_2 ||\mathbf{z}_t||_2 + \lambda_3 ||\mathbf{Z}_s||_F$. Please see Table 1 for details regarding hyperparameters λ_1, λ_2, λ_3. All sub-networks were trained jointly.

Dataset. We considered several CHD types that cause abnormal arrangement and anatomy of the heart chambers and great vessels. Namely, we considered ventricular septal defects (VSD), which involve a hole in the septum that separates the heart's lower chambers; atrial ventricular septal defects (AVSD), characterized by both atrial and ventricular septal openings; Tetralogy of Fallot (ToF), a complex condition comprising four specific heart defects: narrowed pulmonary outflow tract, VSD, overriding aorta, where the aorta is positioned over both the left and right ventricles, and right ventricular hypertrophy. Additionally, we also considered d-transposition of great arteries (TGA), where the aorta and pulmonary artery are reversed; and pulmonary atresia (PuA), discontinuity of the pulmomary artery. The training and testing data comes from the public "imageCHD" dataset [21] that includes 110 CT images, the corresponding diagnosis (CHD type) and ground truth segmentations of 7 cardiac structures. Since the "imageCHD" dataset was initially designed for classification, the provided ground truth segmentations were extremely noisy, posing a significant challenge

Table 1. Implementation details. We selected hyperparameters based on a validation set (n = 5) split from the training set. We then trained the final models on the entire training set for 300 epochs, sampling 32768 points from ground truth SDF per iteration.

Network Design Details (D/W represents the depth and width of the MLPs used)									
	Type Encoder	Type Decoder	Deformation Net				Correction Net	SDF Decoder	Positional Encoding
	D/W	z_t dim.	D/W	D/W	Solver	Integration steps	Z_s dim.	D/W	D/W	Num. of func.
Ours	6/512	64	6/512	3/512	Euler	5	2x2x2x128	6/512	N/A	6
Ours(Disp)				N/A	N/A					
Ours(Vec)					Euler	5	512			
DeepSDF		N/A							8/512	

Training Details (Models were trained and tested on a NVIDIA A40 GPU.)							
	Loss equations	λ values	Learning Rate	Augmentation	Training time	Test time (for 128^3 SDF)
Ours	$L_{total} = \lambda_1 \lVert \mathbf{s} - \mathbf{s}_{gt} \rVert_2 + \lVert \mathbf{s}_f - \mathbf{s}_{gt} \rVert_2 + \lambda_2 \lVert \mathbf{z}_t \rVert_2 + \lambda_3 \lVert \mathbf{Z}_s \rVert_F$	1.5/1e-3/1e-4	1e-4, reduce by 0.9 every 10 epochs	20 copies per sample; B-spline (std=3 vox); Rotation (±5°); Shear (± 0.1)	14 hrs	7.28s
Ours(Disp)	$L_{total} = \lambda_1 \lVert \mathbf{s} - \mathbf{s}_{gt} \rVert_2 + \lVert \mathbf{s}_f - \mathbf{s}_{gt} \rVert_2 + \lambda_2 \lVert \mathbf{z}_t \rVert_2 + \lambda_3 \lVert \mathbf{Z}_s \rVert_F + \lambda_4 L_{disp}$	1.5/1e-3/1e-4/5e-3			13 hrs	6.61s
Ours(Vec)	$L_{total} = \lambda_1 \lVert \mathbf{s} - \mathbf{s}_{gt} \rVert_2 + \lVert \mathbf{s}_f - \mathbf{s}_{gt} \rVert_2 + \lambda_2 \lVert \mathbf{z}_t \rVert_2 + \lambda_3 \lVert \mathbf{Z}_s \rVert_F$	1.5/1e-3/1e-4			13 hrs	7.32s
DeepSDF	$L_{total} = \lVert \mathbf{s} - \mathbf{s}_{gt} \rVert_2 + \lambda_1 \lVert \mathbf{z}_t \rVert_2 + \lambda_2 \lVert \mathbf{Z}_s \rVert_F$	1e-3/1e-4			12 hrs	6.52s

in accurately differentiating the true defects from noise. Therefore, we manually processed the segmentations under the supervision of a radiology expert to remove noise and improve consistency with diagnostic information (Fig. 2.) We have finished processing 51 segmentations that covered 29, 14, 12, 5, and 5 instances of VSD, AVSD, ToF, TGA, PuA, and 5 cases that have normal heart chambers with abnormal pulmonary venous connections (APVR). We used 35 images for training, and 16 images for testing (at least 2 per type).

3   Experiments and Results

Baselines: To our knowledge, our approach is the first to enable conditioned synthesis of cardiac anatomies with different pathological topologies. However, to verify our design choices, we compared the following adapted baselines and variations. *Conditioned DeepSDF:* We adapted DeepSDF [14] to input a type code in addition to a shape code, and predict the SDF values of points based on the point coordinates concatenated with the two latent codes. *Direct Displacement Prediction:* We then compared results using the invertible NODE network against those using an MLP ($\mathcal{D}_{disp}$) to directly predict point displacements between points in the type and shape spaces. We included regularization loss for deformation smoothness, $L_{disp} = ||\nabla \mathcal{D}_{disp}(\mathbf{Z_s}(\mathbf{p}_s), \mathbf{p}_s)||_2^2$. *Position-independent encoding:* We also compared the reconstruction accuracy of using position-dependent shape latent vectors against those using the same shape latent vector for all points. We used the same method to obtain the latent codes and the same positional encoding for all baselines to ensure a fair comparison. Please see Table 1 for implementation details.

Learning the CHD Type-Specific Anatomies. We first visualize the cardiac structures extracted from the learned SDFs corresponding to various CHD types. As Fig. 3 shows, we successfully capture the typical abnormalities given the CHD type as the input, namely, a hole between the ventricles for VSD, a hole connecting the atrium and ventricles in AVSD, the missing PA outlet in PuA, the transposed aorta and PA in TGA and the shifted aortic opening and the narrowed PA trunk in ToF. We observed that certain topological characteristics within the cardiac anatomies, such as the presence of fibrous tissue forming a ring-like structure around the aortic root on the myocardium, were not always maintained in the learned type-specific templates. We note that these features are not indicative of CHD abnormalities and we suspect they may stem from inconsistencies in the ground truth anatomies. In fact, similar topological variations can be observed in the ground truth anatomies, as demonstrated in the normal case depicted in Fig. 2.

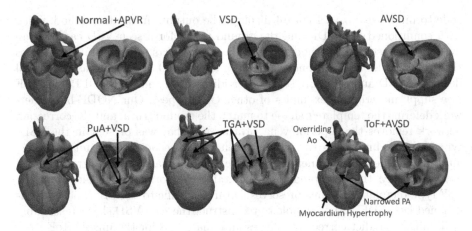

Fig. 3. Learned CHD type-specific cardiac anatomies captured the typical abnormalities given the CHD type as highlighted by arrows.

Representing Seen and Unseen Shapes. We then evaluate how well our trained network can represent both the training (seen) and testing (unseen) CHD shapes. We followed [14] to reconstruct the unseen shapes. Namely, for all baselines, we froze the network parameters and only optimized the shape latent

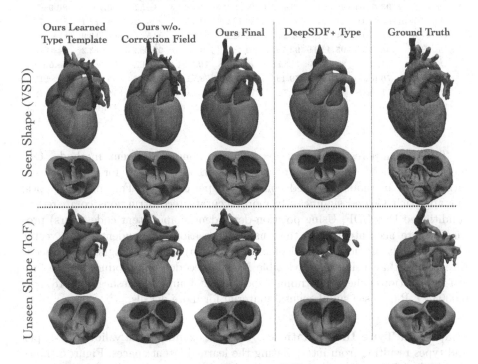

Fig. 4. Example reconstruction results for a seen VSD case and an unseen ToF case. The first and the second rows show the WH and Myo reconstruction, respectively.

code to minimize L_{total}. Figure 4 displays the outputs from our method, along with conditioned DeepSDF and the ground truth, for a seen VSD case and an unseen ToF case. Our method successfully reconstructs type-specific geometries that capture the typical abnormalities for each CHD type, such as the septal defect for VSD and the more severe overriding aorta and narrowed PA for ToF (see suppl. material for examples of other CHD types). Our NODE-based network deforms the template's shape to match the ground truth, and the correction network removes the superior vena cava branch that was present in the VSD-type template but not in the ground truth. When reconstructing unseen cases, although our deformation network matches the template with the ground truth less accurately, the final prediction maintains the correct topologies from the ToF type template and thus preserves features specific to ToF. In contrast, conditioned DeepSDF, although able to reconstruct the seen VSD shape reasonably well, fails to predict correct topologies and anatomies for the unseen ToF case. This comparison verifies the advantage of using disentangled representations of type and shape for modeling CHDs.

Table 2. Dice scores for whole heart (WH) and individual cardiac structures in reconstructed geometries compared to different baselines for seen (S) and unseen (U) cases.

	Methods	WH	LV	RV	LA	RA	Myo	Ao	PA
S	Ours	**93.5**±0.6	95.2±1	93.4±2.4	**93.2**±1.8	**94.8**±1.3	93±2.2	**90.1**±3.6	**86.9**±7.2
	Ours (Disp)	92.9±0.6	94.7±1.2	92.7±2.8	92.7±2	94.4±1.4	92.6±2.2	88.9±4.1	85.9±7.5
	Ours (Vec)	92.9±0.8	94.6±1.3	92.2±3.7	92.5±2	94.2±1.5	92.5±2.4	89.4±3.9	85.9±8
	DeepSDF	93.1±0.7	**95.4**±0.9	**93.7**±1.6	92.4±2	94.3±1.3	**93.2**±2.3	89.8±3.2	84.8±7.4
U	Ours	76±2.7	83.6±3.5	78.8±6	78.8±3.8	**79.5**±4.1	78±4.2	55.2±11.5	**58.5**±16.5
	Ours (Disp)	**76.6**±3.1	**84.7**±3.5	**79.1**±5.8	**79.2**±4.2	78.7±5.3	**80**±3.6	**55.6**±15.3	55.8±15.6
	Ours (Vec)	75.1±3.8	82.2±4.8	77.7±6	78.1±6.4	77.2±7.5	77.7±4.1	53.8±16.1	55.5±18.8
	DeepSDF	69.8±4.6	79.9±5.2	77.3±6.4	68.2±8.4	72.8±6	74.3±6.9	43±15.3	32.1±22.3

Table 2 compares reconstruction accuracy among different methods. Our method produced slightly more accurate reconstructions for seen shapes, although all methods achieved comparable performance. For unseen shapes, our approach yielded significantly more accurate reconstructions compared to conditioned DeepSDF. Using position-dependent shape latent code (Ours) produced more accurate results than using a position-independent shape vector (Vec). Directly predicting the displacements (Disp) produced slightly more accurate results than using the invertible NODE decoder but produced more face self-intersections when deforming type-specific template meshes (0.0018% vs 0.0003%). Please see supplemental material for more details.

Shape and Type Interpolations. We then examined the validity of shapes and types resulting from interpolating the learned latent spaces. Figure 5 shows the results of linearly interpolating the shape latent space or type latent space between a patient with normal heart chambers and APVR (N+APVR) and

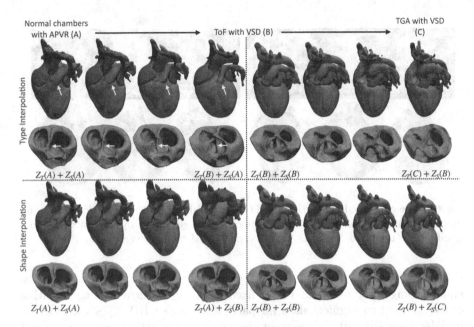

Fig. 5. Reconstruction results when interpolating the latent type or shape code from patients A to B, and B to C. Annotations show the latent code used for interpolation, e.g. $Z_T(A)$ and $Z_S(B)$ indicate using the type code from patient A and the shape code from patient B, respectively. White arrows highlight smooth interpolations that represent different severities of VSD and narrowed PA.

a ToF patient, and between patients with ToF and TGA, respectively. When fixing the shape latent code and interpolating type code from N+APVR to ToF, the APVR connection gradually breaks up, and the VSD and narrowed PA featured in ToF appear, whereas the overall shape of the heart remains roughly fixed. This demonstrates that our method can represent CHD anatomies between the states present during training. For instance, our method could be applied to generate models with varying severity of ToF or VSDs for a specific patient by interpolating the type codes between healthy and diseased states. When interpolating between two disease diagnoses, TGA and ToF, our method was also able to produce a smooth transition between the anatomies of the two types. When fixing the type code and interpolating the shape code, our method preserved the type-specific anatomies while only deforming the shape, thanks to the diffeomorphic mapping of cardiac anatomies between the type space and the shape space. This experiment further demonstrates the advantage of our disentangled representation in conveniently changing the CHD type for a specific patient and altering the shape of the heart for a specific CHD type.

CHD Type-Controlled Shape Synthesis. Since the type and shape representations of cardiac geometries with CHDs are disentangled, we can generate

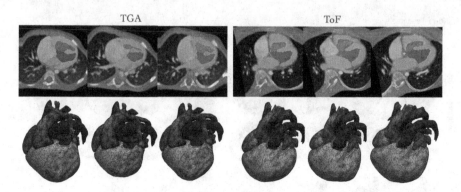

Fig. 6. Type-controlled generation for TGA and ToF. Top: Augmentation results on image-segmentation pairs. Bottom: Cardiac meshes generated by deforming a type-specific mesh template. Textures were mapped based on mesh vertex IDs.

new cardiac shapes for specific CHD types by sampling the shape latent space while keeping the type latent vector fixed. Such type-controlled shape synthesis can be used to augment data for particularly rare CHD types with limited available data (Fig. 6). To augment the image-segmentation pairs, we applied a generated deformation to the original image and segmentation. Furthermore, our method can generate meshes of specific CHDs for use in computational simulations of cardiac function. A single template mesh can be created from the type-specific signed distance function (SDF) and then deformed into different shapes in a diffeomorphic manner, thanks to our use of NODE. The generated meshes within a CHD type have semantic correspondence, as indicated by the texture map created based on global vertex IDs, and share the connectivity of the type-specific template mesh (Fig. 6).

4   Conclusion

We proposed a novel deep-learning approach that learns a type- and shape-disentangled representation of cardiac geometries for major CHD types. Given a CHD-type diagnosis, our approach accurately captures the unique cardiac anatomical abnormalities corresponding to the particular CHD and can generate CHD-type specific shapes that accurately preserve key anatomical features of each diagnosis. Our approach implicitly represented type-specific anatomies of the heart using neural SDFs and learned an invertible deformation for the shape. We demonstrated a better ability to reconstruct unseen shapes with correct typologies using this disentangled design compared to prior methods. We also demonstrated the ability to augment image-segmentation pairs for rarer CHD types and generate cardiac meshes with consistent vertex connectivity. In the future, we plan to verify our approach on a larger dataset, which includes a wider range of CHD types and shapes. Additionally, we will explore applying this approach to facilitate DL-based segmentation, classification, and computational simulations of cardiac function for CHD patients.

References

1. Marsden, A.L., Feinstein, J.: Computational modeling and engineering in pediatric and congenital heart disease. Current Opin. Pediatr. **27**, 587–596 (2015)
2. Attar, R., et al.: 3D cardiac shape prediction with deep neural networks: simultaneous use of images and patient metadata. ArXiv abs/1907.01913 (2019)
3. Beetz, M., et al.: Interpretable cardiac anatomy modeling using variational mesh autoencoders. Frontiers Cardiovasc. Med. **9**, 983868 (2022)
4. Chen, Q., Merz, J., Sanghi, A., Shayani, H., Mahdavi-Amiri, A., Zhang, H.: UNIST: unpaired neural implicit shape translation network. In: 2022 IEEE/CVF Conference on Computer Vision and Pattern Recognition (CVPR), pp. 18593–18601 (2021)
5. Chen, T.Q., Rubanova, Y., Bettencourt, J., Duvenaud, D.K.: Neural ordinary differential equations. In: Neural Information Processing Systems (2018)
6. Deng, Y., Yang, J., Tong, X.: Deformed implicit field: modeling 3d shapes with learned dense correspondence. IN: 2021 IEEE/CVF Conference on Computer Vision and Pattern Recognition (CVPR), pp. 10281–10291 (2020)
7. Dou, H., Ravikumar, N., Frangi, A.F.: A conditional flow variational autoencoder for controllable synthesis of virtual populations of anatomy. ArXiv abs/2306.14680 (2023)
8. Fonseca, C.G., et al.: The cardiac atlas project-an imaging database for computational modeling and statistical atlases of the heart. Bioinformatics **27**, 2288–2295 (2011)
9. Govil, S., et al.: Biventricular shape modes discriminate pulmonary valve replacement in tetralogy of Fallot better than imaging indices. Sci. Rep. **13**, 2335 (2023)
10. Hermsen, J.L., Roldán-Alzate, A., Anagnostopoulos, P.V.: Three-dimensional printing in congenital heart disease. J. Thoracic Disease **12**, 1194–1203 (2020)
11. Hui, K.H., Li, R., Hu, J., Fu, C.W.: Neural template: topology-aware reconstruction and disentangled generation of 3d meshes. In: 2022 IEEE/CVF Conference on Computer Vision and Pattern Recognition (CVPR), pp. 18551–18561 (2022)
12. Liu, Y., Chen, S., Zühlke, L., Black, G., Choy, M., Li, Keavney, B.: Global birth prevalence of congenital heart defects 1970–2017: updated systematic review and meta-analysis of 260 studies. Int. J. Epidemiol. **48**, 455–463 (2019)
13. Marelli, A., Ionescu-Ittu, R., Mackie, A., Guo, L., Dendukuri, N., Kaouache, M.: Lifetime prevalence of congenital heart disease in the general population from 2000 to 2010. Circulation **130**, 749–756 (2014)
14. Park, J.J., Florence, P.R., Straub, J., Newcombe, R.A., Lovegrove, S.: Deepsdf: learning continuous signed distance functions for shape representation. In: 2019 IEEE/CVF Conference on Computer Vision and Pattern Recognition (CVPR), pp. 165–174 (2019)
15. Piazzese, C., Carminati, M.C., Pepi, M., Caiani, E.G.: Statistical shape models of the heart: applications to cardiac imaging (2017)
16. Raju, A., et al.: Deep implicit statistical shape models for 3d medical image delineation. ArXiv abs/2104.02847 (2021)
17. Rodero, C., et al.: Linking statistical shape models and simulated function in the healthy adult human heart. PLoS Comput. Biol. **17**, e100851 (2021)
18. Suinesiaputra, A., et al.: Statistical shape modeling of the left ventricle: myocardial infarct classification challenge. IEEE J. Biomed. Health Inform. **22**, 503–515 (2018)
19. n: Tancik, M., et al.: Fourier features let networks learn high frequency functions in low dimensional domains. In: NeurIPS (2020)

20. Wiesner, D., Suk, J., Dummer, S., Svoboda, D., Wolterink, J.M.: Implicit neural representations for generative modeling of living cell shapes. In: International Conference on Medical Image Computing and Computer-Assisted Intervention (2022)
21. Xu, X., et al.: Imagechd: A 3d computed tomography image dataset for classification of congenital heart disease. ArXiv abs/2101.10799 (2020)
22. Yang, J., Wickramasinghe, U., Ni, B., Fua, P.: Implicitatlas: learning deformable shape templates in medical imaging. In: 2022 IEEE/CVF Conference on Computer Vision and Pattern Recognition (CVPR), pp. 15840–15850 (2022)

Automated Coronary Vessels Segmentation in X-ray Angiography Using Graph Attention Network

Haorui He[1(✉)], Abhirup Banerjee[1,2(✉)] (iD), Robin P. Choudhury[2], and Vicente Grau[1] (iD)

[1] Institute of Biomedical Engineering, Department of Engineering Science, University of Oxford, Oxford, UK
haorui.he@st-hughs.ox.ac.uk
[2] Division of Cardiovascular Medicine, Radcliffe Department of Medicine, University of Oxford, Oxford, UK
abhirup.banerjee@eng.ox.ac.uk

Abstract. Accurate coronary vessels segmentation from invasive coronary angiography (ICA) is essential for diagnosis and treatment planning for patients with coronary stenosis. Current machine learning-based approaches primarily utilise convolutional neural networks (CNNs), which heavily focus on the local vessels features and ignore the geometric structures such as the shapes and directions of vessels. This limits the machine understandability of ICA images and creates a bottleneck for improvements of computer-generated segmentation quality, including unstable generalisation ability in low contrast areas and disconnection in vascular structures. To address these issues, we propose a fusion of Graph Attention Network (GAT) and CNN to assist in the learning of global geometric information during coronary vessels segmentation. We train and evaluate the proposed method on a large-scale ICA dataset and demonstrate that combining GAT into a unified network yields improved segmentation performance. Additionally, we utilise specific metrics to demonstrate the achieved improvements, as they offer greater potential for future research and exploration.

Keywords: Coronary Vessel Segmentation · X-ray Coronary Angiography · Convolutional Neural Network · Graph Neural Network

1 Introduction

Cardiovascular diseases (CVDs) have stayed the dominant cause of death globally for the past decade and took away approximately 19.1 million lives in 2020 [16,17]. Among CVDs, coronary vessel stenosis is a major contributor as it's the primary cause of heart attacks and other complications. Thus, it is crucial to

Supplementary Information The online version contains supplementary material available at https://doi.org/10.1007/978-3-031-52448-6_20.

develop automated methods for sorting out CVD related problems, with a primary focus on precise and swift diagnosis. Invasive coronary angiography (ICA) is considered to be the gold standard imaging modality for the assessment of stenosis [9]. Current research on coronary vessels segmentation from the ICA has focused on pixel-wise manipulation such as convolutional neural networks (CNNs) to isolate vessels from the complex background for identifying stenosis automatically. CNN based models are mainly variants of U-Net [13] or fully convolutional network (FCN) [11]. The earliest fully automatic segmentation of vascular structures focused primarily on the retinal vessels, and it was shown in [12] that computer generated results can achieve human comparable delineation quality. Different CNN methods have been developed for coronary vessels segmentation with techniques including multi-scale feature extraction, cross-domain learning, and semi-supervised learning [4,6,22].

However, as the mechanism of CNN is learning local features and assigning pixel-wise labels on a regular grid, it losses global information of the vessels, which can be used to preserve the integrity of the vascular structure. This information can be learned by graph neural network (GNN) if vessels can be represented by smaller patches (nodes). In [8], a GNN was proposed for utilising the spectral connectivity by using its Laplacian matrix. Attention mechanism can also be applied in GNN [19] to create a graph attention network (GAT), which expands the use of GNN to spatial domain with more flexibility. Recently, GNN is mostly used for various computer vision tasks including semantic segmentation [20] and image recognition [18]. With CNN extracting information in the spatial domain, the fusion of CNN and GNN is inevitable and has been explored in the field of hyperspectral image classification [7,14]. However, it is rarely found for vessel segmentation, as conventional superpixel generation methods can not be applied directly on ICA images and also blocks back-propagation path.

To address this issue, in this work, we propose a novel neural network architecture with a fusion of graph attention network (GAT) and CNN for segmentation of coronary vessels in ICA, which explicitly observes the vascular structural information to optimise the segmentation performance. We have trained and evaluated our proposed model on a large ICA dataset of 323 samples, included additional metrics during evaluation to consider structure completeness, and demonstrated that the predicted vessel segmentation surpasses the base-line performance. The model carries the potential to be applied as additional component to other networks for extrapolating structural information.

2   Materials and Methods

This section introduces the experimental data, neural network architecture, training process, and the evaluation metrics. For convenience, we denote x as the ICA image for network input, y_{UNet} as the segmentation output, and G_{UNet} as the gold standard for segmentation, where $x \in [0, 255]^{C \times H \times W}$ and $y_{UNet}, G_{UNet} \in \{0, 1\}^{1 \times H \times W}$ with C, H, and W representing the number of channels, heights, and widths, respectively. The output of the GAT and the gold standard are labeled as y_{GAT} and G_{GAT} respectively, where $y_{GAT}, G_{GAT} \in \{0,1\}^{1 \times N}$ with N representing the number of generated nodes.

2.1 Study Population

The dataset used for this research is acquired from a database from Renji Hospital of Shanghai Jiao Tong University [5]. The raw data is four-frame ICA sequence, and only the third frame is manually segmented. There are 323 samples and all resized to shape $H \times W = 512 \times 512$. We take the third frame and its gold standard for every sample and keep the original train-test split ($N_{train} = 173$, $N_{val} = 82$, and $N_{test} = 68$). As all channels of the ICA have identical values, C is collapsed from 3 to 1 for more efficient training.

2.2 Overview of Network Architecture

The proposed model is described in Fig. 1 with a GAT block docked onto the four skip-connection of the UNet. It takes x as input and generates two outputs y_{UNet} and y_{GAT}. The whole network is optimised on two loss functions L_{UNet} and L_{GAT}. G_{GAT} for the calculation of L_{GAT} is generated on-the-fly from G_{UNet} using the graph construction method explained below.

UNet Setup. The UNet alone is a simple baseline model for image segmentation with VGG convolution blocks fitted. For convenience, in Fig. 1, the convolution blocks in the encoder of the UNet are coloured purple, and the convolution blocks in the decoder are coloured orange. Each block performs sequential operations consisting of convolution, batch normalisation, and rectified linear unit (ReLU) activation twice in succession. The output feature maps of the top four purple blocks are not only fed into the skip-connection, but also up-sampled to the original input shape for concatenation. Then this concatenated feature map is passed through a fully connected layer for the preparation of GAT block's input. To the decoder side of the skip-connection, the feature map GAT has generated is fused back as an extra feature for the decoder blocks. The loss function L_{UNet} is an elastic interaction-based loss function from [10] that has been proven to be preserving vessel connectivity in [6] better than a simple binary cross-entropy (BCE)-Dice loss. Mathematically, it can be expressed as:

$$L_{UNet} = \frac{1}{8\pi} \int_{\mathbb{R}^2} dxdy \int_{\mathbb{R}^2} \frac{\nabla T(x,y) \cdot \nabla T(x',y')}{|r|} dx'dy' \qquad (1)$$

where $T(x,y) = G_s + \alpha H(\phi)$. G_s is the vessel boundary in the gold standard and can be obtained by convolving the ground truth y_{UNet} with a 2D Gaussian function. $H(\phi)$ is a regularised Heaviside function that represents the vessel boundary in the generated segmentation. x and y here represent the column and row in the image grid $H \times W$. When L_{UNet} is minimised, the boundary in prediction approaches the boundary in gold standard and closes any disconnected branches by closing the boundary.

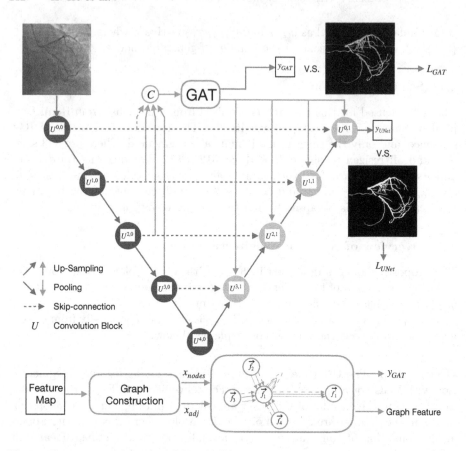

Fig. 1. The proposed fusion model with GAT and CNN for vessel delineation from the ICA images. The CNN is a simple 5-layer UNet with the connections between blocks shown in red arrows. (Color figure online)

2.3　Graph Neural Network

The implemented Graph Neural Network (GNN) and all its connections to the UNet are shown in blue colour in Fig. 1 and illustrated in detail at the bottom of the figure. The GNN can be divided into two components, the Graph Construction part and the GAT part, which we have elaborated separately.

Graph Construction. To fit the input type of GNN, graphs must be generated, which are made up of vertices and edges. In order to create the GNN that can be fitted onto a CNN for vessel segmentation from the complex background of the ICA images, the graphs are converted from concatenated output feature maps of the encoder blocks using the methods inspired from Shin et al. [15].

Node Construction. Firstly, the feature map in the original image grid of size $H \times W$ is divided into smaller sub-grid of size $2^\gamma \times 2^\gamma$ as $H = W = 512 = 2^9$ to create $(9/\gamma)^2$ number of sub-grids. In every sub-grid, the pixel with the highest value is selected as a node. In all-zero sub-grids, the pixel of the left top corner of the center 2×2 grid is selected as a node. The node's normalised pixel intensity are recorded and regarded as the third feature on top of the node location (row and column) of the original image grid.

Edge Construction. This edge construction is only applicable to nodes that are not extracted from an all-zero sub-grid. By pinning one node, edges are created between that node to every other node that lies within a certain geodesic distance threshold ω. After iterating through all the nodes, the graph construction is complete with an adjacent matrix generated from these edges.

Graph Attention Network. The Graph Attention Network (GAT) used in this architecture is applied from Veličković et al. [19], which has been widely used in a large number of spatial graph analysis. The nodes features x_{nodes} and adjacent matrix x_{adj} generated using the above methods are the inputs of GAT. Functionally, attention scores are calculated from node features and masked based on the adjacent matrix. The output of GAT, y_{GAT}, is the probability of predicting whether a node belongs to a vessel sub-grid or all-zero background sub-grid. For GAT to learn from the gold standard G_{GAT}, a weighted node cross entropy loss is used to take the class imbalance into account. Denoting $y_{GAT,i}$ and $G_{GAT,i}$ as i^{th} node from the total number of N nodes, the loss of that node can be expressed as:

$$L_{GAT} = -(1 - \beta)G_{GAT,i} \log y_{GAT,i} - \beta(1 - G_{GAT,i}) \log(1 - y_{GAT,i}) \quad (2)$$

where $\beta = \frac{1}{N} \sum_{i=1}^{N} G_{GAT,i}$. Averaging this value calculated for each node will yield the loss for GNN. The graph feature map from the penultimate attention layer of GAT is fusing back to the skip-connection for the UNet to gain graph information. The fusion is a concatenation followed by a reshape to fit the skip-connections at different scales as the graph construction method ensures it can be simply sized back to a regular grid.

2.4   Training Process

The UNet and GAT blocks cannot be trained with a single optimiser, as the loss is not able to back-propagate through the graph construction algorithm. Due to the efficiency of this algorithm, the generation of x_{nodes} and x_{adj} is limited to every k iterations. When x_{nodes} and x_{adj} are not updating, GAT is frozen with the old version of these two variables. After obtaining y_{UNet} from the network, denoising is applied to remove small pixel clusters. Both G_{UNet} and y_{UNet} are morphologically skeletonised for evaluation of structural completeness.

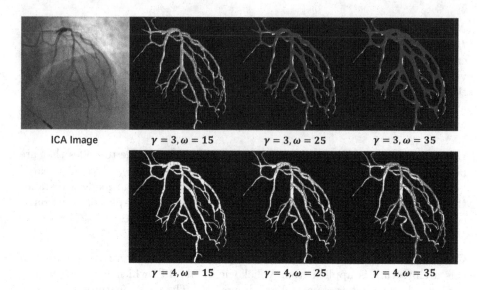

Fig. 2. Visualisation of nodes (green) and edges (blue) of G_{GAT} constructed using the graph construction method with different parameter values in overlap with G_{UNet}. The nodes in non-vessel area forms a regular shape as they are all selected at the same location from their sub-grids. γ is larger than 2, as small numerical values of γ can not extrapolate long-range features for segmentation. (Color figure online)

2.5 Evaluation Metrics

To evaluate the performance of the proposed segmentation method, three conventional metrics, namely Recall, Precision, and Dice coefficient, are used. The evaluation of skeletonised results is performed on two metrics, namely completeness C_r and correctness C_p, which are the buffered versions of Recall and Precision on the skeleton respectively and were originally used for quantifying geometry preservation in curvilinear structures in [21].

3   Experiments and Results

3.1 Experimental Settings

Training of the network is performed on an NVIDIA V100 GPU. We applied random 90°, 180°, or 270° rotation with a probability of 0.7 to both x and G_{UNet}. The Adam optimiser is used for both UNet and GNN, with $\beta_1 = 0.85$ and weight decay $= 10^{-5}$. Learning rate for UNet is set to 10^{-4}, and for GNN is 10^{-5}. The other three key hyper-parameters α, γ, and ω are set to 0.35, 4, and 25, respectively. A demonstration of how the values of γ and ω will influence the generated graph is shown in Fig. 2. It is clear that no edge is created in background area, and the number of edges increases with the increase in ω. Additionally, with higher γ value, the nodes become coarser. Thus, if the connection between nodes clearly depicts the structure of the vascular tree, it is more computationally efficient to use a higher γ with less number of nodes constructed. In Fig. 2, graphs created when ($\gamma = 3, \omega = 15$), ($\gamma = 4, \omega = 25$), and ($\gamma = 4, \omega = 35$) are all well presenting the vessel structure, but ($\gamma = 4, \omega = 25$) yields the least number of nodes and edges, which justifies the selections of these values for our current study.

Table 1. Comparison of segmentation and skeletonisation performance.

Model	Dice (%)	Recall (%)	Precision (%)	C_r(%)	C_p(%)
UNet	79.80 ± 3.46	76.54 ± 6.67	$\mathbf{83.34 \pm 4.76}$	79.59 ± 7.62	$\mathbf{74.52 \pm 6.69}$
GAT-UNet (k = 20)	79.93 ± 4.57	79.24 ± 6.88	81.28 ± 5.81	80.55 ± 8.05	73.46 ± 7.68
GAT-UNet (k = 10)	$\mathbf{80.50 \pm 4.55}$	$\mathbf{81.95 \pm 6.94}$	79.72 ± 5.99	$\mathbf{81.30 \pm 8.10}$	72.69 ± 7.56

Values represent mean ± standard deviation.

3.2   Segmentation Results

To evaluate the performance gain of using the fusion of GAT into the UNet model, we train the proposed GAT-UNet with two settings: $k = 20$ and $k = 10$. When the value of k is set to a lower number, it results in more frequent updates to the weights in a GAT model. A baseline model is built using the UNet alone with the setup mentioned in Sect. 2.2. The loss function is kept as L_{UNet}, same as the proposed method. The batch sizes of all models are set to 4 for a fair comparison. As reported in Table 1, it can be seen that the proposed GAT-UNet with $k = 10$ achieves the highest performance in terms of the Dice coefficient, recall, and C_r on the test dataset. It surpasses the base UNet by over 5% in terms of recall, implying it misses fewer correct vessels than the UNet even after the use of connectivity-preserving loss function. During training, the GAT-UNet with $k = 10$ converges its loss before 80 epochs, but simple UNet takes over 400 epochs for convergence. The precision of GAT-UNet is lower than the UNet; however, it can be attributed to the missingness of small or thin vascular structures in the gold standard. For example, in the larger white box for the first example in Fig. 3, the bottom red arrow points to two distal branches that are not segmented in the gold standard but delineated by the proposed network. The proposed GAT-UNet with $k = 10$ captures the most complete vessel structure with more false positives (in blue) that elongates from the end of those true positive vessels (in green).

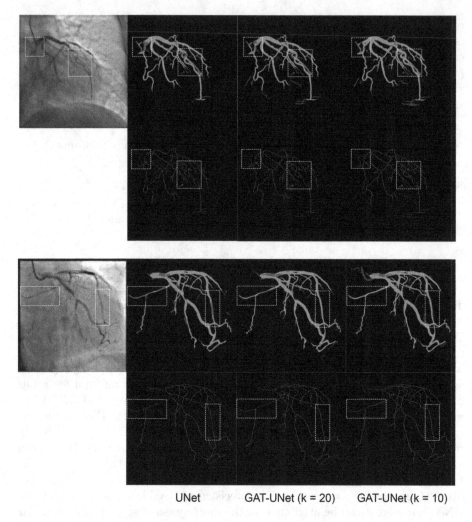

UNet   GAT-UNet (k = 20)   GAT-UNet (k = 10)

Fig. 3. Qualitative evaluation of segmentation and skeletonisation outputs from 3 different networks. True positive (TP) is shown in green, while false positive (FP) and false negative (FN) are presented in blue and red, respectively. (Color figure online)

4   Discussion and Conclusion

In this work, we have developed a novel fused network combining the GNN with CNN for automated coronary vessel segmentation from the ICA images. The qualitative and quantitative results demonstrate the improvement in segmentation and skeletonisation performances over the UNet with connectivity-preserving loss. The faster convergence of the GAT-UNet shows that the GAT block acts like an extra optimiser to guide the proposed network through training with structural information.

Since the GAT-UNet with $k = 10$ obtains the highest recall, the drop from its recall to the buffered skeleton recall C_r is not necessarily a concern. If C_r is higher than recall, it often implies that the missed parts of the vessels tend to sit in thick or main vessels, so that once the vessels are skeletonised, the missing bit takes less portion of the gold standard. This problem is most severe in the simple UNet, as visible from its C_r which is about 2.5% times higher than its recall as well as the qualitative results for both samples in Fig. 3. Particularly, in the white boxes of the second example in Fig. 3, it is observed that the GAT-UNet with $k = 10$ has the least false negative (red) area for both vessels and skeletons. The proposed network also ensures less disconnection in the vascular structure where the red arrow points in the first example of Fig. 3; whereas the same area in baseline UNet's result is filled with fragmented vascular regions. Additionally, it can be found that the improvement in qualitative results are constant and steady from UNet to GAT-UNet with $k = 10$ in all these white box regions.

Based on our experimental analyses, it is visible that the fusion of the GAT blocks in the UNet structure has been able to identify and delineate accurately even the thin vascular structures. Although it is clear that lower value of k (more frequent for GAT weights update) can lead to even better results, the current graph construction algorithm is too time consuming, as it traverses all the elements in the feature map and will increase the training time significantly if lower k is used. Our future research will aim to further improve the performance with more compact design for graph structure and explore the efficacy of this amalgamation for tasks such as semantic segmentation and downstream motion correction [1,3] and 3D coronary tree reconstruction [2].

Acknowledgments. The authors acknowledge the use of the facilities and services of the Institute of Biomedical Engineering (IBME), Department of Engineering Science, University of Oxford and the use of the University of Oxford Advanced Research Computing (ARC) facility, http://dx.doi.org/10.5281/zenodo.22558, in carrying out this work. AB is a Royal Society University Research Fellow and is supported by the Royal Society Grant No. URF\R1\221314. The works of AB, RPC, and VG were partially supported by the British Heart Foundation (BHF) Project under Grant PG/20/21/35082. The work of VG was supported by the CompBioMed 2 Centre of Excellence in Computational Biomedicine (European Commission Horizon 2020 research and innovation programme, grant agreement No. 823712).

References

1. Banerjee, A., Choudhury, R.P., Grau, V.: Optimized rigid motion correction from multiple non-simultaneous x-ray angiographic projections. In: Deka, B., Maji, P., Mitra, S., Bhattacharyya, D.K., Bora, P.K., Pal, S.K. (eds.) PReMI 2019. LNCS, vol. 11942, pp. 61–69. Springer, Cham (2019). https://doi.org/10.1007/978-3-030-34872-4_7

2. Banerjee, A., Galassi, F., Zacur, E., De Maria, G.L., Choudhury, R.P., Grau, V.: Point-cloud method for automated 3D coronary tree reconstruction from multiple non-simultaneous angiographic projections. IEEE Trans. Med. Imaging **39**(4), 1278–1290 (2020)

3. Banerjee, A., Kharbanda, R.K., Choudhury, R.P., Grau, V.: Automated motion correction and 3d vessel centerlines reconstruction from non-simultaneous angiographic projections. In: Pop, M., et al. (eds.) STACOM 2018. LNCS, vol. 11395, pp. 12–20. Springer, Cham (2019). https://doi.org/10.1007/978-3-030-12029-0_2

4. Fan, J., Yang, J., Wang, Y., et al.: Multichannel fully convolutional network for coronary artery segmentation in x-ray angiograms. IEEE Access **6**, 44635–44643 (2018)

5. Hao, D., et al.: Sequential vessel segmentation via deep channel attention network. Neural Netw. **128**, 172–187 (2020)

6. He, H., Banerjee, A., Beetz, M., Choudhury, R.P., Grau, V.: Semi-supervised coronary vessels segmentation from invasive coronary angiography with connectivity-preserving loss function. In: 2022 IEEE 19th International Symposium on Biomedical Imaging (ISBI), pp. 1–5 (2022)

7. Hong, D., Gao, L., Yao, J., Zhang, B., Plaza, A., Chanussot, J.: Graph convolutional networks for hyperspectral image classification. IEEE Trans. Geosci. Remote Sens. **59**(7), 5966–5978 (2020)

8. Kipf, T.N., Welling, M.: Semi-supervised classification with graph convolutional networks (2017)

9. Kočka, V.: The coronary angiography - an old-timer in great shape. Cor Vasa **57**(6), e419–e424 (2015)

10. Lan, Y., Xiang, Y., Zhang, L.: An elastic interaction-based loss function for medical image segmentation. In: Martel, A.L., et al. (eds.) MICCAI 2020. LNCS, vol. 12265, pp. 755–764. Springer, Cham (2020). https://doi.org/10.1007/978-3-030-59722-1_73

11. Long, J., Shelhamer, E., Darrell, T.: Fully convolutional networks for semantic segmentation. In: Proceedings of the IEEE Conference on Computer Vision and Pattern Recognition, pp. 3431–3440 (2015)

12. Maninis, K.-K., Pont-Tuset, J., Arbeláez, P., Van Gool, L.: Deep retinal image understanding. In: Ourselin, S., Joskowicz, L., Sabuncu, M.R., Unal, G., Wells, W. (eds.) MICCAI 2016. LNCS, vol. 9901, pp. 140–148. Springer, Cham (2016). https://doi.org/10.1007/978-3-319-46723-8_17

13. Ronneberger, O., Fischer, P., Brox, T.: U-Net: convolutional networks for biomedical image segmentation. In: Navab, N., Hornegger, J., Wells, W.M., Frangi, A.F. (eds.) MICCAI 2015. LNCS, vol. 9351, pp. 234–241. Springer, Cham (2015). https://doi.org/10.1007/978-3-319-24574-4_28

14. Shen, Y., Zhou, B., Xiong, X., Gao, R., Wang, Y.G.: How GNNs facilitate CNNs in mining geometric information from large-scale medical images (2022)

15. Shin, S.Y., Lee, S., Yun, I.D., Lee, K.M.: Deep vessel segmentation by learning graphical connectivity. Med. Image Anal. **58**, 101556 (2019)

16. Tsao, C.W., et al.: Heart disease and stroke statistics-2022 update: a report from the American heart association. Circulation **145**(8), e153–e639 (2022)

17. Vaduganathan, M., Mensah, G.A., Turco, J.V., Fuster, V., Roth, G.A.: The global burden of cardiovascular diseases and risk. J. Am. Coll. Cardiol. **80**(25), 2361–2371 (2022)

18. Vasudevan, V., Bassenne, M., Islam, M.T., Xing, L.: Image classification using graph neural network and multiscale wavelet superpixels. Pattern Recogn. Lett. **166**, 89–96 (2023)

19. Veličković, P., Cucurull, G., Casanova, A., Romero, A., Liò, P., Bengio, Y.: Graph attention networks (2018)
20. Xie, G.S., Liu, J., Xiong, H., Shao, L.: Scale-aware graph neural network for few-shot semantic segmentation. In: 2021 IEEE/CVF Conference on Computer Vision and Pattern Recognition (CVPR), pp. 5471–5480 (2021)
21. Youssef, R., Ricordeau, A., Sevestre-Ghalila, S., Benazza-Benyahya, A.: Evaluation protocol of skeletonization applied to grayscale curvilinear structures. In: 2015 International Conference on Digital Image Computing: Techniques and Applications (DICTA), pp. 1–6 (2015)
22. Zhang, J., Gu, R., Wang, G., Xie, H., Gu, L.: SS-CADA: a semi-supervised cross-anatomy domain adaptation for coronary artery segmentation. 2021 IEEE 18th International Symposium on Biomedical Imaging (ISBI), pp. 1227–1231 (2021)

Inherent Atrial Fibrillation Vulnerability in the Appendages Exacerbated in Heart Failure

Shaleka Agrawal[1] ⬡, Joseph Ashby[1] ⬡, Jeiyun Bai[1], Fan Feng[1], Xue J. Cai[2],
Joseph Yanni[2], Caroline B. Jones[4] ⬡, Sunil J. R. J. Logantha[2] ⬡, Akbar Vohra[2],
Robert C. Hutcheon[4], Antonio F. Corno[2], Halina Dobrzynski[2] ⬡,
Robert S. Stephenson[3] ⬡, Mark Boyett[5] ⬡, George Hart[2], Jonathan Jarvis[4],
Bruce Smaill[1], and Jichao Zhao[1](✉) ⬡

[1] Auckland Bioengineering Institute, University of Auckland, Auckland, New Zealand
j.zhao@auckland.ac.nz
[2] University of Manchester, Manchester, UK
[3] University of Birmingham, Birmingham, UK
[4] Liverpool John Moores University, Liverpool, UK
[5] University of Bradford, Bradford, UK

Abstract. Atrial fibrillation (AF) frequently accompanies heart failure (HF), however, the causal mechanism underlying their atrial electrophysiological substrates remains unclear. In the present study, we evaluated the effects of abnormal anatomical characteristics on the electrophysiology of rabbit atria with HF. Micro-CT images from adult New Zealand white rabbit hearts (n = 4 HF and n = 4 control) were acquired. Novel imaging methods were used to reconstruct atrial myofiber architecture at a high resolution of 21 μm^3/voxel for quantitative analysis of the structural remodelling. Effects of this structural remodelling on the vulnerability to atrial re-entrant waves was analysed using computer simulation. Reconstructed data showed increased chamber lumen and an uneven reduction in wall thickness across the appendages in HF. Anatomically, myofibers in epicardial walls of the appendages were identified to be circumferential, perpendicular to the pectinate muscles (PMs). The relative ratio of average PM thickness to the atrial wall was larger in HF vs. control (right atrial appendages: 3.5 versus 2.7 and left atrial appendages: 4.4 versus 3.7, p < 0.001). Furthermore, the uncoupled myofiber orientation between the PMs and atrial wall was verified using confocal microscopy at a spatial resolution of 0.2 μm^3. Computer simulations suggested (1) uncoupled myofiber orientation of the PMs and the atrial wall may increase the vulnerability to AF; and (2) decreased atrial thickness and dilated chambers may amplify the unstable substrates leading to re-entry formation in HF. Our ex-vivo to in-silico results demonstrate that uncoupled myofiber orientation in the atria is an important component of the structural remodelling, facilitating the development and maintenance of AF in HF.

Keywords: Atrial fibrillation · heart failure · atrial appendage · computer modelling · myofiber architecture

O. Camara et al. (Eds.): STACOM 2023, LNCS 14507, pp. 220–229, 2024.
https://doi.org/10.1007/978-3-031-52448-6_21

1   Introduction

Heart failure (HF) and atrial fibrillation (AF) are increasingly prevalent and associated with high morbidity, mortality, and healthcare cost. They are closely inter-related with similar risk factors and shared pathophysiology [1, 2]. The sequence of developing AF then HF, or vice versa, may impact prognosis [3]. On the one hand, AF is known to contribute to the development of HF via several mechanisms [4]. On the other hand, HF developed before AF also causes severe remodelling and worsened outcomes [3]. In general, AF is known to lead to loss of atrial systole, and reduced left ventricular (LV) filling which results in a reduction in cardiac output by 25% [4]. During AF, high heart rate also increases the likelihood of LV dysfunction, leading to a reversible form of cardiomyopathy [5–7]. During HF, impaired atrial systole causes atrial enlargement, leading to AF which also causes further atrial dilation [7]. The enlargement of the left atrium (LA) is independently known to lead to right atrium (RA) enlargement and dysfunction [8]. These studies partly illustrate the complex interactions of HF and AF, however, the role of the atrial appendages in these mechanisms has not been well characterised. Hence, exploring the mechanisms underlying the genesis of AF under HF in the atria at different physiological and pathological levels is of the utmost importance.

In recent years, imaging techniques have facilitated better understanding of cardiac disorder mechanisms. Micro-computed tomography (micro-CT) is an advanced imaging technique that allows the acquisition of high-resolution three-dimensional (3D) micrometer-scale images, leading to improved cardiac phenotyping in animals and humans [10, 11]. Iodine-enhanced micro-CT distinguishes cardiac tissue types [12, 13], allowing assessment of clustered myocyte orientation [11, 13]. Coupled with modeling tools, these advanced imaging modalities provide deeper insights into disease mechanisms.

In this study, we used two imaging modalities to quantitatively analyse the structural remodelling of right and left atrial appendages (RAA and LAA). Micro-CT reconstructed images from HF and control atria were loaded into a heart-specific computer model to investigate the effects of structural remodelling on atrial electrical function.

2   Methods

2.1   Experimental Animal Model

Eight adult male New Zealand white rabbits (4 sham-operated control and 4 HF subjects, weighing between 2.5 to 3 kg and aged approximately 3 months) were used to create an experimental model of congestive HF, after aortic valve destruction and thoracic aortic banding, as described in previous publication [14]. Animal care and experimental protocol complied with the Animals Act (Scientific Procedures 1986 and subsequent amendments) under 40/3135.

2.2 Imaging

Micro-CT
After euthanisation, hearts were perfusion-fixed in situ, maintaining normal atrial filling. They were then immersed in phosphate-buffered saline and stained with iodine potassium iodide (I_2KI) in aqueous solution for 3 to 5 days (n = 4 control and n = 4 HF) [11]. Micro-CT images of the hearts ($\approx$ 21μm^3 voxel resolution) were captured using the Metris X-Tec 320 KV bay system at the Henry Moseley X-Ray imaging facility (University of Manchester, United Kingdom) (Fig. 1).

Confocal Microscopy
Atrial appendages from four control adult, male New Zealand white rabbits were preserved in 4% paraformaldehyde, flash-frozen in liquid nitrogen, and stored at –80 °C. Thin strips (200 μm) of transmural tissue were cryo-sectioned from the junction of the atrial wall and pectinate muscles. These sections were then stained with wheat germ agglutinin, mounted, and imaged using an OLYMPUS FV1000 confocal microscope ($\approx$ 0.2 μm^3 pixel resolution) at the Biomedical Imaging Research Unit, University of Auckland [15].

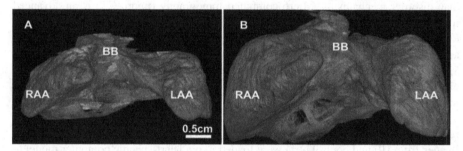

Fig. 1. Anterior views of typical reconstructed rabbit atria. A, Control and **B,** HF. There was a marked increase in overall atrial epicardial dimensions in HF with an evident dilatation of right and left atrial appendages (RAA and LAA). BB – Bachmann's bundle.

2.3 Analysis of AF Substrates

Image Processing and Segmentation
In Amira 5.33, images were visualized, and 2D slices were down-sampled to create an isotropic image volume using a series of 3D image processing steps, including:1) 3D interpolation and smoothing, 2) sequential 3D erosion and dilation operations to fill small holes in epicardial or endocardial surfaces and remove disconnected "islands", and 3) removal of any remaining unconnected regions. To enclose the atrial chambers, surfaces were added for the superior vena cava, inferior vena cava, coronary sinus (CS), pulmonary veins, mitral valve, and tricuspid valve. For atrial appendage segmentation, LAA was identified as a region between the lateral and anterior walls of the left

atrium [16]. RAA was defined as the anterolateral triangular section of the right atrium, demarcated by the crista terminalis (CT) on the endocardium [18, 19].

Extraction of Myofiber Orientation

Structure tensors were constructed from image volumes and then smoothed [19]. 3D gradient information and local fiber alignment were identified using eigen analysis [20]. The fiber fields were also smoothed, and inclination angle of fiber axis were defined [21].

Modelling Electrical Activity

Spread of electrical activation in myocardial tissue was simulated by solving the mono-domain reaction-diffusion equation on a 3D voxel-based finite difference grid ($50 \times 50 \times 50$ μm^3 resolution) and solved using Euler method. To introduce electrical hetero-geneity, a method previously outlined by Seemann [22] and Aslanidi [23] was employed. Courtemanche, Ramirez, and Nattel's model [24] was modified based on voltage clamp recordings from rabbit [25], preserving regional variation in action potential morphol-ogy matching experimental sheep data [26]. Axially anisotropic electrical properties were assigned to connected atrial myocyte bundles and regions characterised by ordered myofiber organisation. Tissue conductivities along and transverse to the myofibers PMs and crista terminalis (CT), were set to 11.7 μS and 0.9 μS, respectively, resulting in conduction velocities (CVs) of $\approx$ 1.3 m/s and $\approx$ 0.6 m/s along and across the CT and PMs. CV in the rest of the atrial appendage was $\approx$ 0.75 m/s. The estimated CV range for working atrial myocardium was between 0.32 m/s to 1.03 m/s [27, 28].

Pooled t-testing was performed, as the two sample sets were of the same size. All results are represented as mean $\pm$ SD, with p < 0.05 being statistically significant.

3 Results

3.1 Effects of HF on Atrial Geometry

HF animals showed evidences of congestive HF with reduced ejection fraction [14, 28], implying structural remodelling in the atria. To investigate possible remodelling of the atrial appendages, micro-CT images were used to measure 1) tissue and cavity volumes; 2) wall and lumen areas; and 3) radial wall thickness. As is shown in Fig. 2A, in HF the cavity volume of atrial chambers was 3–4 times bigger, while tissue volume was only increased by ~ 2 times (p = 0.108). Especially, RAA chambers were dilated more than LAA and it was also observed that the RA was more prone to collapse during imaging. Therefore, cross-sectional areas of walls and lumens of RAA/LAA were measured (Fig. 2B). Under HF, cross-sectional wall area was increased by 30–40%, while luminal area was increased up to 300%. RAA had greater lumen increase, whereas LAA had more wall increase. In detail, RAA lumen increased from 0.04 cm^2 in control to 0.11 cm^2 in HF (~2.5 times; p = 0.0001), and RAA wall enlarged from 0.08 cm^2 to 0.12 cm^2 (~1.5 times; p = 0.005). LAA lumen expanded from 0.28 cm^2 to 0.34 cm^2 (~1.2 times; p = 0.008), while LAA wall was increased from 0.17 cm^2 to 0.24 cm^2 (~1.4 times; p = 0.0003). In the atrial appendages, the thickness of walls and PMs of RAA/LAA was estimated (Fig. 2C). The thickness ratio of PM to atrial wall in the RAA region increased from 2.7 (1.44:0.53 = 2.7) in the control condition to 3.51 for the HF condition (1.30:0.37 = 3.51; p = 0.002), and in the LAA region from 3.65 (2.45:0.67 = 3.65) to 4.44 (2.23:0.50 = 4.44; p = 0.0001).

3.2 Myofiber Architecture in Atrial Appendages

The fiber tracking technique was used to trace and characterise major bundles in the atrial appendages for both study groups. Analysis of the fiber angle data at millimeterscale, provided local myofiber orientation maps of the appendages, Fig. 3A–B. Figures 3C–D demonstrates that the fiber architecture of atrial appendage anatomy was mostly preserved across different groups, although exact location and orientation of the bundles varied between hearts. Architecture of RAA and LAA were dominated by the PMs originating from the CT (Figs. 3A–B). This organisation was unaffected by atrial dilatation under the HF condition. Fiber tracking at the superior margin of the LAA demonstrated dominant epicardial fiber orientation to be circumferential, while the PMs ran vertically to the mitral annulus (Figs. 3B–D). Further investigation in the RAA showed a discontinuous diffusion of PM fibers into the atrial free wall, as shown in Figs. 4A–C. This was further confirmed via high-resolution confocal imaging. As is shown in Fig. 5, myocyte arrangement in the atrial wall (Fig. 5A) vertical to PM (Fig. 5B) and the discontinuous arrangement of myocytes at the junction between the PM and the atrial free wall (Fig. 5C).

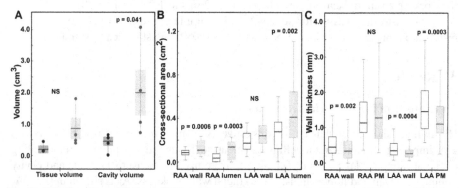

Fig. 2. Effects of heart failure (HF) on the atrial structure. A, Changes of tissue and cavity volumes under HF (red), as compared to control (gray). n = 4 in both groups. **B,** Cross-sectional areas of walls and lumens for right and left atrial appendage (RAA and LAA), respectively. Six equally spaced planes were selected from each heart. n = 24 in both groups. **C,** Radial wall thickness was estimated at the regions of RAA wall, the pectinate muscles (PMs) of RAA (RAA PM), LAA wall, and LAA PM. 36 regions were randomly selected from each heart. n = 144 in both groups. Individual data were plotted together with mean and standard error. NS – Not significant.

3.3 Vulnerability to Re-entry in Atrial Appendages

Upon applying S_1-S_2 pacing protocol to the superior margin of the PMs on the epicardium, a unidirectional conduction block was observed. A unidirectional block is a prerequisite for re-entrant waves as depicted in Fig. 6. This was particularly prominent in the RAA wall. A sharp diffusion of the electrical wave at the junctions of the appendage wall and multiple PMs led to the formation of re-entry circuits, especially in the RAA. This susceptibility to re-entry was amplified under HF, correlated with the increased atrial wall thinning and dilated atrial chambers.

4 Discussion

The role of appendage remodelling has a great pathophysiological significance in. AF [29], however there is relatively little information available. This study presents new information regarding the structure and function of atrial appendages, by combining advanced imaging techniques with computer simulation tools to understand the underlying mechanism of HF-induced AF. We have observed atrial dilatation, a greater reduction in atrial wall thickness compared to associated PM, and increased cross-section area of appendage wall and appendage lumen in the HF atria. The uncoupled myofiber orientation between the PMs and atrial wall was identified to increase the vulnerability to AF.

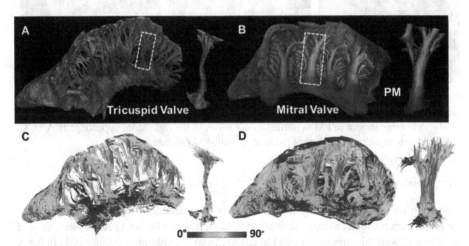

Fig. 3. Myofiber structure in control right and left atrial appendages (RAA and LAA) of control viewed from within atrial cavities. A, Volume-rendered 3D reconstructions of RAA, **B,** LAA with isolated typical pectinate muscles (PMs), indicated with white rectangles. Corresponding orientations of myofiber bundles were displayed in **C,** and **D**. The colour spectrum indicates fiber inclination angle with respect to the valve plane, where blue is in-plane and red is normal to the plane.

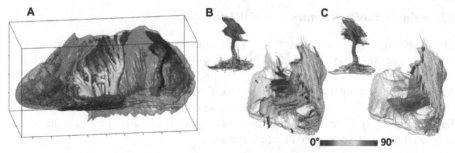

Fig. 4. High-resolution reconstruction of myofiber paths in pectinate muscles (PMs) and adjacent atrial wall from control. A, Fiber tracking was applied to segmented regions of the right atrial appendage (RAA), where the PM of interest is shown in red, with green, and blue representing the inferior and superior arterial wall respectively. **B,** A typical isolated PM, and **C,** Associated atrial wall with their respective myofiber paths. Colour spectrum indicates fiber inclination angle with respect to the valve plane, where blue is in-plane and red is normal to the plane.

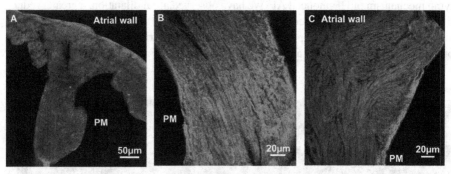

Fig. 5. Representative confocal images of the atrial wall, pectinate muscle (PMs) and its junction from control atria. A, Myocyte arrangement in the atrial wall is in-plane with the valve. **B,** Myocyte arrangement in PM is normal to the valve, in the right atrial appendage (RAA). **C,** Discontinuous myocyte arrangement in the atrial wall-PM junction.

4.1 Atrial Substrates in HF

A striking finding of this study is that while both appendages were dilated, however, the relative increase in cavity compared to tissue mass was higher in RAA, with greater reduction in wall thickness compared to LAA. Possibly explaining thinner PMs in RAA compared to LAA. This suggests that RAA was more prone to significant structural remodelling, and subsequent electrical remodelling. RA enlargement has been shown to be independently associated with AF susceptibility with preserved ejection fraction (EF), however, there is comparatively little information on RA remodelling under reduced EF [9, 32].

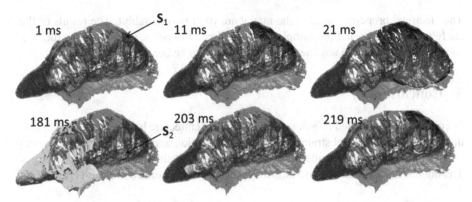

Fig. 6. Vulnerability to re-entrant waves in control. In the right atrial appendage (RAA), an S_1 stimulus ($t = 1$ ms) was applied to initiate a bidirectional conduction wave ($t = 11$ ms and 21 ms). An S_2 stimulus ($t = 181$ ms) led to a unidirectional conduction block ($t = 203$ ms and 219 ms), a condition important for the genesis of re-entrant atrial arrhythmias. Depolarised, refractory and non-depolarised regions are rendered in red, yellow and blue, respectively.

4.2 *In-Silico* Analysis Shows HF Substrates Favour AF Susceptibility

Simulations primarily demonstrated the susceptibility to re-entry due to significant fiber orientation change between the PMs and adjacent atrial walls. This orientation disparity has previously been observed in human hearts [18, 34], however, this is the first study where their effects were studied *in-silico*. In our study, this complex myofiber architecture at the superior margin of RAA, was accompanied by thinning of PMs and atrial walls under HF, potentially explaining their propensity to develop AF. A similar trend was not observed in the LAA, possibly due to comparatively reduced atrial wall and PM junction thinning.

Previous studies have also identified sustained re-entry substrates in HF due to delays in activation times, caused as a consequence of wall thinning, or as a result of the formation of a unidirectional block in the CT [32], but the predominance of each mechanism is poorly understood [33]. Abnormalities in calcium homeostasis are another hallmark of HF, there is a downregulation of intracellular calcium levels which is governed by ryanodine receptors and sarcoplasmic reticulum Ca^{2+}-ATPase [34]. This decrease in intracellular calcium may also contribute to the perpetuation of AF [35].

Our observations highlight the significance of underlying structural abnormalities at the junction of PMs and atrial wall, causing atrial flutter even in patients with grossly normal atria [36]. These structural abnormalities are potentially magnified in HF, predisposing subjects with HF to AF. This multi-scale study of these microscopic substrates has unveiled the importance of atrial appendages in the onset of AF in HF.

Image-based computer models are an important platform to better understand the mechanisms responsible for new-onset and sustained AF. However, spatial dependence of electrical properties of the tissue, and subsequent changes due to HF, must also be incorporated into the model to enable comprehensive analyses. Data quantifying the changes in electrical properties of atrial tissue for rabbits, with and without HF, is not available in literature, though this issue is currently being addressed in our laboratory.

The electrical properties used in the model are from normal rabbit. The results of this study are promising, but the sample size would need to be increased and the methods tested in humans before any medical applications can be considered.

5 Conclusion

In conclusion, uncoupled orientation of myofiber architecture between PMs and epicardial layers provides a substrate for re-entrant activation and the instability of re-entry may be amplified in HF by alterations in tissue mass distribution in dilated appendages. Treatments that forestall heart dilation may prevent atrial dilation and benefit HF patients.

References

1. Mulder, B.A., Rienstra, M., Van Gelder, I.C., et al.: Update on management of atrial fibrillation in heart failure: a focus on ablation. BMJ J. Hear. **108**(6), 422–428 (2022)
2. Carlisle, M.A., Fudim, M., DeVore, A.D., et al.: Heart failure and atrial fibrillation, like fire and fury. JACC Heart. Fail. **7**(6), 447–456 (2019)
3. Smit, M.D., Moes, M.L., Maass, A.H., et al.: The importance of whether atrial fibrillation or heart failure develops first. Eur. J. Heart Fail. **14**(9), 1030–1040 (2014)
4. Deedwania, P.C., Lardizabal, J.A.: Atrial fibrillation in heart failure: a comprehensive review. Am. J. Med. **123**(3), 198–204 (2010)
5. Nerheim, P., Birger-Botkin, S., Piracha, L., et al.: Heart failure and sudden death in patients with tachycardia-induced cardiomyopathy and recurrent tachycardia. Circulation **110**(3), 247–252 (2004)
6. Delgado, V., Bax, J.J.: Atrial functional mitral regurgitation: from mitral annulus dilatation to insufficient leaflet remodeling. Circ. Cardiovasc. Imaging, **10**(3), 6239 (2017)
7. Gopinathannair, R., Chen, L., Chung, M., et al.: Managing atrial fibrillation in patients with heart failure and reduced ejection fraction a scientific statement from the American heart association. Circ. Arrhythmia Electrophysiol. **14**, 688–705 (2021)
8. Ikoma, T., Obokata, M., Okada, K., et al.: Impact of right atrial remodeling in heart failure with preserved ejection fraction. J. Card. Fail. **27**(5), 577–584 (2021)
9. Lombardi, C.M., Zambelli, V., Botta, G., et al.: Postmortem micro-CT of small fetuses and hearts. Ultrasound Obstet. Gynecol. **44**(5), 600–609 (2014)
10. Hutchinson, J.C., Arthurs, O., Ashworth, M., et al.: Clinical utility of postmortem microcomputed tomography of the fetal heart: diagnostic imaging vs macroscopic dissection. Ultrasound Obstet. Gynecol. **47**(1), 58–64 (2016)
11. Stephenson, R.S., Jones, C.B., Guerrero, R., et al.: High-Resolution contrast-enhanced micro-CT to identify the cardiac conduction system in congenitally malformed hearts: valuable insight from a hospital archive. JACC Cardiovasc. Imaging **11**(11), 1706–1712 (2018)
12. Simcock, I.C., Hutchinson, J., Shelmerdine, S., et al.: Investigation of optimal sample preparation conditions with potassium triiodide and optimal imaging settings for microfocus computed tomography of excised cat hearts. Am. J. Vet. Res. **81**(4), 326–333 (2020)
13. Aslanidi, O.V., Nikolaidou, T., Zhao, J., et al.: Application of micro-computed tomography with iodine staining to cardiac imaging, segmentation, and computational model development. IEEE Trans. Med. Imaging **32**(1), 8–17 (2013)
14. Corno, A. F., Cai, X., Jones, C. B., et al.: Congestive heart failure: experimental model. Front. Pediatr. **1**, 68060 (2013)

15. Agrawal, S., Ralugun, G., Ashton, J., et al.: Structural basis of atrial arrhythmogenesis in metabolic syndrome. In: Computing in Cardiology Conference, vol. 46, (2019)
16. Barbero, U., Ho, S.Y.: Anatomy of the atria : a road map to the left atrial appendage. Circ. Arrhy. Electro. 28(4), 347–354 (2017)
17. Ueda, A., McCarthy, K.P., Sánchez-Quintana, D., et al.: Right atrial appendage and vestibule: further anatomical insights with implications for invasive electrophysiology. Europace 15(5), 728–734 (2013)
18. Ho, S.Y., Sánchez-Quintana, D.: The importance of atrial structure and fibers. Clin. Anatomists 22(1), 52–63 (2009)
19. Weickert, J.: Anisotropic diffusion in image processing (1998). B.G. Teubner Stuttgart
20. Axelsson, M., Svensson, S.: 3D pore structure characterisation of paper. Pattern Anal. Appl. 13(2), 159–172 (2010)
21. Butters, T.D., Aslanidi, O.V., Zhao, J.: A novel computational sheep atria model for the study of atrial fibrillation. Interface Focus 3(2), 20120067 (2013)
22. Seemann, G., Höper, C., Sachse, F.B., et al.: Heterogeneous three-dimensional anatomical and electrophysiological model of human atria. Philos. Trans. R. Soc. Math. Phys. Eng. Sci. 364(1843), 1465–1481 (2006)
23. Aslanidi, O.V., Colman, M., Stott, J., et al.: 3D virtual human atria: a computational platform for studying clinical atrial fibrillation. Prog. Biophys. Mol. Biol. 107(1), 156–168 (2011)
24. Courtemanche, M., Ramirez, R.J., Nattel, S.: Ionic mechanisms underlying human atrial action potential properties: insights from a mathematical model. Am. J. Physiol. Heart Circulatory Physiol. 275(1), 301–321 (1998)
25. Shannon, T.R., Wang, F., Puglisi, J., et al.: A mathematical treatment of integrated Ca dynamics within the ventricular myocyte. Biophys. J. 87(5), 3351–3371 (2004)
26. Gray, R.A., Pertsov, A.M., Jalife, J.: Incomplete reentry and epicardial breakthrough patterns during atrial fibrillation in the sheep heart. Circ. 94(10), 2649–2661 (1996)
27. Zhao, J., Butters, T., Zhang, H., et al.: An image-based model of atrial muscular architecture: effects of structural anisotropy on electrical activation. Circ. Arrhy. Electro. 5(2), 361–370 (2012)
28. Logantha, S., Cai, X., Yanni, J., et al.: Remodeling of the Purkinje network in congestive heart failure in the rabbit. Circ. Hear. Fail. 14(7), E007505 (2021)
29. Yu, H.T., Lee, J.S., Kim, T.H., et al.: Advanced left atrial remodeling and appendage contractile dysfunction in women than in men among the patients with atrial fibrillation: potential mechanism for stroke. J. Am. Heart Assoc. 5(7), e003361 (2016)
30. Melenovsky, V., Hwang, S.J., Redfield, M.M., et al.: Left atrial remodeling and function in advanced heart failure with preserved or reduced ejection fraction. Circ. Heart Failure 8(2), 295–303 (2015)
31. Beigel, R., Wunderlich, N., Ho, S., et al.: The left atrial appendage: anatomy, function, and noninvasive evaluation. JACC Cardiovasc. Imaging 7(12), 1251–1265 (2014)
32. Siddiqui, A.U., Daimi, S., Gandhi, K., et al.: Crista terminalis, musculi pectinati, and taenia sagittalis: anatomical observations and applied significance. ISRN Anat. 2013, 1–6 (2013)
33. Zhao, Q., Zhang, H., Tang, Y., et al.: Relationship between autonomic innervation in crista terminalis and atrial arrhythmia. J. Cardiovasc. Electro. 20(5), 551–557 (2009)
34. Luo, M., Anderson, M.E.: Mechanisms of altered Ca^{2+} handling in heart failure. Circ. Res. 113(6), 690–708 (2013)
35. Fong, S.P.T., Agrawal, S., Gong, M., et al.: Modulated calcium homeostasis and release events under atrial fibrillation and its risk factors: A Meta-Analysis. Front. Cardiovasc. Med. 702 (2021)
36. Mizumaki, K., Fujiki, A., Nagasawa, H., et al.: Relation between transverse conduction capability and the anatomy of the crista terminalis in patients with atrial flutter and atrial fibrillation: analysis by intracardiac echocardiography. Circulation 66(12), 1113–1118 (2002)

Two-Stage Deep Learning Framework for Quality Assessment of Left Atrial Late Gadolinium Enhanced MRI Images

K M Arefeen Sultan[1,2]([✉]), Benjamin Orkild[1,3,4], Alan Morris[1],
Eugene Kholmovski[5,6], Erik Bieging[5,7], Eugene Kwan[3,4], Ravi Ranjan[3,4,7],
Ed DiBella[3,5], and Shireen Elhabian[1,2]

[1] Scientific Computing and Imaging Institute, University of Utah,
Salt Lake City, UT, USA
[2] Kahlert School of Computing, University of Utah, Salt Lake City, UT, USA
u1419693@utah.edu
[3] Department of Biomedical Engineering, University of Utah,
Salt Lake City, UT, USA
[4] Nora Eccles Harrison Cardiovascular Research and Training Institute,
University of Utah, Salt Lake City, UT, USA
[5] Department of Radiology and Imaging Sciences, University of Utah,
Salt Lake City, UT, USA
[6] Department of Biomedical Engineering, Johns Hopkins, Baltimore, MD, USA
[7] Division of Cardiology, University of Utah, Salt Lake City, UT, USA

Abstract. Accurate assessment of left atrial fibrosis in patients with atrial fibrillation relies on high-quality 3D late gadolinium enhancement (LGE) MRI images. However, obtaining such images is challenging due to patient motion, changing breathing patterns, or sub-optimal choice of pulse sequence parameters. Automated assessment of LGE-MRI image diagnostic quality is clinically significant as it would enhance diagnostic accuracy, improve efficiency, ensure standardization, and contributes to better patient outcomes by providing reliable and high-quality LGE-MRI scans for fibrosis quantification and treatment planning. To address this, we propose a two-stage deep-learning approach for automated LGE-MRI image diagnostic quality assessment. The method includes a left atrium detector to focus on relevant regions and a deep network to evaluate diagnostic quality. We explore two training strategies, multi-task learning, and pretraining using contrastive learning, to overcome limited annotated data in medical imaging. Contrastive Learning result shows about 4%, and 9% improvement in F1-Score and Specificity compared to Multi-Task learning when there's limited data.

Keywords: Self-supervision · Multi-task learning · Image Quality Assessment

1 Introduction

Atrial fibrillation (AF) is currently the most common cardiac arrhythmia in the United States, with 3 to 5 million people affected, and is expected to affect more

O. Camara et al. (Eds.): STACOM 2023, LNCS 14507, pp. 230–239, 2024.
https://doi.org/10.1007/978-3-031-52448-6_22

than 12 million by 2030 [1]. It has been shown that atrial fibrosis is closely linked to the development and recurrence of AF disease after treatment [2,3]. Currently, catheter ablation is a popular treatment for AF, targeting and eliminating the areas of the heart (i.e., *fibrotic* tissues) responsible for irregular electrical signals by creating targeted lesions or *scars* in these regions. Hence, fibrosis quantification plays a crucial role in guiding catheter ablation procedures. However, the rate of success of catheter ablation is relatively low, with over 40% of patients returning to AF within 1.5 years of ablation [4]. Therefore, it is imperative to understand and address the shortcomings of the treatment of AF.

Late Gadolinium Enhancement (LGE) MRI is a widespread technology used to image and quantify myocardial fibrosis and scarring. LGE-MRI can be performed in atrial fibrillation subjects prior to a catheter ablation treatment to provide the patient's atrial geometry and fibrosis pattern [5,6]. The specific geometry and fibrosis patterns of patients are derived from the LGE-MRI images, which can be used for pre-ablation planning or for creating patient-specific simulations [7,8]. However, LGE-MRI images exhibit variability in quality, with diagnostic accuracy affected by factors such as noise, resolution, intensity level, and patient-related characteristics [9–11].

The clinical significance of quality assessment in LGE-MRI scans for fibrosis quantification is important as it enhances diagnostic accuracy, ablation planning, and treatment guidance. By discarding poor-quality images, clinicians can base their decisions on more reliable results. However, manual quality assessment is laborious and prone to errors making it non-scalable. Automating this process can optimize workflows and can save resources. This automation necessitates the identification of image features predictive of diagnostic quality, a task where deep learning can be instrumental. However, the effectiveness of deep learning depends on the availability of a substantial amount of annotated images, as it has a significant appetite for labeled data. Manual annotation of LGE datasets is a laborious and time-consuming expert-driven task, leading to a scarcity of labeled data. In this paper, we propose a two-stage deep-learning approach that is inspired by the mental process of a radiologist manually evaluating the diagnostic quality of an LGE MRI image for fibrosis quantification. The proposed method is specially designed to mitigate the limited training data scenario in LGE-MRI quality assessment, leveraging contrastive learning for pretraining and multi-task learning for regularization. The contributions of this paper are summarized as follows.

- Introducing a segmentation network to identify relevant left atrium slices in the LGE-MRI scan instead of relying on manual selection of the slices.
- Leveraging a multi-task learning framework to learn quality assessment of atrial fibrosis for the LGE datasets jointly with identification of the atrial blood pool.
- Showcasing the impact of label supervision in contrastive learning to promote learning a discriminative representation in the embedding space.
- Benchmarking the effectiveness of the two approaches using a limited labeled dataset.

2   Related Works

Several automatic methods have been proposed to assess MRI image quality in various anatomical sites. Xu et al. [12] proposed a mean teacher model with ROI consistency to assess the image quality of fetal brain MRI. Liao et al. [13] used a 2D U-Net to jointly assess image quality and segment fetal brain MRIs. Although they labeled the slices as only good/bad, the assessment of image quality encompasses several interconnected factors. This method also required manual labeling of each training MRI slice, which is tedious and time-consuming.

To address the limitation of extensive labeling image datasets, various self-supervised learning methods have been introduced. Chen et al. [14] proposed using contrastive loss by maximizing the agreement between augmented views of the same image and minimizing the agreement between different images. These learned representations are then applied to downstream tasks, such as image classification and object detection. While this method leverages a vast number of negative samples to learn useful representations, it still relies on a considerable amount of unlabeled data. Khosla et al. [15] proposed label supervision within contrastive loss allowing for more efficient representation learning within limited labeled data.

3   Methods

We introduce a two-stage deep-learning approach that emulates the cognitive process of a radiologist manually assessing the diagnostic quality of LGE MRI images for fibrosis quantification. The method consists of two stages: (1) left atrium detection stage and (2) quality assessment stage. Figure 1 depicts the proposed two-stage approach.

3.1   Left Atrium Detection Stage

The primary objective of the LA detection stage is to identify the specific slices within the LGE MRI scan that contain the left atrium. This information is then utilized in the second stage to focus on the relevant region of interest, disregarding any background artifacts that might otherwise interfere with the quality assessment task. Here, we employ a UNet model [16] to generate segmentation masks for the left atrial blood pool in the MRI scan. The predicted masks undergo a sigmoid function to transform pixel values within the range of 0 to 1. Subsequently, we apply a threshold parameter, t, to determine the minimum probability required for a pixel to be considered as part of the left atria. If any pixel value is larger than the threshold, we classify the corresponding slices as containing the left atrium, which is then utilized in the next stage. We exclude the slices where there is no left atrium detected.

3.2   Quality Assessment Stage

This stage incorporates a deep network that effectively maps the image slices to a diagnostic quality score. Here, we assess the effectiveness of two training strategies in addressing the challenge of limited annotated data for the quality assessment task: multi-task learning and pretraining using contrastive learning. We extract the features from a pre-trained network, specifically ResNet34 [17]. The pre-trained weights are based on the Imagenet dataset. After that, we project the embedding space to a latent space using three attribute classifier modules. Each attribute classifier focuses on a fine-grained attribute that is relevant to the quality assessment task.

Image Quality Attributes: We propose the myocardium nulling, sharpness, and enhancement of aorta and valve attributes that are clinically relevant to the diagnostic quality of fibrosis assessment of LGE-MRIs. *Myocardium nulling* compares the intensity of the left ventricular (LV) myocardium to the left ventricular blood pool. A score of 1 means the intensity of the LV myocardium is higher than that of the blood pool, while a score of 5 means the intensity of the LV myocardium is well-nulled and similar to that of the signal-free background. *Sharpness* reflects the amount of blurring in the borders of the LA and other anatomical structures. A score of 1 means there is a severe blurring of the cardiac chambers, while a score of 5 means the edges of the cardiac chambers are well-defined. The third attribute is the *enhancement of fibrous tissue - aorta and valve*. When the wall of the aorta and the cardiac valves show enhancement, this implies the scan also has good quality for detection of fibrosis in the left atrium.

For *fibrosis quality assessment*, experts also score quality of fibrous tissues. A score of 5 defines a high contrast between enhanced fibrous structures and blood pool, whereas 1 defines the absence of enhanced fibrous structures.

These three attributes, along with fibrosis assessment in the left atrium, were all given scores from 1–5 by trained observers and were transformed to binary scores: non-diagnostic and diagnostic (see Sect. 4.1). Next, we discuss the training strategies into 3 subsections, Baseline QA, Multi-Task QA, and pretraining using supervised contrastive learning.

3.2.1   Baseline QA

Attribute Classifier Module: The objective of the attribute classifier submodule is to use image features extracted from the image encoder to classify the three attributes, myocardium nulling, sharpness, and enhancement of aorta and valve, into non-diagnostic and diagnostic. Let these attributes be denoted as a vector $\mathbf{a} = [a_{mn}, a_s, a_{eat}]$ where a_{mn}, a_s, and a_{eat} denote the score of myocardium nulling, sharpness, and enhancement of aorta and valve, and the score $a_* \in \{0, 1\}$. These 3 attribute classifiers are trained using BCE loss, described in Eq. 1.

QA Module: The purpose of this module is to predict the quality of fibrosis assessment. Since the 3 attributes correlate highly with fibrosis assessment as

Table 1. Correlation of three attributes with the main measure of interest - quality of fibrosis assessment.

	Myocardium nulling	Sharpness	Enhancement of aorta and valve
Quality of fibrosis assessment	0.74	0.79	0.76

shown in Table 1, the module for Quality Assessment (QA), shown in Fig. 1, concatenates the output of these 3 attribute classifiers discussed above and then predicts $y_{\hat{qa}}$. The network is trained by minimizing a binary cross-entropy (BCE) loss, $\mathcal{L}_{qa}$, which combines the attribute loss and QA loss. Then the supervised loss of Baseline QA is defined by

$$\mathcal{L}_{qa} = \text{BCE}([a_{mn}, a_s, a_{eft}, y_{qa}], [\hat{a_{mn}}, \hat{a_s}, \hat{a_{eft}}, y_{\hat{qa}}]) \tag{1}$$

where $\hat{a_*}$, and $y_{\hat{qa}}$ are the prediction of the network.

3.2.2 Multi-task QA

Decoder Module: The Decoder module is responsible for transforming the embedding space of the encoder into a segmentation mask. The goal of this module is to segment the blood pool, which helps to provide discriminative features for the scoring task. With the segmentation of the blood pool and the QA network, the overall architecture focuses on the area of the left atrium, shown in Fig. 2, where the scoring plays an important role. The segmentation loss is defined by

$$\mathcal{L}_{seg} = \text{DICE}(\mathbf{M}, \hat{\mathbf{M}}) \tag{2}$$

where $\mathbf{M}$ is the groundtruth LA segmentation mask, and $\hat{\mathbf{M}}$ is the network-generated mask.

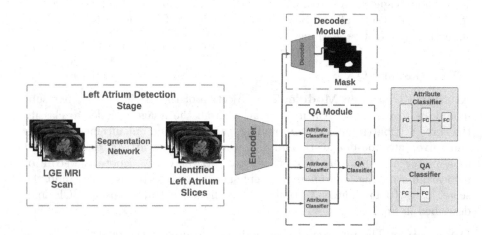

Fig. 1. Two-stage deep learning architecture.

Overall, the loss for training the Multi-Task QA network is defined by

$$\mathcal{L} = \mathcal{L}_{qa} + \mathcal{L}_{seg} \tag{3}$$

3.2.3 Pre-training Using Supervised Contrastive Learning

To partially address this problem, we run another experiment of pretraining the encoder by utilizing supervised contrastive learning [15]. The motivation behind this approach is to enhance our model's representation learning capabilities so that the same class representation comes closer and pushes the representations of different classes apart. The loss is defined by:

$$\mathcal{L}_{\text{sup}} = \sum_{i \in I} \frac{-1}{|P(i)|} \sum_{p \in P(i)} \log \frac{\exp(z_i \cdot z_p / \tau)}{\sum_{a=1}^{2N} \mathbb{1}_{[a \neq i]} \exp(z_i \cdot z_a / \tau)} \tag{4}$$

Here, $P(i)$ is the set of all positives in the augmented view batch corresponding to the anchor (i). For each anchor i, there is 1 positive pair and $2(N-1)$ negative pairs. z_* denotes the projected embedding space of the encoder. For our objective, we optimize the loss function below

$$\mathcal{L}_s^* = \mathcal{L}_{sup_{mn}} + \mathcal{L}_{sup_s} + \mathcal{L}_{sup_{eat}} + \mathcal{L}_{sup_{qa}} \tag{5}$$

where $\mathcal{L}_{sup_{mn}}$, $\mathcal{L}_{sup_s}$, $\mathcal{L}_{sup_{eat}}$, and $\mathcal{L}_{sup_{qa}}$ denotes the supervised contrastive loss of myocardium nulling, sharpness, enhancement of aorta and valve, and quality for fibrosis assessment. After that, we freeze the encoder weights and perform downstream task of supervised learning using the QA module.

4 Results

4.1 Dataset

Our dataset includes 196 scans of labeled data for the QA task and 900 scans that have the blood pool segmentations. All of the scans were acquired as in [3], with a resolution of $1.25 \times 1.25 \times 2.5 \, \text{mm}^3$, approximately 15 min after gadolinium administration, with a 3D ECG-gated, respiratory navigated gradient echo inversion recovery pulse sequence. The 196 scans were divided and scored by experts. These 196 scans have a class imbalance problem because most scans are in the 2 to 4 range. To address this problem, we have transformed the scores of all attributes, including the fibrosis assessment score, into two different labels: diagnostic and non-diagnostic. Scans with a score of ≥ 3 are designated as diagnostic, denoted 1, while less than 3 is non-diagnostic and denoted as 0.

4.2   Data Preprocessing and Augmentation

The dataset was split into train, test, and validation sets. The test set contained 20 patient scans. The remaining scans were divided into training and validation sets in a 90:10 training-to-validation ratio. Each scan was a stack of 2D slices of axial view that was selected by the first stage in Fig. 1 to contain the left atrium. Images were resized to 128×128 using linear interpolation. Since we are assessing image quality, we used geometric transformations, Random flip, perspective transform, shift, scale, and rotate, which were applied with a probability of 0.5 each during training. All data were normalized before being passed through the network.

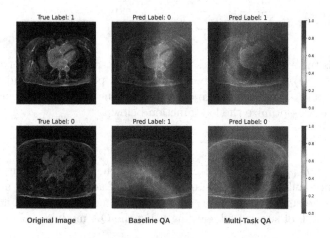

Fig. 2. Result of HiResCAM [18] between our Baseline QA and Multi-Task QA model output on 2 different test scans. We show the critical cases where our QA model fails to focus on (blue color) the relevant areas for scoring, whereas, on the contrary, the Multi-Task QA model is able to do so. The "True Label" depicts the ground truth label: diagnostic (1) and non-diagnostic (0), whereas the "Pred Label" denotes the model prediction. A red arrow in the original image shows the location of the left atrium.

4.3   Summary of Experiments

During training, the model was validated using MSE error across all slices on the validation set and later on evaluated on the test set. For the test set, we performed score predictions across all slices and subsequently report the mode for each scan. The performance is measured by Precision, Recall, F1-Score, and Specificity. The details of the training are given below.

Left Atrium Detection Training: We trained the U-Net network of this stage on the 900 segmentation masks of the blood pool. The network was trained using

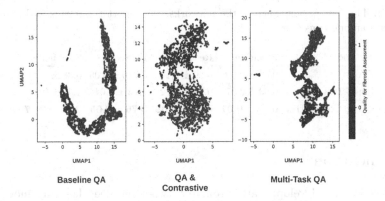

Fig. 3. UMAP [19] Visualization of embedding space representation of encoder of the three models. We report the 1st iteration's visualization among the 5 iterations.

the Adam optimizer with a learning rate of $\eta = 0.001$. The batch size was set to 128. We then used this network to predict the relevant slices of 196 scans for the later stage.

Baseline QA Training: For this model, we only considered the QA module. The network was trained for 40 epochs using the Adam optimizer with a learning rate of $\eta = 0.001$ and with an early stopping criteria of patience 7. The batch size was set to 128. A cosine annealing learning rate scheduler was used to reduce the learning rate throughout training, bringing stable optimization and faster convergence.

Multi-task QA Training: For this model, we considered both the QA and Decoder module in our architecture. The network was trained using the same configuration as the QA Training strategy. To interpret that the Multi-Task QA model focuses on the relevant areas, we report the gradient-based visual explanation from Draelos et al. [18] in Fig. 2.

QA & Contrastive Training: For this model, we follow the 2-step process discussed by Khosla et al. [15]. In the first step, we pre-train the encoder for 100 epochs with a batch size of 512, since contrastive learning benefits from more negative samples during training. We used LARS optimizer [20] with a learning rate of $\eta = 0.5$. In the 2nd step, we freeze the encoder and trained the QA module with the same configuration as the QA Training strategy discussed above. Since gradient-based explanation is not suitable for contrastive learning due to the fact that contrastive learning does not require explicit class labels during training, therefore we reported the embedding representation by utilizing UMAP [19] approach presented in Fig. 3. We can see that the QA & Contrastive based model formed more tight clusters compared to other methods, thus forming cluster alignment. The results of the experiments are shown in Table 2.

Table 2. The test performance of the networks. We report the means and standard deviations among runs here.

Method	Precision	Recall	F1-Score	Specificity
QA	0.60 ± 0.05	0.90 ± 0.15	0.710 ± 0.05	0.38 ± 0.20
Multi-Task QA	0.63 ± 0.07	0.90 ± 0.12	0.73 ± 0.04	0.44 ± 0.24
QA & Contrastive	$\mathbf{0.65 \pm 0.07}$	$\mathbf{0.94 \pm 0.08}$	$\mathbf{0.76 \pm 0.05}$	$\mathbf{0.48 \pm 0.17}$

5   Conclusion

In this study, we developed an automated two-stage deep learning model for quality assessment scoring of LGE MRI images using limited labeled data. Our approach includes a segmentation network to identify relevant left atrium slices, and two strategies: Multi-task learning (reconstruction and quality assessment) and supervised contrastive learning (quality assessment). Multi-task learning benefited from segmentation information, improving quality assessment accuracy through knowledge transfer. Supervised contrastive learning effectively learned informative representations from both contrastive learning and label supervision, yielded the most promising results for quality assessment among all other methods. In conclusion, our research provides valuable insights for quality assessment of the left atrium in LGE MRI images with limited labeled data. Hybrid approaches, combining both techniques, could lead to even more robust and accurate quality assessment models for LGE MRI image analysis and other medical imaging domains, addressing data scarcity challenges in medical imaging advancements.

Acknowledgements. The National Institutes of Health supported this work under grant numbers R01HL162353.

References

1. Colilla, S., Crow, A., Petkun, W., Singer, D.E., Simon, T., Liu, X.: Estimates of current and future incidence and prevalence of atrial fibrillation in the U.S. adult population. Am. J. Cardiol. **112**(8), 1142–1147 (2013)
2. ElMaghawry, M., Romeih, S.: DECAAF: emphasizing the importance of MRI in AF ablation. Glob. Cardiol. Sci. Pract. **2015**, 8 (2015)
3. Marrouche, N.F., et al.: Association of atrial tissue fibrosis identified by delayed enhancement MRI and atrial fibrillation catheter ablation: the DECAAF study. JAMA **311**(5), 498–506 (2014)
4. Verma, A., et al.: Approaches to catheter ablation for persistent atrial fibrillation. New England J. Med. **372**(19), 1812–1822 (2015). https://doi.org/10.1056/NEJMoa1408288
5. Oakes, R.S., et al.: Detection and quantification of left atrial structural remodeling with delayed-enhancement magnetic resonance imaging in patients with atrial fibrillation. Circulation **119**(13), 1758–1767 (2009)

6. Caixal, G., et al.: Accuracy of left atrial fibrosis detection with cardiac magnetic resonance: correlation of late gadolinium enhancement with endocardial voltage and conduction velocity. Europace **23**(3), 380–388 (2021)
7. Lange, M., Kwan, E., Dosdall, D.J., MacLeod, R.S., Bunch, T.J., Ranjan, R.: Case report: personalized computational model guided ablation for left atrial flutter. Front. Cardiov. Med. **9**, 893752 (2022)
8. McDowell, K.S., et al.: Methodology for patient-specific modeling of atrial fibrosis as a substrate for atrial fibrillation. J. Electrocardiol. **45**(6), 640–645 (2012)
9. Gräni, C., et al.: Comparison of myocardial fibrosis quantification methods by cardiovascular magnetic resonance imaging for risk stratification of patients with suspected myocarditis. J. Cardiovasc. Magn. Reson. **21**, 1–11 (2019)
10. Flett, A.S., et al.: Evaluation of techniques for the quantification of myocardial scar of differing etiology using cardiac magnetic resonance. JACC: Cardiov. Imaging **4**(2), 150–156 (2011)
11. Spiewak, M., et al.: Comparison of different quantification methods of late gadolinium enhancement in patients with hypertrophic cardiomyopathy. Eur. J. Radiol. **74**(3), e149–e153 (2010)
12. Xu, J., et al.: Semi-supervised learning for fetal brain MRI quality assessment with ROI consistency. In: Martel, A.L., et al. (eds.) MICCAI 2020. LNCS, vol. 12266, pp. 386–395. Springer, Cham (2020). https://doi.org/10.1007/978-3-030-59725-2_37
13. Liao, L., et al.: Joint image quality assessment and brain extraction of fetal MRI using deep learning. In: Martel, A.L., et al. (eds.) MICCAI 2020. LNCS, vol. 12266, pp. 415–424. Springer, Cham (2020). https://doi.org/10.1007/978-3-030-59725-2_40
14. Chen, T., Kornblith, S., Norouzi, M., Hinton, G.: A simple framework for contrastive learning of visual representations. In: International Conference on Machine Learning, pp. 1597–1607. PMLR (2020)
15. Khosla, P., et al.: Supervised contrastive learning. In: Advances in Neural Information Processing Systems, vol. 33, pp. 18661–18673 (2020)
16. Ronneberger, O., Fischer, P., Brox, T.: U-Net: convolutional networks for biomedical image segmentation. In: Navab, N., Hornegger, J., Wells, W.M., Frangi, A.F. (eds.) MICCAI 2015. LNCS, vol. 9351, pp. 234–241. Springer, Cham (2015). https://doi.org/10.1007/978-3-319-24574-4_28
17. He, K., Zhang, X., Ren, S., Sun, J.: Deep residual learning for image recognition. In: Proceedings of the IEEE Conference on Computer Vision and Pattern Recognition, pp. 770–778 (2016)
18. Draelos, R.L., Carin, L.: Use HIResCAM instead of Grad-CAM for faithful explanations of convolutional neural networks. arXiv preprint arXiv:2011.08891 (2020)
19. McInnes, L., Healy, J., Melville, J.: Umap: uniform manifold approximation and projection for dimension reduction. arXiv preprint arXiv:1802.03426 (2018)
20. You, Y., Gitman, I., Ginsburg, B.: Large batch training of convolutional networks. arXiv preprint arXiv:1708.03888 (2017)

Automatic Landing Zone Plane Detection in Contrast-Enhanced Cardiac CT Volumes

Lisette Lockhart[1]($\boxtimes$), Xin Yi[1], Nathan Cassady[2], Alexandra Nunn[1], Cory Swingen[2], and Alborz Amir-Khalili[1]

[1] Circle Cardiovascular Imaging, Calgary, AB, Canada
{lisette.lockhart,xin.yi,alexandra.nunn,alborz.amir-khalili}@circlecvi.com
[2] Boston Scientific, Marlborough, MA, USA
{nathan.cassady,cory.swingen}@bsci.com

Abstract. Left atrial appendage closure (LAAC) is a common procedure whereby a device is implanted to prevent blood clots in the heart from entering the bloodstream. Selecting an appropriate occlusion device size is essential for the procedure's success, which involves detecting the device landing zone from a contrast-enhanced computed tomography angiography (CTA) image, and measuring its diameter for sizing. Automating landing zone contour detection is challenging due to complexity of locating a 2D contour in a 3D volume. In this paper, we automate landing zone plane detection using convolutional neural networks (CNNs). We reformulate plane detection as a volumetric heatmap regression problem, for which CNNs are well-suited. Our proposed approach can accurately detect the center of the landing zone as well as the cross-section orientation necessary for measuring device size. Compared to other segmentation-based methods, our approach removes the need for costly annotations. The proposed approach can be applied to any volumetric plane detection or pose estimation problem, which we also demonstrate in the context of aortic valve plane detection in 4D flow phase-contrast magnetic resonance angiography (PCMRA) volumes.

Keywords: Left Atrial Appendage Closure · Plane Detection · Pose Estimation

1 Introduction

For patients with non-valvular atrial fibrillation, left atrial appendage closure (LAAC), whereby an occlusion device is implanted via catheter into the left atrial appendage (LAA), has become a standard procedure when oral anticoagulants cannot be used [1, 8]. The landing zone is defined as the LAA cross-section where the device will be implanted. The measurements (e.g. diameter) of the landing zone are used to select an appropriate device size for implantation.

Since most manual landing zone contouring time is spent navigating to the appropriate LAA cross-section from standard multiplanar reformation (MPR)

O. Camara et al. (Eds.): STACOM 2023, LNCS 14507, pp. 240–249, 2024.
https://doi.org/10.1007/978-3-031-52448-6_23

views, we focus on automating landing plane detection to improve LAAC planning workflow. From the detected LAA pose, the landing zone can be contoured manually or automatically using real-time 2D segmentation methods.

An efficient approach for plane detection is useful in several medical image analysis applications. For example, midsagittal plane detection in head magnetic resonance (MR) volumes for surgical brain planning [24], valve plane detection in cardiac computed tomography angiography (CTA) volumes for mitral, aortic, and tricuspid valve repair/replacement [2], and detection of standard planes in fetal ultrasound (US) to assess fetal brain development [13]. It could be especially useful for time-resolved acquisitions that require processing of many volumes, e.g. valve tracking in 4D flow cardiac MR acquisitions to measure blood flow [9], or in multiphase cardiac CTA acquisitions to measure heart valve function [22].

In this paper, we perform automated landing zone centroid localization and plane detection. Our contributions are as follows:

1. We remove the need for dense 3D segmentation annotations and use only sparse 2D contour annotations.
2. We reformulate the centroid localization and plane detection problem as a volumetric landmark heatmap regression.
3. We validate the approach on private datasets of CTA volumes acquired for LAAC planning and PCMRA volumes acquired for aortic valve disease.

2   Related Work

Most work on automation for LAAC planning has focused on LAA segmentation, then landing zone detection and device size measurements from the LAA volume [11,12,20,21]. A more direct approach for landing zone detection [18] in CTA volumes segmented the LAA in two sections corresponding to proximal and distal regions of the landing zone plane. A major limitation of methods that utilize LAA segmentation, either as initialization to an image processing method or as part of a deep learning approach, is the time and cost to generate 3D segmentation annotations. Many samples are required to capture the high variability of LAA size, shape, and morphology. However, automating 2D contouring in a 3D volume with deep learning approaches is extremely challenging due to the high imbalance of foreground (i.e. contour) to background voxels. We therefore focus on automating detection of the landing zone plane, which can make use of established convolutional neural network (CNN) architectures.

Similarly in 4D flow MR imaging, most works focus on aorta segmentation [3,6,10,17,19]. Plane placement was performed without segmentation in [9] by sampling 3D patches with 5 channels, predicting scalar values for probability of containing a vessel and plane parameters, then aggregating results across all patches to get the final predicted vessel plan, and adjusting the plane center according to maximum through-plane velocity. Our approach removes the need for inferring on many patches and outputs an accurate plane center directly.

CNNs have been utilized to detect standard planes in medical image volumes. In [13], an iterative transformation network detected the transformation of an initial 2D plane to a standard plane of interest in 3D US volumes. In [24], a deep CNN utilizing Hough space was proposed for plane detection in CT volumes for brain surgical planning. These methods do not localize the structure of interest unless three intersecting planes are detected.

In this work, we instead reformulate plane detection as a 3D landmark heatmap regression problem. Deep neural networks have shown success for plane detection [4,14,23,26,27] and pose estimation [5,15,16,25] via heatmap regression.

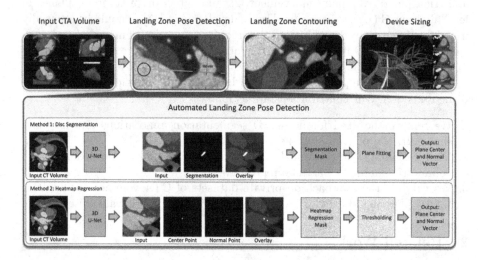

Fig. 1. Overview of the LAAC planning procedure depicting the process of data loading, LAA landing zone pose detection, contouring of the landing zone, and preoperative simulation of the LAAC device fit after the appropriate size is determined. The landing zone contour is shown in green. Our automated landing zone pose detection solutions are presented illustrating two proposed deep learning approaches that employ a U-Net style architecture as a backbone. (Color figure online)

3 Methodology

The LAAC planning pipeline is shown in Fig. 1, where a CTA volume is fed as input. The landing zone detection is the part we aim to automate with the proposed methods. The contouring and device sizing steps shown may be performed using traditional image processing techniques and fall outside this paper's scope. We automate landing zone detection with popular deep CNNs using only sparse landing zone contour annotations. We explore volumetric segmentation and heatmap regression for plane detection, extracting the landing zone contour centroid and plane coordinates from each output for comparison.

3.1 Method 1: Disc Segmentation

A disc can be formulated by filling in the landing zone contour in the landing zone plane. We aim to segment this disc as the objective of the first approach.

The voxels inside the disc were treated as foreground and remaining voxels to background. The landing zone centroid is computed as the centroid of predicted foreground voxels, and the plane coordinates is derived by fitting a plane to the foreground. To address the foreground-background class imbalance, the contour points were dilated and connected into a disc using morphological operations. A volumetric convolutional network (3D U-Net [7]) was trained to segment the landing zone disc using cross-entropy and soft dice loss functions.

3.2   Method 2: Heatmap Regression

In this approach, we define two landmark points: the landing zone contour center and the normal point that is of a fixed distance d to the center point along the normal direction of the plane pointing distally, towards the LAA tip. We further applied a linear weighting scheme to the neighbouring pixels of the landmarks. The heatmap ground truth of the regression is two 3D volumes that contain a sphere with a radius r located around a landmark. The values inside the sphere are determined by distance to the sphere center with a boundary value of 0 and normalized to 1 at the center. A 3D U-Net was trained to regress the heatmaps using mean squared error. Detected points were determined by argmax across each output channel. The landing zone plane was then reconstructed from these points with simple vector math.

A faster variation of this method was implemented using a two-stage network. The first stage performs heatmap regression of the contour center on a lower resolution of the full input volume. A second stage performs heatmap regression of contour center and normal point on a single higher resolution cropped patch of the input volume centered around the landmark predicted in the first stage.

4   Experiments

4.1   Dataset Overview

The landing zone dataset consists of 273 CTA volumes collected from various sites across the U.S., provided by Boston Scientific. For each CTA volume, an LAAC planning expert verified that contrast agent was visible in the LAA structure then contoured the ground truth landing zone manually. A total of 153 (58.6%), 51 (19.6%), and 57 (21.8%) CTA volumes were used for training, validation, and testing, respectively.

The aortic valve plane dataset consists of 200 4D flow cases collected from a private site. Cases were either bicuspid aortic valve, tricuspid aortic valve, or control. Magnitude and velocity components were combined and averaged over time to form 3D PCMRA volume as in [3]. Since only manual aorta segmentation annotations were available for this dataset, the aortic valve plane was computed from aorta segmentation by first computing the skeleton, locating its tip closest to the left ventricle as the center point, then taking the closest point to 10 mm along the skeleton as the normal point. 140 (70%), 30 (15%), and 30 (15%) PCMRA volumes were used for training, validation, and testing, respectively.

4.2   Implementation Details

In the disc segmentation and single-stage heatmap regression experiments, the contour points were dilated twice with a 3×3 structuring element before filling along the plane, giving an empirically chosen thickness of 5 voxels. The 3D U-Net [7] consisted of five downsampling blocks, each with three dilated convolutions, ReLU activation, and batch normalization layers. There were six channels at the top of the U-Net, which doubled at every downsampled block. Striding and upsampling decreased and increased the feature maps size, respectively. Each network has a final convolutional layer of two output channels with either an argmax (disc segmentation) or ReLU activation (heatmap regression) layer.

All networks presented here, unless stated otherwise, were trained with a batch size of 4 and ADAM optimizer with initial learning rate of 0.001 for 10,000 (disc segmentation) or 5,000 (heatmap regression) epochs. The following augmentations were applied during training: mirror (along all axes), translation ([–20, 20] voxels along each axis), rotation ([–50°, 50°]), zoom ([0.8, 1.2] volume size), and gamma intensity (power of [0.25, 2]). Volumes were center-cropped, resampled to $128 \times 128 \times 128$ voxels with isotropic voxel resolution of 1.5 mm/voxel, and then fed as network input.

For the landing zone disc segmentation experiment, foreground and background cross-entropy loss was weighted by 1000 and 1, respectively. Network weights at the epoch with the highest landing zone disc dice score on the validation set were saved and used for testing.

For the single-stage heatmap regression experiments, the distance d between the landing zone center point and normal point was 10 mm and the heatmap sphere radius r was 10 voxels. Foreground (any nonzero voxel) and background mean squared error loss was weighted by 1000 and 1, respectively. Translation augmentations were not applied during training. Network weights at the epoch with the lowest L2 distance between predicted and ground truth landmark points on the validation dataset were saved and used for testing.

In the two-stage heatmap regression experiments, the first stage 3D U-Net had the following modifications from the single-stage version: four downsampling blocks, four channels at the first layer, one output channel, heatmap sphere radius r of 6 voxels, and foreground and background mean squared error loss weighting of 10 and 1. The second stage 3D U-Net used foreground and background mean squared error loss weighting of 2 and 1. Volumes were center-cropped (first stage) or cropped to the predicted center point (second stage), resampled to $64 \times 64 \times 64$ voxels with isotropic voxel resolution of 3.0 mm/voxel (first stage) or 1.5 mm/voxel (second stage), and then fed as network input. The two networks were trained separately then evaluated end-to-end.

Networks were implemented in TensorFlow 1.13.1 with CUDA 10 and trained on an NVIDIA TITAN Xp GPU with 12GB of memory.

4.3   Experimental Results

We evaluate the performance of the approaches using distance metrics, angle error, and run time. The definition of the metrics are as follows:

- **Centroid Distance:** L2 distance between the predicted and ground truth landing zone contour centroid voxel coordinates.
- **Out-of-Plane Distance:** L2 distance between the predicted centroid and the ground truth plane.
- **Contour Points Distance:** average L2 distance between each ground truth landing zone contour point and the predicted plane.
- **Angle Error:** angle between ground truth and predicted plane normal vectors.
- **Run Time:** time needed to run the network, compute the centroid and plane fitting (for disc segmentation method), and compute the normal vector (heatmap regression method) on CPU (AMD Threadripper 2920x processor).

Table 1. Results for disc segmentation and single-stage and two-stage heatmap regression for landing zone and aortic valve plane detection. Best results in bold.

Metric	Disc Segmentation Mean ± Std	Single-stage Mean ± Std	Two-stage Mean ± Std
Landing Zone Plane			
Centroid Dist. (mm)	4.52 ± 3.32	6.47 ± 10.82	**3.81 ± 3.43**
Out-of-Plane Dist. (mm)	3.47 ± 3.24	5.08 ± 8.29	**2.49 ± 2.60**
Contour Points Dist. (mm)	4.77 ± 1.92	6.15 ± 10.74	**3.37 ± 2.51**
Angle Error (°)	37.14 ± 13.88	20.89 ± 11.30	**19.15 ± 12.48**
Run Time (s)	1.86 ± 0.26	1.98 ± 1.00	**0.74 ± 0.23**
Aortic Valve Plane			
Centroid Dist. (mm)	-	9.91 ± 13.27	**9.34 ± 11.71**
Out-of-Plane Dist. (mm)	-	7.40 ± 10.55	**6.59 ± 8.91**
Angle Error (°)	-	22.25 ± 14.52	**22.01 ± 14.29**
Run Time (s)	-	1.73 ± 0.11	**0.26 ± 0.16**

For landing zone detection, defining acceptance criteria for distance and angle errors is challenging due to the variable LAA anatomy. Although, given the size of the LAA, Centroid Distance errors < 10 mm can be generally considered as having localized a valid region within the LAA. In the disc segmentation method, the centroid distance was 4.52 mm. However, the fitted plane angle was high in most cases. In the heatmap regression method, the centroid distance was higher at 6.47 mm, but the angle error was significantly lower than the disc segmentation method. Both methods require only a single network pass and have a run time of < 2 s per volume. The two-stage heatmap regression requires two network passes, but has a significantly faster run time of 0.74 s per volume. By narrowing in on a region-of-interest, all distance errors were reduced. However, the angle error remains similar to single-stage heatmap regression.

For aortic valve plane detection, the distance and angle errors were similar between single-stage and two-stage heatmap regression methods and higher than landing zone detection. However, run time was drastically reduced from 1.73 s to 0.26 s using the two-stage approach.

5 Discussion

The disc segmentation approach can, in theory, directly extract the landing zone contour from the network output (i.e. the perimeter of the predicted disc). However, segmentation errors can cause the disc to be warped such that not all the points are along a plane, disconnected such that the full contour cannot be extracted, have variable thickness, or oversegment the disc along the LAA depth, as shown in Fig. 2.

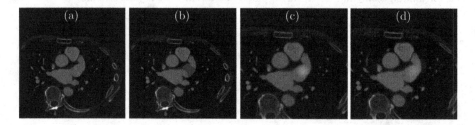

Fig. 2. Results of the disc segmentation and single-stage heatmap regression approaches for landing zone detection. (a–b): an example depicting a failure mode in disc segmentation with the ground truth overlay shown in red (a) and the output prediction overlay shown in blue (b). This example shows clear over-segmentation of the LAA landing zone, which contributes to poor landing zone detection performance. (c–d): results of the single-stage heatmap regression network with the ground truth (c) and the predicted (d) heatmaps overlaid; contour center and normal point heatmaps are color-coded as red and green, respectively, in both images. (Color figure online)

The disc segmentation resulted in high plane angle error, as shown in Table 1. The trained network did not accurately segment the disc, with an average dice score of only 16.53%. This is likely due to the lack of discernible features at the landing zone location. This uncertainty results in the segmented disc being thicker than the training data. As a result, the regressed plane orientation is almost arbitrary, despite the centroid being correctly estimated. Formulating the disc segmentation problem as a disc regression task may improve the results. Also, more robust post-processing methods (e.g. weighted regression) could be applied to improve the plane fitting, but would increase run time.

The heatmap regression method can fully describe the plane with two sets of voxel coordinates, including the point along the plane corresponding to the structure of interest (i.e. the landing zone centroid). The key advantage of this pose estimation with the heatmap regression method is its ability to correctly determine the direction of the normal vector; such that it points towards the LAA tip. This is important as this direction vector is used for visualization and downstream processing (e.g. measuring the LAAC device depth, as in Fig. 1). Beyond application in LAAC, this capability is required where the direction of the plane is important (e.g. valve plane pose estimation for flow quantification).

Unlike the disc segmentation approach, heatmap regression had severe outliers with Centroid Distance ranging from 37.98–60.76 mm. This was caused by a failure mode in the heatmap regression method where other heart structures of similar appearance (e.g. left upper pulmonary vein (LUPV)) cause high uncertainty resulting in foreground activation across the entire image. Using the two-stage approach, the likelihood of mistaking the LUPV for the landing zone was reduced given the smaller field-of-view in the second stage.

In the proposed heatmap regression formulation, the landing zone center point is defined to be the middle of the ground truth LAA landing zone contour and the normal point is defined relative to the center point. This is different than most landmark regression problems, where salient image features (e.g. points along object boundaries) are localized, so both local (image intensity) and regional (object structure) information can be utilized for regression. It is interesting that the network learned to regress landmarks without significant local intensity variations for guidance. This suggests that the CNN might be capable of detecting these landmarks using a combination of other nearby salient features (e.g., the left circumflex artery and the coumadin ridge) without annotations. Image regions contributing to the regression output merits further investigation. However, the high angle error even for the two-stage approach indicates difficulty predicting an accurate normal point relative to the center point.

Although the aorta has a more standard morphology than LAA, the distance and angle errors were both high. This could be caused by lower resolution and contrast of PCMRA volumes. However, the distance errors were not reduced by cropping to a smaller RoI as in the landing zone experiments. Since the pulmonary valve is located close to the aortic valve, it is likely the network has difficulty distinguishing these two structures. This supports the suspected heatmap regression failure mode of detecting another heart structure with similar appearance. Training a network to detect the pulmonary and aortic valves concurrently may help to identify the valve of interest. Outputting one channel per landmark enables simple extraction of the landmark coordinates from the volume. However, the run time and memory would increase significantly as more landmarks are localized.

6   Conclusions

We propose two approaches for landing zone plane detection from CTA volumes that utilize simple landing zone contour annotations instead of costly 3D segmentations. Disc segmentation can accurately detect the landing zone centroid, but not the plane, as evident by high angle errors. On the other hand, heatmap regression can accurately detect the landing zone centroid and plane. By implementing heatmap regression as a two-stage approach, the run time is suitable for intraoperative clinical workflows. As such, this method can automate the tedious process of device sizing for LAAC planning and is a viable solution for other automated plane detection and pose estimation tasks. The approach could be improved to increase learning capacity of the plane normal vector as well as extended to predict multiple planes of interest, which is common in LAAC planning and 4D flow image analysis. It could be extended to also contour the structure of interest along the detected plane.

References

1. Alkhouli, M., Ellis, C.R., Daniels, M., Coylewright, M., Nielsen-Kudsk, J.E., Holmes, D.R.: Left atrial appendage occlusion: current advances and remaining challenges. JACC: Adv. **1**, 100136 (2022)
2. Astudillo, P., et al.: Automatic detection of the aortic annular plane and coronary ostia from multidetector computed tomography. J. Intervent. Cardiol. 2020 (2020)
3. Berhane, H., et al.: Fully automated 3d aortic segmentation of 4d flow MRI for hemodynamic analysis using deep learning. Magn. Reson. Med. **84**(4), 2204–2218 (2020)
4. Blansit, K., Retson, T., Masutani, E., Bahrami, N., Hsiao, A.: Deep learning-based prescription of cardiac MRI planes. Radiol.: Artif. Intell. **1**(6), e180069 (2019)
5. Bulat, A., Tzimiropoulos, G.: Human pose estimation via convolutional part heatmap regression. In: Leibe, B., Matas, J., Sebe, N., Welling, M. (eds.) ECCV 2016. LNCS, vol. 9911, pp. 717–732. Springer, Cham (2016). https://doi.org/10.1007/978-3-319-46478-7_44
6. Bustamante, M., Viola, F., Engvall, J., Carlhäll, C.J., Ebbers, T.: Automatic time-resolved cardiovascular segmentation of 4d flow MRI using deep learning. J. Magn. Reson. Imaging **57**(1), 191–203 (2023)
7. Çiçek, Ö., Abdulkadir, A., Lienkamp, S.S., Brox, T., Ronneberger, O.: 3D U-Net: learning dense volumetric segmentation from sparse annotation. In: Ourselin, S., Joskowicz, L., Sabuncu, M.R., Unal, G., Wells, W. (eds.) MICCAI 2016. LNCS, vol. 9901, pp. 424–432. Springer, Cham (2016). https://doi.org/10.1007/978-3-319-46723-8_49
8. Collado, F.M.S., et al.: Left atrial appendage occlusion for stroke prevention in nonvalvular atrial fibrillation. J. Am. Heart Assoc. **10**(21), e022274 (2021)
9. Corrado, P.A., Seiter, D.P., Wieben, O.: Automatic measurement plane placement for 4D Flow MRI of the great vessels using deep learning. Int. J. Comput. Assist. Radiol. Surg. **17**(1), 199–210 (2022)
10. Fujiwara, T., et al.: Segmentation of the aorta and pulmonary arteries based on 4d flow MRI in the pediatric setting using fully automated multi-site, multi-vendor, and multi-label dense u-net. J. Magn. Reson. Imaging **55**(6), 1666–1680 (2022)
11. Jin, C., et al.: Left atrial appendage segmentation using fully convolutional neural networks and modified three-dimensional conditional random fields. IEEE J. Biomed. Health Inform. **22**(6), 1906–1916 (2018)
12. Leventić, H., et al.: Left atrial appendage segmentation from 3D CCTA images for occluder placement procedure. Comput. Biol. Med. **104**, 163–174 (2019)
13. Li, Y., et al.: Standard plane detection in 3D fetal ultrasound using an iterative transformation network. In: Frangi, A.F., Schnabel, J.A., Davatzikos, C., Alberola-López, C., Fichtinger, G. (eds.) MICCAI 2018. LNCS, vol. 11070, pp. 392–400. Springer, Cham (2018). https://doi.org/10.1007/978-3-030-00928-1_45

14. Lian, C., et al.: Multi-task dynamic transformer network for concurrent bone segmentation and large-scale landmark localization with dental CBCT. In: Martel, A.L., et al. (eds.) MICCAI 2020. LNCS, vol. 12264, pp. 807–816. Springer, Cham (2020). https://doi.org/10.1007/978-3-030-59719-1_78

15. Luo, Z., Wang, Z., Huang, Y., Wang, L., Tan, T., Zhou, E.: Rethinking the heatmap regression for bottom-up human pose estimation. In: CVPR, pp. 13264–13273 (2021)

16. Malik, J., et al.: Handvoxnet: deep voxel-based network for 3d hand shape and pose estimation from a single depth map. In: CVPR, pp. 7113–7122 (2020)

17. Marin-Castrillon, D.M., et al.: Segmentation of the aorta in systolic phase from 4d flow MRI: multi-atlas vs. deep learning. Magn. Reson. Mater. Phys. Biol. Med., 1–14 (2023)

18. Michiels, K., Heffinck, E., Astudillo, P., Wong, I., Mortier, P., Bavo, A.M.: Automated MSCT analysis for planning left atrial appendage occlusion using artificial intelligence. J. Interv. Cardiol. 2022 (2022)

19. Montalt-Tordera, J., et al.: Automatic segmentation of the great arteries for computational hemodynamic assessment. J. Cardiovasc. Magn. Reson. **24**(1), 1–14 (2022)

20. Morais, P., et al.: Fast segmentation of the left atrial appendage in 3-d transesophageal echocardiographic images. IEEE Trans. Ultrason. Ferroelectr. Freq. Control **65**(12), 2332–2342 (2018)

21. Morais, P., et al.: Semiautomatic estimation of device size for left atrial appendage occlusion in 3-D TEE images. IEEE Trans. Ultrason. Ferroelectr. Freq. Control **66**(5), 922–929 (2019)

22. Ortuño, J.E., et al.: Automatic estimation of aortic and mitral valve displacements in dynamic CTA with 4d graph-cuts. Med. Image Anal. **65**, 101748 (2020)

23. Payer, C., Štern, D., Bischof, H., Urschler, M.: Regressing heatmaps for multiple landmark localization using CNNs. In: Ourselin, S., Joskowicz, L., Sabuncu, M.R., Unal, G., Wells, W. (eds.) MICCAI 2016. LNCS, vol. 9901, pp. 230–238. Springer, Cham (2016). https://doi.org/10.1007/978-3-319-46723-8_27

24. Qin, C., et al.: Ideal midsagittal plane detection using deep hough plane network for brain surgical planning. In: Wang, L., Dou, Q., Fletcher, P.T., Speidel, S., Li, S. (eds.) MICCAI 2022. LNCS, vol. 13437, pp. 585–593. Springer, Cham (2022). https://doi.org/10.1007/978-3-031-16449-1_56

25. Wan, C., Probst, T., Van Gool, L., Yao, A.: Dense 3d regression for hand pose estimation. In: CVPR, pp. 5147–5156 (2018)

26. Yang, D., et al.: Automatic vertebra labeling in large-scale 3D CT using deep image-to-image network with message passing and sparsity regularization. In: Niethammer, M., et al. (eds.) IPMI 2017. LNCS, vol. 10265, pp. 633–644. Springer, Cham (2017). https://doi.org/10.1007/978-3-319-59050-9_50

27. Zhang, H., Li, Q., Sun, Z.: Joint voxel and coordinate regression for accurate 3d facial landmark localization. In: 2018 24th International Conference on Pattern Recognition (ICPR), pp. 2202–2208. IEEE (2018)

A Benchmarking Study of Deep Learning Approaches for Bi-Atrial Segmentation on Late Gadolinium-Enhanced MRIs

Yongyao Tan$^{(\boxtimes)}$, Fan Feng, and Jichao Zhao

Auckland Bioengineering Institute, University of Auckland, Auckland, New Zealand
ytan828@aucklanduni.ac.nz

Abstract. Atrial segmentation from late gadolinium-enhanced magnetic resonance imaging (LGE-MRI) provides essential information for patient stratification and targeted ablation treatment for patients with atrial fibrillation (AF). Automatic segmentation based on deep learning approaches, particularly U-Nets architectures, has been widely used to extract the left atrium (LA) automatically in order to overcome manual segmentation limitations. Unfortunately, a lack of an objective benchmarking study evaluated the efficacy of these approaches including recent development. In this work, we performed a comparative study for bi-atrial segmentation with seven U-net architecture variants. We evaluated the models on segmentation of both LA and right atrium (RA) using the 100 3D LGE-MRI dataset from the University of Utah. Moreover, we explored the domain generalization ability using independently acquired 11 LGE-MRIs from Waikato Hospital, New Zealand. Extensive experiments demonstrate that nnU-Net achieved competitive performance over the other state-of-the-art algorithms, such as ResU-Net, U-Net ++, Attention U-Net, and Swin UNETR, achieving an average Dice accuracy of 91.5%, an average surface distance of approximately 1 mm, and a 95% Hausdorff Distance of about 4.5 mm for LA and RA cavities on the Utah dataset. The results show that nnU-Net, a plain U-Net architecture, is sufficient for atrial segmentation, and the state-of-the-art algorithms are potentially unnecessary for this particular task, given no significant performance improvements were observed between most of the tested architectures. We hope this work will also provide insights into new model development for medical image segmentation.

Keywords: Atrial Segmentation · Atrial Fibrillation · Convolutional Neural Network · LGE-MRI

1 Introduction

Atrial fibrillation (AF) is the most common sustained heart rhythm disturbance encountered in the clinic [1]. Structural remodeling of the left atrial (LA) and right atrial (RA) chambers provides reliable information on the evaluation and progression of AF, thus enhancing the effectiveness of AF ablation potentially [2–5]. Late gadolinium-enhanced MRI (LGE-MRI) is widely used to define fibrosis and scars in patients with AF [6].

© The Author(s), under exclusive license to Springer Nature Switzerland AG 2024
O. Camara et al. (Eds.): STACOM 2023, LNCS 14507, pp. 250–258, 2024.
https://doi.org/10.1007/978-3-031-52448-6_24

Hence, segmentation of both LA and RA from LGE-MRIs is vital to improve the understanding and treatment stratification of AF based on structural analysis of the segmented 3D geometry. However, automatic segmentation remains challenging due to large variations of the atrial shape and the intrinsic thin wall thickness.

Recently, due to the success of deep learning methods, specifically, U-Nets adopted in medical image segmentation, a wide range of U-Net variants and improvements have been developed and have achieved promising performance on LA segmentation [7–9]. Despite the extensive research in LA segmentation, a lack of established study has been conducted to evaluate the effect of these U-Net architects on both LA and RA cavities and bi-atrial wall segmentation performance, as well as test its efficiency in an independently acquired clinical dataset.

In this study, a comprehensive comparison was performed using seven U-Net variants to obtain an understanding of their effects on segmentation performance, including Two-stage 2D ResU-Net [10, 11], 2D ResU-Net [12], 2D U-Net + + [13], 2D Attention U-Net [14], 3D Attention U-Net [14], 3D nnU-Net [15], and Swin UNETR [16]. To ensure that the comparison is not biased, two test datasets containing data captured from different centers were used to further evaluate the robustness of the model.

2   Methods

2.1   Data and Pre-processing

Two MRI datasets were used in all experiments: LGE-MRIs from the University of Utah and from Waikato Hospital.

Utah Dataset: The Utah dataset consists of 100 3D LGE MR images acquired from 41 patients at a spatial resolution of 0.625 mm × 0.625 mm × 1.25 mm using either a 1.5 T Avanto or 3.0 T Verio clinical whole body scanner. All 3D LGE-MRI scans contain 44 slices along the Z direction, each with an XY spatial size of 640 × 640 pixels or 576 × 576 pixels. The dataset contains manual segmentation of RA and LA cavities, the epicardium of the RA and LA, and septum. The LA segmentations were provided by the University of Utah alongside the LGE-MRIs. The RA segmentations were manually labeled by our team based on the same protocols used for LA segmentation. Firstly, the RA endocardium was defined by tracing the RA blood pool in each slice of the LGE-MRI manually. The shape of RA endocardium was then adjusted manually according to the boundary of the RA epicardium. Next, the septum, the region of tissue connecting the RA and LA, was manually traced such that the epicardial surfaces of the LA and RA joined together.

Waikato Dataset: The Waikato dataset was captured from 11 AF patients at the Waikato Hospital. The acquisition and labeling protocols for this dataset are the same as in the Utah dataset, resulting in the same spatial resolution and definition of the segmentation.

2.2   Architectures

We performed a comparison of seven U-Net architecture variants, including four 2D networks, Two-stage ResU-Net, ResU-Net, U-Net + +, Attention U-Net, and three 3D networks, Attention U-Net, nnU-Net, and Swin UNETR. All networks were trained to perform atrial semantic segmentation of the LGE-MRI scans. For the comparison in this study, each 2D networks consist of four down-sampling layers with 32 filters for all convolutions in the first layer, subsequently doubled after each CONV layer with a stride of 2 and a kernel size of 5 and followed by halved at each up-sampling layer. Each convolution block contains a convolution layer, followed by a batch normalization and a parametric rectified linear unit activation layer. 20% dropout was applied at every layer for regularization. The Two-stage ResU-Net consists of two 2D ResU-Net used in a sequential manner. The first CNN performed localizing and cropping a focused sub-region to construct an approximate segmentation of the atria. The second CNN then performed targeted prediction on the sub-region to produce slice-by-slice regional segmentation. For 3D networks, we employed different architectures, since nnU-net is a self-configuring algorithm and Swin UNETR is a transformer-based method. The nnU-net architecture consists of 6 layers in both encoder and decoder with 32 filters in the first layer, subsequently doubled in each up-sampling layer as the other architectures, however, it is constrained by a max number of features 320. For 3D Attention U-Net, we used the same architecture as 2D Attention U-Net. The final output layer of all networks

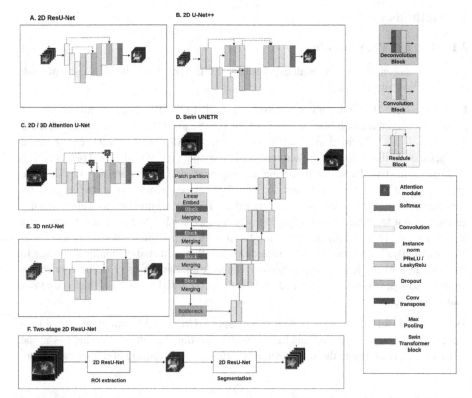

Fig. 1. Summary of the U-Net-based architectures compared in this study (A) ResU-Net, (B) U-Net++, (C) 2D/3D Attention U-Net, (D) Swin UNETR, (E) 3D nnU-Net, (F) Two-stage 2D ResU-Net. Each colored block represents particular layers as per the legend.

contains a 1×1 convolution layer followed by a softmax activation layer, the output of each network consists of class probabilities corresponding to each pixel in the original input image. A visual summary of each architecture is shown in Fig. 1.

2.3 Implementation Details

We implemented all models in PyTorch and MONAI. All models were trained using the AdamW optimizer with a cosine annealing learning rate scheduler with an initial learning rate of 1e-4. Dice cross-entropy loss was used as the training objective function. Each volume was first pre-processed independently using contrast limited adaptive histogram equalization (CLAHE) to enhance the contrast of the image. For 2D networks, all 3D images were divided into 2D slices as network inputs and then were cropped into a unified size of 272×272 centering at the heart region, with an intensity normalization via Z-score. We used online data augmentation strategies such as random rotation, Horizontal and Vertical flips. For 3D models, we used a patch resolution of $32 \times 320 \times 256$ for nnU-Net and Attention U-Net models, and $32 \times 96 \times 96$ for Swin UNETR for computational reasons. For inference, we used a sliding window approach with an overlap portion of 0.5 between the neighboring patches. We did not use any pre-trained weights for the transformer backbone in Swin UNETR. Since it did not demonstrate any performance improvements. In addition, online data augmentation strategies were applied such as random flip in axial, sagittal, and coronal views and random scale and shift intensity. For self-adapting nnUnet architecture, it also employs extensive augmentation strategies, such as gaussian noise transform, gamma Transform and simulate low-resolution Transform, and additional post-proposing steps, such as gaussian computation and test time augmentation.

We randomly split the Utah data into training (25 subjects), validation (5 subjects), and testing (11 subjects) subsets in experiments. Since each patient had multiple 3D LGE-MRIs for pre and post-ablation, the patient-level splits were performed so that the same patient was not included in the different dataset partitions. To evaluate the model generalization ability, we also included the 11 LGE MRIs from Waikato as test dataset. All hyper-parameters were selected based on the performance on the validation set.

2.4 Evaluation Metrics

We use Dice score, Average Surface Distance (ASD), and 95% Hausdorff Distance (HD) to evaluate the accuracy of segmentation in our experiments. For a given semantic class, let G and P denote the ground truth and prediction values, G' and P' represent ground truth and prediction surface point sets, and $d(g', p')$ indicates the Euclidean distance between the two points g' and p'. The Dice score, ASD, and HD metrics are defined as follows

$$Dice(G, P) = \frac{2\sum_{i=1}^{I} G_i P_i}{\sum_{i=1}^{I} G_i + \sum_{i=1}^{I} P_i} \tag{1}$$

$$ASD(G', P') = \frac{1}{2}\left(\frac{\sum_{g' \in G} min_{p' \in P}(g', p')}{\sum_{g' \in G} 1} + \frac{\sum_{p' \in P} min_{g' \in G}(g', p')}{\sum_{p' \in P} 1}\right) \tag{2}$$

$$HD(G', P') = max\{max_{g' \in G}min_{p' \in P}(g', p'), max_{p' \in P}min_{g' \in G}(p' - g')\} \quad (3)$$

The 95% HD uses the 95th percentile of the distances between ground truth and prediction surface point sets.

3   Results and Discussion

3.1   Quantitative Evaluations

Tables 1 and 2 provide a summary of the results for the Utah dataset, and Waikato dataset respectively. For the Utah dataset, 3D nnU-Net (plain Unet architecture) achieves the highest performance. Specifically, for LA and RA cavities, nnU-Net outperforms the second-best baselines by 1.1%, and 1.3% respectively, in terms of Dice score. Furthermore, in the segmentation of thin atrial wall, nnU-Net also outperforms the closest baseline by 0.5% on average in terms of Dice score. Similarly, for the unseen Waikato dataset, nnU-Net has considerably better performance, outperforming other networks by a large margin over all semantic classes in terms of Dice score, Average Surface Distance, and Hausdorff Distance. The four 2D networks (Two-stage ResU-Net, ResU-Net, U-Net + +, and Attention U-Net) have similar performance on both datasets, in which the two-stage network demonstrates a slightly better performance than other one-stage 2D models in terms of Dice score. This indicates that despite the increased architectural complexity, no notable performance differences were observed among these models. These findings further highlight the importance of model configuration introduced by nnU-Net and details in method configuration have more impact on performance than architecture variants. Compared to 2D models, 3D models perform slightly better on the Utah dataset in terms of distance-based metrics (Average Surface Distance and Hausdorff Distance). However, the performance for Swin-UNetr and 3D Attention U-Net degraded dramatically on the Waikato dataset. Overall, our experiments in both datasets demonstrate superior performance of 3D nnU-Net over other U-Net variants. Specifically, 3D nnU-Net achieves better segmentation accuracy. For the unseen waikato dataset, 3D nnU-Net can still outperform other models with efficient generalization ability.

3.2   Qualitative Evaluations

Qualitative atrial segmentation comparisons for the Utah dataset are presented in Fig. 2. nnU-Net shows improved segmentation performance for RA cavity in row 1, while other models confuse RA with other tissues. In row 3, nnU-Net demonstrates higher boundary segmentation accuracy as it accurately identifies the boundaries for RA cavity. In comparison to 2D models, 3D models show improved performance in capturing the fine-grained details of LA cavity, demonstrated in rows 4 and 5, which indicates that 3D models capture better spatial context.

3.3   Model and Computational Complexity

In Table 3, we present the number of parameters, and averaged inference time of the models in Utah benchmarks. 2D methods such as ResU-Net, U-Net++, and Attention U-Net have 26.54M, 28.6M, and 21.88M parameters respectively. 3D nnU-Net outperforms

Table 1. Quantitative comparisons of the segmentation performance for the Utah dataset.

Metrics	RA + LA wall			RA cavity			LA cavity		
	DS	ASD (mm)	HD95 (mm)	DS	ASD (mm)	HD95 (mm)	DS	ASD (mm)	HD95 (mm)
Two-stage 2D ResU-Net	0.667	1.228	7.208	0.898	3.115	12.65	0.906	1.879	9.545
2D ResU-Net	0.628	0.96	4.692	0.894	1.488	6.004	0.906	1.465	7.116
2D U-Net++	0.642	1.148	5.136	0.893	1.753	7.986	0.907	1.659	7.084
2D Attention U-Net	0.631	1.089	6.174	0.894	2.349	9.065	0.908	1.41	6.431
3D Attention U-Net	0.638	**0.651**	4.634	0.885	1.098	**4.857**	0.908	**0.654**	4.411
Swin-UNetr	0.628	0.826	4.715	0.885	**1.088**	5.365	0.904	1.038	5.227
3D nnU-Net	**0.672**	0.762	**4.052**	**0.909**	1.304	5.483	**0.921**	0.882	**3.546**

DS, Dice score; ASD, Average Surface Distance; HD95, 95% Hausdorff Distance; LA/RA, left/right atrium.

Table 2. Quantitative comparisons of the segmentation performance of the Waikato dataset.

Metrics	RA + LA wall			RA cavity			LA cavity		
	DS	ASD (mm)	HD95 (mm)	DS	ASD (mm)	HD95 (mm)	DS	ASD (mm)	HD95 (mm)
Two-stage 2D ResU-Net	0.588	1.894	12.62	0.789	4.483	20.68	0.849	2.332	9.943
2D ResU-Net	0.558	1.377	8.717	0.824	2.926	13.54	0.791	3.056	17.52
2D U-Net++	0.52	1.899	11.21	0.739	4.373	19.05	0.677	3.634	19.63
2D Attention U-Net	0.533	1.106	9.732	0.805	3.001	11.37	0.771	2.546	16.69
3D Attention U-Net	0.445	1.779	24.17	0.611	5.699	46.47	0.749	2.417	26.02
Swin-UNetr	0.413	1.084	18.18	0.486	1.853	18.13	0.751	1.394	22.42
3D nnU-Net	**0.642**	**0.766**	**6.838**	**0.867**	**1.519**	**6.673**	**0.879**	**1.176**	**6.298**

DS, Dice score; ASD, Average Surface Distance; HD95, 95% Hausdorff Distance; LA/RA, left/right atrium.

these models by a large margin in both datasets, however, this is at the cost of drastically increased model complexity as well as slower inference time. In addition, Swin-UNetr

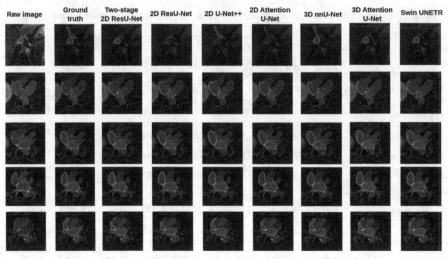

Fig. 2. Segmentation results on the Utah dataset, obtained using different deep learning models, including four 2D networks: Two-stage ResU-Net (3rd column), ResU-Net (4rd column), U-Net++ (5rd column), Attention U-Net (6rd column), and three 3D networks: nnU-Net (7rd column), Attention U-Net (8rd column) and Swin UNETR (9rd column). The first column is the five representative raw 2D LGE-MRIs from a patient with AF and the second column shows the ground truth labels. Red and blue show the LA and RA cavities respectively while green is the atrial wall.

is a moderate-sized model with 62.19M parameters and has the second-highest average inference time after nnUNet.

Table 3. Comparison of the number of parameters and averaged inference time for various models in Utah experiments.

Models	#Params (M)	Inference Time (s)
2D ResU-Net	26.54	0.35
2D U-Net++	28.6	1.34
2D Attention U-Net	21.88	0.5
3D Attention U-Net	23.62	3.05
3D nnU-Net	44.58	45.57
Swin-UNetr	62.19	14.63

4   Conclusion

In this study, seven different U-Net variants were compared for their segmentation performance on the Utah dataset, and we also investigated the generalization ability of models using an unseen dataset from Waikato Hospital. All tested U-Net architectures provide excellent performance for atrial segmentation, and the results suggest that minimal differences were observed between most of the tested architectures and state-of-the-art methods and other architectural changes potentially do not provide a clear benefit for this application. Overall, nnU-Net is an optimal choice for atrial segmentation. This suggests that nnU-Net, a plain U-Net, is sufficient for this particular task. The findings in this study can help to provide a well-informed starting point, saving time and cost in experimentations in future studies while also guiding model selection towards simpler and faster models. While segmentation accuracy is a critical metric, it is not the only one that should be considered when comparing these architectures. Demonstrating model interpretability with methods such as saliency maps is a necessary consideration for adoption into clinical practice. In addition, the performance of atrial wall is lower than other semantic classes and the generalization ability on unseen datasets still has more room to improve. Further research might explore further on model interpretability, and improving performance on atrial wall segmentation and generalization ability.

Acknowledgements. This work was funded by the Heath Research Council of New Zealand and Heart Foundation New Zealand. We would like to acknowledge The NIH/NIGMS Center for Integrative Biomedical Computing (CIBC) at The University of Utah for providing the LGE-MRIs and segmentation labels and Dr. Martin K. Stiles in Waikato Clinical School, The University of Auckland, for providing the LGE-MRIs from Waikato Hospital.

References

1. Chugh, S.S., et al.: Worldwide epidemiology of atrial fibrillation: a Global Burden of Disease 2010 Study. Circulation **129**(8), 837–847 (2014)
2. Njoku, A., et al.: Left atrial volume predicts atrial fibrillation recurrence after radiofrequency ablation: a meta-analysis. Ep Europace. **20**(1), 33–42 (2018)
3. Hopman, L.H., et al.: Right atrial function and fibrosis in relation to successful atrial fibrillation ablation. Eur. Heart J.-Cardiovasc. Imaging **24**(3), 336–345 (2023)
4. Ranjan, R., et al.: Gaps in the ablation line as a potential cause of recovery from electrical isolation and their visualization using MRI. Circ. Arrhythmia Electrophysiol. **4**(3), 279–286 (2011)
5. Takagi, T., et al.: Impact of right atrial structural remodeling on recurrence after ablation for atrial fibrillation. J. Arrhythmia **37**(3), 597–606 (2021)
6. Siebermair, J., Kholmovski, E.G., Marrouche, N.: Assessment of left atrial fibrosis by late gadolinium enhancement magnetic resonance imaging: methodology and clinical implications. JACC Clin. Electrophysiol. **3**(8), 791–802 (2017)
7. Li, L., Zimmer, V.A., Schnabel, J.A., Zhuang, X.: Medical image analysis on left atrial LGE MRI for atrial fibrillation studies: a review. Med. Image Anal. **1**(77), 102360 (2022)
8. Wu, M., Ding, W., Yang, M., Huang, L.: Multi-depth boundary-aware left atrial scar segmentation network. In: Challenge on Left Atrial and Scar Quantification and Segmentation, 18 September 2022, pp. 16–23. Springer, Cham (2022). https://doi.org/10.1007/978-3-031-317 78-1_2
9. Li, L., Zimmer, V.A., Schnabel, J.A., Zhuang, X.: AtrialJSQnet: a new framework for joint segmentation and quantification of left atrium and scars incorporating spatial and shape information. Med. Image Anal. **1**(76), 102303 (2022)

10. Xiong, Z., et al.: A global benchmark of algorithms for segmenting the left atrium from late gadolinium-enhanced cardiac magnetic resonance imaging. Med. Image Anal. 1(67), 101832 (2021)

11. Xiong, Z., Fedorov, V.V., Fu, X., Cheng, E., Macleod, R., Zhao, J.: Fully automatic left atrium segmentation from late gadolinium enhanced magnetic resonance imaging using a dual fully convolutional neural network. IEEE Trans. Med. Imaging 38(2), 515–524 (2018)

12. He, K., Zhang, X., Ren, S., Sun, J.: Deep residual learning for image recognition. In: Proceedings of the IEEE Conference on Computer Vision and Pattern Recognition, pp. 770–778 (2016)

13. Zhou, Z., Rahman Siddiquee, M.M., Tajbakhsh, N., Liang, J.: Unet++: a nested u-net architecture for medical image segmentation. In: Stoyanov, D., et al. (ed.) DLMIA/ML-CDS -2018. LNCS, vol. 11045, pp. 3–11. Springer, Cham (2018). https://doi.org/10.1007/978-3-030-008 89-5_1

14. Oktay, O., et al.: Attention u-net: Learning where to look for the pancreas. arXiv preprint arXiv:1804.03999 (2018)

15. Isensee, F., Jaeger, P.F., Kohl, S.A., Petersen, J., Maier-Hein, K.H.: NnU-Net: a self-configuring method for deep learning-based biomedical image segmentation. Nat. Methods 18(2), 203–211 (2021)

16. Hatamizadeh, A., Nath, V., Tang, Y., Yang, D., Roth, H.R., Xu, D.: Swin unetr: swin transformers for semantic segmentation of brain tumors in MRI images. In: International MICCAI Brainlesion Workshop 27 September 2021, pp. 272–284. Springer, Cham (2021). https://doi.org/10.1007/978-3-031-08999-2_22

CMRxRecon Challenge

Fill the K-Space and Refine the Image: Prompting for Dynamic and Multi-Contrast MRI Reconstruction

Bingyu Xin[1]([✉]), Meng Ye[1], Leon Axel[2], and Dimitris N. Metaxas[1]

[1] Department of Computer Science, Rutgers University, Piscataway, NJ 08854, USA
[2] Department of Radiology, New York University, New York, NY 10016, USA
{bx64,my389,dnm}@cs.rutgers.edu
https://github.com/hellopipu/PromptMR

Abstract. The key to dynamic or multi-contrast magnetic resonance imaging (MRI) reconstruction lies in exploring inter-frame or inter-contrast information. Currently, the unrolled model, an approach combining iterative MRI reconstruction steps with learnable neural network layers, stands as the best-performing method for MRI reconstruction. However, there are two main limitations to overcome: firstly, the unrolled model structure and GPU memory constraints restrict the capacity of each denoising block in the network, impeding the effective extraction of detailed features for reconstruction; secondly, the existing model lacks the flexibility to adapt to variations in the input, such as different contrasts, resolutions or views, necessitating the training of separate models for each input type, which is inefficient and may lead to insufficient reconstruction. In this paper, we propose a two-stage MRI reconstruction pipeline to address these limitations. The first stage involves *filling the missing k-space data*, which we approach as a physics-based reconstruction problem. We first propose a simple yet efficient baseline model, which utilizes adjacent frames/contrasts and channel attention to capture the inherent inter-frame/-contrast correlation. Then, we extend the baseline model to a prompt-based learning approach, **PromptMR**, for all-in-one MRI reconstruction from different views, contrasts, adjacent types, and acceleration factors. The second stage is to *refine the reconstruction* from the first stage, which we treat as a general video restoration problem to further fuse features from neighboring frames/contrasts in the image domain. Extensive experiments show that our proposed method significantly outperforms previous state-of-the-art accelerated MRI reconstruction methods.

Keywords: MRI reconstruction · Prompt-based learning · Dynamic · Multi-contrast · Two-stage approach

1 Introduction

Cardiovascular disease, including conditions such as coronary artery disease, heart failure, and arrhythmias, remains the leading cause of death globally. Car-

O. Camara et al. (Eds.): STACOM 2023, LNCS 14507, pp. 261–273, 2024.
https://doi.org/10.1007/978-3-031-52448-6_25

diac magnetic resonance (CMR) imaging is the most accurate and reliable non-invasive technique for accessing cardiac anatomy, function, and pathology [13]. In the field of accelerated MR imaging (MRI) reconstruction, unrolled networks have achieved state-of-the-art performance. This is attributed to their ability to incorporate the known imaging degradation processes, the undersampling operation in k-space, into the network and to learn image priors from large-scale data [2,16]. As transformers have become predominant in general image restoration tasks [9,18], there is a noticeable trend towards incorporating transformer-based denoising blocks into the unrolled network [2], which enhances reconstruction quality. However, the adoption of transformer blocks concurrently increases the network parameters and computational complexity. The stacking of denoising blocks, in an unrolled manner, further exacerbates this complexity, making the network training challenging. Therefore, one challenging question is how to design efficient denoising blocks within an unrolled model while fully leveraging the k-space information. Another challenge arises from the versatility of MRI, which enables the acquisition of multi-view, multi-contrast, multi-slice, and dynamic image sequences, given specific clinical demands. While there is a prevailing trend towards designing all-in-one models for natural image restoration [7,12], existing MRI reconstruction models cannot offer a unified solution for diverse input types. We thus endeavor to address these challenges with the following contributions:

- Firstly, we propose a simple yet efficient convolution-only baseline model for MRI reconstruction, which outperforms previous state-of-the-art methods on two public multi-coil MRI reconstruction tasks, the CMRxRecon and fastMRI knee image reconstruction.
- Then, by extending our baseline model with prompt-based learning, we are the first to propose an all-in-one approach, **PromptMR**, for multi-view/-contrast and dynamic MRI reconstruction.
- Lastly, we extend our approach to address the capacity limitations of unrolled models, by proposing a two-stage MRI reconstruction pipeline. In the first stage we solve a physics-based inverse problem in k-space domain to fill the missing k-space data, and in the second stage we solve a video restoration problem in the image domain to further refine the MRI reconstruction.

2   Preliminaries

Consider reconstructing a complex-valued MR image x from the multi-coil under-sampled measurements y in k-space, such that,

$$y = Ax + \epsilon, \tag{1}$$

where A is the linear forward complex operator which is constructed based on multiplications with the sensitivity maps S, application of the 2D Fourier transform F, while it under-samples the k-space data with a binary mask M; ϵ is the

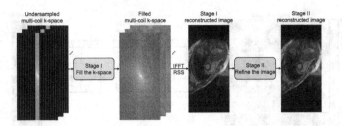

Fig. 1. The proposed two-stage MRI reconstruction pipeline. The first stage solves a physics-based inverse problem to fill the missing k-space data, which are then transformed to the image domain by the inverse Fast Fourier Transformation (IFFT) and root-sum-of-squares (RSS) is applied to get the first-stage reconstructed image. The second stage solves a general denoising problem to further refine the image reconstruction result.

acquisition noise. According to compressed sensing theory [1], we can estimate x by formulating an optimization problem:

$$\min_x \frac{1}{2}||y - Ax||_2^2 + \lambda R(x), \tag{2}$$

where $||y - Ax||_2^2$ is the data consistency term, $R(x)$ is a sparsity regularization term on x (e.g., total variation) and λ is a hyper-parameter which controls the contribution weights of the two terms. E2E-VarNet [16] solves the problem in Eq. 2 by applying an iterative gradient descent method in the k-space domain. In the t-th step, the k-space is updated from k^t to k^{t+1} using:

$$k^{t+1} = k^t - \eta^t M(k^t - y) + G(k^t), \tag{3}$$

where η^t is a learned step size and G is a learned function representing the gradient of the regularization term R. We can unroll the iterative updating algorithm to a sequence of sub-networks, where each cascade represents an unrolled iteration in Eq. 3. The regularization term is applied in the image domain:

$$G(k) = F(\mathcal{E}(\mathbf{D}(\mathcal{R}(F^{-1}(k))))), \tag{4}$$

where $\mathcal{R}(x_1, ..., x_N) = \sum_{i=1}^{N} \hat{S}_i^* x_i$ is the reduce operator that combines N coil images $\{x_i\}_{i=1}^N$ via estimated sensitivity maps $\{\hat{S}_i\}_{i=1}^N$, $\hat{S}_i^*$ is the complex conjugate of $\hat{S}_i$, and $\mathcal{E}(x) = (\hat{S}_i x, ..., \hat{S}_N x)$ is the expand operator that computes coil images from image x. Therefore, the linear forward operator A is computed as $A = MF\mathcal{E}$. $\mathbf{D}$ is a denoising neural network used to refine the complex image. $\hat{S} = \text{SME}(y_{\text{ACS}})$ is computed by a sensitivity map estimation (SME) network from the low-frequency region of k-space y_{ACS}, called the Auto-Calibration Signal (ACS), which is typically fully sampled. The final updated multi-coil k-space is converted to the image domain by applying an inverse Fourier transform followed by a root-sum-of-squares (RSS) method reduction [14] for each pixel.

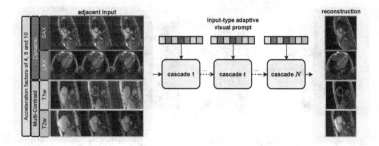

Fig. 2. Overview of PromptMR in Stage I: an all-in-one unrolled model for MRI reconstruction. Adjacent inputs, depicted in image domain for visual clarity, provide neighboring k-space information for reconstruction. To accommodate different input varieties, the input-type adaptive visual prompt is integrated into each cascade of the unrolled architecture to guide the reconstruction process.

3   Method

We propose a two-stage pipeline for dynamic and multi-contrast MRI reconstruction, as shown in Fig. 1. Below, we give more details of each stage.

3.1   Stage I: Filling the K-Space

The center of the k-space preserves image contrast, and the periphery of the k-space contains edge information. In the first stage, we fill the missing k-space data constrained by the existing k-space acquisition and learned image priors.

Baseline Model. We follow the implementation of E2E-VarNet [16] to construct an unrolled model in Stage I. Inspired by the adjacent slice reconstruction (ASR) method [2], which learns inter-slice information by jointly reconstructing a set of adjacent slices instead of relying on a single k-space to be reconstructed, we devise the following new method. We generalize ASR to adjacent k-space reconstruction along any dimension, e.g., temporal/slice/view/contrast dimension, and the updating formula of Eq. 3 is improved as follows:

$$k_{adj}^{t+1} = k_{adj}^t - \eta^t M(k_{adj}^t - y_{adj}) + G(k_{adj}^t), \qquad (5)$$

where $k_{adj}^t = [k_{c-a}^t, ..., k_{c-1}^t, k_c^t, k_{c+1}^t, ..., k_{c+a}^t]$ is the concatenation of the central k-space k_c^t with its $2a$ adjacent k-spaces along a specific dimension. To efficiently extract features from adjacent inputs, we design a Unet-style network [15] with channel attention [3,4], namely CAUnet, for both the denoising network D and the sensitivity map estimation network, as shown in Appendix A.1. The CAUnet has a 3-level encoder-decoder structure. Each level consists of a DownBlock, UpBlock, and corresponding skip connection. The architecture integrates a BottleneckBlock for high-level semantic feature capturing and employs Channel Attention Blocks (CABs) within each block. The overall unrolled architecture is shown in Appendix A.2.

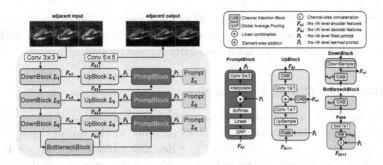

Fig. 3. Overview of the PromptUnet architecture in PromptMR, featuring a 3-level encoder-decoder design. Each level comprises a DownBlock, UpBlock and Prompt-Block. The PromptBlock in the i-th level encodes input-specific context into fixed prompt P_i, producing adaptively learned prompt $\hat{P}_i$. These prompts, across multiple levels, integrate with decoder features $F_{d,i}$ in the UpBlocks to allow rich hierarchical context learning.

PromptMR. Considering various image types (e.g., different views, different contrasts) with different adjacent types (e.g., dynamic, multi-contrast) under different undersampling rates (e.g., ×4, ×8, ×10), instead of training separate models for each specific input, we propose to learn an all-in-one unified model for all possible adjacent inputs. The image structure remains consistent for multi-contrast adjacent input, while only the contrast varies. Conversely, the contrast remains constant for dynamic adjacent input, but the image structure shifts. To achieve effective performance on diverse input types, the unified model should be able to encode the contextual information conditioned on the input type. Inspired by the recent development of visual prompt learning [5,6] and prompt learning-based image restoration method [12], we introduce PromptMR, an all-in-one approach for MRI reconstruction, as illustrated in Fig. 2. While PromptMR retains the same unrolled architecture of the basline model, it extends CAUnet to PromptUnet by integrating PromptBlocks to learn input-type adaptive prompts and then interact with decoder features in the UpBlocks at multiple levels, to enrich the input-specific context, as shown in Fig. 3. The PromptBlock at i-th level takes features $F_{d,i} \in \mathbb{R}^{H_f \times W_f \times C_f}$ from the decoder and N_p-components fixed prompt $P_i \in \mathbb{R}^{N_p \times H_p \times W_p \times C_p}$ as input. Then, $F_{d,i}$ are processed by a global average pooling (GAP) layer, followed by a linear layer and a softmax layer to generate the normalized prompt weights $\{\omega_{ij}\}_{j=1}^{N_p}$. These weights linearly combine the N_p prompt components as $\sum_{j=1}^{N_p} \omega_{ij} P_{ij}$, which is then interpolated to match the spatial dimension of $F_{d,i}$, before going through a 3×3 convolution layer to generate the input-type adaptive prompt $\hat{P}_i$. The process in the PromptBlock can be summarized as:

$$\hat{P}_i = \mathrm{Conv}_{3\times3}(\mathrm{Interp}(\textstyle\sum_{j=1}^{N_p} \omega_{ij} P_{ij})), \quad \omega_i = \mathrm{Softmax}(\mathrm{Linear}(\mathrm{GAP}(F_{d,i}))) \quad (6)$$

The generated prompts by the PromptBlocks at multiple levels can learn hierarchical input-type contextual representations, which are integrated with the decoder features to guide the all-in-one MRI reconstruction.

3.2 Stage II: Refining the Image

After the first stage, the missing k-space data has been filled, and image aliasing artifacts have been largely removed. However, due to the unrolled nature and memory limitations, the capability of the denoising block we can use is constrained, which may prevent the full exploration of dynamic and multi-contrast information. In stage II, we further explore the inter-frame/-contrast coherence in the image domain for multi-frame/-contrast feature aggregation by using a powerful restoration model, ShiftNet [8], as the refinement network. This network employs stacked Unets and grouped spatio-temporal shift operations to expand the effective receptive fields. Details of the ShiftNet are not covered here, since it is not the core part of this paper, and ShiftNet can be replaced by any state-of-the-art video restoration model.

4 Experiments

In this section, we first provide experimental details and results of our proposed method on the CMRxRecon dataset. We use SSIM, PSNR, and NMSE to compare the performance of different reconstruction methods under various acceleration factors ($\times 4$, $\times 8$, $\times 10$). Then, we conduct extensive ablation studies of our proposed method and also benchmark on another large-scale MRI dataset, the fastMRI multi-coil knee dataset. For experiments on fastMRI dataset, we refer readers to the Appendix B.

4.1 CMRxRecon Dataset

The CMRxRecon Dataset [17] includes 120 cardiac MRI cases of fully sampled dynamic cine and multi-contrast raw k-space data obtained on 3 T magnets. The dynamic cine images in each case include short-axis (SAX), two-chamber (2-CH), three-chamber (3-CH), and four-chamber (4-CH) long-axis (LAX) views. Typically 5–10 slices were acquired for SAX cine, while a single slice was acquired for each LAX view. The cardiac cycle was segmented into 12–25 phases with a temporal resolution of 50 ms. The multi-contrast cardiac MRI in each case is in the SAX view, which contains 9 T1-weighted (T1w) images conducted using a modified look-locker inversion recovery (MOLLI) sequence and 3 T2-weighted (T2w) images performed using T2-prepared FLASH sequence.

The shape of each k-space data is [time phases/contrasts, slices, coils, readouts, phase encodings]. All data were compressed into 10 virtual coils. We splited the cases in an 8 : 2 ratio, resulting in 14, 964 dynamic images and 6, 516 multi-contrast images for training, and 2, 940 dynamic images and 1, 272 multi-contrast images for testing.

Table 1. Comparison of $\mathrm{NMSE}(\times 10^{-2})/\mathrm{PSNR}/\mathrm{SSIM}$ of different MRI reconstruction methods on CMRxRecon dataset under ×10 acceleration. The best and second best results are highlighted in red and blue colors, respectively.

Stage	Task	Method	Cine		Mapping	
			SAX	LAX	T1w	T2w
I	One-by-One	E2E-Varnet [16]	1.6/42.05/0.9744	2.1/39.93/0.9673	1.5/43.12/0.9800	1.4/41.20/0.9777
		HUMUS-Net-L [2]	1.3/42.96/0.9791	2.0/40.07/0.9689	1.3/43.85/0.9832	1.1/42.39/0.9824
		Baseline (Ours)	1.1/43.68/0.9814	1.9/40.38/0.9705	1.2/44.14/0.9839	0.9/43.12/0.9845
	All-in-One	PromptIR [12]	2.5/40.16/0.9659	2.7/38.62/0.9581	2.3/41.10/0.9726	1.4/41.10/0.9784
		PromptMR (Ours)	1.1/45.58/0.9865	1.2/43.72/0.9836	1.0/46.84/0.9899	0.7/46.24/0.9903
II	PromptMR+ShiftNet [8]		0.7/45.63/0.9866	0.9/43.76/0.9837	0.7/47.04/0.9903	0.5/46.33/0.9905

4.2 Results

We assessed the performance of our proposed baseline model, PromptMR, and two-stage reconstruction pipeline using the CMRxRecon dataset. In the first stage, we compared the E2E-VarNet [16] and HUMUS-Net-L [2] with our baseline in a one-by-one setup, in which we trained four separate models from scratch for SAX/LAX/T1w/T2w reconstruction task, respectively. Then we compared our PromptMR and PromptIR [12] in an all-in-one configuration. In the second stage, we deployed ShiftNet to refine the images reconstructed by PromptMR. In our experiment, we minimize the SSIM loss between the target image and the reconstructed image; all unrolled models consist of 12 cascades, except for HUMUS-Net-L, which only has 8 cascades due to its large parameter size; we trained networks using AdamW [10] optimizer with a weight decay of 0.01 for 12 epochs; the learning rate was set as 2×10^{-4} for the first 11 epochs and 2×10^{-5} for the last epoch.

The results are shown in Table 1. Notably, our baseline model outperforms E2E-VarNet and HUMUS-Net-L across all tasks. Moreover, our PromptMR demonstrates significant enhancement in the all-in-one setup when compared to the baseline model trained for individual tasks. PromptIR performs poorly due to the fact that it is not tailored to account for the MRI forward model. The refinement in the second stage offers a marginal boost to the SSIM, but provides considerable improvements for NMSE and PSNR. The qualitative results are shown in Fig. 4. More qualitative comparisons can be found in Appendix C. These qualitative comparisons show that our method can recover more finer details for small anatomical structures on the reconstructed images.

4.3 Ablation Study

Single MRI Reconstruction Task. We started with an ablation study on two single MRI reconstruction tasks, dynamic cine SAX image reconstruction and multi-contrast T1-weighted image reconstruction, both under ×10 acceleration, to investigate the impact of adjacent reconstruction and prompt module in the proposed PromptMR. We changed the number of adjacent images to 1,

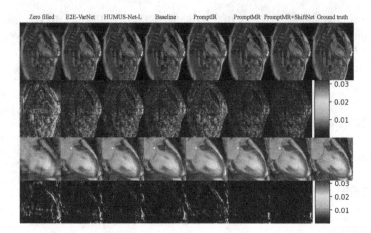

Fig. 4. The reconstruction results and absolute error maps of different methods for LAX 2-CH cine image of case P101 under ×10 acceleration factors. The bottom two rows show the zoomed area. Red arrows show the difference in recovery of the mitral valve structure for different reconstruction methods. (Color figure online)

Table 2. Impact of the adjacent input number and the PromptBlock in PromptMR for two single MRI reconstruction tasks: dynamic cine SAX and multi-contrast T1-weighted (T1w) reconstruction under ×10 acceleration.

# of adj	PromptBlock	SAX PSNR/SSIM	T1w PSNR/SSIM
1	✓	43.19/0.9798	44.36/0.9845
3	✓	43.96/0.9822	44.78/0.9856
5	✓	43.87/0.9820	44.75/0.9856
5	✗	43.68/0.9814	44.14/0.9839

3, and 5, where '1' indicates the absence of adjacent input. The results, shown in Table 2, underscore the utility of incorporating adjacent input to enhance the reconstruction quality. Moreover, the inclusion of PromptBlocks proves beneficial for individual MRI reconstruction tasks.

All-In-One MRI Reconstruction Task. To investigate the impact of the PromptBlock in the all-in-one MRI reconstruction task, we trained both our baseline and PromptMR model using all possible input data in the CMRxRecon dataset. As depicted in Table 3, the integration of the PromptBlock into our baseline model enables PromptMR to achieve significant improvements across all individual reconstruction tasks. We also used t-SNE [11] to visualize the learned prompts in the 12-th cascade at multiple decoder levels from different types of data in the test set. Figure 5 shows that the prompts can learn to encode discriminative information for different input types at lower levels.

Table 3. Impact of PromptBlock in all-in-one task. Results are reported on the CMRxRecon dataset under ×10 acceleration.

Method	SAX PSNR/SSIM	LAX PSNR/SSIM	T1w PSNR/SSIM	T2w PSNR/SSIM
Baseline (Ours)	43.97/0.9825	42.11/0.9786	44.90/0.9862	44.45/0.9874
PromptMR (Ours)	45.58/0.9865	43.72/0.9836	46.84/0.9899	46.24/0.9903

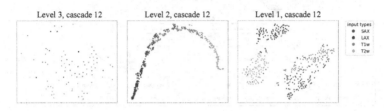

Fig. 5. Visualization of the learned prompts at each decoder level in the 12-th cascade in PromptMR using t-SNE.

5 Conclusion

In this work, we introduce a robust baseline model for MRI reconstruction that utilizes neighboring information of adjacent k-space. To accommodate various input types, adjacent configurations, and undersampling rates within a unified model, we enhance our baseline with prompt-based learning blocks, creating an all-in-one MRI reconstruction model, **PromptMR**. Finally, to overcome the model capacity constraints of unrolled architectures, we propose a second stage of image refinement to delve deeper into the adjacent information, which is particularly useful when immediate reconstruction latency is not a priority.

Appendix

A   Details of the Baseline Model

A.1   CAUnet

(See Fig. 6).

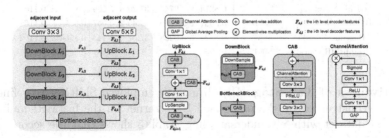

Fig. 6. Overview of the CAUnet architecture in the proposed baseline model.

A.2   Unrolled Model Architecture

(See Fig. 7).

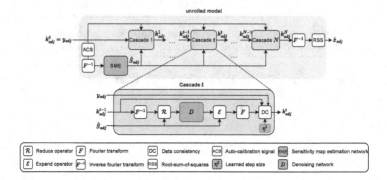

Fig. 7. Overview of the unrolled model architecture for both our baseline model and PromptMR. The primary distinction is in the denoiser D and sensitivity map estimation (SME) networks: the baseline employs CAUnet, whereas PromptMR utilizes PromptUnet. Each cascade represents an updating step in Eq. 5 in the main text. The red module indicates the learnable part in the unrolled model.

B   Experiments on FastMRI Multi-Coil Knee Dataset

Benchmark on FastMRI Multi-Coil Knee Dataset. To assess the performance of our proposed method across different anatomies, we benchmarked it on another large-scale MRI reconstruction dataset, the fastMRI multi-coil knee dataset [19]. Since the online evaluation platform for the fastMRI test set is unavailable[1], we divided the original 199 validation cases into 99 for validation and 100 for testing. The results of other methods are reported using their officially pretrained models. As presented in Table 4, our models outperform all previous state-of-the-art methods, without significantly increasing the number of network parameters compared to E2E-Varnet.

Table 4. Performance of state-of-the-art accelerated MRI reconstruction techniques on the fastMRI knee multi-coil ×8 test dataset. The best and second best results are highlighted in red and blue colors, respectively.

Method	# of params	NMSE($\times 10^{-2}$)($\downarrow$)	PSNR($\uparrow$)	SSIM($\uparrow$)
E2E-Varnet [16]	30M	0.8690 ± 0.9279	37.30 ± 4.925	0.8936 ± 0.1157
HUMUS-Net [2]	109M	0.8974 ± 0.9743	37.20 ± 5.009	0.8946 ± 0.1162
HUMUS-Net-L [2]	228M	0.8587 ± 0.9930	37.45 ± 5.067	0.8955 ± 0.1161
Baseline (ours)	48M	0.8321 ± 0.9258	37.57 ± 5.143	0.8964 ± 0.1162
PromptMR (ours)	80M	0.8034 ± 0.9249	37.78 ± 5.281	0.8983 ± 0.1167

Effectiveness of Two-Stage Pipeline. We employed ShiftNet to refine the images reconstructed by the pretrained E2E-Varnet on the fastMRI multi-coil knee test dataset with ×8 undersampling. Table 5 shows that the second-stage

---

[1] https://github.com/facebookresearch/fastMRI/discussions/293.

refinement substantially improves the reconstruction quality, which implies that the multi-slice information in the fastMRI dataset might not be comprehensively utilized by the single-stage unrolled model.

Table 5. Effectiveness of the second-stage image refinement on the fastMRI knee multi-coil ×8 test dataset.

Stage	Method	# of params	NMSE($\times 10^{-2}$)($\downarrow$)	PSNR($\uparrow$)	SSIM($\uparrow$)
I	E2E-Varnet [16]	30M	0.8690 ± 0.9279	37.30 ± 4.925	0.8936 ± 0.1157
II	ShiftNet [8]	2M	0.8415 ± 0.9131	37.46 ± 4.973	0.8953 ± 0.1157

C Additional Qualitative Results on CMRxRecon Dataset

(See Fig. 8).

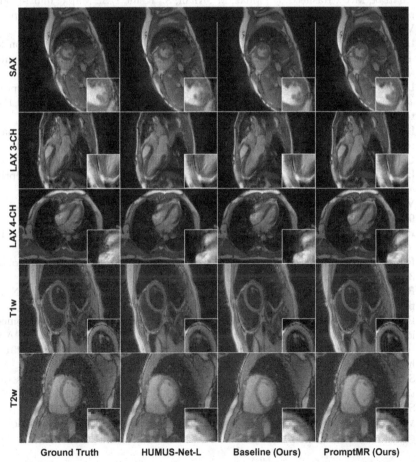

Ground Truth   HUMUS-Net-L   Baseline (Ours)   PromptMR (Ours)

Fig. 8. Visual comparison of reconstructions from the CMRxRecon dataset with ×10 acceleration. PromptMR can recover fine details (highlighted in red box) on reconstructed images that other state-of-the-art methods may miss. (Color figure online)

References

1. Donoho, D.L.: Compressed sensing. IEEE Trans. Inf. Theory **52**(4), 1289–1306 (2006)
2. Fabian, Z., Tinaz, B., Soltanolkotabi, M.: Humus-net: hybrid unrolled multi-scale network architecture for accelerated MRI reconstruction. Adv. Neural. Inf. Process. Syst. **35**, 25306–25319 (2022)
3. Hu, J., Shen, L., Sun, G.: Squeeze-and-excitation networks. In: Proceedings of the IEEE Conference on Computer Vision and Pattern Recognition, pp. 7132–7141 (2018)
4. Huang, Q., Yang, D., Wu, P., Qu, H., Yi, J., Metaxas, D.: MRI reconstruction via cascaded channel-wise attention network. In: 2019 IEEE 16th International Symposium on Biomedical Imaging (ISBI 2019), pp. 1622–1626. IEEE (2019)
5. Jia, M., et al.: Visual prompt tuning. In: European Conference on Computer Vision. pp. 709–727. Springer (2022). https://doi.org/10.1007/978-3-031-19827-4_41
6. Khattak, M.U., Rasheed, H., Maaz, M., Khan, S., Khan, F.S.: Maple: multi-modal prompt learning. In: Proceedings of the IEEE/CVF Conference on Computer Vision and Pattern Recognition, pp. 19113–19122 (2023)
7. Li, B., Liu, X., Hu, P., Wu, Z., Lv, J., Peng, X.: All-in-one image restoration for unknown corruption. In: Proceedings of the IEEE/CVF Conference on Computer Vision and Pattern Recognition, pp. 17452–17462 (2022)
8. Li, D., et al.: A simple baseline for video restoration with grouped spatial-temporal shift. In: Proceedings of the IEEE/CVF Conference on Computer Vision and Pattern Recognition, pp. 9822–9832 (2023)
9. Liang, J., Cao, J., Sun, G., Zhang, K., Van Gool, L., Timofte, R.: Swinir: image restoration using swin transformer. In: Proceedings of the IEEE/CVF International Conference on Computer Vision, pp. 1833–1844 (2021)
10. Loshchilov, I., Hutter, F.: Decoupled weight decay regularization. arXiv preprint arXiv:1711.05101 (2017)
11. Van der Maaten, L., Hinton, G.: Visualizing data using t-sne. J. Mach. Learn. Res. **9**(11) (2008)
12. Potlapalli, V., Zamir, S.W., Khan, S., Khan, F.S.: Promptir: prompting for all-in-one blind image restoration. arXiv preprint arXiv:2306.13090 (2023)
13. Rajiah, P.S., François, C.J., Leiner, T.: Cardiac MRI: state of the art. Radiology **307**(3), 223008 (2023)
14. Roemer, P.B., Edelstein, W.A., Hayes, C.E., Souza, S.P., Mueller, O.M.: The NMR phased array. Magn. Reson. Med. **16**(2), 192–225 (1990)
15. Ronneberger, O., Fischer, P., Brox, T.: U-Net: convolutional networks for biomedical image segmentation. In: Navab, N., Hornegger, J., Wells, W.M., Frangi, A.F. (eds.) MICCAI 2015. LNCS, vol. 9351, pp. 234–241. Springer, Cham (2015). https://doi.org/10.1007/978-3-319-24574-4_28
16. Sriram, A., et al.: End-to-end variational networks for accelerated MRI reconstruction. In: Martel, A.L., et al. (eds.) MICCAI 2020. LNCS, vol. 12262, pp. 64–73. Springer, Cham (2020). https://doi.org/10.1007/978-3-030-59713-9_7
17. Wang, C., et al.: Cmrxrecon: an open cardiac MRI dataset for the competition of accelerated image reconstruction. arXiv preprint arXiv:2309.10836 (2023)

18. Zamir, S.W., Arora, A., Khan, S., Hayat, M., Khan, F.S., Yang, M.H.: Restormer: eDfficient transformer for high-resolution image restoration. In: Proceedings of the IEEE/CVF Conference on Computer Vision and Pattern Recognition, pp. 5728–5739 (2022)
19. Zbontar, J., et al.: fastmri: An open dataset and benchmarks for accelerated MRI. arXiv preprint arXiv:1811.08839 (2018)

Learnable Objective Image Function for Accelerated MRI Reconstruction

Artem Razumov[✉] and Dmitry V. Dylov

Skolkovo Institute of Science and Technology, Moscow, Russia
arazumov96@gmail.com

Abstract. Magnetic Resonance Imaging (MRI) provides strong contrast for soft tissues but requires long acquisition times, oftentimes resulting in the motion artifacts. Recent advancements in MRI reconstruction from undersampled data rely on compressed sensing (CS) and deep learning (DL) techniques, allowing for significant scan acceleration while maintaining the image quality nearly at the fully sampled level.

In this study, we propose to use the convolutional neural networks (CNN), such as U-net, to parametrize the objective function employed in the compressed sensing optimization problems. By doing so, our aim is to avoid unrealistic reconstructions often associated with traditional DL-based image reconstruction techniques.

To validate the proposed method, we used the CMRxRecon dataset containing cardiac raw k-space data. The results demonstrate realistic reconstruction of anatomical structures, outperforming classical CS reconstruction methods.

Keywords: compressed sensing · computer vision · deep learning

1 Introduction

Magnetic Resonance Imaging (MRI) has had a significant impact on medicine due to its excellent soft-tissue contrast and the absence of ionizing radiation. However, even today, MRI remains known for its long data acquisition times, which can range from 15 to 60 min to complete a single scan [4]. To reduce such long scan times, the compressed sensing (CS) methodology has been developed [3,4,9,14]. CS involves an incomplete sampling of the raw k-space data that an MR machine can acquire, with the consequent compensation of the artifacts caused by undersampling by reconstruction algorithms. These recovery algorithms use a prior knowledge about the data: the image sparsity in some domain [17].

The publication of the open fastMRI dataset [8,10,18] has resulted in a a notable resurgence of interest to this problem. The deep learning community eagerly unholstered the arsenal of available reconstruction methods, ranging from basic U-Net-like models, to those that incorporate specifics of the imaging process [11,12], to super-resolution methods [2]. The deep learning models proved superior to the classical reconstruction models in compressed sensing, ultimately resulting in their adoption by the MR industry.

O. Camara et al. (Eds.): STACOM 2023, LNCS 14507, pp. 274–282, 2024.
https://doi.org/10.1007/978-3-031-52448-6_26

The goal of this work is to propose a further development of CS methodology by additional learnable prior knowledge [7], capable of restricting the accelerated reconstructions to realistic high-quality scans of preserved diagnostic value.

2   Methods

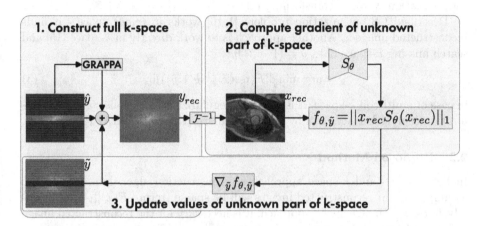

Fig. 1. Inference of the proposed method: a U-net like CNN is a part of compressed sensing reconstruction. Objective function $f_{\theta,\tilde{y}}$ is parametrized by parameters θ of the CNN. Note that the gradient of x_{rec} also depends on S_θ, i.e., the CNN is also used during the gradient computation. In the case of multicoil data, $y_{rec} = \hat{y} + \tilde{y} + y_{grappa}$, where y_{grappa} is the prediction of missing k-space data by GRAPPA convolution [5].

This section outlines how to approach under-sampled magnetic resonance imaging (MRI) reconstruction problems by first formulating the problem, describing classical methods for reconstruction, and then proposing our novel algorithm based on compressed sensing and learnable image objective function.

2.1   Compressed Sensing

We begin with notation of fully sampled image x, fully sampled multicoil k-space y. The under sampled k-space $\hat{y}$ can be modeled with sampling operator M, coil sensitivity operator S and Fourier transform operator $\mathcal{F}$:

$$\begin{cases} y = \mathcal{F}Sx \\ \hat{y} = My \end{cases} \tag{1}$$

The task it to find fully sampled image x (or fully sampled k-space data y) by given under sampled k-space data $\hat{y}$. Note that this approach can be generalized to single-coil k-space where S operator is just identity function.

The CS approach based on prior knowledge that real-world images are sparse [9], that is, l_1-norm of image (or total-variation, or l_1-norm of wavelet transform of image) should be less for fully sampled image than under sampled image. Thereby, CS solves reconstruction as following optimization problem:

$$x_{rec} = \arg \min_{x_{rec}} ||\hat{y} - M\mathcal{F}Sx||_2 + \lambda ||F(x)||_1 \qquad (2)$$

where λ - regularization parameter, F - some image transform (identity function, total variation, wavelet transform, etc.)

Described CS optimization problem [1,16] works in *image* space - it finds reconstructed image x. Another approach is to work directly in k-space [15] and search missing k-space data $\tilde{y} = (1 - M)y$:

$$\tilde{y} = \arg \min_{\tilde{y}} ||F(\text{rss}(\mathcal{F}^{-1}(\hat{y} + \tilde{y})))||_1 \qquad (3)$$

in case of multi-coil k-space, it is worth to regularize $\tilde{y}$ to be consistent with calibration central k-space data by adding regularization with GRAPPA operator.

2.2   Proposed Method

In both *image*- and k-space approaches reconstruction algorithms rely on the assumption that reconstructed images will be sparse (term $||F(x)||_1$). Additionally, in cases of image-space approach, it is necessary for the reconstructed image to be consistent with recorded signals (term $||\hat{y} - M\mathcal{F}Sx||_2$), while in cases of k-space approach and multi-coil data, consistency must also be maintained with calibration data [15]. However, these assumptions may not suffice when dealing with highly accelerated data, which prompts us to propose extending prior knowledge about such data by utilizing a CNN (see Fig. 1).

We propose to extend k-space CS approach by using U-net like CNN model $S_\theta(\cdot)$ with learnable parameters θ:

$$\begin{cases} y_{grappa} = G(\hat{y}) \\ x_{rec} = \text{rss}(\mathcal{F}^{-1}(\hat{y} + y_{grappa} + \tilde{y})) \\ \tilde{y} = \arg \min_{\tilde{y}} ||x_{rec}S_\theta(x_{rec})||_1 \end{cases} \qquad (4)$$

where x_{rec} is root-sum-of-squares reconstruction, y_{grappa} is missing k-space data, predicted by GRAPPA kernels.

In case of singlecoil data, proposed approach simplified to:

$$\begin{cases} x_{rec} = \text{rss}(\mathcal{F}^{-1}(\hat{y} + \tilde{y})) \\ \tilde{y} = \arg \min_{\tilde{y}} ||x_{rec}S_\theta(x_{rec})||_1 \end{cases} \qquad (5)$$

These optimization problems are solved by Adam algorithm with numerical gradient computation. The parameters θ of U-net model $S_\theta(\cdot)$ are optimized by Adam algorithm by solving following optimization problem:

$$\theta = \arg \min_\theta \mathbb{E}||x - x_{rec}||_1 \qquad (6)$$

where x_{rec} is reconstructed image.

Algorithm 1: Learnable objective function training

Data: $Y \in \mathbb{C}^N$ - full k-space data set, M - undersampling operator
Result: θ - parameters of U-net model $S_\theta(\cdot)$
initialize $\theta, \theta_{m_1}, \theta_{m_2}$;
for *training epoch* **do**
    for $y \leftarrow sample(Y)$ **do**
        $\hat{y} \leftarrow My$ $\triangleright$ undersampling k-space data ;
        $\tilde{y} \leftarrow 0$ $\triangleright$ init unknown k-space data ;
        $\tilde{y}_{m_1} \leftarrow 0$ $\triangleright$ init 1st moment;
        $\tilde{y}_{m_2} \leftarrow 0$ $\triangleright$ init 2nd moment;
        for $i \leftarrow 0$ **to** *niter* **do**
            $x_{rec} \leftarrow \text{rss}(\mathcal{F}^{-1}(\hat{y} + \tilde{y}))$;
            $f_{\theta,\tilde{y}} \leftarrow \|x_{rec} \cdot S_\theta(x_{rec})\|_1$;
            $\text{grad}_{\tilde{y}} \leftarrow \nabla_{\tilde{y}} f_{\theta,\tilde{y}}$;
            $\tilde{y}, \tilde{y}_{m_1}, \tilde{y}_{m_2} \leftarrow \text{AdamStep}(\text{grad}_{\tilde{y}}, \tilde{y}_{m_1}, \tilde{y}_{m_2})$;
        end
        $x_{rec} \leftarrow \text{rss}(\mathcal{F}^{-1}(\hat{y} + \tilde{y}))$;
        $\mathcal{L} \leftarrow \|x_{rec} - x\|_1$;
        $\text{grad}_\theta \leftarrow \nabla_\theta \mathcal{L}$;
        $\theta, \theta_{m_1}, \theta_{m_2} \leftarrow \text{AdamStep}(\text{grad}_\theta, \theta_{m_1}, \theta_{m_2})$;
    end
end

---

3   Experiments

The short-axis cine MRI dataset from CMRxRecon Challenge (Cine SAX) with the raw 2D k-space with about 5–10 slices and 12 timesteps was used. Due to computational complexity, only first timestemp was used for training.

For this study, we used data from the CMRxRecon Challenge that included raw 2D k-space. We only used the first timestamp for training our models due to computational complexity of training procedure. To compare with other methods, we employed Berkeley Advanced Reconstruction Toolbox's (BART) l_1-wavelet FISTA reconstruction algorithm on both single-coil and multi-coil data sets. For singlecoil, reconstruction was obtained with `bart pics -R W:1:2:0.01 -S`. In case of multicoil data, sensitivity map was evaluated by `bart ecalib -m1 -S -a -W`.

We then evaluated the reconstructed images using SSIM, PSNR image metrics. Additionally, segmentation U-net model was trained on fully sampled images and provided labels of cardiac cavities. DICE score performance of pretrained segmentation model on reconstructed images was also used for scoring.

For training U-net for proposed reconstruction algorithm, we used simple U-net model [13] with 32 channels and 4 pool layers but with instance norm ans SiLU activations. The optimization for finding reconstructed images was performed by Adam algorithm [6] with learning rage 1e–3 and *eps* equal to 1e–2 for better training stability and $\beta_1 = 0.9$ and $\beta_2 = 0.999$. Optimization of U-net parameters was also performed by Adam with learning rate 1e–3.

4   Results

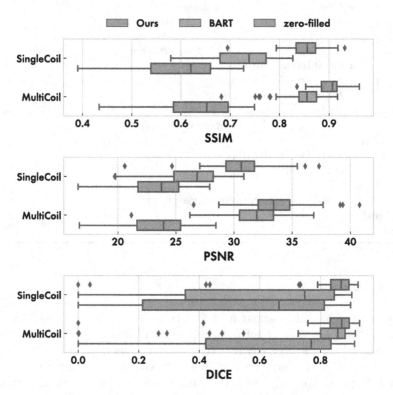

Fig. 2. Image quality and segmentation scores on reconstructed images. Note DICE variance for zer-filled and BART reconstructed images

The comparison of image metrics and segmentation scores on a validation dataset is demonstrated in Fig. 2. Our proposed method outperforms BART as indicated by high SSIM, PSNR image metrics, as well as DICE segmentation scores. The line plot in Fig. 3 also shows that our proposed method has less correlation with aliasing artifacts caused by undersampling than does the BART algorithm, but has larger varience than BART.

The proposed method demonstrates a lower gain in SSIM score comapring singlecoil and multicoil data. The mean SSIM score of the proposed method on single coils was 0.85 while it was 0.91 for multi-coils (see Fig. 2). On the other hand, BART has lower scores but larger differences between single and multi-coil SSIM scores compared to our proposed method (mean SSIM score of 0.72

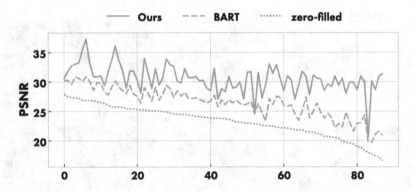

Fig. 3. PSNR scores from the validation dataset. The scores are plotted in descending order on the zero-filled reconstruction. It is worth noting that for BART models, the PSNR scores tend to decrease along with the zero-filled scores. In contrast, our proposed model maintains a relatively consistent level of PSNR scores regardless of the change in zero-filled scores. However, our method exhibits higher variance than BART

for single coils vs. mean SSIM score of 0.85 for multi-coils), indicating that our proposed method has slightly smaller gain from usign multicoil data. This is supported by Fig. 4, which shows that the BART reconstruction between single and multicoil images are noticeably different while our proposed method has a small difference in terms of overall scores but still performs better than BART on both types of images.

In some cases (as shown in Fig. 5), the image metrics between our proposed method and BART are not very different but there may be instances where BART's reconstruction does not fully restore or blurred regions, which can prevent a segmentation model from correctly identifying cardiac cavities. This is likely due to blurring of borders in certain areas that affect the accuracy of the segmentation on BART's reconstruction compared to our proposed method where sharp and distinguishable cardiac cavity borders are present.

5   Discussion

As shown in Fig. 2 and 3, our proposed method outperforms BART when both methods are iteratively reconstructing images by solving minimization problem - but with different objective function for each method: l_1-norm of wavelet transform of image $||W(x_{rec})||_1$ for BART vs $||x_{rec} \cdot S_\theta(x_{rec})||_1$ - l_1-norm parametrized by $S_\theta(\cdot)$ U-net model for ours method). This supports the idea that additional prior knowledge can be incorporated into the objective function to improve compressed sensing approach and reconstruct an image closer to a fully sampled one. Additionally, visually, our proposed method has sharper and more detailed reconstruction compared to BART on ×10 undersampling, without introducing new artifacts or non-existent image features (which is common for straightforward deep learning image reconstruction models). Therefore, similar to many

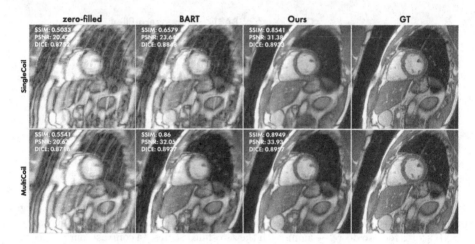

Fig. 4. Comparison of different reconstruction methods on singlecoil and multicoil *k*-space. SSIM and PSNR scores are evaluated for demonstrated cropped region.

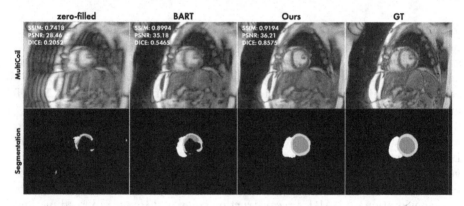

Fig. 5. Segmentation performance of different reconstruction methods multicoil *k*-space.

compressed sensing techniques, our proposed method solves the reconstruction problem by solving minimization problem, but also takes advantage of deep learning models by incorporating additional prior knowledge into its objective function.

Limitations. Although the requirements for our proposed method inference match those of straightforward deep learning models, its training process involves significantly higher computational and memory costs. This is due to the computation of second order gradients during model parameter optimization, which can make training more unstable and requires storing large computational graph for all iterations related to prediction of reconstructed images. As a result, we were limited in batch size of 4 and number of reconstruction iterations (5 iterations) for Tesla V100 with 32 Gb GPU memory.

Conclusion. In this paper, we present a technique for undersampled MRI reconstruction based on a learnable objective function. We validated the effectiveness of our proposed approach using the CMRxRecon dataset and demonstrated its superiority over classical compressed sensing techniques by achieving reconstructions that are closer to fully sampled images, thanks in part to the incorporation of deep learning models into the objective function.

References

1. Beck, A., Teboulle, M.: A fast iterative shrinkage-thresholding algorithm for linear inverse problems. SIAM J. Imaging Sci. **2**(1), 183–202 (2009)
2. Belov, A., Stadelmann, J., Kastryulin, S., Dylov, D.V.: Towards Ultrafast MRI via Extreme k-Space Undersampling and Superresolution. In: de Bruijne, M., Cattin, P.C., Cotin, S., Padoy, N., Speidel, S., Zheng, Y., Essert, C. (eds.) MICCAI 2021. LNCS, vol. 12906, pp. 254–264. Springer, Cham (2021). https://doi.org/10.1007/978-3-030-87231-1_25
3. Candès, E.J., Romberg, J., Tao, T.: Robust uncertainty principles: exact signal reconstruction from highly incomplete frequency information. IEEE Trans. Inf. Theory **52**(2), 489–509 (2006)
4. Debatin, J.F., McKinnon, G.C.: Ultrafast MRI. Springer, Heidelberg (1998). https://doi.org/10.1007/978-3-642-80384-0
5. Griswold, M.A., et al.: Generalized autocalibrating partially parallel acquisitions (grappa). Magn. Reson. Med. Off. J. Int. Soc. Magn. Reson. Med. **47**(6), 1202–1210 (2002)
6. Kingma, D.P., Ba, J.L.: Adam: a method for stochastic optimization (2015)
7. Kuzmina, E., Razumov, A., Rogov, O.Y., Adalsteinsson, E., White, J., Dylov, D.V.: Autofocusing+: noise-resilient motion correction in magnetic resonance imaging. In: International Conference on Medical Image Computing and Computer-Assisted Intervention, pp. 365–375. Springer, Heidelberg (2022), https://doi.org/10.1007/978-3-031-16446-0_35
8. Liu, R., Zhang, Y., Cheng, S., Luo, Z., Fan, X.: A deep framework assembling principled modules for CS-MRI: unrolling perspective, convergence behaviors, and practical modeling. IEEE Trans. Med. Imaging **39**(12), 4150–4163 (2020)
9. Lustig, M., Donoho, D., Pauly, J.M.: Sparse MRI: the application of compressed sensing for rapid MR imaging. Magn. Reson. Med. **58**, 1182–1195 (2007)
10. Oh, G., et al.: Unpaired deep learning for accelerated MRI using optimal transport driven cycleGAN. IEEE Trans. Comput. Imag. **6**, 1285–1296 (2020)
11. Pezzotti, N., et al.: An adaptive intelligence algorithm for undersampled knee MRI reconstruction. IEEE Access **8**, 204825–204838 (2020)
12. Razumov, A., Rogov, O., Dylov, D.V.: Optimal MRI undersampling patterns for ultimate benefit of medical vision tasks. Magn. Reson. Imaging **103**, 37–47 (2023)
13. Ronneberger, O., Fischer, P., Brox, T.: U-net: convolutional networks for biomedical image segmentation. In: Navab, N., Hornegger, J., Wells, W.M., Frangi, A.F. (eds.) MICCAI 2015. LNCS, vol. 9351, pp. 234–241. Springer, Cham (2015). https://doi.org/10.1007/978-3-319-24574-4_28

14. Song, L., Zhang, J., Wang, Q.: MRI reconstruction based on three regularizations: total variation and two wavelets. Biomed. Signal Process. Control **30**, 64–69 (2016)
15. Uecker, M., et al.: Espirit-an eigenvalue approach to autocalibrating parallel MRI: where sense meets grappa. Magn. Reson. Med. **71**(3), 990–1001 (2014)
16. Yang, J., Zhang, Y., Yin, W.: A fast alternating direction method for tvl1-l2 signal reconstruction from partial Fourier data. IEEE J. Sel. Topics Signal Process. **4**(2), 288–297 (2010)
17. Ye, J.C.: Compressed sensing MRI: a review from signal processing perspective. BMC Biomed. Eng. **1**(1), 1–17 (2019)
18. Zbontar, J., et al.: fastMRI: an open dataset and benchmarks for accelerated MRI (2019)

Accelerating Cardiac MRI via Deblurring Without Sensitivity Estimation

Jin He, Weizhou Liu, Yun Tian[(⊠)], and Shifeng Zhao

School of Artificial Intelligence, Beijing Normal University, Beijing, China
tianyun@bnu.edu.cn

Abstract. Reducing acquisition time in Cardiac Magnetic Resonance Imaging (MRI) brings several benefits, such as improved patient comfort, reduced motion artifacts and increased doctors' work efficiency, but it may lead to image blurring during the reconstruction process. In this paper, we propose a new method for restoring blurry cardiac MRI images caused by under-sampling, treating it as an image deblurring problem to achieve clear reconstruction, and ensuring consistency with training by using a simple modified input during inference. A U-Net network architecture which initially designed for natural image deblurring, has been adapted to effectively discern the differences between blurred and clear MRI images, eliminating the need for sensitivity estimation. Moreover, to address the inconsistency between training on local patches and testing on the entire image, we propose a partial overlap cropping approach during inference time, effectively resolving this discrepancy. We evaluated our method using the cardiac MRI dataset from the CMRxRecon challenge, revealing its potential to reduce acquisition time while preserving high image quality in cardiac MRI, even under highly under-sampled conditions. Importantly, this achievement was attained in a coil-agnostic manner, enabling us to achieve favorable results on both multi-coil and single-coil data.

1 Introduction

Cardiac magnetic resonance imaging (MRI) has emerged as a crucial imaging modality for diagnosing cardiac diseases, owing to its superior soft tissue contrast, non-radiation and non-invasive nature. In addition to its diagnostic capabilities, it has numerous other important applications, such as assessing heart function, researching the development of new therapies, and monitoring the progression of heart diseases and the efficacy of treatments. However, the lengthy scanning time of cardiac MRI can cause discomfort to patients and lead to motion artifacts in the images. Therefore, there is a pressing need for high-quality, fast imaging from highly under-sampled k-space (raw data). Designing an effective approach for accelerating cardiac MRI to meet these criteria remains an open challenge.

Accelerated MRI or parallel imaging (PI), which is a type of MRI reconstruction technique, utilizes multiple coils to acquire data simultaneously, producing high-quality images with shorter scan times. Coil sensitivity, based on

O. Camara et al. (Eds.): STACOM 2023, LNCS 14507, pp. 283–292, 2024.
https://doi.org/10.1007/978-3-031-52448-6_27

size, shape, and arrangement of the coils, allows for image reconstruction with fewer samples than conventional MRI. Reconstruction compensates for k-space subsampling using knowledge of coil sensitivities and sampling trajectory.

Traditional PI methods like SENSE [1] and GRAPPA [2] are common in clinical settings, with each coil's sensitivity map weighting the acquired data. SENSE applies Fourier inverse transform to under-sampled k-space for image recovery, while GRAPPA estimates fully-sampled data then uses Fourier transform. Other methods include iterative self-consistent parallel imaging reconstruction (SPIRiT) [3], parallel imaging using eigenvector maps (ESPIRiT) [4], and annihilating filter-based low-rank Hankel matrix (ALOHA) [5]. Although these methods are commonly employed by medical device manufacturers, they can be time-consuming. Moreover, When the k-space data is highly under-sampled, such as with acceleration factors of 4, 8, or 10, these methods may not perform well and can result in significant image artifacts.

Recent advances in deep neural networks, include convolutional neural networks [6,7] and recursive neural networks [8,9], offers a promising solution to address this issue, which has sparked further research in Accelerated MRI. This methods can capture more global information or high/low-frequency information, but they come with computationally complex issues. And more importantly, they require the prior knowledge of coil sensitivities, which can be susceptible to coil configuration discrepancies and estimation errors. Therefore, these approaches may not be suitable when dealing with multi-coil data that has errors in coil sensitivity or lacks coil sensitivity information, or when only single-coil data is available.

In this work, we observe that under-sampling leads to blurring in the zero-filled images obtained from parallel imaging, even at much lower sampling rates than the Nyquist sampling theorem. To that end, a deblurring method is introduced into medical MRI image reconstruction, which has the advantages of both accurate reconstruction speed and complete coil-agnosticism over the aforementioned method. The approach can handle data without given coil sensitivities, including single-coil data.

To the best of our knowledge, this is the first work to apply a natural image deblurring network to MRI reconstruction. Specifically, the k-space data is preprocessed into a 2D image format before feeding it into the network, which enables the network to adapt to both k-space and image-domain datasets simultaneously. We introduce a network that employs a U-Net structure from coarse to fine, augmented with convolution blocks to capture global and frequency information. Notably, because of the inconsistency of the distribution in training (with cropped patches from images) and inference (with the full-resolution image), we propose a strategy for test-time cropping of images intelligently adapts to the available GPU memory, boosting performance without the need for retraining or fine-tuning.

2   Method

We apply a deblurring method to the blurred cardiac MRI images obtained from k-space transformation. The overall framework is illustrated in Fig. 1. We provide a rationale for treating the reconstruction task as a deblurring problem. To this end, we introduce a multi-input multi-output U-Net (MIMO-UNet) network architecture. This architecture is a classical network structure used for natural image deblurring, offering lower computational complexity compared to the Transformer network structure and the ability to adapt to medical images. Additionally, a new ResBlock is incorporated into the network, which captures both long-term and short-term interactions while integrating low-frequency and high-frequency residual information. Next, to solve the inconsistency, we illustrate our approach for dividing images into overlapping windows of uniform size during the test phase. These windows are then fed into the network for processing, after which they are reassembled to reconstruct an image that matches the original size.

2.1   Problem Formulation

Let I_f denote the fully-sampled image, I_u represent the observed under-sampled image, and ε indicate the noise. In addition, $\mathcal{F}$ represents the Fourier Transform matrix, $\mathcal{F}^{-1}$ represents the Inverse Fourier Transform matrix, $\mathcal{M}$ represents the under-sampling mask, and $\odot$ denotes the element-wise multiplication operation. The MRI reconstruction process, as referenced in [6], can be formulated as $\mathcal{F}(I_u) = \mathcal{M} \odot \mathcal{F}(I_f) + \varepsilon$. Based on this formulation, the following equation can be derived:

$$I_u = \mathcal{F}^{-1}(\mathcal{M}) * I_f + \mathcal{F}^{-1}(\varepsilon) \tag{1}$$

On the other hand, as referenced in [14], a blurred image is denoted as:

$$I_b = \Phi(I_s; \theta) \tag{2}$$

where Φ represents the image blur function, θ is a parameter vector, and I_s is the latent sharp version of the blurred image I_b. The objective of image deblurring is to recover a sharp image by finding the inverse of the blur function, denoted as $I_{db} = \Phi^{-1}(I_b; \theta)$, where Φ^{-1} is the deblurring model, and I_{db} represents the deblurred image, which serves as an estimate of the latent sharp image I_s. Specifically, the motion blur process can be modeled as a convolution process:

$$I_b = K * I_s + \theta \tag{3}$$

where K represents the blur kernel and θ is additive Gaussian noise. When we represent the blur process in Eq. 2 using Eq. 3, θ corresponds to the blur kernel and Gaussian noise, while Φ corresponds to the convolution and sum operator. This equation exhibits similarity to Eq. 1.

The above content indicates that a motion deblur network architecture can be applied for MR reconstruction. Image deblurring is a well-studied problem in the field of computer vision, and the literature in this domain is extensive and diverse. We leave for future work to inspect the most performing architecture. Nonetheless, by borrowing from and adapting image deblurring methods, new solutions and improvements can be provided for reconstruction problems.

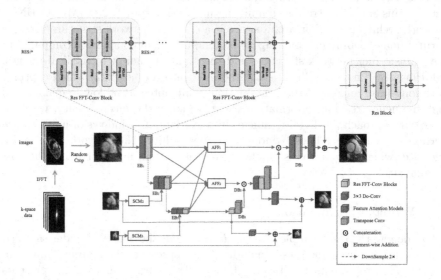

Fig. 1. Overall Framework of Training Network.

2.2   Training Network

Network Structure. MIMO-UNet [10] is composed of encoding blocks (EB), decoding blocks (DB), shallow convolutional modules (SCM), feature attention modules (FAM), and asymmetric feature fusions (AFF), utilizing a single U-Net network to simulate a multi-level cascaded U-Net network for single-image deblurring. SCM extracts features from down-sampled blurry images, forming a residual structure with input image and fine-tuning combined features. FAM combines SCM output with previous EB, emphasizing or suppressing features and learning spatial/channel importance. AFF integrates EB output, enabling multi-scale feature integration via convolutional layers, enhancing information flow across architectural scales.

The Deep Residual Fourier Transform (DeepRFT) [11] framework is based on the design of MIMO-UNet. In MIMO-UNet, all ResBlocks are replaced by Res FFT-Conv Blocks, which consist of two streams for residual learning: (1) The FFT-Conv stream performs a 1×1 convolution after transforming spatial feature maps into the frequency domain, enabling low-frequency and high-frequency learning. (2) The regular convolution stream focuses on local details and is better suited for learning high-frequency differences. Moreover, to achieve significant deblurring performance improvement, the Depthwise Over-parameterized Convolutional layer (DO-Conv) [12] is employed instead of conventional convolution operations. DO-Conv introduces performance gains for image deblurring without introducing additional parameters.

Loss Functions. We utilize three loss functions for our network: Charbonnier Loss, Edge Loss, and L1 loss for frequency domain images. In our context, $k \in \{0, ..., K - 1\}$ represents the k-th level image in DeepRFT, I denotes the k-th level reconstructed image, and $\tilde{I}$ represents the k-th level ground truth image. Additionally, the constant value ε is empirically set to 10^{-3}.

(1) Charbonnier Loss: $\mathcal{L}_{char} = \sqrt{||\tilde{I} - I||^2 + \varepsilon^2)}$

(2) Edge Loss: $\mathcal{L}_{edge} = \sqrt{||\Delta\tilde{I} - \Delta I||^2 + \varepsilon^2)}$, where Δ denotes the Laplacian operator.

(3) FFT Loss: $\mathcal{L}_{fft} = ||FT(\tilde{I}) - FT(I)||_1$, where FT denotes the FFT operation.

Our final loss then becomes:

$$\mathcal{L}_{total} = \sum_{k=0}^{K-1} \mathcal{L}_{char}^k + \gamma_1 \mathcal{L}_{edge}^k + \gamma_2 \mathcal{L}_{fft}^k$$

where γ_1 and γ_2 are hyperparameters, and γ_1 is set to 0.05 and γ_2 is set to 0.01 empirically.

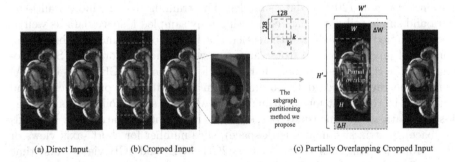

(a) Direct Input   (b) Cropped Input   (c) Partially Overlapping Cropped Input

Fig. 2. Several methods for processing input images during inference time. On the left side of each image pair is the input image, and on the right side is the reconstructed image. (a) represents the input image in which the distribution is inconsistent with the training data. (b) corresponds to the input image obtained by cropping without any overlap, where Blue arrows show artifacts from direct cropping, causing visible lines. Our proposed approach (c) involves partial overlap cropping, which intuitively yields better deblurring effects. (Color figure online)

2.3   Inference Time

Compared to the methods (a) and (b) in Fig. 2, we propose a partial overlap cropping approach to process the input image (Fig. 2c). Each image with a size of $H \times W$ is divided into several 128×128 sub-images, where adjacent sub-images share the same area and width in the overlapping regions. We denote the overlap width as k, and the width and height of the image after padding as W' and H', respectively. The additional width and height introduced by the padding are represented by ΔW and ΔH. We enforce a constraint that each pair of adjacent sub-images can overlap at most once, resulting in a total of the following number of sub-images for a given image:

$$\frac{W' - k}{128 - k} \cdot \frac{H' - k}{128 - k}$$

where $H' = H + \Delta H$ with $\Delta H \equiv 128 - k \bmod 128$, and $W' = W + \Delta W$ with $\Delta W \equiv 128 - W \bmod 128$. It can be easily proven that the additional time complexity caused by the cropping-overlapping technique is $O(\frac{W'H'}{64^2})$. Since the length of the images in this dataset is on the order of hundreds, each image is divided into 8~12 sub-images.

3   Experiments

3.1   Dataset

The CMRxRecon challenge dataset [15] comprises 120 training data, 60 validation data, and 120 test data samples. The training data includes multi-coil data and single-coil data, each comprising fully sampled k-space data as well as under-sampled k-space data with acceleration factors of 4, 8, and 10. The dataset consists of images captured from various views, including short-axis (SA), two-chamber (2CH), three-chamber (3CH), and four-chamber (4CH) long-axis (LA) views. The raw k-space data obtained from the scanner is pre-processed and converted to the *.mat format using MATLAB. The single-coil data is complex k-space data with dimensions (kx, ky, sz, t) representing matrix size in x-axis (k-space), matrix size in y-axis (k-space), slice number for short -axis views or slice group for long-axis views (such as 2CH, 3CH, and 4CH views), and time frame. On the other hand, the multi-coil data with dimensions (kx, ky, sc, sz, t) is compressed to 10 virtual coils to facilitate standardization and optimize storage efficiency. Here, sc represents the coil array number, which is set to 10.

3.2   Implementation Details

Data Preprocessing. During the training phase, we begin by converting the raw k-space data into image domain data with dimensions (kx, ky, sz, t) using the inverse fast Fourier transform (IFFT). Then, we combine all types of data, including single-coil and multi-coil data with acceleration factors of 4, 8, and

10, to create the training set for the network. Next, we split each image data into individual images of size (kx, ky). Subsequently, we apply data augmentation techniques to these two-dimensional images, including flipping, rotation, gamma correction, and random cropping into blocks of size 128×128. These augmentation methods play a vital role in enhancing the network's robustness and generalization capability.

In the testing phase, we also partition the input images, which have undergone IFFT transformation and possess dimensions of (kx, ky, sz, t), into distinct data subsets with dimensions of (kx, ky). However, in contrast to the data shuffling process employed during training, we employ the number of time frames, t, as the batch size. This choice is predicated on our observation that the size of t in the dataset remains constant, thereby ensuring that it remains within the storage capacity of the GPU. As detailed in Sect. 2.2, there exists a disparity between the training and testing distributions. Consequently, we adopt the aforementioned methodology and divide each (kx, ky) image into multiple 128×128 patches with partial overlap. These patches are subsequently inputted into the pre-trained network, and the resultant outputs are cropped and reassembled to reconstruct the original image.

Table 1. The quantitative metric results of the mean values for the multi-coil and single-coil datasets under three acceleration factors.

Type	Acc.		PSNR	SSIM	RMSE
Single-coil	Sax	4×	32.1724	0.8756	0.0181
		8×	30.4539	0.8506	0.0265
		10×	30.1156	0.8446	0.0286
	Lax	4×	29.3903	0.8202	0.0278
		8×	28.0868	0.7921	0.0380
		10×	27.5125	0.7799	0.0430
	Total		**29.6219**	**0.8272**	**0.0303**
Multi-coil	Sax	4×	35.298	0.9229	0.0089
		8×	32.6465	0.8909	0.0166
		10×	31.9710	0.8811	0.0189
	Lax	4×	32.1404	0.8789	0.0148
		8×	29.8762	0.8401	0.0251
		10×	29.1551	0.8247	0.0294
	Total		**31.8478**	**0.8731**	**0.0189**

Training Details. This study was conducted using PyTorch on two NVIDIA Tesla V100 (32 GB) GPU workstations. Taking inspiration from the MPR-Net [13] and DeepRFT, we employed the following network training hyperparameters: optimizer (Adam) and a learning rate that gradually decays to

1×10^{-6} using the cosine annealing strategy. Considering the CMRxRecon dataset, where the images had a shortest side smaller than 256 but larger than 128, we adjusted the patch size to 128×128. Moreover, taking into account the adapted GPU size and the dataset differences compared to DeepRFT, we set the batch size to 64, the initial learning rate for 3×10^{-4}, and training for 800 epochs to accommodate the GPU capabilities and dataset size.

Evaluation Metrics. In the testing phase, the image reconstruction process is capable of running on a relatively small GPU workstation, such as an 8 GB GPU, which demonstrates the memory-friendly nature of our method. To evaluate the quality of the reconstructed images, the challenge utilized three objective metrics for all the images: SSIM (Structural Similarity Index Measure), PSNR (Peak Signal-to-Noise Ratio), and RMSE (Root Mean Square Error). These metrics provide quantitative measures to assess the fidelity and similarity of the reconstructed images compared to their ground truth counterparts.

During the competition phase, we faced limitations in accessing the ground truth images corresponding to the validation set, and no separate test set was provided. As a result, our evaluation of the network's performance was solely based on the validation set. Therefore, all experimental data presented in this paper represent the performance of our method exclusively on the validation set. Following the competition requirements, we employed specific strategies during the validation phase. To facilitate ranking, we use only the central 2 slices of the images and considered the first 3 time frames. Additionally, the original images are cropped by discarding the outer 1/6 of the image.

3.3   Results

Quantitative Metrics Results. We present the reconstruction results of both multi-coil and single-coil datasets, under-sampled at $4\times, 8\times$, and $10\times$, in Table 1. Taking into account the outcomes of other contestants, it is evident that our method achieves outstanding performance not only on the multi-coil dataset but also on the single-coil dataset, approaching a closer resemblance to the ground truth. Certainly, the limitation of our method lies in its inability to achieve the same level of performance when applied to single-coil datasets compared to multi-coil datasets. Additionally, it does not attain the top position in terms of performance when evaluated on multi-coil datasets.

Furthermore, based on current information, our method holds the top position on the online leaderboard for the single-coil dataset in the challenge. We observe that the objective performance of single-coil data is generally inferior to that of multi-coil data, mainly due to the ability of multi-coil data to capture more spatial information and local details. However, in specific application scenarios such as fast imaging requirements, specific regions of interest, particularly the heart, and cost-effectiveness considerations, single-coil data still holds its advantages. Our method effectively leverages these advantages and enables single-coil data to achieve excellent reconstruction results in such scenarios.

Table 2. Compare the effects of inputting original images, cropped images, and the inference time strategy we proposed for image processing. The last two row shows the average time required for reconstructing a raw k-space data under the three strategies.

Metric	Direct input	Cropped input	Ours
PSNR	31.5196	31.5877	31.8478
SSIM	0.8724	0.8682	0.8731
RMSE	0.0201	0.0314	0.0303
Time (s) - Multi-coil	46.63	48.15	51.10
Time (s) - Single-coil	19.90	22.10	24.87

Table 3. Comparing the impact of model size on performance. Furthermore, when processing the data without using any method, simply reading and writing the multi-coil data requires 31.08 s per person.

Metric	DeepRFT-small	DeepRFT	DeepRFT-plus
PSNR	30.8899	31.4084	31.8478
SSIM	0.8552	0.8630	0.8731
RMSE	0.0368	0.0331	0.0303
FLOPs (G)	0.54	0.56	0.58
Params (M)	5.42	10.18	24.47
Time (s/p) - Multi-coil	32.62	41.77	51.10
Time (s/p) - Single-coil	11.15	15.72	24.87

Ablation Study. As shown in Table 2, our proposed method achieved the best performance in terms of the evaluation metrics for processing the test-time images, thereby validating the effectiveness of our approach. Moreover, we conducted an analysis in Table 3 to investigate the influence of model size on the reconstruction results, and the results confirmed the optimality of the selected model size. Furthermore, we demonstrated the fast execution speed of our method, as presented in Table 3, where the average time required to reconstruct a raw k-space data range from 32 to 51 s.

4   Conclusion

In this study, we introduce a coil-agnostic framework that treats accelerated MRI reconstruction as a deblurring problem and addresses the issue of inconsistent training/inference input distributions. In experiments, our proposed framework outperforms many competitors and exhibits outstanding performance on single-coil data, contributing to enhanced patient experiences and improved diagnostic outcomes. In ablation experiments, our proposed pipeline demonstrates a small parameter count and fast reconstruction speed. Future research directions include exploring the effectiveness of more deblurring-related frameworks and incorporating frequency-domain images or constructing networks with 3D or even entire cine sequences as inputs to integrate more information.

References

1. Pruessmann, K.P., Weiger, M., Scheidegger, M.B., Boesiger, P.: SENSE: Sensitivity encoding for fast MRI. Magn. Reson. Med. **42**(5), 952–962 (1999)
2. Griswold, M.A., et al.: Generalized autocalibrating partially parallel acquisitions (GRAPPA). Magn. Reson. Med. **47**(6), 1202–1210 (2002)
3. Lustig, M., Pauly, J.M.: SPIRiT: iterative self-consistent parallel imaging reconstruction from arbitrary k-space. Magn. Reson. Med. **64**(2), 457–471 (2010)
4. Uecker, M., et al.: ESPIRiT-an eigenvalue approach to autocalibrating parallel MRI: where SENSE meets GRAPPA. Magn. Reson. Med. **71**(3), 990–1001 (2014)
5. Lee, J., Jin, K.H., Ye, J.C.: Reference-free single-pass EPI Nyquist ghost correction using annihilating filter-based low rank Hankel matrix (ALOHA). Magn. Reson. Med. **76**(6), 1775–1789 (2016)
6. Lyu, J., et al.: Region-focused multi-view transformer-based generative adversarial network for cardiac cine MRI reconstruction. Med. Image. Anal. **85**, 102760 (2023)
7. Lv, J., Wang, P., Tong, X., Wang, C.: Parallel imaging with a combination of sensitivity encoding and generative adversarial networks. Quant. Imaging Med. Surg. **10**(12), 2260–2273 (2020)
8. Qin, C., Schlemper, J., Caballero, J., Price, A.N., Hajnal, J.V., Rueckert, D.: Convolutional recurrent neural networks for dynamic MR image reconstruction. IEEE Trans. Med. Imaging **38**(1), 280–290 (2019)
9. Qin, C., et al.: Complementary time-frequency domain networks for dynamic parallel MR image reconstruction. Magn. Reson. Med. **386**(6), 3274–3291 (2021)
10. Cho, S.-J., Ji, S.-W. , Hong, J.-P., Jung, S.-W., Ko, S.-J.: Rethinking coarse-to-fine approach in single image deblurring. In: 2021 IEEE/CVF International Conference on Computer Vision (ICCV), pp. 4621–4630 (2021)
11. Mao, X., Liu, Y., Liu, F., Li, Q., Shen, W., Wang, Y.: Intriguing findings of frequency selection for image deblurring. In: Proceedings of the AAAI Conference on Artificial Intelligence, vol. 37, pp. 1905–1913 (2023)
12. Cao, J., et al.: DO-conv: depthwise over-parameterized convolutional layer. IEEE Trans. Image Process. **31**, 3726–3736 (2022)
13. Mehri, A., Ardakani, P.B., Sappa, A.D.: MPRNet: multi-path residual network for lightweight image super resolution. In: 2021 IEEE Winter Conference on Applications of Computer Vision (WACV), pp. 2703–2712 (2021)
14. Zhang, K., et al.: Deep image deblurring: a survey. Int. J. Comput. Vision **130**(9), 2103–2130 (2022)
15. Wang, C., Lyu, J., Wang, S., et al.: CMRxRecon: an open cardiac MRI dataset for the competition of accelerated image reconstruction. arXiv preprint arXiv:2309.10836 (2023)

T1/T2 Relaxation Temporal Modelling from Accelerated Acquisitions Using a Latent Transformer

Michael Tänzer[1,2] , Fanwen Wang[1,2(✉)] , Mengyun Qiao[1] , Wenjia Bai[1] ,
Daniel Rueckert[1,3] , Guang Yang[1,2] , and Sonia Nielles-Vallespin[1,2]

[1] Imperial College London, Exhibition Road, London SW7 2AZ, UK
[2] Royal Brompton and Harefield Hospital, Sydney Street, London SW3 6NP, UK
fanwen.wang@imperial.ac.uk
[3] Technische Universität München (TUM), Arcisstraße 21, 80333 München, Germany

Abstract. Quantitative cardiac magnetic resonance T1 and T2 mapping enable myocardial tissue characterisation but the lengthy scan times restrict their widespread clinical application. We propose a deep learning method that incorporates a time dependency *Latent Transformer* module to model relationships between parameterised time frames for improved reconstruction from undersampled data. The module, implemented as a multi-resolution sequence-to-sequence transformer, is integrated into an encoder-decoder architecture to leverage the inherent temporal correlations in relaxation processes. The presented results for accelerated T1 and T2 mapping show the model recovers maps with higher fidelity by explicit incorporation of time dynamics. This work demonstrates the importance of temporal modelling for artifact-free reconstruction in quantitative MRI.

Keywords: Quantitative MRI · Deep learning · MRI reconstruction

1 Introduction

Magnetic resonance imaging (MRI) is a crucial non-invasive tool for assessing tissue properties and functions, with applications across medical disciplines. In cardiac imaging, quantitative T1 and T2 mapping provide insights into myocardial composition and structure, enabling characterisation of cardiomyocytes [7, 8, 10]. However, long scan times limit the clinical utility of cardiac T1 and T2 mapping. Although undersampled acquisitions offer a means to accelerate scans, they often lead to artifacts and errors, not only in the reconstructed images but also to a greater extent in the computed T1/T2 maps.

Recent deep learning approaches have shown promise for reconstructing high-quality maps from highly accelerated scans. Encoder-decoder models like

---

M. Tänzer and F. Wang—Equal contribution.
D. Rueckert, G. Yang and S. Nielles-Vallespin—Co-last authors.

O. Camara et al. (Eds.): STACOM 2023, LNCS 14507, pp. 293–302, 2024.
https://doi.org/10.1007/978-3-031-52448-6_28

AUTOMAP [16] leveraged deep convolutional neural networks to learn efficient representations directly from undersampled k-space and mapping targets. Schlemper et al. [11] explored the ability of cascaded CNN to learn the spatial-temporal correlations from multi-coil undersampled cardiac cine MRI. Qin et al. [9] exploited the spatiotemporal correlations using recurrent network for dynamic multi-coil cardiac cine data. Lyu et al. [4] divided temporal cine MRI data into several views and used a video-Transformer [14] model to capture spatial and temporal relationship. However, most existing methods disregard dependencies between parameterised time frames. As relaxation processes induce temporal correlations, explicitly modelling time structure is essential for accurate reconstruction.

We introduce an innovative deep reconstruction model that introduces a temporal dependency module to effectively capture inter-frame relationships within encoded sequences. The module is seamlessly integrated into an encoder-decoder architecture by modifying the latent vectors in the skip connections of the encoder-decoder model to better exploit the temporal correlation. By incorporating temporal dynamics, the proposed model aims to significantly enhance reconstructions derived from accelerated cardiac T1 and T2 mapping scans, regardless of the number of time points or mapping methodologies employed. This could facilitate accurate analyses of myocardial tissue properties from faster, patient-friendly scans. We present preliminary validation of our technique for accelerated cardiac T1 and T2 mapping against state-of-the-art methods and gold-standard fully-sampled acquisitions.

2   Materials and Methods

2.1   Data Acquisition

We used both single and multi-coil T1 and T2 mapping data from the MICCAI 2023 CMRxRecon training dataset [12]. The T1 mapping data was acquired using a Modified Look-Locker Inversion recovery (MOLLI) sequence [6] with nine frames of variable T1 weighting in short-axis view at end-diastole. For each subject, between five and seven slices were collected with a slice thickness of 5.0 mm. The matrix size of each T1-weighted frame was 144×512 with an in-plane spatial resolution of $1.4 \times 1.4 \, \mathrm{mm}^2$. For the multi-coil data, the coils were compressed to 10 virtual coils. The T2 mapping data was acquired using a T2-prepared (T2prep) FLASH sequence with three T2 weightings and with geometrical parameters identical to the T1 mapping acquisition.

2.2   Data Processing

Both single and multi-coil T1 and T2 mapping data from the 120 healthy subjects in the training dataset were randomly split into 80% for training, 10% for validation and 10% for testing. We pre-processed the provided k-space data by scaling it to a range where the model could perform optimally. Specifically, we multiplied all k-space data by a fixed value of 10^2 to bring the magnitude of the images values approximately into the $[0, 1]$ range. This transformation is reversed before computing the mappings.

During training, the data was also augmented using a random undersampling mask for every subject. The random mask was generated by selecting every k^{th} line starting from line s, where k is the acceleration factor and s is a randomly sampled integer between 0 and k. As in the original acquisition, the random mask preserved the central 24 lines of k-space. This allowed the model to exhibit greater adaptability to minor variations in the acquisition protocol, thereby providing a valuable and realistic data augmentation strategy.

2.3   Model

Latent Transformer. The proposed model, the Latent Transformer (LT), employs an encoder-decoder architecture with shared encoding-decoding blocks across all time-frames (Fig. 1). Embedded within each skip connection between an encoding layer and the corresponding decoding layer is an LT block, which enables modelling of dependencies across frames before passing signals to the decoder layers (Fig. 1, E). Specifically, there is a unique LT block for each layer of the main encoder-decoder network. Each LT block utilises multi-layer and multi-head self-attention (Fig. 1, C and D) to compute the updated latent code as a weighted linear combination of itself and the latent codes of the other time-frames in a pixel-wise manner (Fig. 1, A and B). The LT blocks are key to exploit temporal correlations within the data to aid reconstruction performance, as they allow information from each frame to affect the reconstruction of all other frames.

Single-Coil. The proposed model architecture consists of two complementary components as illustrated in Fig. 1:

- An encoder-decoder network that serves as the main artifact removal model. In our experiments, we utilise a U-Net architecture as a strong baseline model. The goal of this encoder-decoder network is to remove large-scale artifacts from the input MRI frames.
- The latent transformer model to explicitly capture inter-frame dependencies.

Finally, the predictions from the main encoder-decoder model and the LT model are combined to produce the final reconstructed output frames. By fusing the outputs this way, the model leverages both general artifact removal capabilities and inter-frame dependencies for enhanced MRI reconstruction.

Multi-coil Model. Similar to the single-coil model, the multi-coil artifact-removal architecture consists of a main encoder-decoder network and LT model tailored for multi-coil data. To reduce computational complexity, the LT is applied on the coil-combined complex image rather than each coil individually.

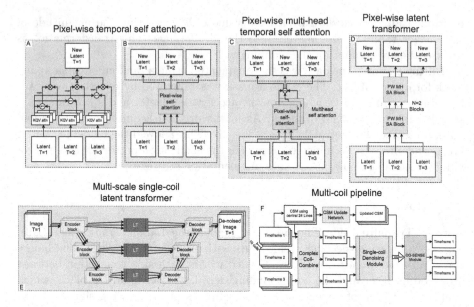

Fig. 1. Diagram representing the architecture for the single-coil (E) and multi-coil (F) tasks. The figure also shows how the latent transformer is implemented using a pixel-wise self-attention mechanism (A, B) in a multilayer and multi-head fashion (D). The figure shows a case where three time-frames are available, but the method extends seamlessly to any number of time-frames with no modification.

Coil sensitivity maps (CSMs) C are extracted using an iterative approach [3] with smoothing based on the central 24 lines of undersampled k-space $x \in \mathbb{C}^{W \times M}$ among all W coils. The undersampled multi-coil data is multiplied by the conjugate CSMs to maintain complex information as $\hat{x} \in \mathbb{C}^{W}$ before input to the model.

The multi-coil reconstruction can be formulated as:

$$\hat{x}_{rec} = \arg\min_{x}(\|Ex - y\|_2^2 + \lambda \|\hat{x} - S(\hat{x}; \theta)\|_2^2) \tag{1}$$

$$E = M \cdot F \cdot \hat{C} \tag{2}$$

where x is the multi-coil complex image and $y \in \mathbb{C}^{W \times M}$ is the acquired multi-coil k-space data. E represents the operator combining the undersampling mask M, Fourier transform F, and updated sensitivity maps $\hat{C}$. $S(\hat{x}; \theta)$ denotes the single-coil based deep neural network with parameters θ. We separate the optimisation into conjugate gradient SENSE [5] reconstruction and neural network-based reconstruction, iteratively updating $\hat{x}$.

$$\hat{x}_{rec} = (E^H E + \lambda I)^{-1}(E^H y + \lambda S(x; \theta)) \tag{3}$$

where $\hat{x}_{rec}$ is calculated with fixed θ parameters in the network.

An additional CSM update module is integrated to improve the original C under the supervision of $\hat{x}_{rec}$ for a better SENSE reconstruction. The CSM C initialised by iterative method [3] works as a warm start:

$$\hat{C} = N(C; \beta) \tag{4}$$

$N(C; \beta)$ is the network to update the CSM with parameters of β. It consisted of four single-scale convolutional layers with kernel size of 3×3 followed by ReLU and a fifth layer with only 2D convolution.

2.4   Assessments

The results reported in this work were computed by comparing the model outputs with fully-sampled reconstructions on a fixed test set. We reported quantitative metrics including root-mean-square-error (RMSE), normalised mean-square-error (NMSE), peak signal-to-noise ratio (PSNR), and structural similarity index (SSIM). To assess the impact on the downstream mapping task, these metrics were also calculated for the estimated T1 and T2 parameter maps.

To match the evaluation protocol used in the CMRxRecon challenge, the reconstructed images and parameter maps were cropped to a region of interest before metric computation. The quantitative assessment was specifically focused on the most clinically relevant region by retaining only the central 50% of rows and central 33% of columns.

3   Results

3.1   Single-Coil Reconstruction

To evaluate the proposed model, we conducted experiments for both the T1 and T2 mapping tasks using acceleration factors 4, 8, and 10. In this section, we compared two model configurations:

- U-Net: A baseline U-Net architecture that serves as a standard encoder-decoder network.
- U-Net + Latent Transformer: The proposed model combining the baseline encoder-decoder model with the latent-transformer module to exploit inter-frame dependencies.

Table 1 summarises the results across both tasks for the three considered acceleration factors. Results demonstrate the ability of the latent transformer to effectively model the temporal correlations and improve the reconstruction.

Figure 2 also qualitatively compares the reconstruction produced using Zero-filling, a U-Net model and the proposed U-Net + LT model.

Table 1. Single-coil model results for both the reconstructed MR acquisitions and the computed T1 and T2 maps. The table compares a U-Net based artifact-removal process with a pipeline that used a Latent Transformer-enhanced U-Net at its core. Our report underlined the best model for a given mapping task and acceleration factor.

Model	PSNR ↑	SSIM ↑	NMSE ↓	RMSE ↓
U-Net T1 4×	30.160	0.811	0.028	<u>4.35E–05</u>
U-Net T1 8×	28.158	0.766	0.044	5.58E–05
U-Net T1 10×	27.661	0.762	0.048	5.91E–05
U-Net T2 4×	28.940	0.827	0.023	2.81E–05
U-Net T2 8×	27.350	0.802	0.032	3.44E–05
U-Net T2 10×	<u>27.246</u>	<u>0.808</u>	<u>0.032</u>	3.50E–05
U-Net T1 4× + LT	<u>30.184</u>	<u>0.813</u>	0.028	4.36E–05
U-Net T1 8× + LT	<u>28.431</u>	<u>0.774</u>	<u>0.040</u>	<u>5.39E–05</u>
U-Net T1 10× + LT	<u>27.933</u>	<u>0.769</u>	<u>0.045</u>	<u>5.71E–05</u>
U-Net T2 4× + LT	<u>29.067</u>	<u>0.831</u>	<u>0.022</u>	2.76E–05
U-Net T2 8× + LT	<u>27.469</u>	<u>0.806</u>	<u>0.031</u>	3.39E–05
U-Net T2 10× + LT	27.210	0.807	0.033	3.51E–05

Model	Map PSNR ↑	Map SSIM ↑	Map NMSE ↓	Map RMSE ↓
U-Net T1 4×	19.709	0.600	0.024	134.219
U-Net T1 8×	18.593	0.534	0.031	152.737
U-Net T1 10×	18.548	0.524	0.031	153.266
U-Net T2 4×	11.884	0.433	0.505	64.175
U-Net T2 8×	11.838	0.406	0.511	64.483
U-Net T2 10×	11.860	0.410	0.508	64.350
U-Net T1 4× + LT	<u>19.834</u>	<u>0.607</u>	<u>0.023</u>	<u>132.514</u>
U-Net T1 8× + LT	<u>18.778</u>	<u>0.538</u>	<u>0.029</u>	<u>149.393</u>
U-Net T1 10× + LT	<u>18.592</u>	<u>0.526</u>	<u>0.030</u>	<u>152.493</u>
U-Net T2 4× + LT	<u>11.910</u>	<u>0.439</u>	<u>0.502</u>	<u>63.977</u>
U-Net T2 8× + LT	<u>11.898</u>	<u>0.413</u>	<u>0.504</u>	<u>64.056</u>
U-Net T2 10× + LT	<u>11.903</u>	<u>0.414</u>	<u>0.503</u>	<u>64.017</u>

3.2   Multi-coil Reconstruction

To evaluate the proposed model for multi-coil MRI reconstruction, we compared two model configurations built upon MoDL [1] framework:

- MoDL: The baseline MoDL model using fixed coil sensitivity maps and a standard single-scale network architecture. This serves as the standard MoDL implementation.
- MoDL + Proposed Model: An enhanced MoDL pipeline incorporating our proposed single-coil model architecture with latent transformers and learnable coil sensitivity maps.

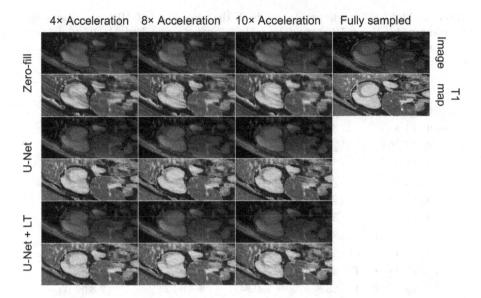

Fig. 2. Qualitative comparison between zero-filling reconstruction, U-Net-based reconstruction and the proposed U-Net + LT model. The figure shows both the reconstructed images and the T1 mapping associated with the shown slice.

Experiments were conducted on the multi-coil T1 and T2 mapping datasets. Quantitative metrics compare the two MoDL-based approaches to analyse the benefits of the proposed model enhancements, including the latent transformer's ability to exploit inter-frame dependencies.

The results in Table 2 summarise the reconstruction performance for the two models across both mapping tasks. We observed consistent improvements from the proposed techniques for integrating learnable coil sensitivity estimation and advanced single-coil modelling into the MoDL framework.

4   Discussion

The introduction of the Latent Transformer (LT) module demonstrates clear improvements for the vast majority of single coil results across all analysed metrics, tasks, and acceleration factors, for both images and derived T1 and T2 maps (Table 1). For multi-coil data, the improvements from the LT blocks are more modest. In particular, the LT addition degrades performance on reconstructed images but improves results for the computed T1 and T2 maps (Table 1). The difference in behaviour between the two tasks likely arises from two concurring causes. First, the network works as a regularisation term. The weighting between the data consistency layer of the CG-SENSE and network may even downgrade for heavy networks. When the network becomes too complex or contains too many parameters, it may prioritise fitting the training data over maintaining consistency with the acquired data, leading to smaller weighting and reduced

Table 2. Multi-coil model results for both the reconstructed MR acquisitions and the computed T1 and T2 maps. The table compares a standard MoDL model with our proposed method for artifact-removal. Our report underlined the best model for a given mapping task and acceleration factor.

Model	PSNR ↑	SSIM ↑	NMSE ↓	RMSE ↓
MoDL T1 4×	34.790	0.894	0.026	2.85E–05
MoDL T1 8×	31.534	0.855	0.025	3.80E–05
MoDL T1 10×	29.904	0.840	0.030	4.57E–05
MoDL T2 4×	33.896	0.911	0.010	1.63E–05
MoDL T2 8×	29.937	0.870	0.018	2.54E–05
MoDL T2 10×	29.205	0.867	0.021	2.79E–05
MoDL T1 4× + LT	34.460	0.887	0.028	2.97E–05
MoDL T1 8× + LT	30.763	0.837	0.028	4.15E–05
MoDL T1 10× + LT	29.762	0.839	0.032	4.67E–05
MoDL T2 4× + LT	32.866	0.899	0.011	1.83E–05
MoDL T2 8× + LT	28.853	0.847	0.023	2.89E–05
MoDL T2 10× + LT	28.005	0.844	0.028	3.22E–05
Model	Map PSNR ↑	Map SSIM ↑	Map NMSE ↓	Map RMSE ↓
MoDL T1 4×	22.346	0.719	0.017	105.874
MoDL T1 8×	19.934	0.623	0.021	131.508
MoDL T1 10×	20.004	0.611	0.022	129.568
MoDL T2 4×	12.365	0.577	0.455	60.838
MoDL T2 8×	11.968	0.480	0.497	63.645
MoDL T2 10×	11.946	0.473	0.499	63.805
MoDL T1 4× + LT	22.035	0.739	0.015	110.417
MoDL T1 8× + LT	20.413	0.647	0.023	124.659
MoDL T1 10× + LT	20.012	0.613	0.022	129.470
MoDL T2 4× + LT	12.247	0.547	0.447	61.585
MoDL T2 8× + LT	12.114	0.494	0.481	62.617
MoDL T2 10× + LT	11.977	0.483	0.496	63.532

effect of the LT module. This can result in degraded performance and lower accuracy in image reconstruction. From Table 2, the proposed method outperform the original MoDL at higher acceleration factors, indicating the potential of the networks correcting for severe undersampling artifacts. Lighter variants on U-net with attention across channels may be taken into consideration. For the CSM update module, we also tried using the CSM generated on the fully-sampled k-space as a hard constrain to supervise, but got inferior performance. Networks with unrolled manner [13] or an additional J-SENSE module [2] can be incorporated to get a better data consistency performance. Second, the LT module seems to produce some image artifacts but ultimately captures inter-frame dynamics better than simpler models focused solely on de-noising, as evidenced by improved T1 and T2 map estimates. In summary, the proposed LT framework demonstrates clear utility in exploiting temporal correlations, especially for single coil acquisitions, further validating its use for reconstructing highly accelerated MRI data for T1 and T2 mapping.

5   Conclusion

This work proposes a deep learning approach for reconstructing undersampled MRI that incorporates our novel Latent Transformers module to model inter-frame dependencies. Experiments on accelerated cardiac T1/T2 mapping show improved image quality and parameter mapping compared to baseline models, demonstrating the importance of temporal modelling. While promising, limitations remain including the basic U-Net architecture used. For future work, we will explore integrating the latent transformers into more advanced models like Restormer [15] and unrolled networks [11] with better CSM estimation to further boost performance. Overall, this study validates explicitly modelling time correlations with transformers to enable accurate reconstructions from highly accelerated quantitative MRI.

Acknowledgement. We want to show our gratitude to Zimu Huo, who provided support and advice on multi-coil reconstruction. This work was supported by the British Heart Foundation (RG/19/1/34160). Guang Yang was supported in part by the ERC IMI (101005122), the H2020 (952172), the MRC (MC/PC/21013), the Royal Society (IEC/NSFC/211235), the NVIDIA Academic Hardware Grant Program, the SABER project supported by Boehringer Ingelheim Ltd, and the UKRI Future Leaders Fellowship (MR/V023799/1). Fanwen Wang was supported by the UKRI Future Leaders Fellowship (MR/V023799/1). Michael Tänzer was supported by the UKRI CDT in AI for Healthcare (EP/S023283/1).

References

1. Aggarwal, H.K., Mani, M.P., Jacob, M.: MoDL: model-based deep learning architecture for inverse problems. IEEE Trans. Med. Imaging **38**(2), 394–405 (2019). https://doi.org/10.1109/TMI.2018.2865356
2. Arvinte, M., Vishwanath, S., Tewfik, A.H., Tamir, J.I.: Deep J-sense: accelerated MRI reconstruction via unrolled alternating optimization. In: de Bruijne, M., et al. (eds.) MICCAI 2021. LNCS, vol. 12906, pp. 350–360. Springer, Cham (2021). https://doi.org/10.1007/978-3-030-87231-1_34
3. Inati, S.J., Hansen, M.S., Kellman, P.: A fast optimal method for coil sensitivity estimation and adaptive coil combination for complex images. In: Proceedings of the 22nd Annual Meeting of ISMRM, Milan, p. 4407 (2014)
4. Lyu, J., et al.: Region-focused multi-view transformer-based generative adversarial network for cardiac cine MRI reconstruction. Med. Image Anal. **85**, 102760 (2023)
5. Maier, O., et al.: CG-SENSE revisited: results from the first ISMRM reproducibility challenge. Magn. Reson. Med. **85**(4), 1821–1839 (2021)

6. Messroghli, D.R., Radjenovic, A., Kozerke, S., Higgins, D.M., Sivananthan, M.U., Ridgway, J.P.: Modified look-locker inversion recovery (MOLLI) for high-resolution T1 mapping of the heart. Magn. Reson. Med. **52**(1), 141–146 (2004). https://doi.org/10.1002/mrm.20110
7. Messroghli, D.R., et al.: Myocardial T1 mapping: application to patients with acute and chronic myocardial infarction. Magn. Reson. Med. Official J. Int. Soc. Magn. Reson. Med. **58**(1), 34–40 (2007)
8. Moon, J.C., et al.: Myocardial T1 mapping and extracellular volume quantification: a society for cardiovascular magnetic resonance (SCMR) and CMR working group of the european society of cardiology consensus statement. J. Cardiovasc. Magn. Reson. **15**(1), 1–12 (2013)
9. Qin, C., et al.: Complementary time-frequency domain networks for dynamic parallel MR image reconstruction. Magn. Reson. Med. **86**(6), 3274–3291 (2021)
10. Sado, D.M., et al.: Identification and assessment of Anderson-Fabry disease by cardiovascular magnetic resonance noncontrast myocardial T1 mapping. Circ. Cardiovasc. Imaging **6**(3), 392–398 (2013)
11. Schlemper, J., Caballero, J., Hajnal, J.V., Price, A.N., Rueckert, D.: A deep cascade of convolutional neural networks for dynamic MR image reconstruction. IEEE Trans. Med. Imaging **37**(2), 491–503 (2017)
12. Wang, C., et al.: Cmrxrecon: an open cardiac MRI dataset for the competition of accelerated image reconstruction (2023)
13. Wang, Z., et al.: A faithful deep sensitivity estimation for accelerated magnetic resonance imaging. arXiv preprint arXiv:2210.12723 (2022)
14. Yan, S., et al.: Multiview transformers for video recognition. In: Proceedings of the IEEE/CVF Conference on Computer Vision and Pattern Recognition, pp. 3333–3343 (2022)
15. Zamir, S.W., Arora, A., Khan, S., Hayat, M., Khan, F.S., Yang, M.H.: Restormer: efficient transformer for high-resolution image restoration. In: Proceedings of the IEEE/CVF Conference on Computer Vision and Pattern Recognition, pp. 5728–5739 (2022)
16. Zhu, B., Liu, J.Z., Cauley, S.F., Rosen, B.R., Rosen, M.S.: Image reconstruction by domain-transform manifold learning. Nature **555**(7697), 487–492 (2018)

T1 and T2 Mapping Reconstruction Based on Conditional DDPM

Yansong Li, Lulu Zhao, Yun Tian[✉], and Shifeng Zhao

School of Artificial Intelligence, Beijing Normal University, Beijing, China
tianyun@bnu.edu.cn

Abstract. Cardiac magnetic resonance imaging (CMR) has emerged as a crucial imaging modality for the diagnosis of cardiac diseases. T1 and T2 mapping are essential techniques for detecting cardiomyopathies. However, the imaging speed is noticeably slow and conventional mapping models often struggle to produce accurate results when the imaging process is compromised. To overcome this limitation, accelerated mapping techniques have been developed to reduce motion artifacts and enhance image quality. In this study, we propose a novel reconstruction method based on a conditional denoising diffusion probabilistic model (CDDPM). By utilizing accelerated mapping as a conditioning factor and iteratively applying a denoising process, we generate refined T1 and T2 maps from initially corrupted data consisting of pure Gaussian noise. The experimental results of the CMR Reconstruction Challenge demonstrate the effectiveness of our proposed method. Objective indicators show significant improvements, indicating enhanced image quality. Furthermore, our method successfully improves the texture quality of the images, providing more detailed and accurate information for cardiomyopathy diagnosis.

Keywords: CMR reconstruction · T1/T2 mapping · Diffusion model

1 Introduction

Cardiac magnetic resonance imaging (CMR) is a valuable non-invasive imaging technique with excellent soft tissue contrast for diagnosing heart diseases [1]. However, its inherent slow imaging speed and susceptibility to motion artifacts have been limitations. In recent years, there has been significant interest in accelerating CMR image acquisition and reconstruction.

One promising approach is the reconstruction of CMR images from highly undersampled k-space data, obtained during the scan. Advanced reconstruction algorithms that leverage sparsity in the Fourier domain can generate high-quality CMR images from fewer k-space samples. This reduces scan time, improves patient comfort, and mitigates motion artifacts, leading to more accurate diagnosis.

Accelerated CMR image reconstruction techniques, such as compressed sensing, parallel imaging, and deep learning-based methods, exploit image structure

O. Camara et al. (Eds.): STACOM 2023, LNCS 14507, pp. 303–313, 2024.
https://doi.org/10.1007/978-3-031-52448-6_29

and temporal redundancies to achieve high-fidelity reconstructions with minimal data acquisition. These advancements enable faster and more robust CMR examinations, facilitating early detection, accurate diagnosis, and treatment planning for cardiovascular conditions.

Parallel imaging (PI) is a CMR reconstruction technique that utilizes multiple coils to simultaneously obtain data and generate high-quality images in a shorter scanning time. PI can perform image reconstruction with fewer samples by reconstructing k-space undersampling using coil sensitivity. Traditional PI techniques, including SENSE [2], GRAPPA [3], SPIRiT [4], and ESPIRiT [5], have already been used clinically. However, these methods may cause image artifacts when k-space data is highly undersampled.

In the field of accelerating CMR image reconstruction, deep learning plays a crucial role and has achieved good results [6–8]. Among them, the generative model is more commonly applied [9,10]. Generative models are statistical models that learn the distribution of a dataset and generate new data that resembles the original dataset. By utilizing generative models, we can reconstruct high-quality CMR images from highly undersampled k-space data, thereby improving diagnostic accuracy and patient experience.

The choice of a specific generative model depends on the characteristics of the dataset and the requirements of the task at hand. In recent years, popular generative models include Variational Autoencoders (VAEs) [11], Generative Adversarial Networks (GANs) [12], and the highly successful Diffusion Models [13,14] in the field of Mediacal image generation.

Ho et al. [13] proposed the denoising diffusion probabilistic model (DDPM), which uses a variational inference approach to estimate the parameters of a diffusion process. Song et al. [14] proposed Noise-conditioned Score Networks (NCSNs), which rely on a maximum likelihood-based estimation approach. They utilize the score function of the loglikelihood of the data to estimate the parameters of the diffusion process. Jalal et al. [15] proposed Compressed Sensing with Generative Models (CSGM). CSGM trains the score-based generative models on CMR to utilize them as prior information for the inversion pathway in reconstructing realistic CMR data from undersampled CMR in a posterior sampling scheme. Chung et al. [16] proposed a score-based diffusion framework that solves the inverse problem for image reconstruction from accelerated CMR scans. They train a continuous time-dependent score function with denoising score matching. At the inference stage, they iterate between the numerical SDE solver and data consistency step to achieve reconstruction.

In this work, we propose a novel reconstruction method based on Conditional Denoising Diffusion Probability Model (CDDPM) to improve the accuracy and quality of T1 and T2 mapping in CMR. Our approach utilizes accelerated mapping as a conditional factor and incorporates an iterative denoising process to refine T1 and T2 mappings from the initially damaged data. By employing this method, we are able to enhance image quality, reduce motion artifacts, and provide more detailed and accurate information for the diagnosis of cardiomyopathy. We have demonstrated the effectiveness of the proposed method in the dataset provided by the CMR Reconstruction Challenge [17], with significant improvements in both objective indicators and texture quality of the reconstructed images.

2   Method

The proposed framework is illustrated in Fig. 1. By adding noise to the original data x_0 (fully sampled CMR) and predicting noise through a conditional network with conditional y (accelerated CMR), the reconstructed image is restored from the noise x_T.

2.1   Conditional Denoising Diffusion Probability Model

DDPM uses Variational inference [18] to define a forward Markovian diffusion process q that gradually adds Gaussian noise to a high-resolution image x_0 over T iterations:

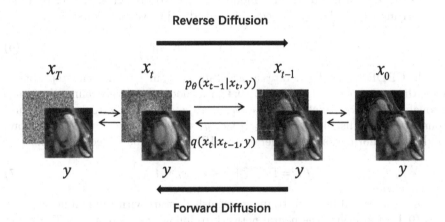

Fig. 1. Framework for CMR reconstruction based on CDDPM.

$$q\left(x_t \mid x_{t-1}\right) = \mathcal{N}\left(x_t; \sqrt{1 - \beta_t} \cdot x_{t-1}, \beta_t \cdot \mathbf{I}\right), \forall t \in \{1, 2, \ldots, T\} \qquad (1)$$

where the parameters $T, \beta_1, \beta_2, \ldots, \beta_T \in [0, 1)$ represent the number of diffusion steps and the variance schedule across diffusion steps, respectively. $\mathbf{I}$ is the identity matrix and $\mathcal{N} \sim (x; \mu, \sigma)$ represents the normal distribution of mean μ and covariance σ. Considering $\alpha_t = 1 - \beta_t$ and $\bar{\alpha}_t = \prod_{s=0}^{T} \alpha_s$ one can directly sample an arbitrary step of the noised latent conditioned on the input x_0 as follows:

$$x_t = \sqrt{\bar{\alpha}_t} x_0 + \sqrt{1 - \bar{\alpha}_t} \epsilon \qquad (2)$$

Leveraging the above definitions, approximating a reverse process to get a sample from $q(x_0)$. To this end, we can parameterize this reverse process by starting at $p_\theta(x_T) = \mathcal{N} \sim (x_T; \mu, \sigma)$ as follows:

$$p_\theta(x_{0:T}) = p(x_T) \prod_{t=1}^{T} p_\theta(x_{t-1} \mid x_t) \tag{3}$$

$$p_\theta(x_{t-1} \mid x_t) = \mathcal{N}(x_{t-1}; \mu_\theta(x_t, t), \Sigma_\theta(x_t, t) \mathbf{I}) \tag{4}$$

To train this model in a way that allows $p(x_0)$ to learn the true data distribution $q(x_0)$, we can optimize the following variational bound on negative log-likelihood:

$$\mathbb{E}\left[-\log p_\theta(x_0)\right] \leq \mathbb{E}_q\left[-\log p(x_T) - \sum_{t \geq 1} \log \frac{p_\theta(x_{t-1} \mid x_t)}{q(x_t \mid x_{t-1})}\right] = -L_{\text{VL-B}} \tag{5}$$

Ho et al. [34] found it better not to directly parameterize $\mu_\theta(x_t, t)$ as a neuralnetwork, but rather to train a model $\epsilon_\theta(x_t, t)$ to predict ϵ. Hence, by reparameterizing Eq. 6, they proposed a simplified objective as follows:

$$\mathrm{L}(\theta) := \mathrm{E}_{t,x_0,\epsilon}\left[\|\epsilon - \epsilon_\theta(x_t, t)\|^2\right] \tag{6}$$

The CDDPM model generates a target image x_0 in T refinement steps. Starting with a pure noise image $x_T \sim \mathcal{N}(0; \mathbf{I})$, the model iteratively refines the image through successive iterations($x_{T-1}, x_{T-2}, \ldots, x_0$) according to learned conditional transition distributions $p_\theta(x_{t-1} \mid x_t, y)$ such that $x_0 \sim p(x \mid y)$. Therefore, conditional information y (accelerated CMR) needs to be added in Eq. 6:

$$\mathrm{L}(\theta) := \mathrm{E}_{t,x_0,\epsilon}\left[\|\epsilon - \epsilon_\theta(x_t, y, t)\|^2\right] \tag{7}$$

Once the neural network training is finished, start with random noise x_T to $\mathcal{N} \sim (0; \mathbf{I})$, and utilize the neural network to follow the current step T. Predict the noise at step $T-1$ using x_T and compute x_{T-1} based on the predicted mean and variance. Continue this iterative process until $t = 0$ to generate the data x_0 (reconstructed CMR).

2.2   Network Structure

As shown in Fig. 2, our method takes U-Net as backbone. U-Net is particularly well-suited for noise prediction in the diffusion process due to its compatibility with the image's dimensionality. The architecture of U-Net, with its encoder-decoder structure, allows it to capture and model the complex relationships between input images and their corresponding noise patterns. This capability is crucial in scenarios where noise characteristics depend on the specific features and structures within the image. By leveraging U-Net's hierarchical representation learning capabilities, it becomes proficient at learning and predicting noise patterns that align with the image content, ultimately delivering accurate and meaningful noise predictions.

U-Net is composed of three main blocks: the residual blocks, the attention blocks, and the time embedding as illustrated in Fig. 2.

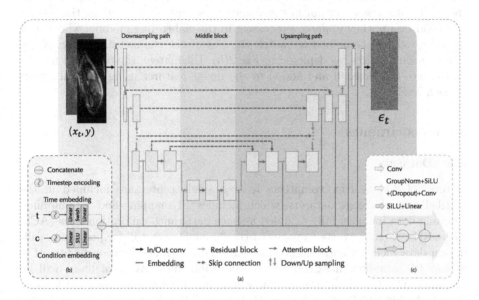

Fig. 2. Architecture of the U-Net used in CDDPM. (a) Overall U-Net structure. (b) Time and condition embedding module. (c) Detailed illustration of the Residual block.

Residual blocks: Inspired by residual learning, the residual blocks serve as a fundamental block of U-Net that help in capturing and preserving essential image features during the encoding and decoding phases. Comprising multiple convolutional layers and skip connections, these blocks facilitate the reuse of low-level features and enable the network to effectively learn residual representations.

Attention blocks: The attention blocks are designed to enhance the model's ability to focus on relevant image regions while suppressing irrelevant or noisy regions. They employ mechanisms like spatial or channel-wise attention to selectively emphasize or suppress certain parts of the feature maps. By incorporating attention mechanisms, U-Net can allocate more resources to crucial image features, thereby enhancing its ability to handle complex patterns and variations.

Time embedding: The Time embedding is specifically employed in scenarios involving input data with temporal dimensions, such as video or sequence data. It encodes the temporal information using specialized techniques. This module allows U-Net to capture temporal dependencies and exploit temporal context during the diffusion process, resulting in more accurate noise prediction over time.

Both the downsampling and upsampling path consist of 30 layers, with 5 of them containing the downsampling or upsampling modules. The downsampling path incorporates 4 attention blocks, the middle block has 1 attention block, and the upsampling path includes 4 attention blocks. The U-Net takes as input a noisy sample denoted as x_t., a conditional state represented as y, and a time step t. It produces an output in the shape of a sample. For the given time step t and condition y, they undergo processing in the embedding module, where they are adjusted and combined to match the size of the channel dimension of the input image features, as depicted in Fig. 2(b). Ultimately, the temporal feature vectors are broadcasted and added to the image features using residual blocks at each level.

3   Experiments

3.1   Datasets

A total of 300 healthy volunteers from a single center are included in the CMRxRecon challenge datasets, which includes T1 mapping and T2 mapping. Training data includes fully sampled k-space data, auto-calibration lines (ACS, 24 lines) and reconstructed images in the *.mat format will be provided. Validation data includes undersampled k-space data with acceleration factors of 4, 8, and 10, sampling mask, and auto-calibration lines (ACS, 24 lines) will be provided. The test dataset is similarly structured, with undersampled k-space data featuring acceleration factors of 4, 8, and 10, the sampling mask, auto-calibration lines (ACS, 24 lines), and the associated reconstructed images. The raw k-space data obtained from the scanner is pre-processed and is subsequently converted to the *.mat format using MATLAB. Furthermore, the multi-coil data is compressed to 10 virtual coils for standardization and storage efficiency.

3.2   Data Preprocessing

During the training phase, the raw k-space data is initially transformed into the image domain using the inverse fast Fourier transform (IFFT), resulting in data with dimensions represented as (kx, ky, sz, t). To create the training set for the neural network, both single-coil and multi-coil data are used, and acceleration factors of 4, 8, and 10 are combined into a single dataset.

We utilize function h to combine the multi-coil k-space data into a unified real-valued image. The merging procedure involves applying an inverse 2D Fourier transform to each coil and subsequently performing a sum-of-squares (SOS) operation. Through SOS calculation, all the coils are effectively combined into a singular real-valued image.

$$SOS\left(x_1, ..., x_N\right) = \sum_{n=1}^{N} |x_n|^2 \tag{8}$$

where $x_1,..., x_N$ are the images from the N coils.

After combining the data, each image is split into individual images of size (kx, ky). To enhance the diversity of the training data, we apply data augmentation techniques to these two-dimensional images. One common augmentation technique is random cropping, where the images are randomly cropped into patches of size 256 × 256. By cropping to a smaller size, the network training process can be faster without affecting image quality.

The same process is also applied to the test set. The under-sampled k-space data in the test set, along with the corresponding sampling mask and auto-calibration lines, are transformed into the image domain using the IFFT. This results in smaller-sized images, which are then processed by the network to predict the reconstructed images.

After obtaining the predicted patch-level images, they are reassembled to their original size. This reassembly process involves stitching the overlapping patches together and handling any potential artifacts or inconsistencies at the patch boundaries. By reassembling the patches, we obtain the final reconstructed images for the test set. These reconstructed images can be evaluated and compared with the ground truth or reference images to assess the performance of the network in terms of CMR reconstruction. This patch-based approach allows for efficient processing and prediction on large images, while also enabling the evaluation of the network's performance on the entire image domain.

3.3 Training Details

Before employing random cropping augmentations, the input data undergoes a normalization process. The model is implemented using PyTorch and trained on a workstation equipped with an NVIDIA Tesla V100 GPU with 32 GB of memory. The model comprises a total of 52,090,177 parameters, and it was trained from scratch, without utilizing any pre-trained model. Training of the model is carried out for 4 * 24 h using the Adam optimizer, with diffusion step set to 1000.

3.4 Evaluation Metrics

The evaluation metrics primarily used in this competition are Peak Signal-to-Noise Ratio (PSNR), Structural Similarity Index (SSIM), and Root Mean Square Error (RMSE). These metrics provide quantitative measures to assess the performance of different methods in terms of image quality and similarity to ground truth.

During the competition phase, since the ground truth or reference images are not available for the test set, the evaluation of the methods can only be conducted using the validation set. The validation set serves as a proxy for assessing the generalization and effectiveness of the proposed approaches.

Participants are expected to submit their predictions for the test set, and these predictions will be evaluated based on the metrics mentioned above using the corresponding ground truth images from the validation set. This evaluation process allows for comparing and ranking the performance of different methods in terms of noise reduction or image restoration without relying on the ground truth data.

The PSNR metric measures the difference between the predicted images and the ground truth images in terms of their signal power, thereby providing an indication of the reconstruction accuracy. Meanwhile, the SSIM metric evaluates the structural similarity between the predicted and ground truth images, considering factors including luminance, contrast, and structural content. Lastly, the RMSE metric quantifies the average difference between the predicted and ground truth images, providing an overall assessment of the reconstruction error.

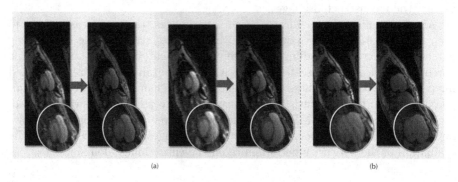

(a)   (b)

Fig. 3. 4× accelerated reconstruction for Validation Set P001. (a) T1 mapping. (b) T2 mapping.

4   Result

Table 1 presents the reconstruction outcomes from both the multi-coil and single-coil datasets T1/T2, considering undersampling factors of 4, 8, and 10. It's important to note that we utilize a U-Net directly on both the single-coil and combined multi-coil, rather than incorporating an additional step to estimate coil sensitivities. By omitting the use of coil sensitivities, our approach centers on the inherent spatial and temporal correlations present in the data to reconstruct high-quality images. The results demonstrate that the reconstruction of multi-coil data outperforms that of single-coil data. Remarkably, the neural network is trained to learn how to map the aliased coil images to the combined unaliased images [19]. This learning process empowers the exploit the sensitivity information embedded within the data for de-aliasing, without the need for explicit coil sensitivity maps. As a result, our strategy facilitates a more generalized and robust reconstruction process that is not reliant on specific coil sensitivity profiles.

The reconstruction results in Table 1 demonstrate the effectiveness of our method across different undersampling factors and datasets. The achieved image quality is evaluated based on metrics such as PSNR, SSIM, and RMSE, providing

Table 1. The quantitative metric results of the values for T1/T2 mapping of multi-coil and single-coil with different acceleration factors.

Modality	Coil	Acc	PSNR	SSIM	RMSE
T1 mapping	single	4×	30.1389	0.8476	0.1393
		8×	28.8884	0.8361	0.1335
		10×	27.1265	0.7911	1.3116
	multi	4×	32.4824	0.8609	0.1307
		8×	28.1413	0.8453	0.1427
		10×	27.1353	0.8223	0.1528
T2 mapping	single	4×	30.3452	0.8325	0.1331
		8×	29.3465	0.8102	0.1448
		10×	27.7631	0.7891	0.1591
	multi	4×	32.7234	0.8631	0.1343
		8×	29.4661	0.8497	0.1282
		10×	28.8552	0.8314	0.1522

quantitative measures of the reconstruction accuracy, structural similarity, and overall error.

The absence of coil sensitivity information in our approach simplifies the reconstruction pipeline and reduces computational complexity. This makes our method more practical and applicable to various imaging scenarios, especially those where access to coil sensitivity maps may be limited or unavailable.

Overall, the results presented in Table 1 highlight the potential of our coil-insensitive approach in achieving accurate and high-fidelity reconstructions, even under challenging undersampling conditions.

Figure 3 show cases the reconstruction details of different regions in T1 mapping using 4× accelerated sampling. Despite the accelerated sampling, the reconstructed T1 mapping exhibit remarkable details of various anatomical regions.

5   Conclusion

To enhance the accuracy and image quality of T1 and T2 mapping in CMR for the detection of cardiomyopathies, we propose a novel reconstruction method based on CDDPM. By leveraging accelerated mapping as a conditioning factor and employing an iterative denoising process, the proposed method effectively generates refined T1 and T2 maps from initially corrupted data containing pure Gaussian noise. The results from our experiments, conducted as part of the CMR Reconstruction Challenge, demonstrate the effectiveness of the proposed method. Objective indicators reveal significant improvements, indicating a notable enhancement in image quality. Importantly, the proposed method not only improves the accuracy of T1 and T2 mapping but also enhances the texture quality of the images. This enhancement provides more detailed and accurate information for the diagnosis of cardiomyopathies. A drawback associated with this method is its relatively slow sampling during the training phase. However, considering that our approach strategically decouples the computationally intensive training steps from the more efficient inference steps, the extended training process is typically acceptable. This strategy enables the rapid acquisition of high-quality images immediately after scanning, making it highly practical for real-world applications. In the future, we will focus on accelerating the training process and ensuring image quality.

References

1. Hundley, W.G., Bluemke, D.A., Finn, J.P., et al.: ACCF/ACR/AHA/NASCI/ SCMR 2010 expert consensus document on cardiovascular magnetic resonance: a report of the American College of Cardiology Foundation Task Force on Expert Consensus Documents. J. Am. Coll. Cardiol. **55**(23), 2614–2662 (2010)
2. Pruessmann, K.P., Weiger, M., Scheidegger, M.B., et al.: SENSE: sensitivity encoding for fast MRI. Magn. Reson. Med. Official J. Int. Soc. Magn. Reson. Med. **42**(5), 952–962 (1999)
3. Griswold, M.A., Jakob, P.M., Heidemann, R.M., et al.: Generalized autocalibrating partially parallel acquisitions (GRAPPA). Magn. Reson. Med. Official J. Int. Soc. Magn. Reson. Med. **47**(6), 1202–1210 (2002)
4. Lustig, M., Pauly, J.M.: SPIRiT: iterative self-consistent parallel imaging reconstruction from arbitrary k-space. Magn. Reson. Med. **64**(2), 457–471 (2010)
5. Uecker, M., Lai, P., Murphy, M.J., et al.: ESPIRiT-an eigenvalue approach to autocalibrating parallel MRI: where SENSE meets GRAPPA. Magn. Reson. Med. **71**(3), 990–1001 (2014)
6. Qin, C., Schlemper, J., Caballero, J., et al.: Convolutional recurrent neural networks for dynamic MR image reconstruction. IEEE Trans. Med. Imaging **38**(1), 280–290 (2018). https://doi.org/10.1109/TMI.2018.2863670
7. Qin, C., Duan, J., Hammernik, K., et al.: Complementary time-frequency domain networks for dynamic parallel MR image reconstruction. Magn. Reson. Med. **86**(6), 3274–3291 (2021). https://doi.org/10.1002/mrm.28917
8. Lyu, J., Sui, B., Wang, C., et al.: DuDoCAF: dual-domain cross-attention fusion with recurrent transformer for fast multi-contrast MR imaging. In: Wang, L., Dou, Q., Fletcher, P.T., Speidel, S., Li, S. (eds.) MICCAI 2022. LNCS, vol. 13436, pp. 474–484. Springer, Cham (2022). https://doi.org/10.1007/978-3-031-16446-0_45
9. Lyu, J., Li, G., Wang, C., et al.: Region-focused multi-view transformer-based generative adversarial network for cardiac cine MRI reconstruction. Med. Image Anal. **85**, 102760 (2023). https://doi.org/10.1016/j.media.2023.102760
10. Lyu, J., Tong, X., Wang, C.: Parallel imaging with a combination of SENSE and generative adversarial networks (GAN). Quant. Imaging Med. Surg. **10**(12), 2260–2273 (2020). https://doi.org/10.21037/qims-20-518
11. Kingma, D.P., Welling, M.: Auto-encoding variational bayes. In: Proceedings of International Conference on Learning Representations, ICLR, pp. 1–14. Springer, Cham (2014)
12. Goodfellow, I., Pouget-Abadie, J., Mirza, M., et al.: Generative adversarial networks. Commun. ACM **63**(11), 139–144 (2020)
13. Ho, J., Jain, A., Abbeel, P.: Denoising diffusion probabilistic models. Adv. Neural. Inf. Process. Syst. **33**, 6840–6851 (2020)

14. Song, Y., Ermon, S.: Generative modeling by estimating gradients of the data distribution. In: Advances in Neural Information Processing Systems, vol. 32 (2019)
15. Jalal, A., Arvinte, M., Daras, G., et al.: Robust compressed sensing MRI with deep generative priors. Adv. Neural. Inf. Process. Syst. **34**, 14938–14954 (2021)
16. Chung, H., Ye, J.C.: Score-based diffusion models for accelerated MRI. Med. Image Anal. **80**, 102479 (2022)
17. Wang, C., Lyu, J., Wang, S., et al.: CMRxRecon: an open cardiac MRI dataset for the competition of accelerated image reconstruction. arXiv preprint arXiv:2309.10836 (2023)
18. Blei, D.M., Kucukelbir, A., McAuliffe, J.D.: Variational inference: a review for statisticians. J. Am. Stat. Assoc. **112**(518), 859–877 (2017)
19. Kwon, K., Kim, D., Park, H.W.: A parallel MR imaging method using multilayer perceptron. Med. Phys. **44**(12), 6209–6224 (2017)

k-t CLAIR: Self-consistency Guided Multi-prior Learning for Dynamic Parallel MR Image Reconstruction

Liping Zhang⬤ and Weitian Chen(✉)⬤

CUHK Lab of AI in Radiology (CLAIR), Department of Imaging and Interventional Radiology, The Chinese University of Hong Kong, Shatin, Hong Kong
lpzhang@link.cuhk.edu.hk, wtchen@cuhk.edu.hk
https://github.com/lpzhang/ktCLAIR

Abstract. Cardiac magnetic resonance imaging (CMR) has been widely used in clinical practice for the medical diagnosis of cardiac diseases. However, the long acquisition time hinders its development in real-time applications. Here, we propose a novel self-consistency guided multi-prior learning framework named *k-t* CLAIR to exploit spatiotemporal correlations from highly undersampled data for accelerated dynamic parallel MRI reconstruction. The *k-t* CLAIR progressively reconstructs faithful images by leveraging multiple complementary priors learned in the *x-t*, *x-f*, and *k-t* domains in an iterative fashion, as dynamic MRI exhibits high spatiotemporal redundancy. Additionally, *k-t* CLAIR incorporates calibration information for prior learning, resulting in a more consistent reconstruction. Experimental results on cardiac cine and T1W/T2W images demonstrate that *k-t* CLAIR achieves high-quality dynamic MR reconstruction in terms of both quantitative and qualitative performance.

Keywords: Cardiac MRI reconstruction · Deep learning · Unrolled neural networks · Parallel imaging · Compressed sensing

1 Introduction

Cardiac magnetic resonance imaging (CMR) has gained widespread adoption in the diagnosis of cardiac diseases, owing to its exceptional soft-tissue contrast, non-invasive nature, and high spatial resolution. However, the acquisition procedure is inherently time-consuming due to the requirement of repeated acquisition of multiple heartbeat cycles. The prolonged duration can lead to potential patient discomfort and introduce motion artifacts in the resulting images.

Undersampling k-space is effective for accelerating the process, but direct reconstruction using incomplete data results in poor signal-to-noise ratio (SNR)

Supplementary Information The online version contains supplementary material available at https://doi.org/10.1007/978-3-031-52448-6_30.

and severe artifacts due to the violation of the Nyquist sampling theorem. Parallel imaging (PI) techniques [5,14] have been widely adopted to achieve acceleration in MR acquisition. However, their speedup rates are often constrained by hardware limitations. The combination of compressed sensing (CS) techniques with PI methods [8,10–13] has shown promise in enabling rapid imaging at high acceleration rates and improving the quality of image reconstruction for accelerated dynamic MRI. However, CS-based methods often rely on the assumption that the desired image has a sparse representation and incoherent undersampling artifacts in a known transform domain.

Deep learning (DL)-based reconstruction approaches have gained significant popularity in recent years, surpassing the limitations of traditional CS techniques. These methods utilize neural networks, either directly or iteratively, to address the inverse problem in either the image domain [1,6,21,27] or the k-space domain [2,7]. However, a common issue with these approaches is the excessive smoothing of reconstructed images. To tackle this challenge, recent advancements have focused on cross-domain models [3,4,18,26,29], which leverage multiple sources of prior knowledge to achieve high-quality image reconstruction while preserving sharp details. However, most existing methods in the literature primarily focus on static MRI and do not adequately address the challenges of accelerated dynamic MRI reconstruction, which requires capturing crucial spatial-temporal information. Some recent approaches [16,17,21] have emerged to exploit temporal correlations between dynamic MR frames in the spatiotemporal domain. However, these methods have mainly focused on single-coil cardiac MR reconstruction, with only a limited number of methods available for multi-coil cardiac MR reconstruction [9,15]. As a result, there is a pressing need for more efficient and effective deep learning models to further advance fast cardiac MR.

In this work, we propose k-t CLAIR, a self-consistency guided multi-prior learning framework for accelerated dynamic parallel MRI reconstruction. It leverages unrolled neural networks to exploit spatiotemporal correlations by learning complementary priors in different domains: spatiotemporal (x-y-t space), frequency-temporal (k_x-k_y-t space), and spatial-temporal frequency (x-y-f space). Self-consistency learning is enforced in the k-t domain using an end-to-end data-driven approach. Additionally, we introduce a frequency fusion layer to coordinate feature learning across all priors, facilitating faithful dynamic MRI reconstruction. Experimental results on highly undersampled cardiac cine, T1 weighted (T1W) acquisitions for T1 mapping, and T2 weighted (T2W) acquisitions for T2 mapping demonstrate the superior performance of our proposed method in reconstructing high-quality dynamic images across various accelerations.

2   Methods

2.1   Dynamic Parallel MRI Problem Formulation

In Parallel MRI, a complex-valued MR image sequence denoted as $\mathbf{m} \in \mathbb{C}^N$ in the x-y-t space is simultaneously encoded by N_c receiver coils. The acquired

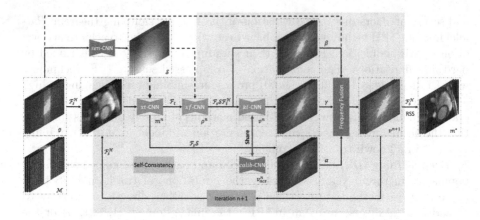

Fig. 1. Illustration of the overall architecture of the proposed k-t CLAIR.

k-space measurement, represented as $\tilde{\mathbf{v}} \in \mathbb{C}^{N_c N}$ in the k_x-k_y-t space, can be expressed as:

$$\tilde{\mathbf{v}} = \mathcal{M}\mathcal{F}_s\mathcal{S}\mathbf{m}, \tag{1}$$

where the forward process involves sequential operators of coil sensitivity map projection $\mathcal{S}$, spatial Fourier transform $\mathcal{F}_s$, and sampling pattern $\mathcal{M}$. The reconstruction of $\mathbf{m}$ from $\tilde{\mathbf{v}}$ is an ill-posed inverse problem. While conventional CS-MRI methods [11] utilize predefined prior knowledge to solve the inverse problem, we formulate it as a learnable multi-prior optimization problem:

$$\min_{\mathbf{m}} \|\mathcal{A}\mathbf{m} - \tilde{\mathbf{v}}\|_2^2 + \lambda_{\mathrm{xt}}\mathcal{R}_{\mathrm{xt}}(\mathbf{m}; \theta_{\mathrm{xt}}) + \lambda_{\mathrm{xf}}\mathcal{R}_{\mathrm{xf}}(\mathcal{F}_t\mathbf{m}; \theta_{\mathrm{xf}}) + \lambda_{\mathrm{kt}}\mathcal{R}_{\mathrm{kt}}(\mathcal{F}_s\mathbf{m}; \theta_{\mathrm{kt}}), \tag{2}$$

where $\mathcal{R}_{\mathrm{xt}}(\mathbf{m}; \theta_{\mathrm{xt}})$, $\mathcal{R}_{\mathrm{xf}}(\mathcal{F}_t\mathbf{m}; \theta_{\mathrm{xf}})$, and $\mathcal{R}_{\mathrm{kt}}(\mathcal{F}_s\mathbf{m}; \theta_{\mathrm{kt}})$ are data-adaptive priors with learnable parameters θ_{xt}, θ_{xf}, and θ_{kt} to regularize the data in the x-t, x-f, and k-t spaces, respectively. $\mathcal{F}_t$ and $\mathcal{F}_s$ represent the Fourier transform operators along the temporal (t) and spatial (x-y) dimensions of the image sequence $\mathbf{m}$, respectively. The parameters λ_{xt}, λ_{xf}, and λ_{kt} control the balance between the impact of the imposed prior regularization and the fidelity to the acquired data.

2.2 k-t CLAIR for Dynamic Parallel MRI Reconstruction

We propose k-t CLAIR, as shown in Fig. 1, to tackle the optimization problem with multiple priors in Eq. (2) through an iterative approach. k-t CLAIR exploits spatiotemporal correlations by iteratively learning complementary priors in the x-t, (x-f), and k-t domains while enforcing self-consistency learning in the k-t domain. Specifically, the proposed approach consists of four steps of prior learning within each iteration: (1) an image enhancement prior learning step removes aliasing artifacts in the x-t domain using a xt-CNN; (2) a dynamic temporal prior learning step captures dynamic details in the x-f domain through a xf-CNN; (3) a k-space restoration prior learning step restores high-frequency information in the *k-t* domain using a *kt*-CNN; and (4) a self-consistency prior learning step enforces calibration data consistency in the *k-t* domain via a *calib*-CNN. Furthermore, we introduce a frequency fusion block to merge and coordinate the feature learning processes of all the priors. Notably, in our approach, we also incorporate the joint learning of coil sensitivity maps using a *sen*-CNN, inspired by the work [28]. This joint learning of coil sensitivity maps complements the multi-prior learning framework of *k-t* CLAIR and enhances the overall reconstruction process for dynamic parallel MRI.

Image Enhancement Prior in the *x-t* Domain. We begin by leveraging the spatiotemporal correlations present in the coil-combined image sequence in the *x-t* domain during each iteration. This is achieved by learning a data-adaptive image prior from the training data using an *xt*-CNN. Our objective is to generate artifact-free images and restore the overall anatomical structure from highly degraded images, enabling us to provide complete k-space signals for subsequent dynamic feature extraction and high-frequency restoration. Specifically, within each iteration, the *xt*-CNN employs a step of the unrolled gradient descent (GD) algorithm to guarantee fidelity to the acquired data and enforce image constraints on the coil-combined images. At the n-th iteration, the *x-t* domain reconstruction can be expressed as follows:

$$\mathbf{m}^{n+1} = \mathbf{m}^n - \eta^n \big(xt\text{-CNN}^n(\mathbf{m}^n) + \lambda^n \mathcal{S}^H \mathcal{F}_s^H (\mathcal{M} \mathcal{F}_s \mathcal{S} \mathbf{m}^n - \tilde{\mathbf{v}}) \big), \qquad (3)$$

where η^n denotes the learning rate at the n-th iteration. Similar to the approach in [22], we utilize U-Net [19] to learn a highly nonlinear image prior.

Dynamic Temporal Prior in the *x-f* Domain. To further capture dynamic features, we propose to learn a temporal prior in the *x-f* domain. This approach takes advantage of the sparsity property exhibited by the signal in the temporal Fourier domain, arising from the periodic cardiac motion observed in dynamic imaging. Our *xf*-CNN, similar to the *xt*-CNN, utilizes the GD algorithm for data fidelity and capturing temporal dynamic frequency. Compared to methods in [8,15], our direct restoration of temporal dynamic frequencies proves more effective in restoring dynamic motion patterns within our framework. At the n-th iteration, the *x-f* domain reconstruction can be formulated as follows:

$$\rho^{n+1} = \rho^n - \zeta^n \big(xf\text{-CNN}^n(\rho^n) + \lambda^n \mathcal{F}_t \mathcal{S}^H \mathcal{F}_s^H (\mathcal{M} \mathcal{F}_s \mathcal{S} \mathcal{F}_t^H \rho^n - \tilde{\mathbf{v}}) \big), \qquad (4)$$

where ζ^n denotes the learning rate. In our approach, we leverage U-Net [19] to learn a highly nonlinear dynamic temporal prior.

Self-consistency Guided Prior in the *k-t* Domain. Previous approaches [20, 23,29] have shown that leveraging k-space correlations preserves high-frequency information and enhances image quality. We extend this idea by employing a *kt*-CNN model in the *k-t* domain to learn spatio-temporal k-space priors from multi-coil data for high-frequency restoration. The *kt*-CNN consists of four convolutional layers that estimate missing signals based on neighboring data. To ensure more accurate and consistent signal restoration in the *k-t* domain, we incorporate calibration information using a *calib*-CNN. The *calib*-CNN shares the same architecture as the *kt*-CNN and learns scan-specific feature embeddings from auto-calibration signals (ACS). By sharing network parameters between the *calib*-CNN and *kt*-CNN models, we enable the learning of scan-specific features that improve the prediction of missing high-frequency k-space data. This integration of calibration information enhances reconstruction accuracy and fidelity in the *k-t* domain. The update formula for the *n*-th iteration can be expressed as follows:

$$\mathbf{v}_{\text{acs}}^{n+1} = calib\text{-CNN}^n(\mathbf{v}_{\text{acs}}), \quad \mathbf{v}^{n+1} = kt\text{-CNN}^n(\mathbf{v}^n), \tag{5}$$

where $\mathbf{v}_{\text{acs}}$ refers to the ACS data.

Frequency Fusion Block. To leverage the benefits of different priors, a frequency fusion layer is introduced to coordinate the feature learning processes in the *x-t*, *x-f*, and *k-t* domains. The formulation of this block is as follows:

$$\mathbf{v}^{n+1} = \alpha \mathcal{M}\mathcal{F}_s \mathcal{S}\mathbf{m}^n + \beta \mathcal{M}\mathcal{F}_s \mathcal{S}\mathcal{F}_t^H \rho^n + \gamma(1 - \mathcal{M})\mathbf{v}^n, \tag{6}$$

where the coefficients α, β, and γ control the influence of each prior in the fusion process. By adjusting these coefficients, the network can balance the contributions from different priors based on their respective strengths and leverage their complementary information for more accurate and faithful dynamic MRI reconstruction. For simplicity during training, the values of α, β, and γ were initially set to 0.5, 0.5, and 1, respectively. However, further exploration can be conducted by treating these values as learnable parameters.

2.3   Objective Function and Evaluation Metric

For quantitative evaluation, we employ three commonly used reconstruction metrics: Structural Similarity Index (SSIM), Normalized Mean Squared Error (NMSE), and Peak Signal-to-Noise Ratio (PSNR). These metrics allow us to assess the quality and fidelity of the reconstructed data. During the training process, we utilize two loss functions: L_1 loss and SSIM loss. These loss functions are employed to minimize the difference between the reconstructed images and the ground truth, thereby ensuring accurate reconstruction. Additionally, we apply an L_1 loss to constrain the *calib*-CNN model throughout training. This constraint helps maintain consistent embedding of acquisition features in each iteration, thereby preserving the integrity of the acquired data during the reconstruction process. The total losses can be expressed as follows:

$$\mathcal{L} = \lambda_1 \mathcal{L}_{L_1}(m^*, m^G) + \lambda_2 \mathcal{L}_{SSIM}(m^*, m^G) + \lambda_3 \sum_{n=1}^{T} \mathcal{L}_1(v_{\text{acs}}^n, v_{\text{acs}}), \tag{7}$$

where m^* and m^G are the Root Sum of Squares (RSS) reconstruction and reference image, respectively. v_{acs}^n represents the predicted ACS data from the $(n-1)$-th *calib*-CNN using the acquired ACS data v_{acs} as input. λ_1, λ_2, and λ_3 are trade-off parameters set to 1 during training.

3   Experiment and Results

3.1   Data and Implementation Details

We conducted experiments on two Cardiac MRI Reconstruction tasks: accelerated cine and T1/T2 mapping, as part of the CMRxRecon Challenge. The dataset included 300 healthy volunteers from a single center, with 120 training data, 60 validation data, and 120 test data. The training data consisted of fully sampled k-space raw data, while the validation and test data consisted of undersampled k-space data with acceleration factors of 4, 8, and 10, along with a sampling mask and ACS of 24 lines. The ground truth images for the validation set were withheld, and the test data was not accessible to the participants. For more detailed information, please refer to [24,25] and the challenge website[1].

3.2   Implementation Details

We divided the training data into two subsets: 80% for model training and the remaining 20% for model validation. All models were optimized using the Adam optimizer with parameters $\beta_1 = 0.9$ and $\beta_2 = 0.999$, an initial learning rate of $3e^{-4}$, and a batch size of 1. The number of iterations (T) was set to 12 for all rolling-based models. The network architecture remained consistent for both the xt-CNN and xf-CNN models. For the mapping task, the initial feature dimension was set to 64, and for the cine task, it was set to 32. These dimensions were then doubled after each pooling stage to enhance the network's ability to capture critical features. Regarding the kt-CNN and *calib*-CNN models, the feature dimension was specifically set to match the number of coils in each respective task. Networks were trained for 30 epochs for the cine reconstruction task and 50 epochs for the T1/T2 mapping reconstruction task. After 20 and 40 epochs, the learning rates were reduced by a factor of 10. Models were trained using PyTorch Lightning on 2 NVIDIA RTX A6000 GPUs for T1/T2 mapping and 4 Tesla A100 GPUs for the Cine task. Detailed model information for both tasks can be found in the Supplementary Materials.

3.3   Experiments and Results

Reconstruction of Accelerated T1/T2 Acquisitions for T1/T2 Mapping. We conduct a comparative analysis of different methods for reconstruction of accelerated acquisitions of T1- and T2-weighted images in T1 and T2 mapping, respectively. The evaluated methods include Zero-filled, FastMRI U-Net [28], and E2EVarNet [22]. Additionally, we extend the works in [22,28] by incorporating 3D Convolution to capture temporal correlations. Notably, E2EVarNet [22] utilizes a deep learning unrolling network structure for iterative reconstruction in the image domain, while our proposed method is based on an unrolling framework that incorporates multiple prior knowledge in each reconstruction iteration.

---

[1] https://cmrxrecon.github.io.

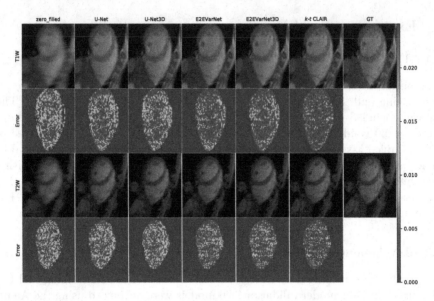

Fig. 2. Reconstruction of 10× accelerated T1W/T2W images and masked error maps.

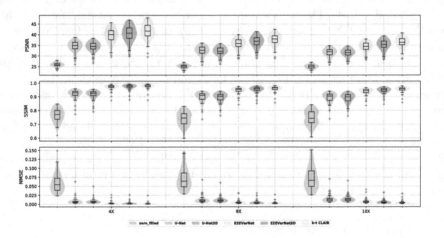

Fig. 3. Performance for T1W/T2W reconstruction on the unseen training data.

We demonstrate the effectiveness of our proposed method by presenting reconstruction results for highly accelerated T1 and T2-weighted images, as shown in Fig. 2. Our method yields reconstructions with enhanced fidelity, accurately capturing rich and intricate structures compared to alternative methods. The reconstruction error maps consistently highlight the superior performance of our approach in restoring highly undersampled images in the ventricle and ventricular myocardium regions. Detailed comparison results on the 20% hold-out training data are presented in Table 1, demonstrating the superiority of our method

Table 1. Quantitative evaluation for cardiac CINE (LAX and SAX) and T1W/T2W images reconstruction on the unseen training data. Note that the results were reported after cropping the middle 1/6 of the original image.

Methods	Acc.	SSIM↑				NMSE↓				PSNR↑			
		Lax	Sax	T1W	T2W	Lax	Sax	T1W	T2W	Lax	Sax	T1W	T2W
Zero-filled	4×	0.6975	0.7601	0.7301	0.7991	0.0677	0.0599	0.0769	0.0417	26.06	27.22	25.92	25.79
	8×	0.6902	0.7328	0.7022	0.7771	0.0756	0.0738	0.0910	0.0507	25.82	26.40	25.17	24.95
	10×	0.6864	0.7284	0.7023	0.7901	0.0788	0.0780	0.0963	0.0509	25.65	26.16	24.93	24.96
	Avg	0.7188		0.7501		0.0721		0.0679		26.26		25.29	
U-Net	4×	0.8503	0.8983	0.9109	0.9247	0.0197	0.0123	0.0106	0.0064	31.24	34.17	35.05	33.74
	8×	0.8262	0.8701	0.8876	0.9053	0.0299	0.0202	0.0151	0.0098	29.62	31.97	33.13	31.66
	10×	0.8152	0.8635	0.8809	0.9055	0.0338	0.0229	0.0186	0.0111	29.04	31.39	32.08	31.11
	Avg	0.8566		0.9025		0.0226		0.0119		31.39		32.79	
U-Net3D	4×	0.8604	0.9076	0.9078	0.9216	0.0175	0.0106	0.0097	0.0065	31.64	34.72	34.92	33.49
	8×	0.8365	0.8798	0.8845	0.9017	0.0266	0.0179	0.0147	0.0100	29.99	32.44	32.94	31.45
	10×	0.8262	0.8728	0.8770	0.9024	0.0305	0.0207	0.0183	0.0112	29.40	31.80	31.95	30.99
	Avg	0.8665		0.8992		0.0202		0.0117		31.82		32.63	
E2EVarNet	4×	0.9569	0.9714	0.9738	0.9661	0.0039	0.0027	0.0035	0.0024	38.16	41.14	41.26	37.94
	8×	0.9016	0.9324	0.9466	0.9423	0.0130	0.0083	0.0071	0.0052	32.96	35.87	36.73	34.43
	10×	0.8849	0.9183	0.9342	0.9378	0.0166	0.0109	0.0103	0.0064	31.92	34.67	34.97	33.52
	Avg	0.9291		0.9501		0.0090		0.0058		35.95		36.48	
E2EVarNet3D	4×	0.9640	0.9759	0.9771	0.9692	0.0031	0.0022	0.0030	0.0020	39.11	42.01	42.34	38.61
	8×	0.9156	0.9429	0.9538	0.9485	0.0106	0.0067	0.0059	0.0042	33.88	36.83	37.78	35.26
	10×	0.9003	0.9297	0.9434	0.9446	0.0138	0.0090	0.0085	0.0052	32.77	35.54	36.00	34.37
	Avg	0.9394		0.9561		0.0074		0.0048		36.86		37.39	
k-t CLAIR	4×	0.9673	0.9777	0.9806	0.9728	0.0027	0.0020	0.0021	0.0015	39.74	42.49	43.57	39.48
	8×	0.9238	0.9493	0.9609	0.9544	0.0090	0.0056	0.0043	0.0032	34.57	37.64	38.90	36.14
	10×	0.9099	0.9372	0.9531	0.9507	0.0118	0.0075	0.0060	0.0040	33.47	36.30	37.31	35.19
	Avg	**0.9454**		**0.9621**		**0.0063**		**0.0035**		**37.54**		**38.43**	

over other approaches for both T1 and T2-weighted images across all acceleration rates and evaluation metrics. Figure 3 further illustrates the effectiveness of the proposed method. Additionally, *k-t* CLAIR demonstrates strong generalization capabilities and achieves promising reconstructions on validation data. This is supported by the detailed quantitative scores obtained from the Challenge Leaderboard[2] for cardiac T1 and T2-weighted reconstruction on the validation set, as shown in Table 2.

Accelerated Cine Images Reconstruction. For the cine task evaluation, we compare our method with Zero-filled, FastMRI U-Net [28], and E2EVarNet [22]. Additionally, we compare our method with U-Net3D and E2EVarNet3D, similar to the mapping task. The reconstruction results for 10× acceleration are shown in Fig. 4, illustrating the faithful reconstructions produced by our method across all views, in comparison to other methods. The reconstruction error maps within the myocardium and chambers consistently demonstrate the superior accuracy of our method in rapid CMR reconstruction. Quantitative results on the 20%

---

[2] https://www.synapse.org/#!Synapse:syn51471091/wiki/622548.

Table 2. Quantitative evaluation of the proposed k-t CLAIR for cardiac CINE (LAX and SAX) and T1W/T2W images reconstruction on the Validation Set.

Acc.	SSIM↑				NMSE↓				PSNR↑			
	Lax	Sax	T1W	T2W	Lax	Sax	T1W	T2W	Lax	Sax	T1W	T2W
4×	0.963	0.976	0.983	0.970	0.003	0.003	0.003	0.003	38.838	41.656	43.801	38.849
8×	0.914	0.945	0.961	0.950	0.011	0.007	0.006	0.006	33.508	36.853	38.961	35.621
10×	0.901	0.932	0.954	0.947	0.014	0.009	0.008	0.007	32.49	35.602	37.225	34.721
Avg.	0.939		0.961		0.008		0.005		36.491		38.196	

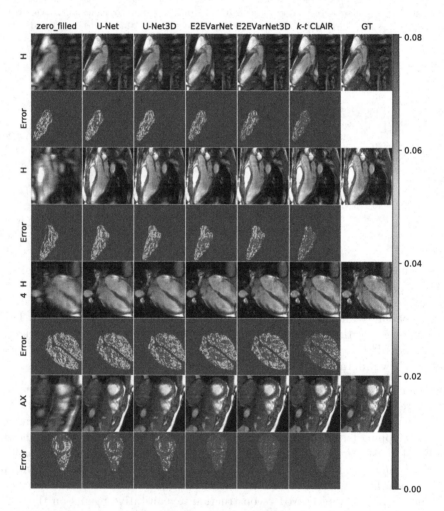

Fig. 4. Reconstruction of 10× accelerated CINE and masked error maps.

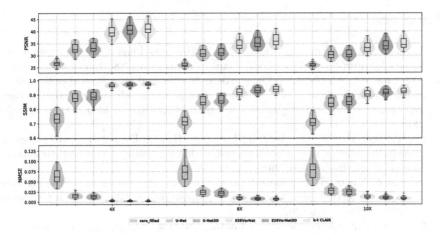

Fig. 5. Performance for CINE reconstruction on the unseen training data.

hold-out training data are presented in Table 1, indicating the outstanding reconstruction quality and improvement achieved by our proposed method across all evaluation metrics and views. The effectiveness of the proposed method is consistently illustrated by the violin plot, as shown in Fig. 5. The performance on the Sax view generally surpasses that of the Lax views at the same acceleration rate, possibly due to the larger amount of available data for the Sax view. Furthermore, our method also performs well in generating reconstructions for cine data in the validation set, as detailed in Table 2.

4   Conclusions

In this paper, we introduced *k-t* CLAIR, a self-consistency guided multi-prior learning framework for faithful dynamic MRI reconstruction. By leveraging multiple priors and incorporating calibration information, *k-t* CLAIR improved the quality and accuracy of dynamic MRI reconstructions. Experimental results demonstrated its effectiveness in achieving high-quality reconstructions for cardiac cine and T1W/T2W images in T1/T2 mapping. In future work, we plan to extend the application of this method to other anatomical structures in dynamic MRI and validate its application in clinical routines.

Acknowledgements. This work was supported by a grant from Innovation and Technology Commission of the Hong Kong SAR (MRP/046/20X); and by a Faculty Innovation Award from the Faculty of Medicine of The Chinese University of Hong Kong.

References

1. Aggarwal, H.K., Mani, M.P., Jacob, M.: MoDL: model-based deep learning architecture for inverse problems. IEEE Trans. Med. Imaging **38**(2), 394–405 (2018)
2. Akçakaya, M., Moeller, S., Weingärtner, S., Uğurbil, K.: Scan-specific robust artificial-neural-networks for k-space interpolation (RAKI) reconstruction: database-free deep learning for fast imaging. Magn. Reson. Med. **81**(1), 439–453 (2019)
3. Eo, T., Jun, Y., Kim, T., Jang, J., Lee, H.J., Hwang, D.: Kiki-net: cross-domain convolutional neural networks for reconstructing undersampled magnetic resonance images. Magn. Reson. Med. **80**(5), 2188–2201 (2018)
4. Fabian, Z., Tinaz, B., Soltanolkotabi, M.: Humus-net: hybrid unrolled multi-scale network architecture for accelerated MRI reconstruction. Adv. Neural. Inf. Process. Syst. **35**, 25306–25319 (2022)
5. Griswold, M.A., et al.: Generalized autocalibrating partially parallel acquisitions (GRAPPA). Magn. Reson. Med. Official J. Int. Soc. Magn. Reson. Med. **47**(6), 1202–1210 (2002)
6. Hammernik, K., et al.: Learning a variational network for reconstruction of accelerated MRI data. Magn. Reson. Med. **79**(6), 3055–3071 (2018)
7. Han, Y., Sunwoo, L., Ye, J.C.: k-space deep learning for accelerated MRI. IEEE Trans. Med. Imaging **39**(2), 377–386 (2020)
8. Jung, H., Sung, K., Nayak, K.S., Kim, E.Y., Ye, J.C.: k-t FOCUSS: a general compressed sensing framework for high resolution dynamic MRI. Magn. Reson. Med. Official J. Int. Soc. Magn. Reson. Med. **61**(1), 103–116 (2009)
9. Küstner, T., et al.: Cinenet: deep learning-based 3D cardiac cine MRI reconstruction with multi-coil complex-valued 4D spatio-temporal convolutions. Sci. Rep. **10**(1), 13710 (2020)
10. Lingala, S.G., Hu, Y., DiBella, E., Jacob, M.: Accelerated dynamic MRI exploiting sparsity and low-rank structure: kt SLR. IEEE Trans. Med. Imaging **30**(5), 1042–1054 (2011)
11. Lustig, M., Donoho, D., Pauly, J.M.: Sparse MRI: the application of compressed sensing for rapid MR imaging. Magn. Reson. Med. Official J. Int. Soc. Magn. Reson. Med. **58**(6), 1182–1195 (2007)
12. Lustig, M., Pauly, J.M.: Spirit: iterative self-consistent parallel imaging reconstruction from arbitrary k-space. Magn. Reson. Med. **64**(2), 457–471 (2010)
13. Otazo, R., Candes, E., Sodickson, D.K.: Low-rank plus sparse matrix decomposition for accelerated dynamic MRI with separation of background and dynamic components. Magn. Reson. Med. **73**(3), 1125–1136 (2015)
14. Pruessmann, K.P., Weiger, M., Scheidegger, M.B., Boesiger, P.: Sense: sensitivity encoding for fast MRI. Magn. Reson. Med. Official J. Int. Soc. Magn. Reson. Med. **42**(5), 952–962 (1999)
15. Qin, C., et al.: Complementary time-frequency domain networks for dynamic parallel MR image reconstruction. Magn. Reson. Med. **86**(6), 3274–3291 (2021)
16. Qin, C., Schlemper, J., Caballero, J., Price, A.N., Hajnal, J.V., Rueckert, D.: Convolutional recurrent neural networks for dynamic MR image reconstruction. IEEE Trans. Med. Imaging **38**(1), 280–290 (2018)
17. Qin, C., Schlemper, J., Duan, J., Seegoolam, G., Price, A., Hajnal, J., Rueckert, D.: k-t NEXT: dynamic MR image reconstruction exploiting spatio-temporal correlations. In: Shen, D., et al. (eds.) MICCAI 2019. LNCS, vol. 11765, pp. 505–513. Springer, Cham (2019). https://doi.org/10.1007/978-3-030-32245-8_56

18. Ran, M., et al.: MD-Recon-Net: a parallel dual-domain convolutional neural network for compressed sensing MRI. IEEE Trans. Radiat. Plasma Med. Sci. **5**(1), 120–135 (2020)
19. Ronneberger, O., Fischer, P., Brox, T.: U-Net: convolutional networks for biomedical image segmentation. In: Navab, N., Hornegger, J., Wells, W.M., Frangi, A.F. (eds.) MICCAI 2015. LNCS, vol. 9351, pp. 234–241. Springer, Cham (2015). https://doi.org/10.1007/978-3-319-24574-4_28
20. Ryu, K., Alkan, C., Choi, C., Jang, I., Vasanawala, S.: K-space refinement in deep learning MR reconstruction via regularizing scan specific spirit-based self consistency. In: Proceedings of the IEEE/CVF International Conference on Computer Vision, pp. 4008–4017 (2021)
21. Schlemper, J., Caballero, J., Hajnal, J.V., Price, A.N., Rueckert, D.: A deep cascade of convolutional neural networks for dynamic MR image reconstruction. IEEE Trans. Med. Imaging **37**(2), 491–503 (2017)
22. Sriram, A., et al.: End-to-end variational networks for accelerated MRI reconstruction. In: Martel, A.L., et al. (eds.) MICCAI 2020. LNCS, vol. 12262, pp. 64–73. Springer, Cham (2020). https://doi.org/10.1007/978-3-030-59713-9_7
23. Sriram, A., Zbontar, J., Murrell, T., Zitnick, C.L., Defazio, A., Sodickson, D.K.: Grappanet: combining parallel imaging with deep learning for multi-coil MRI reconstruction. In: Proceedings of the IEEE/CVF Conference on Computer Vision and Pattern Recognition, pp. 14315–14322 (2020)
24. Wang, C., et al.: Recommendation for cardiac magnetic resonance imaging-based phenotypic study: imaging part. Phenomics **1**, 151–170 (2021)
25. Wang, C., et al.: Cmrxrecon: an open cardiac MRI dataset for the competition of accelerated image reconstruction. arXiv preprint arXiv:2309.10836 (2023)
26. Wang, S., et al.: Dimension: dynamic MR imaging with both k-space and spatial prior knowledge obtained via multi-supervised network training. NMR Biomed. **35**(4), e4131 (2022)
27. Yang, G., et al.: Dagan: deep de-aliasing generative adversarial networks for fast compressed sensing MRI reconstruction. IEEE Trans. Med. Imaging **37**(6), 1310–1321 (2017)
28. Zbontar, J., et al.: fastMRI: an open dataset and benchmarks for accelerated MRI. arXiv preprint arXiv:1811.08839 (2018)
29. Zhang, L., Li, X., Chen, W.: Camp-net: consistency-aware multi-prior network for accelerated MRI reconstruction. arXiv preprint arXiv:2306.11238 (2023)

Cardiac MRI Reconstruction from Undersampled K-Space Using Double-Stream IFFT and a Denoising GNA-UNET Pipeline

Julia Dietlmeier[1]([✉]) [ID], Carles Garcia-Cabrera[2] [ID], Anam Hashmi[2] [ID],
Kathleen M. Curran[3] [ID], and Noel E. O'Connor[1] [ID]

[1] Insight SFI Research Centre for Data Analytics, Glasnevin, Dublin 9, Ireland
{julia.dietlmeier,noel.oconnor}@insight-centre.org
[2] ML-Labs, Dublin City University, Glasnevin, Dublin 9, Ireland
{carles.garciacabrera6,anam.hashmi2}@mail.dcu.ie
[3] School of Medicine, University College Dublin, Dublin, Ireland
kathleen.curran@ucd.ie
http://www.insight-centre.org, http://www.ml-labs.ie/, http://www.ucd.ie/

Abstract. In this work, we approach the problem of cardiac Magnetic Resonance Imaging (MRI) image reconstruction from undersampled k-space. This is an inherently ill-posed problem leading to a variety of noise and aliasing artifacts if not appropriately addressed. We propose a two-step double-stream processing pipeline that first reconstructs a noisy sample from the undersampled k-space (frequency domain) using the inverse Fourier transform. Second, in the spatial domain we train a denoising GNA-UNET (enhanced by Group Normalization and Attention layers) on the noisy aliased and fully sampled image data using the Mean Square Error loss function. We achieve competitive results on the leaderboard and show that the algorithmic combination proposed is effective in high-quality MRI reconstruction from undersampled cardiac long-axis and short-axis complex k-space data.

Keywords: Cardiac MRI · Undersampled k-space · Deep Learning · Denoising UNET

1 Introduction

In clinical diagnostics, Magnetic Resonance Imaging (MRI) is a widely used non-invasive medical imaging modality providing superior soft tissue contrast. Cardiac MRI, in particular, offers the opportunity for the diagnosis of cardiovascular disease. The long acquisition time in MRI remains a major weakness of the approach. The undersampling of k-space by different factors (also known as the compressed sensing (CS) approach [10]) offers accelerated data collection providing discomfort relief for paediatric patients. In addition to the motion artifacts inherently present in cardiac MRI, the process of frequency domain undersampling results in the loss of high-frequency information [11], which translates to a variety of noise and structural aliasing artifacts in the spatial domain, and thus further deteriorates the reconstructed image quality.

A simple and a straightforward approach to denoising and artifact removal is to use a deep learning (DL) model, e.g. a Convolutional Neural Network (CNN), in the spatial domain to map a noisy aliased MR image to a higher-quality aliasing-free image. In fact, deep learning (DL) became an irreplaceable tool in MRI reconstruction and most DL approaches treat it as a denoising problem while some propose to directly complete the missing and corrupted k-space data using specially designed interpolation CNNs [7] or Transformers [8]. Some challenges still remain such as degraded performance of such models due to the high level of noise or limited training data and limited computational resources.

The work reported in this paper was conducted as part of the CMRxRecon challenge, Task 1: Cine reconstruction, and specifically we target high-quality MR image reconstruction in spatial domain. Our contribution is primarily in designing an effective double-stream pipeline to process the undersampled complex k-space data and to reconstruct the CMR images.

2   Related Works

Recently, end-to-end DL approaches have tried to solve the inverse problem of reconstructing MR images from sub-Nyquist sampled data [12]. For example, Wang et al. [13] trained a CNN based model using a large number of existing high quality MR images from downsampled reconstruction images as either an initialization or regularization term in classical CS approaches, to learn fully sampled reconstruction. Kwon et al. (2017), utilized a multilayer perceptron for fast reconstruction of MR images [14]. Lee et al. (2018), trained a residual network for phase and magnitude information separately to perform MR image reconstruction [15]. Numerous studies have incorporated the MRI domain knowledge, such as data consistency in the space to improve the standard deep learning algorithms for CS-MRI [16]. Some research used generative adversarial networks (GANs) [17]. Others used k-space reconstruction neural networks trained on autocalibration signal (ACS) data [18]. Typically, DL based methods require a large amount of training data, so several studies have tried to employ untrained CNNs such as Deep Image prior and Deep encoder and their variants [19–21]. More recently the MR image reconstruction research has shifted to using an *unrolled* framework with DL models because of better reconstruction performance and faster reconstruction times compared to iterative methods [22–24]. However, these approaches still rely on fully sampled scans as the ground truth data while overall it is preferable to reduce reliance on data.

In order to increase data efficiency, recent work has proposed designing data augmentation pipelines specifically suited to accelerated MRI reconstruction with appropriate image-based [25] or acquisition-based, physics-driven transformations [26]. These pipelines would supplement prior proposals that utilize

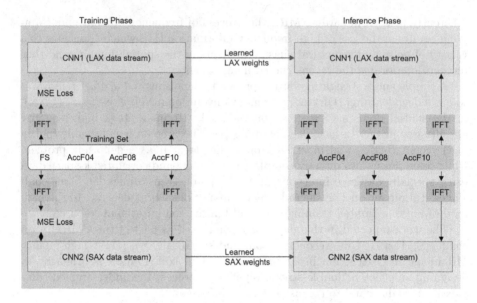

Fig. 1. Architecture of the proposed double-stream cardiac MRI reconstruction pipeline. Our proposed design involves processing the long-axis (LAX) and short-axis (SAX) data streams separately. Also, in the training phase we train the CNNs using the MSE loss function only on the AccFactor10 undersampled data. Inspired by the results from [9], we hypothesize that because the ×10 acceleration factor is the most sparse one it results in strongest structural artifacts. Therefore a model trained on the fully sampled (FS) and AccF10 data can also be applied to the factor ×4 (AccF04) and factor ×8 (AccF08) undersampled data during the inference phase.

prospectively undersampled (unsupervised) data that currently lag in reconstruction performance [21]. However, data augmentation places an additional burden on the neural network by requiring it to learn every conceivable scale of every feature separately, even though it just attempts to approximate equivariance. The outcomes of such acquired equivariance are frequently worse than those of assured equivariance [27]. Therefore, scale-equivariant CNNs have been gaining traction in improving the data efficiency [28,29].

3   Architecture

The original data are a complex k-space with different subsampling (acceleration) factors such as ×4, ×8, and ×10. Subsampling masks and fully sampled k-space are also provided in the training set. A diagram of our processing pipeline is illustrated in Fig. 1. The pipeline operates directly on k-space data by converting frequency domain into spatial domain using the inverse Fourier transform and uses aliased undersampled images and the fully sampled images from the spatial domain to train a CNN.

We process the long-axis (LAX) and short-axis (SAX) k-space data in two separate but identical data streams due to our empirical observation during experimentation that training two models on separate LAX and SAX data yields better results.

3.1 GNA-UNET with Group Normalization and GCT Attention

In the medical imaging domain, a UNET model is probably the most well known supervised deep learning architecture that was initially introduced by Ronneberger et al. in 2015 [3] and since then has found its applications in many downstream tasks such as image segmentation and image to image translation [6]. Furthermore, the concept of UNET is essential to recent Transformer architectures [5] and Denoising Diffusion Probabilistic Models (DDPM) [4]. In a nutshell, it is a symmetric encoder-decoder architecture consisting of contracting and an expanding branches and skip connections that enable sharing of information. Another remarkable property of UNET is that this model is known to be able to perform well with limited training data [31]. Here, we work with 2D slices and build a 2D GNA-UNET with the configuration outlined in Table 1. In particular, we design a GNA-UNET with five downsampling/upsampling stages.

Table 1. GNA-UNET configuration developed in this work.

Module	Shape (in_channels, out_channels)
encoder block	(1, 64)
encoder block	(64, 128)
encoder block	(128, 256)
encoder block	(256, 512)
encoder block	(512, 1024)
conv_block	(1024, 2048)
decoder block	(2048, 1024)
decoder block	(1024, 512)
decoder block	(512, 256)
decoder block	(256, 128)
decoder block	(128, 64)
output	(64,1)

Our GNA-UNET Model is First Enhanced by the Group Normalization (GN) Layers [30]. This is motivated by the fact that we are using a small batch_size = 2 in our experiments. As has been shown in [30], GN outperforms commonly used Batch Normalization (BN) for small batch sizes. In particular, GN divides the channels into groups and computes within each group the mean and variance for normalization. Our Ablation study in Sect. 4.6 demonstrates the effectiveness of GN as compared to BN.

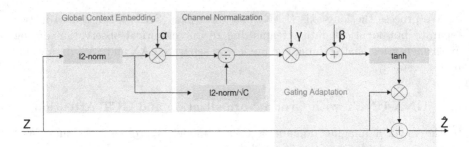

Fig. 2. Simplified diagram of the Gated Channel Transformation (GCT) attention layer. The weight parameter α controls the weight of each channel before the channel normalization. The gating weight and bias, γ and β, are responsible for adjusting the scale of the input feature Z channel-wisely. Image source: adapted from [32].

Second, We Introduce Gated Channel Transformation (GCT) Attention Layers into GNA-UNET. GCT attention has been shown to improve the discriminability of deep CNNs by leveraging the relationship among channels [32]. In particular, we add GCT layers after each convolutional **conv2d** layer and in the skip connections. This configuration has been determined by our Ablation study in Sect. 4.6. The overview diagram of GCT is given in Fig. 2 where α, β and γ are trainable parameters. As shown by the authors, GCT layer inserted far away from the network output, reduces the variance of input features and thus encourages cooperation among channels and avoids excessive activation values or loss of useful features. The GCT consists of *Global Context Embedding* module that exploits global contextual information ooutside the small receptive fields of convolutional layers. The trainable embedding weight α controls the weight of each channel. The *Channel Normalization* module uses l_2-based channel normalization. The scalar $\sqrt{C}$ is used to normalize the scale, with C being the number of channels. The *Gating Adaptation* module is used to adapt the original feature. GCT can facilitate competition and cooperation during the training process by introducing the gating mechanism with *tanh* activation function. Gating weight γ and the gating bias β are trainable parameters.

To regularize our model, we further add **Dropout** layers with Dropout probability $p = 0.25$ in each of five encoder blocks. Without the **Dropout** layers, the model starts overfitting from the very beginning of the training phase.

Thus, our encoder block consists of **conv_block**(in_channels, out_channels) $\rightarrow$ **Dropout**($p = 0.25$) $\rightarrow$ **MaxPooling2d**(2, 2) layers.

A convolutional block **conv_block** consists of the following layers: **conv2d** (kernel_size = 3, padding = 1) $\rightarrow$ **GCT** $\rightarrow$ **GN**(ng) $\rightarrow$ **conv2d** (kernel_size = 3, padding = 1) $\rightarrow$ **GCT** $\rightarrow$ **GN**(ng) $\rightarrow$ **ReLU**. Where **GN**(ng) is a group normalization layer with the number of groups hyperparameter $ng = 8$ determined in Sect. 4.6, and **ReLU** is the Rectified Linear Unit activation function.

The decoder block mirrors the encoder block in reverse and consists of the combination of the transposed convolution layer **ConvTranspose2d** (kernel_size = 2, stride = 2, padding = 0) $\rightarrow$ **conv_block**.

The total number of learnable GNA-UNET parameters is 124,427,137.

3.2 Loss Function

We use the Mean Square Error (MSE) as the objective function to update the model parameters:

$$Loss_{MSE} = \frac{1}{N} \sum_{i=1}^{N} (y_i - \hat{y}_i)^2 \tag{1}$$

where N is the number of pixels in the image, and y_i and $\hat{y}_i$ represent the target and reconstructed images.

4 Methodology

In order to be able to obtain quantitative validation metrics, we perform our experiments on the training and validation sets of CMRxRecon challenge data which contain the long-axis and short-axis fully sampled and undersampled images of 120 and 60 patients, respectively. We do not use the sampling masks.

4.1 Preprocessing

We process 1-channel 2D greyscale images in the spatial domain first and after performing the IFFT, we rescale by the maximum image intensity and then apply the linear scaling transform to map the pixel intensity I into $\bar{I} \in [0, 1]$:

$$\bar{I} = \frac{(I - I_{min})}{(I_{max} - I_{min})} \tag{2}$$

While the IFFT operates on the original k-space resolutions, we resize all reconstructed images to 512×512 pixels as an input to the GNA-UNET model.

4.2 Dataset

A total of 300 healthy volunteers from a single center were included in this study. The released dataset [34] includes 120 volunteers for training data, 60 for validation data and 120 for test data. Training data includes fully sampled k-space data, auto-calibration lines (ACS, 24 lines) and undersampled k-space with acceleration factors $\times 4$, $\times 8$ and $\times 10$.

4.3 Implementation and Training Details

The processing pipeline was implemented in Python 3.9 and the DL open source library Pytorch 2.0.1. All experiments were performed on a desktop computer with the Ubuntu operating system 18.04.3 LTS with the Intel(R) Core(TM) i9-9900K CPU, Nvidia GeForce RTX 2080 Ti GPU, and a total of 62 GB RAM.

We train the GNA-UNET for 300 epochs using an AdamW optimizer [33] with a learning rate of $lr = 0.001$. We use the batch size of 2. No data augmentation was used in the training phase.

4.4   Performance Evaluation Metrics

Mean Square Error (MSE) is the most widely used image quality assessment metric with better values closer to zero. The MSE and the normalized MSE (NMSE) between two images $\hat{y}$ and y are defined as follows:

$$MSE = \frac{1}{MN} \sum_{m=1}^{M} \sum_{n=1}^{N} [\hat{y}(n,m) - y(n,m)]^2 \tag{3}$$

$$NMSE = 1 - \frac{\|y - \hat{y}\|_2}{\|y - \overline{y}\|}, \overline{y} = \frac{1}{N} \sum_i y_i \tag{4}$$

Peak Signal to Noise Ratio (PSNR) is defined as the ratio between the maximum possible signal power and the power of the distorting noise. This ratio between two images is computed in the decibel form as [2]:

$$PSNR = 10 \log_{10}(peakval^2)/MSE \tag{5}$$

where *peakval* is the maximum possible intensity value in an image.

The Structural Similarity Index Measure (SSIM) is a perception-based model that captures the mutual dependencies among adjacent pixels to assess the similarity of two images, such as brightness, contrast and structural properties [1]:

$$SSIM(y,\hat{y}) = \frac{(2\mu_y\mu_{\hat{y}} + c_1)(2\sigma_{y\hat{y}} + c_2)}{(\mu_y^2 + \mu_{\hat{y}}^2 + c_1)(\sigma_y^2 + \sigma_{\hat{y}}^2 + c_2)} \tag{6}$$

where μ_y and $\mu_{\hat{y}}$ represent the mean values of the model output $\hat{y}$ and the target output y, σ_y and $\sigma_{\hat{y}}$ denote the corresponding pixel variance values and $\sigma_{y\hat{y}}$ is the covariance value. In order to stabilize the division, and with $P = \max(y) - \min(y)$, the constants c_1 and c_2 are defined as follows:

$$c_1 = (0.01P)^2, c_2 = (0.03P^2) \tag{7}$$

4.5   Experimental Results

To allow fast experimentation, we do not use all training data provided but construct two training subsets as follows. For subset $\mathbb{S}_1$ (single coil) and subset $\mathbb{M}_1$ (multi coil), we process all timeframes and sz slices in the supplied `cine_lax.mat` and `cine_sax.mat` files. We further randomly subsample each full training set and select 1,000 LAX and 1,000 SAX images for training.

Table 2 shows our best results on the validation set provided by the challenge platform while Table 3 shows validation results obtained by using a simple denoising Autoencoder model (DAE) with the encoder-bottleneck-decoder architecture, i.e. without skip connections. Two large DAE models each with 131,282,057 learnable parameters were trained on 128 × 128 resized LAX and SAX images. The comparison between Tables 2 and 3 shows that GNA-UNET based processing outperforms the DAE model by a large margin. We think that

Table 2. Our best results on the CMRxRecon validation set using GNA-UNET.

Modality/Metric	SSIM ↑	NMSE ↓	PSNR ↑	Modality/Metric	SSIM ↑	NMSE ↓	PSNR ↑
Single_LAX_04	0.5505	0.1514	21.9632	Multi_LAX_04	0.5933	0.1358	22.5673
Single_LAX_08	0.5683	0.1506	22.1463	Multi_LAX_08	0.6121	0.1285	23.3891
Single_LAX_10	0.5677	0.1536	22.0728	Multi_LAX_10	0.6323	0.1399	21.8900
Single_SAX_04	0.6003	0.2004	22.1731	Multi_SAX_04	0.5903	0.2017	22.0045
Single_SAX_08	0.6116	0.1995	22.3188	Multi_SAX_08	0.6520	0.1541	22.9344
Single_SAX_10	0.6153	0.1991	22.3178	Multi_SAX_10	0.6699	0.2087	23.3594
Cine average	0.5856	0.1757	22.1653	Cine average	0.6250	0.1614	22.6908

Table 3. Results on the CMRxRecon validation set (single coil) by using the DAE model (trained for 100 epochs) as a denoising CNN.

Modality/Metric	SSIM ↑	NMSE ↓	PSNR ↑
Single_LAX_04	0.3047	0.7520	15.6542
Single_LAX_08	0.3054	0.7422	15.6861
Single_LAX_10	0.3051	0.7448	15.6745
Single_SAX_04	0.3218	1.1617	14.6089
Single_SAX_08	0.3227	1.1517	14.6522
Single_SAX_10	0.3228	1.1466	14.6672
Cine average	0.3137	0.9498	15.1572

this is due to the absence of encoder-decoder skip connections. The results for the multi coil data are some way off the top of the leaderboard, and we are currently trying to better understand why this might be the case with a view to solving this problem in future work.

Selected qualitative results are shown in Fig. 3. While working with the single coil data on a subset of the training set, we obtained high $PSNR = 35.8596$ dB however when evaluating on the validation set provided by the challenge platform, our best $PSNR = 22.3188$ dB was much lower. This shows that despite the regularization, our model does not yet generalize well to the new unseen data.

Our Results on the Test Set are: $PSNR = 34.4493$ dB, $SSIM = 0.921$ and $NMSE = 0.0563$.

4.6   Ablation Studies

We perform two ablations studies to first demonstrate the effectiveness of the Group Normalization (GN) and second the effectiveness of the Gated Channel Transformation (GCT) attention unit. Both ablation studies were performed on a random subset of the TrainingSet of the CMRxRecon dataset with the 70%-10%-20% non-overlapping training-validation-test split.

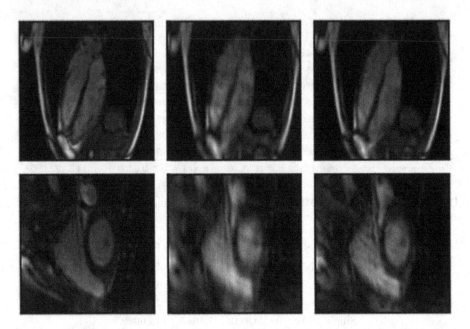

Fig. 3. Selected qualitative results (single coil, patient P002). First row: long-axis images. Second row: short-axis images. **Left:** ground truth; **Middle:** undersampled AccFactor10; **Right:** reconstructed images using our GNA-UNET pipeline.

We tune the hyperparameters of the Group Normalization layers and report the best result based on five training runs. The base model in Table 4 refers to the encoder-decoder structure with skip connections without the Batch Normalization (BN) layers. It can be seen that the model performs poorly without any type of normalization. The addition of BN layers results in 12.66 dB $PSNR$ gain while the addition of GN layers results in 14.56 dB $PSNR$ gain (with $ng = 8$) relative to the base model. Interestingly, the best performing hyperparameter $ng = 8$ is fixed for all convolutional blocks. We have also investigated such configuration as $2-4-8-16-32-64$ in the encoder with the reverse configuration $64-32-16-8-4-2$ in the decoder. This type of configuration resulted in a slightly worse performance. The addition of GCT layers results in a marginal improvement over the use of GN layers as can be seen in Table 5. The position of GCT layers, as well as the normalization norm and the gating function influence performance (Table 5).

Table 4. Ablation study to demonstrate the effectiveness of Group Normalization (GN) layers instead of Batch Normalization (BN) layers. All investigated fixed number of groups (ng) values consistently outperform BN. The best value $ng = 8$ corresponds to 1.9 dB $PSNR$ gain relative to the use of BN.

Configuration	SSIM ↑	MSE ↓	PSNR ↑
base model	0.3759	0.00781	21.2789
base model with Batch Normalization	0.9152	0.0004722	33.9389
base model with Group Normalization			
$ng = 2$	0.9322	0.0003059	35.7680
$ng = 4$	0.9333	0.0003118	35.6037
$ng = 8$	**0.9335**	**0.0002984**	**35.8393**
$ng = 16$	0.9244	0.0003304	35.2098
$ng = 32$	0.9302	0.0003147	35.4652
$ng = 64$	0.9285	0.0003631	34.8299
$ng = 2 - 4 - 8 - 16 - 32 - 64$	0.9257	0.0003041	35.6154
$ng = 64 - 32 - 16 - 8 - 4 - 2$	0.9285	0.0003816	34.7310

Table 5. Ablation study to demonstrate the effectiveness of GCT layers.

Position	SSIM ↑	MSE ↓	PSNR ↑
after each **conv2d** layer	0.9333	0.0002983	35.8570
only in the skip connections	0.9326	0.0003062	35.8225
only in the bottleneck	0.9318	0.0003412	35.4324
after each **conv2d** layer and in the skip connections	**0.9346**	**0.0002969**	**35.8596**

Table 6. Our model information. Performance on a training set shows PSNR, SSIM and MSE values while performance on a validation set shows PSNR, SSIM and NMSE values. Inference time is given per patient and includes LAX and SAX processing time.

Task of participation	Task1: Cine	Use of pre-training	No
University/organization	Insight SFI Centre	Data augmentation	No
Single- or multi-channel	Single	Data standardization	No
Hardware configuration	RTX 2080 Ti GPU	Model parameters	124,427,137
Training time	9 h	Loss function	MSE
Inference time	13.27 s	Physical model	No
Performance on a train. set	(35.86, 0.93, 0.00029)	Use of unrolling	No
Performance on a val. set	(22.17, 0.59, 0.17)	k-space fidelity	No
Docker submitted?	Yes	Model backbone	UNET
Use of segmentation labels	No	Operations	Amplitude

5 Conclusion

In this paper, we report the research carried out as part of our participation in the exciting CMRxRecon challenge, Task 1: Cine reconstruction. While not reaching the top of the leaderboard, we achieved competitive results on the validation set using a double-stream processing pipeline including the IFFT and a denoising GNA-UNET model with Group Normalization and GCT attention layers. We also verified that GNA-UNET outperforms the denoising Autoencoder by a large margin (7.5336 dB $PSNR$ gain). Our ongoing work is centered around the use of generative denoising diffusion probabilistic models and k-space interpolation methods for the cine reconstruction on the dataset provided.

Acknowledgements. This publication has emanated from research conducted with the financial support of Science Foundation Ireland under Grant numbers 18/CRT/6183 and 12/RC/2289_P2.

References

1. Wang, Z., Bovik, A.C., Sheikh, H.R., Simoncelli, E.P.: Image quality assessment: from error visibility to structural similarity. IEEE Trans. Image Process. **13**, 600–612 (2004). https://doi.org/10.1109/TIP.2003.819861
2. Sara, U., Akter, M., Uddin, M.S.: Image quality assessment through FSIM, SSIM, MSE and PSNR-a comparative study. J. Comput. Commun. **7**(3) (2019)
3. Ronneberger, O., Fischer, P., Brox, T.: U-net: convolutional networks for biomedical image segmentation. In: Navab, N., Hornegger, J., Wells, W.M., Frangi, A.F. (eds.) MICCAI 2015. LNCS, vol. 9351, pp. 234–241. Springer, Cham (2015). https://doi.org/10.1007/978-3-319-24574-4_28
4. Yuan, X., et al.: Spatial-frequency U-net for denoising diffusion probabilistic models. arXiv (2023)
5. Hatamizadeh, A., et al.: UNETR: transformers for 3D medical image segmentation. In: IEEE WACV Conference (2022)
6. Kalantar, R., et al.: CT-based pelvic T1-weighted MR image synthesis using UNet, UNet++ and cycle-consistent generative adversarial network (cycle-GAN). Front. Oncol. Front **11**, 665807 (2021)
7. Ding, Q., Zhang, X.: MRI Reconstruction by Completing Under-sampled K-space Data with Learnable Fourier Interpolation. In: Wang, L., Dou, Q., Fletcher, P.T., Speidel, S., Li, S. (eds.) MICCAI 2022. LNCS, vol. 13436, pp. 667–685. Springer, Cham (2022). https://doi.org/10.1007/978-3-031-16446-0_64
8. Zhao, Z., Zhang, T., Xie, W., Wang, Y., Zhang, Y.: K-space transformer for undersampled MRI reconstruction. In: BMVC (2022)
9. Versteeg, E., Klomp, D.W.J., Siero, J.C.W.: Accelerating brain imaging using a silent spatial encoding axis. Magn. Reson. Med. **88**(4), 1785–1793 (2022)
10. Kojima, S., Shinohara, H., Hashimoto, T., Suzuki, S.: Undersampling patterns in k-space for compressed sensing MRI using two-dimensional Cartesian sampling. Radiol. Phys. Technol. **11**(3), 303–319 (2018)
11. McGibney, G., Smith, M.R., Nichols, S.T., Crawley, A.: Quantitative evaluation of several partial Fourier reconstruction algorithms used in MRI. Magn. Reson. Med. **30**(1), 51–9 (1993)

12. Sriram, A., et al.: End-to-end variational networks for accelerated MRI reconstruction. In: Martel, A.L., et al. (eds.) MICCAI 2020. LNCS, vol. 12262, pp. 64–73. Springer, Cham (2020). https://doi.org/10.1007/978-3-030-59713-9_7

13. Wang, S., et al.: Accelerating magnetic resonance imaging via deep learning. In: IEEE 13th International Symposium on Biomedical Imaging (ISBI), pp. 514–517 (2016). https://doi.org/10.1109/ISBI.2016.7493320

14. Kwon, K., Kim, D., Park, H.: A parallel MR imaging method using multilayer perceptron. Med. Phys. **44**(12), 6209–6224 (2017). https://doi.org/10.1002/mp.12600

15. Lee, D., Yoo, J., Tak, S., Ye, J.C.: Deep residual learning for accelerated MRI using magnitude and phase networks. IEEE Trans. Biomed. Eng. **65**(9), 1985–1995 (2018)

16. Hyun, C.M., Kim, H.P., Lee, S.M., Lee, S., Seo, J.K.: Deep learning for undersampled MRI reconstruction. Phys. Med. Biol. **63**(13), 135007 (2018)

17. Yang, G., et al.: DAGAN: deep de-aliasing generative adversarial networks for fast compressed sensing mri reconstruction. IEEE Trans. Med. Imaging **37**(6), 1310–1321 (2018)

18. Akçakaya, M., Moeller, S., Weingärtner, S., Uğurbil, K.: Scan-specific robust artificial-neural-networks for k-space interpolation (RAKI) reconstruction: database-free deep learning for fast imaging. Magn. Reson. Med. **81**(1), 439–453 (2019)

19. Arora, S., Roeloffs, V., Lustig, M.: Untrained modified deep decoder for joint denoising parallel imaging reconstruction. In: ISMRM and SMRT Virtual Conference and Exhibition (2020)

20. Yoo, J., Jin, K.H., Gupta, H., Yerly, J., Stuber, M., Unser, M.: Time-dependent deep image prior for dynamic MRI. IEEE Trans. Med. Imaging **40**(12), 3337–3348 (2021)

21. Darestani, M.Z., Heckel, R.: Accelerated MRI with un-trained neural networks. IEEE Trans. Comput. Imaging **7**, 724–733 (2021)

22. Schlemper, J., Caballero, J., Hajnal, J.V., Price, A.N., Rueckert, D.: A deep cascade of convolutional neural networks for dynamic MR image reconstruction. IEEE Trans. Med. Imaging **37**(2), 491–503 (2017)

23. Muckley, M.J., et al.: Results of the 2020 fastMRI challenge for machine learning MR image reconstruction. IEEE Trans. Med. Imaging **40**(9), 2306–2317 (2021)

24. Ramzi, Z., Chaithya, G.R., Starck, J.-L., Ciuciu, P.: NC-PDNet: a density-compensated unrolled network for 2D and 3D non-cartesian MRI reconstruction. IEEE Trans. Med. Imaging **41**(7), 1625–1638 (2022)

25. Fabian, Z., Heckel, R., Soltanolkotabi, M.: Data augmentation for deep learning based accelerated MRI reconstruction with limited data. In: Proceedings of the 38th International Conference on Machine Learning (ICML), pp. 3057–3067 (2021)

26. Desai, A.D., et al.: Vortex: physics-driven data augmentations for consistency training for robust accelerated MRI reconstruction. In: International Conference on Medical Imaging with Deep Learning (MIDL) (2021)

27. Bekkers, E.J.: B-spline CNNs on lie groups. arXiv preprint arXiv:1909.12057 (2019)

28. Gunel, B., et al.: Scale-equivariant unrolled neural networks for data-efficient accelerated MRI reconstruction. In: Wang, L., Dou, Q., Fletcher, P.T., Speidel, S., Li, S. (eds.) MICCAI 2022. LNCS, vol. 13436, pp. 737–747. Springer, Cham (2022). https://doi.org/10.1007/978-3-031-16446-0_70

29. Wimmer, T., Golkov, V., Dang, H.N., Zaiss, M., Maier, A., Cremers, D.: Scale-equivariant deep learning for 3D data. arXiv preprint arXiv:2304.5864 (2023)

30. Wu, Y., He, K.: Group normalization. In: ECCV (2018)
31. Ali, O., Ali, H., Ayaz Ali Shah, S., Shahzad, A.: Implementation of a modified U-net for medical image segmentation on edge devices. IEEE Trans. Circ. Syst. II: Express Brief (2022)
32. Yang, Z., Zhu, L., Wu, Y., Yang, Y.: Gated channel transformation for visual recognition. In: CVPR (2020)
33. Loshchilov, I., Hutter, F.: Decoupled weight decay regularization. In: ICLR (2019)
34. Wang, C., et al.: CMRxRecon: an open cardiac MRI dataset for the competition of accelerated image reconstruction (2023). https://arxiv.org/abs/2309.10836

Multi-scale Inter-frame Information Fusion Based Network for Cardiac MRI Reconstruction

Wenzhe Ding, Xiaohan Liu, Yong Sun, Yiming Liu, and Yanwei Pang[(✉)]

Tianjin Key Lab of Brain Inspired Intelligence Technology, School of Electrical and Information Engineering, Tianjin University, Tianjin, China
pyw@tju.edu.cn

Abstract. Accelerated cine MRI reconstruction from under-sampled data is paramount in clinical diagnosis. Nonetheless, the existing method falls short in fully harnessing inter-frame information, thereby impeding its overall reconstruction performance. This paper presents a novel multi-scale inter-frame information fusion strategy, aimed at extracting and leveraging the multi-scale features from adjacent multi-frame data to guide the reconstruction process more comprehensively. The proposed framework incorporates several specific encoders for feature extraction from each frame, followed by an information fusion block that effectively combines the multi-scale features of multiple frames. This ensures the effective utilization of supplementary information from multiple frames at varying scales. Moreover, the fused inter-frame information is also utilized in subsequent refinement blocks to perform feature enhancement and guide the reconstruction. Consequently, the introduced multi-scale inter-frame information fusion strategy not only enhances the overall reconstruction performance but also demonstrates high efficiency. Experimental results show that the method can obtain competitive reconstruction results and the metrics of SSIM, NMSE and PSNR can reach 0.9299, 0.0098 and 35.43 respectively averaging over all validation datasets.

Keywords: Deep Learning · Cardiac MRI · Cine Reconstruction

1 Introduction

Magnetic Resonance Imaging (MRI) is an effective medical imaging technique for disease diagnosis which has the advantages of non-invasiveness, non-radiation and high contrast on soft tissues. Cardiac MRI can provide comprehensive information about the structure and function of the heart, and has become an important tool in the diagnosis and treatment of heart diseases. However, due to the limitation of hardware, the required scan time of cardiac MRI is long and involuntary physiological activities such as breathing and heart beating cause motion artifacts. Therefore, accelerated cardiac MRI reconstruction is crucial for clinical practice.

Subsampling k-space data is one of the widely used approaches to accelerate cardiac MRI [1], but images reconstructed from under-sampled k-space data are poor quality and introduce many artifacts. Jung *et al.* [2] proposed k-t FOCUSS to improve the quality of reconstructed images which exploits the sparsity of x-f signals and combines

© The Author(s), under exclusive license to Springer Nature Switzerland AG 2024
O. Camara et al. (Eds.): STACOM 2023, LNCS 14507, pp. 339–348, 2024.
https://doi.org/10.1007/978-3-031-52448-6_32

Compressed Sensing with k-t BLAST/k-t SENSE. Furthermore, many reconstruction methods using deep learning have been proposed including CRNN [3] and GAN [4]. Huang *et al.* [5] developed GRAPPA and took advantage of local correlations along both spatial and temporal dimensions to interpolate the missing data in k-t space. Schlemper *et al.* [6] developed a deep cascade of CNN to learn spatio-temporal correlations for dynamic cardiac MRI reconstruction. Lyu *et al.* [7] exploited Transformer to interact different spatio-temporal information from multiple views. Huang *et al.* [8] applied motion information into the reconstructing process through motion estimation and motion compensation (ME/MC). However, most existing cardiac cine MRI reconstruction algorithms exploit temporal correlations between frames by RNN or 3D convolutions, which suffer from high computational costs.

In this paper, we propose a Multi-Scale Inter-Frame Information Fusion Based Network for cardiac MRI reconstruction, which utilizes multiple successive under-sampled k-space frames to reconstruct the target frame. The network mainly consists of two stages, in which the first stage extracts and fuses multi-scale features from multiple frames to reconstruct the initial target frame, and the second stage refine the initial reconstruction through cascaded refinement blocks. Because of the use of 2D convolutions in whole network, our method is computational-friendly. The experiments prove that our method achieves better reconstruction performance than 2D reconstruction algorithms. This work is done in the context of the Cardiac MRI Reconstruction Challenge (CMRxRecon), where we focus on Task 1 (cine reconstruction).

2 Method

2.1 Overall Architecture

The overall architecture of our proposed method is illustrated in Fig. 1. Given a sequence of $2N + 1$ consecutive under-sampled multi-coil k-space frames $\{K_{t-N}, ..., K_t, ..., K_{t+N}\}$ as input, where K_t denotes the k-space data of the target frame and the other $2N$ frames are reference frames (in our work we set $N = 1$), our goal is to reconstruct the target frame $\hat{I}_t$ that best approaches to the ground truth I_t, where I_t represents the spatial domain image of the target frame. The reconstruction process consists of two stages: the inter-frame information fusion stage and the refinement stage.

First to deal with multi-coil k-space data, we exploit the k-space data of the target frame to estimate sensitivity maps through a simple U-Net [9]. And then the sensitivity maps are utilized to combine or expand the coil dimension. The inter-frame information fusion stage contains four blocks, a coil combination block, an Inter-Frame Features Fusion (IFFF) Net, a coil expansion block and a DC layer. The refinement stage is comprised of three cascaded Refinement Blocks which is based on VarNet [10]. The architecture of each Block is similar to the inter-frame information fusion stage, where the IFFF Net is replaced by IFFE (Inter-Frame Features Enhancement) Net. The Data Consistency (DC) layer is used to replace the reconstructed data with the original data in sampled positions and retain the reconstruction in unsampled positions.

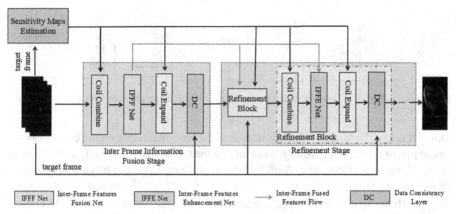

Fig. 1. The overall framework of the proposed Multi-Scale Inter-Frame Information Fusion Based Network.

2.2   Inter-Frame Features Fusion Net

As shown in Fig. 2, in the Inter-Frame Features Fusion Net, we use three encoders to respectively extract the features from three frames. The encoder has 4 layers and each layer follows the structure of two consecutive convolution blocks consisting of a convolution layer, a Instance Normalization layer and a LeakyReLU, in which the kernel size is set to 3. Layers are connected by a down-sampling operation, which is implemented by average pooling. After each down-sampling, the number of feature channels is doubled and the size of feature maps is halved. Therefore, multi-scale features of each frame can be obtained through encoders.

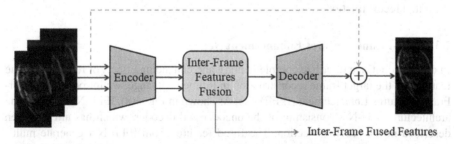

Fig. 2. The detailed architecture of the Inter-Frame Features Fusion Net.

Motivated by the improvement brought by multi-scale features fusion [11, 12], we upsample the multi-scale features extracted from each frame to the original resolution and concatenate them together along the channel dimension, and then the features are fused through a 1 × 1 convolution layer.

Considering adjacent reference frames can provide complementary information to the target frame and improve reconstruction performance, we utilize Transformer to learn similar features between the target and reference frames in Inter-Frame Features Fusion

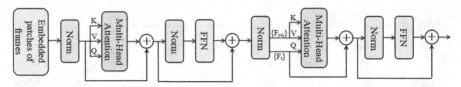

Fig. 3. The architecture of the Inter-Frame Features Fusion Block.

Block. It is not effective to calculate inter-frame attention directly on the corresponding spatial position due to the misalignment among the features of different frames. To solve the problem, we perform inter-frame attention mechanism on the patch-level features. First the three-frame features $\{F_{t+i}\}_{i=-N}^{N} \in R^{B \times (2N+1) \times C \times H \times W}$ are divided into patches $\{F_{t+i}^{k}\}$ with the spatial size p × p, where k represents the patch index and the number of patches is B × H/p × W/p, and then we calculate the frame self-attention and frame cross-attention successively as shown in Fig. 3. The frame self-attention takes all tokens in the three-frame features as Queries, Keys and Values and the computation of attention is limited to each patch. The process of frame cross-attention is similar to frame self-attention, in which the only difference is taking tokens in the target frame feature as Queries. The attention is calculated as Eq. 1.

$$Attention\ (Q, K, V) = softmax\left(\frac{QK^{T}}{\sqrt{d}}\right)V \qquad (1)$$

The decoder takes the fused feature map in the lowest resolution as input and upsamples features through transposed convolutions. Then the up-sampled features are concatenated with fused multi-frame features from encoders in the same resolution to realize the integration of low-resolution features and high-resolution features. Finally a 1 × 1 convolution layer is applied to transform features into the spatial domain and generate the initial reconstruction.

2.3 Inter-Frame Features Enhancement Net

In order to continue to utilize complementary and enhancing effects of reference frame features on the target frame reconstruction in Refinement stage, we introduce the Inter-Frame Features Enhancement (IFFE) Net. As shown in Fig. 4(a), IFFE Net exploits the architecture of U-Net consisting of the encoder and decoder, which has already been described in the previous section. The fused features from IFFF Net generate multi-scale inter-frame features through encoder and then are input into Inter-Frame Features Enhancement Block (IFFEB) together with image features in the same resolution before upsampling in each layer.

In IFFE Block (see in Fig. 4(b)), we first concatenate the image features from ConvBlock and inter-frame fused features in the same resolution along the channel dimension and perform feature fusion through 1 × 1 convolution layer. Then we calculate spatial attention weights and channel attention weights [13] respectively for the fused features $F \in R^{C \times H \times W}$. In channel attention the fused feature maps are aggregated through max pooling and average pooling along the width and height dimension, generating two feature maps with the size of C × 1 × 1. We merge the two output feature maps which

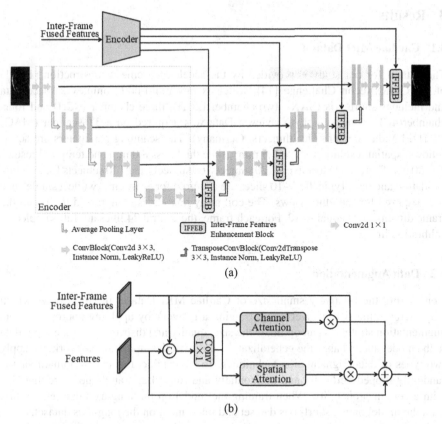

Fig. 4. (a) Illustration of the Inter-Frame Features Guidance Net. (b) The architecture of the Inter-Frame Features Enhancement Block.

are activated by ReLU using element-wise summation and obtain the channel attention map M_C through a Sigmoid. In spatial attention, we first apply max pooling and average pooling along the channel dimension and concatenate them. A 3×3 convolution layer and a Sigmoid function are then used to generate the spatial attention map M_S. Finally we apply element-wise multiplication to the primary fused features F and attention maps and fuse them as Eq. 2.

$$F' = (F \otimes M_S) \oplus (F \otimes M_C) \oplus F \tag{2}$$

where M_S and M_C denote spatial attention map and channel attention map respectively, $\otimes$ denotes element-wise multiplication and $\oplus$ represents element-wise summation. F is the primary fused features and F' is the enhanced features.

3 Results

3.1 Cardiac MRI Dataset

The dataset for our study is provided by the accelerated cine reconstruction (Task1) of the CMRxRecon Challenge [14], which is raw multi-coil complex k-space data and includes short-axis (SAX), two-chamber (2CH), three-chamber (3CH) and four-chamber(4CH) long-axis (LAX) views. Data was acquired on a 3T scanner (MAG-NETOM Vida, Siemens Healthineers, Germany). The scanning parameters are set as follows: spatial resolution 1.5×1.5 mm^2, slice thickness 8.0 mm and temporal resolution 50 ms. There are respectively 120 subjects, 60 subjects and 120 subjects for training, validation and test. Typically 5–10 slices are acquired for SAX cine, while a single slice was acquired for the other views. The coil array number is compressed to 10 and the frame dimension is equal to 12. For each frame, the central 24 lines are full sampled as calibration lines.

3.2 Data Augmentation

Considering the relatively small size of Cardiac MRI Dataset in LAX views which may increase the risk of overfitting, training a network by applying appropriate data augmentation strategies can optimize the size, quality and distribution of the input data to the model and enhance the generalization ability of the model. In our work, we apply two types of data augmentation techniques. To be specific, the spatial domain data is randomly flipped vertically and horizontally and rotated at a angle no more than 45° with a certain probability. When training the model on the long-axis dataset, we first train the model on the short-axis dataset and fine-tune it on the long-axis dataset.

3.3 Implementation Details

Our network is trained for a total of 50 epochs using SSIM loss on NVIDIA Tesla V100 DGXS GPU with 32G memory. The optimizer chosen is Adam. The initial learning rate for training is set to 0.005 and is scheduled to decay by a factor of 0.1 every 40 epochs. The batch size is 1. The number of feature channels of the first convolution layer in the Inter-Frame Features Fusion Net and Inter-Frame Features Enhancement Net is set to 18 and the depth is set to 4. And the number of feature channels of the first convolution layer in sensitivity maps estimation U-Net are set to 8 and the depth is set to 3. The specific model information is shown in Table 1.

Table 1. Model Information (On the training set we evaluate the reconstruction quality of the entire image, and on the validation set we only evaluate the reconstruction quality of one sixth of the middle of image).

Model Information			
Task of participation	Cine reconstruction	Use of pre-training	Use datasets of SAX views to pre-train when training on LAX views datasets
University/organization	Tianjin University	Data augmentation	Randomly flip vertically and horizontally and rotate at a angle no more than 45°
Single-channel or multi-channel	Multi-coil	Data standardization	Z-score standardization
Hardware configuration (GPU, VRAM, CPU, number of cores)	GPU: NVIDIA Tesla V100 DGXS GPU; CPU: Intel(R) Xeon(R) CPU ES-2698 v4 @ 2.20 GHz; number of cores: 20	Model parameter number	27.2M
Training time	SAX: 43.3 h LAX: 13.3 h (50 epochs total time)	Loss function	SSIM loss
Inference time	0.4454 s	Incorporation of a physical model	No
Performance on the training set	SSIM 0.9800 NMSE 0.0055 PSNR 42.29	Use of unrolling	Yes
Performance on the validation set	SSIM 0.9299 NMSE 0.0098 PSNR 35.43	Application of k-space fidelity	Use Data Consistency (DC) layer to replace the reconstructed data with the original data in sampled positions
Docker submitted?	Yes	Model backbone	E2E-VarNet
Use of segmentation labels	No	Amplitude or complex-valued operations	Complex-valued operations

3.4 Validation and Test Results

To assess the quality of reconstructed MR images, three evaluation metrics of Structural Similarity (SSIM), Peak Signal to Noise Ratio (PSNR), and Normalized Mean Squared

Error (NMSE) are adopted. Table 2 shows the validation results of our method for all under-sampling factors (only evaluate one sixth of the middle of reconstructed images). Averaging over six datasets of different accelerations and views, our method can get a reconstruction result with SSIM metric of 0.9299, NMSE metric of 0.0098 and PSNR metric of 35.43 on validation set. Therefore, our method can basically achieve satisfactory reconstruction results both quantitatively and qualitatively. On test sets the SSIM, NMSE and PSNR of our method is respectively 0.984, 0.0067 and 43.79 (evaluate entire reconstructed images).

Table 2. Validation results of our method for all under-sampling factors.

Accelerations	Views	NMSE	PSNR	SSIM
4	SAX	0.0026	41.07	0.9742
	LAX	0.0039	38.14	0.9589
8	SAX	0.0085	35.49	0.9362
	LAX	0.0145	32.34	0.9002
10	SAX	0.0113	34.25	0.9228
	LAX	0.0177	31.31	0.8873

4 Discussion

Cardiac MRI reconstruction can improve the quality of reconstructed images by taking advantage of the similarity among adjacent frames because of the existence of time dimension in acquired k-space data. In this work, different from some algorithms which reconstruct under-sampled MR images in a frame-by-frame manner, we calculate the correlation of features from consecutive frames and merge the complementary information by introducing Inter-Frame Features Fusion (IFFF) Block based on Transformer. In the following stage, we reuse the similar and complementary information among frames to enhance the features and guide the refinement reconstruction. Moreover, due to the use of 2D convolutions in total network, the proposed method is computationally friendly.

The visual results of reconstructions by zero-filled and our method are illustrated in Fig. 5. The error map shows the difference between the reconstructed image and the ground-truth. For easy observation, only one sixth of the middle of reconstructed images (one third of the image size in x-axis and half of the image size in y-axis) is visualized and the error is magnified the error a hundred times. As can be seen, our method makes full use of the complementary information among frames so that reconstructed images are closer to the ground-truth and the error is smaller. Moreover, compared with zero-filled images our method can reconstruct most of the structure with fewer artifacts and get better performance in recovering some fine textures.

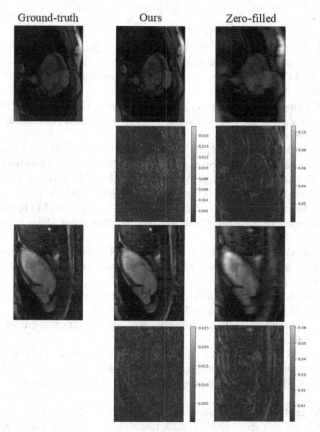

Fig. 5. Visual results of ground-truth and reconstructions by zero-filled and our method on the dataset for the under-sampling factor of 8. The top two rows show reconstructions and error maps in short-axis view and the bottom two rows show reconstructions in long-axis view.

5   Conclusion

In this paper, we propose a Multi-Scale Inter-Frame Information Fusion Based Network for cardiac cine MRI reconstruction, which utilizes 2D network to learn the spatial and temporal correlations. Specifically, the network can respectively extracts features of different scales in consecutive frames and effectively fuses features in each scale. The experimental results demonstrate our method can reconstruct high quality cardiac images more approaching to the ground truth both in short-axis view and in long-axis view.

Acknowledgement. This work was supported by the National Natural Science Foundation of China (Grant No. 52227814).

References

1. Lustig, M., Donoho, D., Pauly, J.M.: Sparse MRI: the application of compressed sensing for rapid MR imaging. Magn. Reson. Med. **58**(6), 1182–1195 (2007)
2. Jung, H., Ye, J.C., Kim, E.Y.: Improved k–t BLAST and k–t SENSE using FOCUSS. Phys. Med. Biol. **52**(11), 3201 (2007)
3. Qin, C., Schlemper, J., Caballero, J., Price, A.N., Hajnal, J.V., Rueckert, D.: Convolutional recurrent neural networks for dynamic MR image reconstruction. IEEE Trans. Med. Imaging **38**(1), 280–290 (2018)
4. Murugesan, B., Raghavan, S.V., Sarveswaran, K., Ram, K., Sivaprakasam, M.: Recon-GLGAN: a global-local context based generative adversarial network for MRI reconstruction. In: Knoll, F., Maier, A., Rueckert, D., Ye, J. (eds.) MLMIR 2019. LNCS, vol. 11905, pp. 3–15. Springer, Cham (2019). https://doi.org/10.1007/978-3-030-33843-5_1
5. Huang, F., Akao, J., Vijayakumar, S., Duensing, G.R., Limkeman, M.: K-t GRAPPA: A k-space implementation for dynamic MRI with high reduction factor. Magn. Reson. Med. **54**(5), 1172–1184 (2005)
6. Schlemper, J., Caballero, J., Hajnal, J.V., Price, A.N., Rueckert, D.: A deep cascade of convolutional neural networks for dynamic MR image reconstruction. IEEE Trans. Med. Imaging **37**(2), 491–503 (2017)
7. Lyu, J., et al.: Region-focused multi-view transformer-based generative adversarial network for cardiac cine MRI reconstruction. Med. Image Anal. **85**, 102760 (2023)
8. Huang, Q., et al.: Dynamic MRI reconstruction with end-to-end motion-guided network. Med. Image Anal. **68**, 101901 (2020)
9. Ronneberger, O., Fischer, P., Brox, T.: U-Net: convolutional networks for biomedical image segmentation. In: Navab, N., Hornegger, J., Wells, W.M., Frangi, A.F. (eds.) MICCAI 2015. LNCS, vol. 9351, pp. 234–241. Springer, Cham (2015). https://doi.org/10.1007/978-3-319-24574-4_28
10. Sriram, A., et al.: End-to-end variational networks for accelerated MRI reconstruction. In: Martel, A.L., et al. (eds.) MICCAI 2020. LNCS, vol. 12262, pp. 64–73. Springer, Cham (2020). https://doi.org/10.1007/978-3-030-59713-9_7
11. Fan, G., Hua, Z., Li, J.: Multi-scale depth information fusion network for image dehazing. In: IEEE/CVF Conference on Computer Vision and Pattern Recognition (CVPR), pp. 7262–7280 (2021)
12. Wang, Z., Peng, Y., Li, D., Guo, Y., Zhang, B.: MMNet: a multi-scale deep learning network for the left ventricular segmentation of cardiac MRI images. Appl. Intell. **52**(5), 5225–5240 (2022)
13. Woo, S., Park, J., Lee, J.Y., Kweon, I.S.: CBAM: convolutional block attention module. In: Ferrari, V., Hebert, M., Sminchisescu, C., Weiss, Y. (eds.) ECCV 2018. LNCS, vol. 11211, pp. 3–19. Springer, Cham (2018). https://doi.org/10.1007/978-3-030-01234-2_1
14. Wang, C., et al.: CMR×Recon: an open cardiac MRI dataset for the competition of accelerated image reconstruction (2023). https://doi.org/10.48550/arXiv.2309.10836

Relaxometry Guided Quantitative Cardiac Magnetic Resonance Image Reconstruction

Yidong Zhao$^{(\boxtimes)}$, Yi Zhang, and Qian Tao

Department of Imaging Physics, Delft University of Technology,
Delft, The Netherlands
{y.zhao-8,y.zhang-43,q.tao}@tudelft.nl

Abstract. Deep learning-based methods have achieved prestigious performance for magnetic resonance imaging (MRI) reconstruction, enabling fast imaging for many clinical applications. Previous methods employ convolutional networks to learn the image prior as the regularization term. In quantitative MRI, the physical model of nuclear magnetic resonance relaxometry is known, providing additional prior knowledge for image reconstruction. However, traditional reconstruction networks are limited to learning the spatial domain prior knowledge, ignoring the relaxometry prior. Therefore, we propose a relaxometry-guided quantitative MRI reconstruction framework to learn the spatial prior from data and the relaxometry prior from MRI physics. Additionally, we also evaluated the performance of two popular reconstruction backbones, namely, recurrent variational networks (RVN) and variational networks (VN) with U-Net. Experiments demonstrate that the proposed method achieves highly promising results in quantitative MRI reconstruction.

Keywords: Caridac MRI · Quantitative mapping · Relaxometry · Image reconstruction

1 Introduction

Quantitative Magnetic Resonance Image (qMRI) has emerged as an indispensable imaging modality in research and clinical applications thanks to its quantitative measurements of tissue properties, such as T_1 and T_2 relaxation times [23]. A common approach for qMRI typically involves a two-step process. Firstly, multiple k-space data of a subject are acquired and reconstructed into a series of weighted images with varying imaging parameters (such as echo times and diffusion weights). Subsequently, the underlying tissue properties are estimated by fitting a signal model to the images [24]. However, acquiring fully sampled k-space data, adhering to the Nyquist criterion, for multiple measurements in qMRI requires a time-consuming endeavor [6]. This extended acquisition process introduces the potential for motion artifacts and leads to patient discomfort due to prolonged scan duration.

O. Camara et al. (Eds.): STACOM 2023, LNCS 14507, pp. 349–358, 2024.
https://doi.org/10.1007/978-3-031-52448-6_33

Motivated by reducing the acquisition time in qMRI while keeping the reconstruction and estimation quality, *accelarated qMRI* has become one of the central research topics in qMRI. The methods in accelerated qMRI can be divided into two categories according to whether the method includes an intermediate step of image reconstruction from k-space data: If so, the method is categorized as *indirect reconstruction*; Otherwise, the method is categorized as *direct reconstruction* where only the parameter mapping estimations are gained [24]. This paper will focus on indirect reconstruction methods since the goals require both reconstructing images and estimating the mapping parameters. However, our proposed method is flexible yet novel, taking self-supervised quantitative mapping as auxiliary constraints to guide the reconstruction.

In conventional qMRI, the acceleration can be achieved by parallel imaging (PI) [9,12] or compressed sensing (CS) [2]. With specific undersampling patterns, PI utilizes multiple radio-frequency (RF) coils simultaneously with individual coil sensitivity maps to reduce the aliasing artifacts. The applications of PI include SENSE [20], which is the original implementation, and GRAPPA [5], which allows auto-calibration when sensitivity maps are missing. Meanwhile, CS-based methods achieve acceleration by using partial sampling in k-space, which loosens the requirement of the Nyquist criterion with sparsity assumption in true signals. Though the CS-based methods are well-established, they still struggle with the design of regularizer functions, which varies across different tissues [17].

Accelerated qMRI reconstruction benefits from recent rapid development in deep learning on both computer vision and inference learning since MRI reconstruction can be formulated as an inverse problem: Given a (partial) measurement in k-space, the target is to recover the image signal as close as possible. Deep learning-based methods for MRI reconstruction can be categorized into *iterative* methods [8,16,25,29,30] and *one-step* methods [11,31]. One-step methods aim to predict the refined reconstructed images from images reconstructed from corrupted k-space [11] or partially sampled k-space measurements directly [31]. Iterative methods leverage neural networks to learn an incremental refining process, analogous to a conventional optimization process. As a first attempt, Yang et al. [29] proposed ADMM-Net: a neural network to parameterize the alternating direction method of multipliers (ADMM), thus solving the inverse problem of MRI reconstruction in combination with CS. Similarly, Hammernik et al. proposed a variational network [8] in an unrolled gradient descent scheme for generalized CS reconstruction. Inspired by a more general idea in meta-learning, recurrent inference machines (RIM) [21] were designed to find a maximum-a-posteriori (MAP) estimation to solve inverse problems when the corresponding forward model is known. This is then applied in MRI reconstruction [16], where in the RIM framework, the parameters are shared across the iterations with internal hidden states instead of individual units for each iteration in previous works. A more recent work, RecurrentVarNet [30], combines variational networks and RIM with a hybrid domain-learning strategy using convolutional neural networks (CNN) to guide the optimization in k-space.

Unlike naive MRI reconstruction, in qMRI reconstruction, the involvement of physical signal models in methodological design can further improve the quality of both reconstruction and parameter mapping estimation since an anatomical correspondence across images is often assumed when designing the acquisition pipeline [10]. Apart from conventional least-square methods, parameter mappings can be estimated by dictionary matching [7,19,24]. Given the reconstructed images or raw k-space measurements, there are several methods utilizing self-supervised learning to estimate parameter mappings [1,14,15]. However, few existing works using [3] address the joint tasks of reconstruction and parameter estimation, while none lie in the deep learning paradigm.

1.1 Contributions

In this work, we tackle the quantitative mapping problem in the CMRxRecon challenge and make the following contributions:

- We introduce a novel quantitative mapping network that learns to mimic MR physics in an unsupervised fashion.
- We evaluate two different CNN architectures for image prior learning in the variational reconstruction network.
- We leverage the quantitative mapping network to guide the reconstruction process, ensuring that the output signal conforms to the MR relaxometry.

2   Methods

2.1   Parallel Imaging

Accelerated MRI. In this paper, we focus on the reconstruction of k_t qMRI baseline images $x \in \mathbb{C}^{k_x \times k_y \times k_t 1}$, given their corresponding k-space measurements $y \in \mathbb{C}^{n_c \times k_x \times k_y \times k_t}$ of n_c coils, where k_x and k_y are image shapes along the frequency and phase encoding axes, respectively. The cth coil measurement y_c is formulated as

$$y_c = U\mathcal{F}S_c x + \epsilon_c, \tag{1}$$

where $\mathcal{F}$ is the 2D Fourier transform operator, S_c denotes the sensitivity map, U characterises the Cartesian under-sampling pattern and ϵ_c represents the measurement noise. The overall forward operator and its adjoint operator in accelerated parallel imaging can be formulated as $\mathcal{A} = U \circ \mathcal{F} \circ \mathcal{E}$ and $\mathcal{A}^* = \mathcal{R} \circ \mathcal{F} \circ U$ [30], where

$$\mathcal{E}(x) = [S_1 x, S_2 x, \cdots, S_{n_c} x], \tag{2}$$

$$\mathcal{R}(y) = \sum_{c=1}^{n_c} S_c^* y_c. \tag{3}$$

---

[1] We use x for both complex and magnitude image for simplicity.

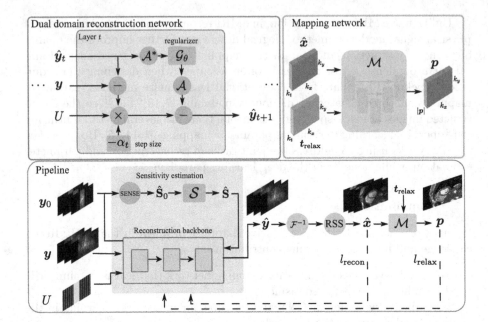

Fig. 1. The reconstruction backbone consists of unrolled gradient descent layers, and the image prior is learned during training by $\mathcal{G}_\theta$. A pre-trained mapping network $\mathcal{M}$ is introduced to predict the quantitative parameters p and guide the reconstruction with MR relaxometry.

Sensitivity Estimation. The sensitivity maps are estimated using the auto-calibration region U_{AC} in k-space which is always sampled in the low-frequency band. The initial sensitivity maps can be estimated as

$$\hat{S}_c^0 = \frac{\mathcal{F}^{-1}U_{AC}\boldsymbol{y}_c}{\text{RSS}\left(\{\mathcal{F}^{-1}U_{AC}\boldsymbol{y}_l\}_{l=1}^{n_c}\right)}, \tag{4}$$

where RSS denotes the root-sum-of-square operator defined in [13]

$$\text{RSS}(\boldsymbol{x}_1, \boldsymbol{x}_2, \cdots, \boldsymbol{x}_{n_c}) = \sqrt{\sum_{c=1}^{n_c} |\boldsymbol{x}_c|^2}. \tag{5}$$

2.2   Relaxometry Guided Reconstruction

Dual-Domain Reconstruction Network. Reconstruction of an image $\boldsymbol{x}$ given its under-sampled parallel imaging measurements $\boldsymbol{y}$ can be formulated as optimizing the Lagrangian $\mathcal{L}(\boldsymbol{x})$ defined as

$$\arg\min_{\hat{\boldsymbol{x}}} \mathcal{L}(\boldsymbol{x}) = \|\mathcal{A}\hat{\boldsymbol{x}} - \boldsymbol{y}\|_2^2 + R(\hat{\boldsymbol{x}}), \tag{6}$$

where the first term constrains the data fidelity via a $2 - norm$ operator and the latter term $R(\cdot)$ is a regularizer to stabilize the solution space. The update rule of gradient-descent for optimization of Eq. 6 is

$$\hat{x}_{t+1} = \hat{x}_t - \alpha_t \frac{\partial \mathcal{L}}{\partial \hat{x}^*}$$
$$= \hat{x}_t - \alpha_t \mathcal{A}^* (\mathcal{A}\hat{x}_t - y) - \mathcal{G}_{\theta_t}(\hat{x}_t), \tag{7}$$

where $\mathcal{G}_{\theta_t}$ is a convolutional network that learns the scaled Wirtinger derivatives $\frac{\partial R}{\partial \hat{x}_t^*}$ of the regularizer $w.r.t.$ the current reconstruction, and α_t is a trainable step scalar. Applying Fourier transform on both sides of Eq. 7 yields

$$\hat{y}_{t+1} = \hat{y}_t - \alpha_t U(\hat{y}_t - y) - \mathcal{F} \circ \mathcal{E} \circ \mathcal{G}_{\theta_t}(\mathcal{R} \circ \mathcal{F}^{-1}\hat{y}_t), \tag{8}$$

in which the data fidelity part is updated in k-space and the regularization is performed in the spatial domain. In this work, we follow [8,30] and use the variational network based on the unrolled update rule defined in Eq. 8 for image construction. Specifically, we investigate and compare two different types of CNNs to learn the regularizer $\mathcal{G}_{\theta_t}$: U-Net [22] and convolutional Gated Recurrent Unit (GRU) blocks as in [30]. Additionally, we employ another U-Net $\mathcal{S}$ to refine the initial sensitivity map estimated in Eq. 4 $\hat{S}_c = \mathcal{S}(\hat{S}_c^0)$. The reconstruction network is trained with a combination of L_1 loss and structural similarity index measure (SSIM) [28]:

$$l_{\text{recon}}(\hat{x}) = \gamma_1 \|\hat{x} - x\|_1 + \gamma_2 \text{SSIM}(\hat{x}, x), \tag{9}$$

where $\hat{x}$ and x denote the predicted image and the fully sampled image, respectively. The architecture of reconstruction backbone layers is shown on the upper left panel in Fig. 1.

Quantitative Mapping Network. Given reconstructed magnitude image $x \in \mathbb{R}^{k_x \times k_y \times k_t}$, quantitative mapping is usually treated as a parameter fitting problem and solved by least squares or dictionary matching [7,19,24]. Voxels in a reconstructed image of good quality should conform to the MR relaxometry and thus have a relatively lower level of fitting error since tissue anatomies are assumed to be spatially aligned. Inspired by this, we propose using the parameter fitting error to guide the reconstruction procedure. However, both least squares and dictionary matching are not differentiable and thus cannot be integrated into the computational graph. To make the mapping procedure differentiable, we propose using a U-Net $\mathcal{M}$ to predict the parameters given the image x and the inversion time or echo time t_{relax}. With the MR relaxation physics known, the network $\mathcal{M}$ is trained in a purely unsupervised fashion, and the physics-informed training loss l_{relax} is defined as:

$$l_{\text{relax}}(x) = \|s(\mathcal{M}(x, t_{\text{relax}}), t_{\text{relax}}) - x\|_1, \tag{10}$$

where $s(\cdot)$ can be either s_{T_1} or s_{T_2} which characterises the signal intensity models for T_1 or T_2 relaxation:

$$s_{T_1}(A, B, T_1^*, t_{\text{relax}}) = \left| A - B \exp\left(-\frac{t_{\text{relax}}}{T_1^*}\right) \right|, \tag{11}$$

$$s_{T_2}(A, T_2, t_{\text{relax}}) = A \exp\left(-\frac{t_{\text{relax}}}{T_2}\right). \tag{12}$$

Note that from Eq. 11, the parameter of interest T_1 is derived by $T_1 = (B/A - 1)T_1^*$.

Joint Reconstruction and Quantitative Mapping. The mapping network $\mathcal{M}$ is pre-trained with the fully sampled images only and is frozen during the reconstruction network training process. The physics-informed loss l_{relax} can then be used to enforce a relaxation physics-informed reconstruction. We also penalize the difference between the mapping prediction on fully sampled image $\mathcal{M}(x)$ and on the predicted image $\mathcal{M}(\hat{x})$, such that the reconstructed image has a consistent parameter map as the fully sampled image. The total training loss is then formulated as:

$$l(\hat{x}) = l_{\text{recon}}(\hat{x}) + \gamma_3 l_{\text{relax}}(\hat{x}) + \gamma_4 \|\mathcal{M}(\hat{x}) - \mathcal{M}(x)\|_1. \tag{13}$$

An overview of the proposed method is illustrated in Fig. 1.

3   Experiments

3.1   Dataset

We conduct the experiments on the CMRxRecon challenge data[2] [27], consisting of 120 subjects for training. Imaging was performed on a Siemens 3T MRI scanner (MAGNETOM Vida) and the multi-coil images were compressed to 10 virtual coils, with acceleration factors 4, 8, and 10. More details on the image acquisition protocol are described in [26]. Data of each subject comprise two different qMRI sequences: the modified-look-locker (MOLLI) [18] sequence with 9 baseline images for T_1-mapping and the T2-prepared (T2prep)-FLASH sequence with 3 baseline images for T_2-mapping. The validation set contains 59 subjects, and the results are evaluated on the official platform.

3.2   Training Configuration

We first train the mapping network $\mathcal{M}$, a U-Net with 256 base filters and 1 pooling layer, by Adam optimizer with an initial learning rate of $\eta_0 = 10^{-4}$ for 200 epochs. And then we freeze $\mathcal{M}$ for training the reconstruction backbone. We

---

[2] https://cmrxrecon.github.io/.

studied two different architectures for the reconstruction: the recurrent variational network (RVN) [30] using GRU units for regularizer and the variational network [8] with U-Net as the regularizer (VN-UNet). The number of unrolled layers was set as 10 for both configurations. The loss function weighting was set as $\gamma_1 = 0.2, \gamma_2 = 0.8$. For VN-UNet, we also study the performance with relaxometry guidance, setting $\gamma_3 = 0.01, \gamma_4 = 0.1$ (VN-UNet-relax). During training, the network is trained by the Adam optimizer with an initial learning rate $\eta_0 = 10^{-3}$ for 400 epochs.

Data augmentation was performed on each individual coil image as in [4], with random rotation $[-45°, 45°]$, translation $[-10\%, 10\%]$, shearing $[-20°, 20°]$ and vertical/horizontal flip. Additionally, we contaminate the k-space by additive Gaussian noise with a random signal-to-noise-ratio (SNR) in $[6.67, +\infty]$. The augmentation was performed with a probability of 40%.

4   Results

4.1   Evaluation Results

We evaluate the three aforementioned configurations: RVN, VN-UNet, and VN-UNet-relax on the validation set. For simplicity, we only list the results on T_1 mapping with acceleration factor $R = 4$ in Table 1. The best performance was achieved by VM-UNet-relax with a PSNR of 42.59 dB. The evaluation results on all acceleration factors of both T_1 and T_2 mapping are shown in Table 2.

Table 1. Ablation of model architecture settings on the T_1 mapping validation set ($R = 4$). The relaxometry guided variational network with U-Nets achieved the best performance.

Model Settings	PSNR (dB) ↑	NMSE ($\times 10^{-3}$) ↓	SSIM (%) ↑
RVN	41.55	3.4	97.53
VN-UNet	42.50	**2.6**	97.91
VN-UNet-relax	**42.59**	**2.6**	**97.93**

4.2   Qualitative Results

We show a few exemplar reconstructed images and their corresponding quantitative maps in Fig. 2. The baseline images with the shortest and longest inversion or echo times are listed.

5   Discussion and Conclusion

Table 1 shows that the VN-UNet achieved better performance than RVN, and the performance is slightly improved by introducing relaxometry guidance. However, only image-based metrics like SSIM are provided during the validation

Table 2. Evaluation results of VN-UNet-relax on the validation set. A higher acceleration factor causes a performance drop.

Dataset	PSNR (dB)	NMSE ($\times 10^{-3}$)	SSIM (%)
T_1-$R = 4$	43.1	2	98.1
T_1-$R = 8$	37.8	7	95.6
T_1-$R = 10$	36.6	9	95.1
T_2-$R = 4$	38.9	3	97.1
T_2-$R = 8$	35.2	6	94.9
T_2-$R = 10$	34.4	7	94.6

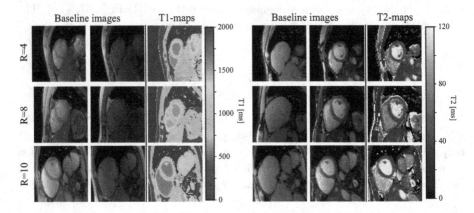

Fig. 2. Qualitative results for T1 and T2 mapping sequences. The baseline images with the shortest and longest inversion/echo times are shown. The proposed method can generate both images and quantitative maps simultaneously. Perceptually, the reconstructed images of all acceleration factors are of good quality.

phase. Therefore, the effect of introducing MR relaxometry-related terms in the loss function needs further investigation because the discrepancy between the predicted and the ground truth quantitative maps can not be evaluated. From Table 2, we observe a performance drop with the increase of acceleration factor. However, it can barely be perceived from the qualitative results shown in Fig. 2.

In conclusion, we proposed a learning-based framework for qMRI reconstruction with a variational network as the reconstruction backbone and introduced an additional mapping network. The proposed framework can output both the baseline images and the mapping result simultaneously. Choosing U-Net as the regularizer achieved better performance, which is further improved by introducing MR relaxometry.

References

1. Barbieri, M., et al.: A deep learning approach for magnetic resonance fingerprinting: scaling capabilities and good training practices investigated by simulations. Phys. Med. **89**, 80–92 (2021)
2. Donoho, D.L.: Compressed sensing. IEEE Trans. Inf. Theory **52**(4), 1289–1306 (2006)
3. Eliasi, P.A., Feng, L., Otazo, R., Rangan, S.: Fast magnetic resonance parametric imaging via structured low-rank matrix reconstruction. In: 2014 48th Asilomar Conference on Signals, Systems and Computers, pp. 423–428. IEEE (2014)
4. Fabian, Z., Heckel, R., Soltanolkotabi, M.: Data augmentation for deep learning based accelerated MRI reconstruction with limited data. In: International Conference on Machine Learning, pp. 3057–3067. PMLR (2021)
5. Griswold, M.A., et al.: Generalized autocalibrating partially parallel acquisitions (GRAPPA). Magn. Reson. Med.: Off. J. Int. Soc. Magn. Reson. Med. **47**(6), 1202–1210 (2002)
6. Haacke, E.M.: Magnetic resonance imaging: physical principles and sequence design (1999)
7. Haaf, P., Garg, P., Messroghli, D.R., Broadbent, D.A., Greenwood, J.P., Plein, S.: Cardiac T1 mapping and extracellular volume (ECV) in clinical practice: a comprehensive review. J. Cardiovasc. Magn. Reson. **18**(1), 1–12 (2017)
8. Hammernik, K., et al.: Learning a variational network for reconstruction of accelerated MRI data. Magn. Reson. Med. **79**(6), 3055–3071 (2018)
9. Heidemann, R.M., et al.: A brief review of parallel magnetic resonance imaging. Eur. Radiol. **13**, 2323–2337 (2003)
10. Huizinga, W., et al.: PCA-based groupwise image registration for quantitative MRI. Med. Image Anal. **29**, 65–78 (2016)
11. Hyun, C.M., Kim, H.P., Lee, S.M., Lee, S., Seo, J.K.: Deep learning for undersampled mri reconstruction. Phys. Med. Biol. **63**(13), 135007 (2018)
12. Larkman, D.J., Nunes, R.G.: Parallel magnetic resonance imaging. Phys. Med. Biol. **52**(7), R15 (2007)
13. Larsson, E.G., Erdogmus, D., Yan, R., Principe, J.C., Fitzsimmons, J.R.: Snr-optimality of sum-of-squares reconstruction for phased-array magnetic resonance imaging. J. Magn. Reson. **163**(1), 121–123 (2003)
14. Liu, F., Feng, L., Kijowski, R.: MANTIS: model-augmented neural network with incoherent k-space sampling for efficient MR parameter mapping. Magn. Reson. Med. **82**(1), 174–188 (2019)
15. Liu, F., Kijowski, R., El Fakhri, G., Feng, L.: Magnetic resonance parameter mapping using model-guided self-supervised deep learning. Magn. Reson. Med. **85**(6), 3211–3226 (2021)
16. Lønning, K., Putzky, P., Sonke, J.J., Reneman, L., Caan, M.W., Welling, M.: Recurrent inference machines for reconstructing heterogeneous MRI data. Med. Image Anal. **53**, 64–78 (2019)
17. Lustig, M., Donoho, D., Pauly, J.M.: Sparse MRI: the application of compressed sensing for rapid MR imaging. Magn. Reson. Med.: Off. J. Int. Soc. Magn. Reson. Med. **58**(6), 1182–1195 (2007)
18. Messroghli, D.R., Radjenovic, A., Kozerke, S., Higgins, D.M., Sivananthan, M.U., Ridgway, J.P.: Modified look-locker inversion recovery (MOLLI) for high-resolution t1 mapping of the heart. Magn. Reson. Med.: Off. J. Int. Soc. Magn. Reson. Med. **52**(1), 141–146 (2004)

19. O'Brien, A.T., Gil, K.E., Varghese, J., Simonetti, O.P., Zareba, K.M.: T2 mapping in myocardial disease: a comprehensive review. J. Cardiovasc. Magn. Reson. **24**(1), 1–25 (2022)
20. Pruessmann, K.P., Weiger, M., Börnert, P., Boesiger, P.: Advances in sensitivity encoding with arbitrary k-space trajectories. Magn. Reson. Med.: Off. J. Int. Soc. Magn. Reson. Med. **46**(4), 638–651 (2001)
21. Putzky, P., Welling, M.: Recurrent inference machines for solving inverse problems. arXiv preprint arXiv:1706.04008 (2017)
22. Ronneberger, O., Fischer, P., Brox, T.: U-net: convolutional networks for biomedical image segmentation. In: Navab, N., Hornegger, J., Wells, W.M., Frangi, A.F. (eds.) MICCAI 2015, Part III. LNCS, vol. 9351, pp. 234–241. Springer, Cham (2015). https://doi.org/10.1007/978-3-319-24574-4_28
23. Seraphim, A., Knott, K.D., Augusto, J., Bhuva, A.N., Manisty, C., Moon, J.C.: Quantitative cardiac MRI. J. Magn. Reson. Imaging **51**(3), 693–711 (2020)
24. Shafieizargar, B., Byanju, R., Sijbers, J., Klein, S., den Dekker, A.J., Poot, D.H.: Systematic review of reconstruction techniques for accelerated quantitative MRI. Magn. Reson. Med. (2023)
25. Sriram, A., et al.: End-to-end variational networks for accelerated MRI reconstruction. In: Martel, A.L., et al. (eds.) MICCAI 2020, Part II. LNCS, vol. 12262, pp. 64–73. Springer, Cham (2020). https://doi.org/10.1007/978-3-030-59713-9_7
26. Wang, C., et al.: Recommendation for cardiac magnetic resonance imaging-based phenotypic study: imaging part. Phenomics **1**, 151–170 (2021)
27. Wang, C., et al.: CMRxRecon: an open cardiac MRI dataset for the competition of accelerated image reconstruction. arXiv preprint arXiv:2309.10836 (2023)
28. Wang, Z., Bovik, A.C., Sheikh, H.R., Simoncelli, E.P.: Image quality assessment: from error visibility to structural similarity. IEEE Trans. Image Process. **13**(4), 600–612 (2004)
29. Yang, Y., Sun, J., Li, H., Xu, Z.: ADMM-CSNet: a deep learning approach for image compressive sensing. IEEE Trans. Pattern Anal. Mach. Intell. **42**(3), 521–538 (2018)
30. Yiasemis, G., Sonke, J.J., Sánchez, C., Teuwen, J.: Recurrent variational network: a deep learning inverse problem solver applied to the task of accelerated MRI reconstruction. In: Proceedings of the IEEE/CVF Conference on Computer Vision and Pattern Recognition, pp. 732–741 (2022)
31. Zhu, B., Liu, J.Z., Cauley, S.F., Rosen, B.R., Rosen, M.S.: Image reconstruction by domain-transform manifold learning. Nature **555**(7697), 487–492 (2018)

A Context-Encoders-Based Generative Adversarial Networks for Cine Magnetic Resonance Imaging Reconstruction

Weihua Zhang[✉], Mengshi Tang, Liqin Huang, and Wei Li

College of Physics and Information Engineering, Fuzhou University, Fuzhou, China
zhang_wh2000@163.com

Abstract. Cine imaging serves as a vital approach for non-invasive assessment of cardiac functional parameters. The imaging process of Cine cardiac MRI is inherently slow, necessitating the acquisition of data at multiple time points within each cardiac cycle to ensure adequate temporal resolution and motion information. Over prolonged data acquisition and during motion, Cine images can exhibit image degradation, leading to the occurrence of artifacts. Conventional image reconstruction methods often require expert knowledge for feature selection, which may result in information loss and suboptimal outcomes. In this paper, we employ a data-driven deep learning approach to address this issue. This approach utilizes supervised learning to compare data with different acceleration factors to full-sampled spatial domain data, training a context-aware network to reconstruct images with artifacts. In our model training strategy, we employ an adversarial approach to make the reconstructed images closer to ground truth. We incorporate loss functions based on adversarial principles and introduce image quality assessment as a constraint. Our context-aware model efficiently accomplishes artifact removal and image reconstruction tasks.

Keywords: Image Reconstruction · Cine MRI · Generative Adversarial Networks · Context Encoder · Deep Learning

1 Introduction

Cine is a non-invasive diagnostic method for acquiring information about the heart and vascular structures. It is commonly employed for evaluating cardiovascular diseases such as myocardial abnormalities, cardiac chamber and valve irregularities, and congenital heart conditions. Cardiac MRI offers several advantages, including the absence of ionizing radiation, making it a safer alternative. It provides high-resolution, high-quality images that offer clear visualization of soft tissue structures and intricate blood flow details. Cine MRI, a specialized subset of MRI, is capable of capturing information regarding organ tissues and their temporal dynamics. This imaging technique is generated by repeatedly scanning a region of interest over a defined time period. Typically, it begins with acquiring single-layer slices of organ tissues, which are then composited into an animated sequence. However, the imaging process for Cine cardiac MRI is relatively

O. Camara et al. (Eds.): STACOM 2023, LNCS 14507, pp. 359–368, 2024.
https://doi.org/10.1007/978-3-031-52448-6_34

slow. It requires the collection of data at multiple time points within each cardiac cycle, often synchronized with electrocardiograms or respiratory signals, to ensure sufficient temporal resolution and motion information. Due to the extended acquisition time, variations in patient heart rate or irregular breathing can lead to motion artifacts, resulting from inconsistencies between data acquired at different time points. Cine MRI rapidly collects a substantial amount of data within a short time frame. Inadequate sampling frequency or data points can lead to incomplete data in k-space, resulting in aliasing artifacts or blurred images. Additionally, magnetic field inhomogeneities can cause changes in signal phase or frequency, leading to chemical shift artifacts or geometric distortion artifacts. Nowadays, accelerated cine scanning can effectively address the aforementioned issues without compromising image quality. This allows for real-time imaging while minimizing artifacts caused by factors such as unstable heart rates or irregular breathing [3]. Accelerated cine is a method used to acquire MRI data by leveraging compressed sensing and parallel imaging techniques. High-speed real-time cine MRI, utilizing compressed sensing and parallel imaging techniques, has demonstrated clinical utility [1]. To enhance MRI quality, conventional image processing methods such as filtering, denoising, and enhancement have been employed. However, these methods often require manual parameter adjustment and may result in the loss of essential details. In recent years, deep learning techniques have been used to directly learn image features and various parameters from data, enabling rapid and effective reconstruction [2]. During the network learning process, employing two-stage networks for domain enhancement has been shown to improve MRI analysis outcomes [6]. In recent years, network models that operate in both the frequency and spatial domains, using a cross-network approach, have been applied for the reconstruction and analysis of cardiac MRI [4, 5].

In this paper, we employ a data-driven deep learning approach to address the challenges of cine MRI reconstruction. The method used involves supervised learning, where different acceleration factors are compared with fully sampled spatial domain data. It treats sampled accelerated images with artifacts as the processing targets and fully sampled k-space data as labels. This is achieved through the training of a convolutional neural network based on a context encoder to reconstruct cine MRI [7]. In our model training strategy, we adopt an adversarial approach to make the reconstructed images closely resemble real images (ground truth) [8]. The combination of this context encoder and adversarial learning proves effective for image reconstruction [10]. We utilize loss functions based on adversarial principles and incorporate image quality assessment as a constraint. Experimental results demonstrate the network's effectiveness in image reconstruction, particularly with the inclusion of image quality constraints. In the validation dataset provided for the competition, the context-aware model successfully accomplishes artifact removal and image reconstruction tasks.

2 Methods

To address the cine MRI reconstruction challenge, we employed a Generative Adversarial Network (GAN) architecture, as illustrated in Fig. 1. Utilizing a U-shaped context encoder network allows for the capture of deeper information while preserving spatial details in the generated images [9]. Initially, the accelerated cine MRI with artifacts is fed into the generative network composed of the context encoder, resulting in the reconstructed MRI. Subsequently, a discriminator is introduced to engage in an adversarial process, discerning between real and fake images [8]. The k-space fully sampled data is utilized as the ground truth, ensuring the reconstructed MRI closely approximates real images. The final loss function incorporates Mean Squared Error (MSE) for image quality evaluation and Structure Similarity Index (SSIM).

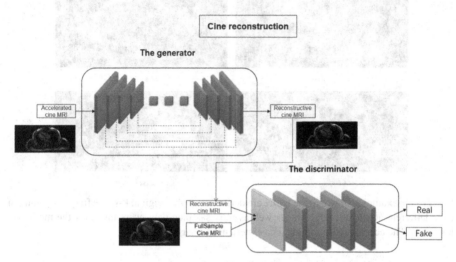

Fig. 1. The network architecture for reconstructing cine MRI in this paper.

2.1 Context Encoder for Reconstructing Cine MRI

The structure of the context encoder follows an encoder-decoder connectivity, and its methodology was introduced for image restoration. It allows predictions of missing regions based on pixel context within the image [7]. Leveraging the concept of the context encoder, it is possible to infer objects containing artifacts or other degraded content based on known structures or attributes of other objects within the image. Not only does this structure enable understanding relationships between different objects in the image, but it also provides an approximate context for the location of a particular structure. In recent years, context encoders have also been applied to segmentation tasks, demonstrating excellent results [9]. In our case, we employ them for cine MRI reconstruction, taking degraded images from accelerated sampling as input. The encoder performs feature analysis and extraction on the input image, resulting in latent deep feature representations. The decoder then translates these low-dimensional latent features into an output image, which is validated against fully sampled k-space data to assess the authenticity of the generated image.

The overall structure of the generator consists of a deep U-shaped network. It takes single-channel images of size 224 × 448 as input, as illustrated in Fig. 2, and proceeds with consecutive downsampling to obtain latent feature representations. These latent features are then reconstructed into images of the original input size using the same number of transposed convolution layers. Each convolution module involves two convolution operations and nonlinear activation, as depicted in Fig. 1. During the training of the context encoder, no pre-training is performed; instead, random initialization parameters are directly employed to learn MRI features.

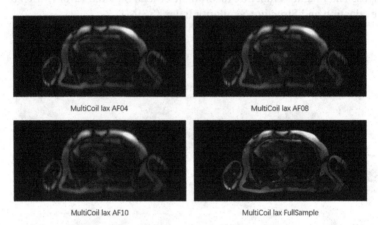

Fig. 2. The examples shown in the figure are derived from the original k-space frequency domain data. Different acceleration factors, as well as fully sampled data, were taken for the multi-coil long-axis. The accelerated sampling data experienced image degradation.

2.2 GAN for Authenticity

Based on the concept of Generative Adversarial Networks (GANs), a setup involves a discriminator and a context encoder engaged in a game. The adversarial approach is an optimization method rooted in game theory, and its essence lies in leveraging competition or collaboration between two or more models to enhance their performance and effectiveness. In the optimization process of the discriminator, fully sampled data is considered real, while generated images are deemed fake. In contrast, during the optimization of the context encoder, generated images are treated as real. The discriminator aims to distinguish between generated images and labeled images, while the generator strives to produce as realistic images as possible. The primary objective of introducing this mechanism is to make the reconstructed images closely resemble reality [8]. The structure of the discriminator is illustrated in Fig. 1.

2.3 Loss Function

To enhance the realism of the generated reconstructed images, we have formulated the following loss function strategy based on the GAN loss function. Specifically, for optimizing the discriminator, we compare the input fully sampled label data with a tensor of the same shape (all pixel values set to 1). We minimize their binary cross-entropy loss function to make the discriminator consider the fully sampled label data as real. Furthermore, we minimize the binary cross-entropy loss function between the reconstructed image and a tensor of the same shape (all pixel values set to 0), considering the cine MRI reconstructed by the context encoder as fake. Through these methods, the discriminator aims to distinguish real MRI from generated MRI as effectively as possible. The sum of these two binary cross-entropy loss functions yields the following overall loss function:

$$D\_R\_Loss = L_{BCE}(D(Y), 1) \tag{1}$$

$$D\_F\_Loss = L_{BCE}(D(G(X)), 0) \tag{2}$$

$$D\_Loss = D\_R\_Loss + D\_F\_Loss \tag{3}$$

In this context, Y represents fully sampled cine MRI, X denotes accelerated sampled cine MRI with artifacts, D stands for the discriminator, G represents the generator, '1' and '0' denote tensors of the same size with all pixel values set to 1 and 0, respectively, and L_{BCE} signifies the binary cross-entropy loss function. D_Loss corresponds to the ultimate discriminator loss.

During the optimization of the generator based on the context encoder, additional loss terms, including mean squared error and structural similarity loss, were incorporated into the original loss function. The generator's loss not only takes into account the realism of the reconstructed MRI images but also considers the disparity between the reconstructed images and the label data. Specifically, the mean squared error loss function is used to quantify the dissimilarity between the reconstructed image and the fully sampled image, as shown below:

$$MSE(Y, G(X)) = (\sum_{i=1}^{n} (Y_i - G(X)_i)^2)/n \tag{4}$$

n represents the number of sample pixels, and a smaller Mean Squared Error (MSE) value between fully sampled and reconstructed images indicates a better reconstruction outcome.

Structural similarity is based on the high structural content in images, characterized by strong pixel correlations within the image. For cine MRI of the heart, which visually contains essential information about organ structures, structural similarity serves as an approximate measure of image distortion by assessing whether the structural information between the reconstructed MRI and fully sampled MRI has changed. This quantifies the degree of similarity between the two images, as described below:

$$SSIM(X, Y) = (2\mu_X\mu_Y + c_1)(2\sigma_{XY} + c_2)/(\mu_X^2 + \mu_Y^2 + c_1)(\sigma_X^2 + \sigma_Y^2 + c_2) \tag{5}$$

The formula compares the structural information of two given images, X and Y. Here, μ_X represents the mean of image X, μ_Y is the mean of image Y, σ_X2 is the variance of X, σ_Y2 is the variance of Y, and σ_{XY} is the covariance between X and Y. c_1 and c_2 are constants used for stability, and the values of SSIM range from -1 to 1. When two images are identical, the SSIM value is 1. The loss function based on SSIM is defined as follows:

$$L_{SSIM} = 1 - SSIM\,(Y,\ G(X)) \tag{6}$$

The loss for the context encoder is defined as a sum of multiple losses, as illustrated below:

$$G\_Loss = L_{BCE}(D(G(X)),\,1) + L_{MSE} + L_{SSIM} \tag{7}$$

During the network training process, an alternating approach is employed, wherein the discriminator is optimized first, followed by the optimization of the generator. Finally, the model of the context encoder is saved for the purpose of reconstructing cine MRI that has undergone degradation due to acceleration.

3 Experiments and Results

3.1 Dataset and Training Protocols

All the data is sourced from the MICCAI Challenge 2023, encompassing both multi-coil and single-coil data in Task 1 Cine reconstruction. The training set comprises cine MRI data with acceleration factors of 4, 8, 10, and full-sampling, each category containing 120 files, with each file containing both long-axis and short-axis data. The validation set exclusively consists of accelerated-sampling data, with each category containing 60 files. These images originate from the Siemens 3T MRI scanner (MAGNETOM Vida). Typically, 5 to 10 slices were acquired for short-axis cine, while only a single slice was acquired for the other views. The data types consist of raw K-space data for short-axis (multi-slices) and long-axis (multi-views) sequences. The cardiac cycle was segmented into 12–25 phases, with a temporal resolution of 50 ms, a spatial resolution of 2.0 × 2.0 mm^2, a slice thickness of 8.0 mm, and a slice gap of 4.0 mm [3].

In practical implementation, we convert the raw k-space data into spatial domain data to facilitate visualization and enable reconstruction tasks using a context encoder. Data augmentation techniques are applied to images, including rotations and flips, to increase sample diversity and allow the model to learn rotational invariance. Each image is then rescaled and interpolated to meet input standards, employing spline interpolation to achieve smoother and more detailed images. After testing, the model converges after approximately 300 epochs, at which point the context encoder can effectively reconstruct MRI scans and remove artifacts. We conduct comparative experiments to demonstrate the effectiveness of incorporating image quality assessment as a constraint and compare it with networks that do not employ adversarial strategies.

3.2 Results

In this section, we will present the results of our experiments, including the reconstructed cine MRI and the evaluation based on the metrics provided by the challenge. We will also conduct comparative experiments (Table 1).

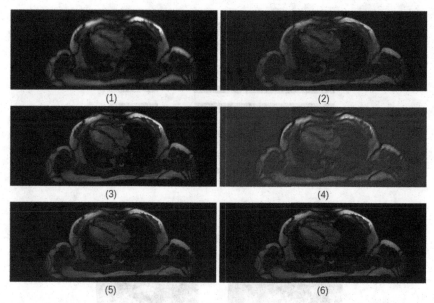

Fig. 3. (1) represents the original image, (2) illustrates the performance of the network used for image restoration in cine MRI reconstruction [7], (3) showcases the results achieved using network with image quality constraints, (4) demonstrates the results obtained with adversarial loss strategy, (5) represents the method employed in this paper, and (6) depicts the effects of using pooling layers.

Table 1. Comparative experiments.

	SSIM	NMSE	PSNR
Figure 3.(2)	0.4959	0.3127	19.6248
Figure 3.(3)	0.7865	0.0685	26.6558
Figure 3.(4)	0.2674	2.3862	10.6332
Figure 3.(5)	0.7988	0.0623	27.3955
Figure 3.(6)	0.7867	0.0733	26.7023

The network employed for image restoration performs relatively poorly in the reconstruction task. Results without the use of adversarial strategies and with pooling are slightly less effective than the method employed in this paper. The network without constraints exhibits the poorest performance. By adopting the approach presented in this paper, image reconstruction can better capture image features and details, resulting in reconstructed images that closely align with the gold standard.

From Fig. 4, it is evident that most instances of image degradation can be improved using the strategy employed in this paper. The optimization results for different image degradation scenarios are shown in Table 2. Cine_PSNR is around 25, Cine_SSIM is approximately 0.78, and Cine_NMSE is roughly 0.16. In summary, it is possible

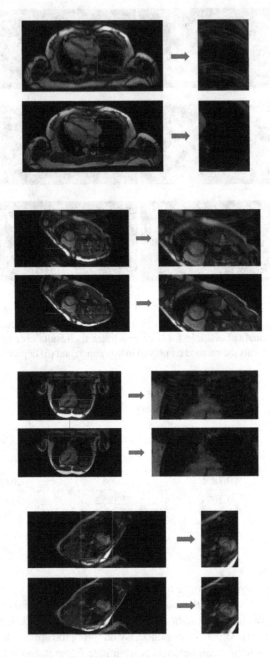

Fig. 4. Visualization of reconstructed results

Table 2. Image evaluation metrics

	SSIM	NMSE	PSNR
MultiCoil_lax_AF04	0.7988	0.0623	27.3955
MultiCoil_lax_AF08	0.7554	0.0863	25.5630
MultiCoil_lax_AF10	0.7590	0.0852	25.5527
MultiCoil_sax_AF04	0.8135	0.1792	25.0799
MultiCoil_sax_AF08	0.7929	0.1852	24.3341
MultiCoil_sax_AF10	0.7628	0.2068	23.5277
SingleCoil_lax_AF04	0.6931	0.1020	24.4874
SingleCoil_lax_AF08	0.6886	0.1070	24.2814
SingleCoil_lax_AF10	0.6932	0.1042	24.2982
SingleCoil_sax_AF04	0.7154	0.2179	23.0952
SingleCoil_sax_AF08	0.7277	0.2071	23.1881
SingleCoil_sax_AF10	0.7120	0.2336	22.6461

to achieve satisfactory image reconstruction results with a relatively small model size compared to image restoration networks, and the speed of reconstructing a single image is also relatively fast.

4 Conclusions

This paper addresses the problem of image degradation in accelerated-sampled cine MRI using a model with a context encoder and a generative adversarial training strategy. The issue is resolved by incorporating constraints based on the relationships between structures in the images and analyzing spatial information within the images. A discriminator network is introduced to engage in a game with the model, resulting in more realistic reconstructed images. Building upon the existing image restoration network, we modify the model structure, introduce constraints, and employ adversarial training to enhance the reconstruction performance. In the Challenge, we have achieved satisfactory results, effectively eliminating most artifacts. However, there are instances of edge blurring and vessel loss in certain regions. Future work will consider optimizing image edge details and employing frequency domain preprocessing to further improve the results.

Acknowledgements. This work was supported by National Natural Science Foundation of China (62271149), Fujian Provincial Natural Science Foundation project (2021J02019).

References

1. Feng, L., Otazo, R., Srichai, M.B., Lim, R.P., Sodickson, D.K., Kim, D.: Highly-accelerated real-time cine MRI using compressed sensing and parallel imaging. J. Cardiovasc. Magn. Reson. **13**(1), 1–2 (2011)

2. Schlemper, J., Caballero, J., Hajnal, J.V., Price, A.N., Rueckert, D.: A deep cascade of convolutional neural networks for dynamic MR image reconstruction. IEEE Trans. Med. Imaging **37**(2), 491–503 (2017)

3. 2023 CMRxRecon Task1: Cine reconstruction. https://cmrxrecon.github.io/Task1-Cine-reconstruction.html. Accessed 04 Aug 2023

4. Nitski, O., Nag, S., McIntosh, C., Wang, B.: CDF-Net: cross-domain fusion network for accelerated MRI reconstruction. In: Martel, A.L., et al. (eds.) MICCAI 2020. LNCS, vol. 12262, pp. 421–430. Springer, Cham (2020). https://doi.org/10.1007/978-3-030-59713-9_41

5. Machado, I., et al.: Quality-aware cine cardiac mri reconstruction and analysis from undersampled k-space data. In: Puyol Antón, E., et al. (eds.) STACOM 2021. LNCS, vol. 13131, pp. 12–20. Springer, Cham (2022). https://doi.org/10.1007/978-3-030-93722-5_2

6. Zhang, X., Yang, X., Huang, L., Huang, L.: Two stage of histogram matching augmentation for domain generalization: application to left atrial segmentation. In: Zhuang, X., Li, L., Wang, S., Wu, F. (eds.) LAScarQS 2022. LNCS, vol. 13586, pp. 60–68. Springer, Cham (2022). https://doi.org/10.1007/978-3-031-31778-1_6

7. Pathak, D., Krahenbuhl, P., Donahue, J., Darrell, T., Efros, A.A.: Context encoders: feature learning by inpainting. In: Proceedings of the IEEE Conference on Computer Vision and Pattern Recognition, pp. 2536–2544 (2016)

8. Goodfellow, I., et al.: Generative adversarial nets. In: Advances in Neural Information Processing Systems, vol. 27 (2014)

9. Gu, Z., et al.: CE-Net: context encoder network for 2D medical image segmentation. IEEE Trans. Med. Imaging **38**(10), 2281–2292 (2019)

10. Huang, C.Y., Chen, O.T.C., Wu, G.Z., Chang, C.C., Hu, C.L.: Ultrasound imaging improved by the context encoder reconstruction generative adversarial network. In: 2018 IEEE International Ultrasonics Symposium (IUS), pp. 1–4. IEEE (2018)

Accelerated Cardiac Parametric Mapping Using Deep Learning-Refined Subspace Models

Calder D. Sheagren[1,2(✉)], Brenden T. Kadota[1,2], Jaykumar H. Patel[1,2], Mark Chiew[1,2], and Graham A. Wright[1,2]

[1] Department of Medical Biophysics, University of Toronto, Toronto, ON, Canada
calder.sheagren@mail.utoronto.ca
[2] Physical Science Platform, Sunnybrook Research Institute, Toronto, ON, Canada

Abstract. Cardiac parametric mapping is useful for evaluating cardiac fibrosis and edema. Parametric mapping relies on single-shot heartbeat-by-heartbeat imaging, which is susceptible to intra-shot motion during the imaging window. However, reducing the imaging window requires undersampled reconstruction techniques to preserve image fidelity and spatial resolution. The proposed approach is based on a low-rank tensor model of the multi-dimensional data, which jointly estimates spatial basis images and temporal basis time-courses from an auxiliary parallel imaging reconstruction. The tensor-estimated spatial basis is then further refined using a deep neural network, trained in a fully supervised fashion, improving the fidelity of the spatial basis using learned representations of cardiac basis functions. This two-stage spatial basis estimation will be compared against Fourier-based reconstructions and parallel imaging alone to demonstrate the sharpening and denoising properties of the deep learning-based subspace analysis.

Keywords: Parametric mapping · Deep subspace learning

1 Introduction

T_1-weighted (T_1w) and T_2-weighted (T_2w) imaging are staple techniques to determine regions of cardiac muscle, fat, fibrosis, edema, and amyloid, among other possible pathologies [5]. However, these are qualitative imaging techniques that rely on hyperintense or hypointense signal intensities to make a diagnosis, the distribution of which can change as a function of acquisition parameters and sequence type. Moreover, diagnosing diffuse disease on contrast-weighted imaging is difficult due to the global impact of the pathology on myocardial signal intensity. To address these issues, parametric mapping has been introduced to reproducibly quantify the underlying tissue parameters T_1 and T_2 at every voxel, providing absolute normative and abnormal values for a given field strength.

Cardiac T_1 and T_2 mapping have been used clinically to diagnose diseases such as acute myocardial infarction, cardiac amyloidosis, Fabry's disease, and

O. Camara et al. (Eds.): STACOM 2023, LNCS 14507, pp. 369–379, 2024.
https://doi.org/10.1007/978-3-031-52448-6_35

iron overload, among other diseases [9]. Additionally, T_1 and T_2 mapping play an integral role in diagnosing acute myocarditis as per the Lake Louise criteria [7]. One challenge with parametric mapping is the long temporal window needed to acquire single-shot contrast-weighted images in the cardiac cycle. This causes a trade-off between intra-shot cardiac motion, spatial resolution, and signal-to-noise ratio (SNR) [13]. To shorten the temporal footprint without sacrificing SNR, parallel imaging techniques such as SENSE and GRAPPA that exploit redundancies between the multi-coil acquisitions have been introduced to accelerate acquisitions by a factor of 2–3 [4,11]. Parallel imaging is widely used in the clinic, but has a practical limit to the acceleration factors supported. To further reduce the temporal window and enable imaging at high heart rates, imaging at higher parallel imaging acceleration factors is needed, requiring more sophisticated undersampled data reconstruction methods to preserve SNR and image quality.

Parametric mapping can be interpreted as a *spatiotemporal* imaging method, where the same spatial volume is imaged at multiple timepoints with different contrasts. In spatiotemporal imaging, both the space and time dimensions are often highly compressible [3,10]. One way to leverage this compressibility is to explicitly factor the spatiotemporal image $X \in \mathbb{C}^{N_x \times N_y \times N_t}$ as a single outer product $X = C \otimes T$. Here, $C \in \mathbb{C}^{N_x \times N_y}$ is a *spatial basis*, and $T \in \mathbb{C}^{N_t}$ is a *temporal basis*. This model is very simplistic and may not capture complex interactions between physiology and motion or contrast changes, so the notion of *partially separable functions* has been introduced to model X as a linear combination

$$X = \sum_{l=1}^{L} a_l C_l \otimes T_l. \tag{1}$$

Now, each basis component $C_l \in \mathbb{C}^{N_x \times N_y}$, $T_l \in \mathbb{C}^{N_t}$, for $l = 0, \ldots, L \in \mathbb{N}$, where L is the chosen rank of the tensor. Decomposing spatial information from temporal dynamics allows for sophisticated reconstruction methods with dimension-specific regularizers, such as spatial sparsity preservation and temporal low-rank preservation [17].

To generalize dimension-specific regularizers, the notion of *deep subspace learning* was introduced to apply deep learning to sub-components of the reconstructed tensor to preserve temporal dynamics while enhancing spatial information [2]. In this paper, we propose a method using Deep learning-Refined sUbspace ModelS (DRUMs).

2   Methods

2.1   Dataset

The CMRxRecon challenge dataset was used to develop this method [16]. Briefly, 300 volunteers were scanned on a 3 T MRI (MAGNETOM Vida, Siemens Healthineers, Germany). Steady-state free precession T_1 maps and gradient-echo T_2 maps were acquired with parallel imaging undersampling factors of 2 (hereafter

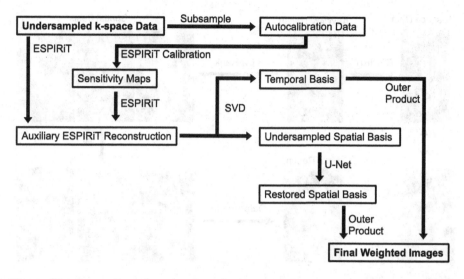

Fig. 1. Visual overview of proposed method. The undersampled k-space data is subsampled into the autocalibration data, which is used for sensitivity map calculation. Undersampled k-space data is used to calculate coil sensitivity maps, which jointly are inputs to an auxiliary ESPIRiT reconstruction. The SVD of the ESPIRiT reconstruction is used to generate spatial and temporal basis sets. The spatial basis is fed through a U-Net designed to remove residual noise and incorporate missing high-order image features. The final basis is combined with the temporal basis to obtain the final weighted images. ESPIRiT: *eigenvalue-based iterative self-consistent parallel imaging reconstruction*; SVD: *singular value decomposition*.

referred to as "fully sampled"), 4, 8, and 10 to reduce the temporal acquisition window during the cardiac cycle. Sequence parameters for T_1 mapping include: FOV = 360 × 307 mm^2, spatial resolution = 1.4 × 1.4 mm^2, slice thickness = 5.0 mm, TR/TE = 2.67 ms/1.13 ms, Partial Fourier = 7/8, 24 autocalibration lines, TI={100, 180, 260, 900, 1000, 1050, 1700, 1800, 2500}ms. Sequence parameters for T_2 mapping include: FOV = 360 × 288 mm^2, spatial resolution = 1.9 × 1.9 mm^2, slice thickness = 5.0 mm, TR/TE = 3.06 ms/1.29 ms, T2 preparation times = {0, 35, 55} ms, Partial Fourier = 6/8, 24 autocalibration lines.

2.2   DRUMs Algorithm

In this section, we will discuss the methods used for the DRUMs reconstruction combining parallel imaging, subspace models, and deep learning-refined spatial bases. For a visual overview of the method, see Fig. 1. For sample images at multiple stages in the method, see Fig. 2. Training and inference code for this model is available at https://github.com/WrightGroupSRI/CMRxRecon.

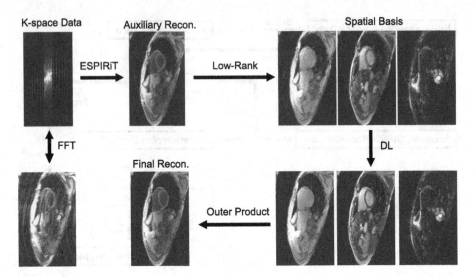

Fig. 2. Sample images at multiple stages of the reconstruction pipeline. Raw k-space data is reconstructed using the ESPIRiT method, and is subsequently decomposed into a spatial and temporal basis. A deep learning network is applied to the spatial basis, which is combined with the temporal basis vectors and singular values to obtain a final reconstruction. The FFT reconstruction is provided in the lower left-hand corner for visual comparison with the proposed method. ESPIRiT: *eigenvalue-based iterative self-consistent parallel imaging reconstruction*; FFT: *fast Fourier transform*.

One set of coil sensitivity maps $\{S_q\}_{q=1}^{10}$ per slice were calculated using the `bart` ESPIRiT calibration (`ecalib`) utility [15]. These sensitivity maps were used for a slice-by-slice, contrast-independent L^1-ESPIRiT reconstruction in the `bart` framework that solves the following inverse problem:

$$X_E = \arg\min_X \frac{1}{2} \sum_{q=1}^{10} \|PFS_q X - y\|_2^2 + \lambda\|\Psi X\|_1. \tag{2}$$

Here, X_E is the reconstructed image, P is the undersampling projection mask operator, F is the Fourier transform, y is the undersampled multi-coil k-space data, $\lambda = 0.01$ is the regularization parameter determined via visual analysis, and Ψ is the spatial wavelet transform. Sparse thresholding in the wavelet domain was selected as a regularizer due to its performance in prior compressed-sensing applications [8]. 100 iterations were used with an eigenvalue-guided step size selection.

X_E was then decomposed into a spatial basis C and a temporal basis T in a slice-by-slice manner using the singular value decomposition. The rank-9 basis for T_1 mapping images was truncated to a rank-3 basis, allowing for a constant rank $L = 3$ for both T_1 and T_2 mapping. T_2 mapping used the full rank of 3 to improve parametric accuracy at the cost of some spatial regularization.

To recover image sharpness lost in the compressed sensing reconstruction and reduce residual statistical noise gained in the ESPIRiT reconstruction, the spatial basis C was input into a deep learning model U_θ. The network was trained to optimize the model weights θ in a fully supervised manner using spatial bases generated from undersampled and fully sampled data. For more details, see Sect. 2.3. After the model was applied, final weighted images were calculated as

$$X = \sum_{l=1}^{L} a_l \left(U_\theta(C_l) \right) \otimes T_l. \tag{3}$$

Parameter maps were fitted during the submission validation process using CMRxRecon-specific code. Briefly, T_1 maps were fitted using a two-step process:

$$X(\text{TI}) = A - B \exp(-\text{TI}/T_1^*), \quad T_1 = \left(\frac{B}{A} - 1 \right) T_1^*. \tag{4}$$

T_2 maps were fitted using a one-step process:

$$X(T_\text{prep}) = A \exp(-T_\text{prep}/T_2), \tag{5}$$

where T_prep denotes the T_2 preparation time.

2.3   Model Training

A U-Net model architecture was used here due to its success in multiple reconstruction and denoising applications [6, 12]. The purpose of the U-Net in this application is to restore high-order spatial information, reduce statistical noise, and reduce aliasing artifacts in the undersampled spatial basis vectors. Our U-Net consists of 4 downsampling and upsampling layers and initializes with 64 convolutional filters. Each downsampling layer level downsamples the spatial features by a factor of 2 and doubles the convolutional filters. The convolutional kernels are sized at 3×3 and were followed by Batch normalization and a rectified linear unit activation function. Spatial basis vectors were split into the real and imaginary components of the complex signal, dephased to a consistent phase, and cropped to a constant size of (128, 128). Basis vectors were concatenated into an image size of $(128, 128 \times 2L)$ to pass all three vectors to the model simultaneously.

Our U-Net learns to minimize the residual between the undersampled spatial basis and fully sampled spatial basis coefficients by varying model weights θ:

$$\hat{\theta} = \arg \min_\theta \mathcal{L}(U_\theta(C) - C, \hat{C}). \tag{6}$$

Here, $\hat{C}$ is the fully sampled spatial basis coefficients. We pass spatial basis coefficients with vectors concatenated along the channel dimension through the network. The dataset was normalized using z-standardization along the channel dimension for undersampled and fully sampled spatial basis coefficients. Here

Table 1. Description of model information for mapping submission.

Task of Participation	Mapping	Pre-Training	None
University/Organization	University of Toronto Sunnybrook Research Institute	Data Augmentation	None
Coil Configuration	Multi-channel	Data Standardization	Z-score
Training Hardware	NVIDIA P100 Pascal 12 GB Intel E5-2683v4 4 × 2.1 GHz	Parameter Number	31,036,800
Validation Hardware	NVIDIA RTX 2060 Super 8 GB Intel i7-6700k 6 × 4.00 GHz		
Training Time	10 h	Loss Function	L^1 + SSIM
Inference Time	3 h	Physical Model	None (Low-rank)
Training Set Performance	See Sect. 3	Use of Unrolling	None
Validation Set Performance	See Table 2	k-space Fidelity	Compressed sensing
Docker Submitted	Yes	Model Backbone	U-Net
Segmentation Labels	None	Operations	Dephased complex-valued

we use a combination of L^1 and Structural Similarity Index (SSIM) losses to help preserve edges in our spatial bases. L^1 loss was chosen to enforce voxelwise consistency with the ground-truth data, and the SSIM loss was chosen due to its versatility in image information and contrasts. The total loss function is

$$\mathcal{L} = 1 - \text{SSIM}\left(U_\theta(C), \hat{C}\right) + \|U_\theta(C) - \hat{C}\|_1. \tag{7}$$

Here, a window size of 5 was used in the SSIM loss.

The U-Net was implemented using the PyTorch deep learning library. The given training data was subdivided. We split the training data into internal training, internal validation, and internal test set sizes of 100, 10 and 10 patients (3252, 318, and 318 slices). The Adam optimizer was used with default parameters and a learning rate of 10^{-3} with a batch size of 16 slices. The learning rate was chosen using a learning rate range test as shown in [14]. Dropout of 0.50 was used to regularize the model to prevent overfitting. A single model for use on all image contrasts and acceleration factors was trained for 500 epochs on an NVIDIA P100 Pascal GPU with 12 GB of memory, which took approximately 10 h. For an overview of the model information, please see Table 1.

2.4 Experiments

Normalized root mean squared error (NRMSE) was compared between fully-sampled images and reconstructed images for patients in the internal testing sub-set of the overall training dataset with provided fully-sampled reference images. This allows for comparison of what reconstruction steps impacted NRMSE the most and what performance differences are present between acceleration factors and image contrasts.

Peak signal to noise ratio (PSNR), normalized mean squared error (NMSE), and structural similarity index metric (SSIM) were measured in the validation

Table 2. Validation set performance on the segmented myocardial ROI. In each cell, the leftmost value denotes the CMRxRecon parallel imaging submission, and the rightmost value denotes our proposed DRUMs method. Numbers in bold refer to best-performing methods. R: acceleration factor.

Metric	$R = 4$	$R = 8$	$R = 10$
T_1 PSNR [dB]	23.00/**31.28**	22.38/**29.10**	22.23/**27.74**
T_2 PSNR [dB]	23.89/**29.45**	23.40/**27.70**	23.50/**27.03**
T_1 SSIM [1]	0.66/**0.85**	0.63/**0.81**	0.63/**0.79**
T_2 SSIM [1]	0.77/**0.87**	0.75/**0.84**	0.77/**0.83**
T_1 NMSE [1]	0.21/**0.03**	0.26/**0.05**	0.27/**0.06**
T_2 NMSE [1]	0.08/**0.02**	0.09/**0.03**	0.09/**0.03**

set between reconstructed images and unseen fully-sampled images. Comparison was performed on the parameter level, with T_1 and T_2 maps serving as inputs for reconstructed and fully-sampled images. Parameter values were measured over a manually-segmented region of interest containing the left ventricular myocardium, left ventricular blood pool, and right ventricular blood pool. Results for the DRUMs algorithm were compared against conventional parallel imaging reconstruction as submitted by CMRxRecon on the validation set leaderboard.

3   Results

The proposed DRUMs method was successfully implemented in Python with a trained neural network. On a CPU, each 3D k-space dataset reconstruction took 100 s, which is broken down into the following: Sensitivity map calculation - 23.35 s; ESPIRiT reconstruction - 75.96 s; SVD and basis fitting - 0.34 s; and U-Net inference - 0.33 s. On a GPU, the entire training dataset was able to be reconstructed within 5 h.

For sample images at various acceleration factors, refer to Fig. 3. NRMSE across all contrasts for the zero-filled Fast Fourier Transform (FFT) reconstruction was 0.081 ± 0.02. NRMSE across all contrasts for the ESPIRiT reconstruction was 0.033 ± 0.02. NRMSE across all contrasts for the low-rank approximation of the ESPIRiT reconstruction was 0.035 ± 0.02, and NRMSE across all contrasts for the full DRUMs reconstruction was $0.030 \pm .01$. We speculate that the low-rank approximation increased the NRMSE due to constraining the data in a subspace. NRMSE for T_1w images was 0.030 ± 0.02, and NRMSE for T_2w images was 0.028 ± 0.01, showing no large difference between the two tissue contrasts. NRMSE for acceleration factors 4, 8, and 10 was 0.023 ± 0.01, 0.031 ± 0.02, and 0.035 ± 0.02, respectively. This is expected, as errors increase with fewer given k-space lines.

In the unseen validation set, metrics were evaluated using the fitted T_1 and T_2 maps in a manually-segmented myocardial ROI. Table 2 contains a compar-

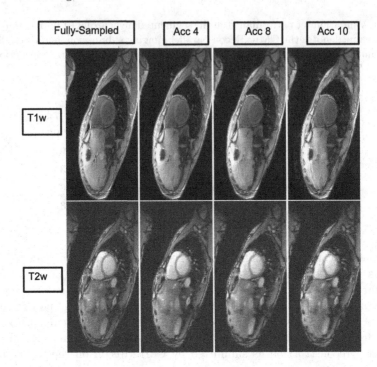

Fig. 3. Sample T_1w and T_2w images reconstructed using the proposed DRUMs method across acceleration factors. Top row: T_1w images. Bottom row: T_2w images. Left column: fully sampled images reconstructed using the ESPIRiT method to preserve signal intensity scaling. Right columns: images reconstructed at acceleration factors of 4, 8, and 10. Images at higher acceleration factors are visually similar to fully-sampled images, with residual blurring and aliasing artifacts more present at higher acceleration factors, particularly in the T_2w images. ESPIRiT: *eigenvalue-based iterative self-consistent parallel imaging reconstruction*

ison of the proposed DRUMs method and the benchmark CMRxRecon parallel imaging reconstruction submission. DRUMs reconstruction outperformed the parallel imaging reconstruction in every metric at every acceleration factor. Our method generally performed better quantitatively in T_2 reconstruction than T_1 reconstruction, with the exception of PSNR, which is surprising given the lack of temporal regularization in our T_2 reconstruction. The parallel imaging reconstruction was also superior in T_2 mapping vs T_1 mapping with the same method for all metrics including PSNR. For sample M_0 and T_2 maps at acceleration 10 from a patient in the validation set with unseen fully sampled data, see Fig. 4.

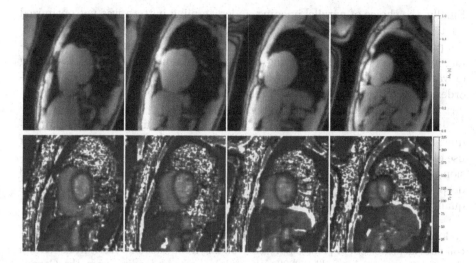

Fig. 4. Sample quantitative results for DRUMs reconstructed images. Results from a single subject from the validation set with no provided ground-truth images are plotted here, with a bicubic interpolation for voxel smoothing. Reconstructed images at an acceleration factor of 10 were fitted to the following equation: $S(t) = M_0 \exp(-T_{\mathrm{prep}}/T_2)$, $M_0 \in [0, 1]$, $T_2 \in [0, 250]$ ms. Top row: proton-density M_0 maps. Bottom row: T_2 maps. Parameter values are reasonable for 3 T acquisition, and good delineation between myocardium and LV blood pool is observed. The RV free wall is visible in some slices, and residual blurring and aliasing artifacts are present around the heart.

4   Discussion and Conclusion

In this paper, we have proposed a deep learning-refined subspace model reconstruction framework that is compatible with cardiac T_1 and T_2 mapping. The subspace formalism and constant rank allows for simple cross-application between T_1 and T_2 mapping. Temporal consistency is enforced using data-driven constraints rather than fitting to predefined modelling functions. Applying the deep learning model to the spatial basis vectors allows for improvements in spatial fidelity while reducing the capability for hallucinating due to the predetermined temporal constraints.

This proposed method has several limitations. First, it was only trained on one T_1 mapping sequence and one T_2 mapping sequence from a single vendor, so while we hope this method can generalize well to unknown sequences or contrasts due to its low-rank nature, this has not been rigorously evaluated in this submission. Second, the T_2 mapping uses the full rank of 3, which improves parametric accuracy at the cost of temporal compressibility. This can be observed in Fig. 3, where T_1w images have improved image quality at higher acceleration values. We will investigate the effects of using rank-2 approximations of T_2w multi-contrast images on overall image quality and parametric robustness.

Finally, the choice of parallel imaging undersampling pattern was not optimized for temporal incoherence of artifacts. Temporally-incoherent artifacts improve the performance of sparsity-based reconstruction methods like compressed sensing, and can be accomplished using time-varying pseudorandom undersampling or variable-density sampling patterns [1]. Generally, choosing different higher-order k-space lines at different contrasts allows for more robust temporal regularization that can admit higher acceleration factors.

In the future, we hope this method can be applied inline on scanners to produce high-quality reconstructions in accelerated sequences that facilitate greater use of advanced reconstruction methods in standard clinical practice. In conclusion, a method combining parallel imaging, temporal low-rank constraints, and deep subspace learning spatial restoration was proposed to improve the image quality of highly accelerated cardiac parametric mapping sequences.

Acknowledgments. This work made use of the Digital Research Alliance of Canada compute facilities. CS, JP, and GW receive funding from Canadian Institutes of Health Research grant number PJT178299. BK and MC receive funding from the Canada Research Chairs Program and NSERC Discovery grant number RGPIN/2023-03410.

References

1. Ahmad, R., Xue, H., Giri, S., Ding, Y., Craft, J., Simonetti, O.P.: Variable density incoherent spatiotemporal acquisition (VISTA) for highly accelerated cardiac MRI. Magn. Reson. Med. **74**(5), 1266–1278 (2015). https://doi.org/10.1002/mrm.25507
2. Chen, Y., Shaw, J.L., Xie, Y., Li, D., Christodoulou, A.G.: Deep learning within a Priori temporal feature spaces for large-scale dynamic MR image reconstruction: application to 5-D cardiac MR multitasking. In: Shen, D., et al. (eds.) MICCAI 2019. LNCS, vol. 11765, pp. 495–504. Springer, Cham (2019). https://doi.org/10.1007/978-3-030-32245-8_55
3. Feng, L., Axel, L., Chandarana, H., Block, K.T., Sodickson, D.K., Otazo, R.: XD-GRASP: golden-angle radial MRI with reconstruction of extra motion-state dimensions using compressed sensing. Magn. Reson. Med. **75**(2), 775–788 (2016)
4. Griswold, M.A., et al.: Generalized autocalibrating partially parallel acquisitions (GRAPPA). Magn. Reson. Med. **47**(6), 1202–1210 (2002)
5. Henningsson, M., Malik, S., Botnar, R., Castellanos, D., Hussain, T., Leiner, T.: Black-blood contrast in cardiovascular MRI. J. Magn. Reson. Imaging **55**(1), 61–80 (2022). https://doi.org/10.1002/jmri.27399
6. Lehtinen, J., et al.: Noise2noise: learning image restoration without clean data. arXiv preprint arXiv:1803.04189 (2018)
7. Luetkens, J.A., et al.: Comparison of original and 2018 lake louise criteria for diagnosis of acute myocarditis: results of a validation cohort. Radiol.: Cardiothorac. Imaging **1**(3), e190010 (2019)
8. Lustig, M., Donoho, D., Pauly, J.M.: Sparse MRI: the application of compressed sensing for rapid MR imaging. Magn. Reson. Med. **58**(6), 1182–1195 (2007). https://doi.org/10.1002/mrm.21391
9. Messroghli, D.R., et al.: Myocardial T1 mapping: application to patients with acute and chronic myocardial infarction. Magn. Reson. Med. **58**(1), 34–40 (2007)

10. Ong, F., et al.: Extreme MRI: large-scale volumetric dynamic imaging from continuous non-gated acquisitions. Magn. Reson. Med. **84**(4), 1763–1780 (2020). https://doi.org/10.1002/mrm.28235

11. Pruessmann, K.P., Weiger, M., Scheidegger, M.B., Boesiger, P.: SENSE: sensitivity encoding for fast MRI. Magn. Reson. Med. **42**(5), 952–962 (1999)

12. Ronneberger, O., Fischer, P., Brox, T.: U-net: convolutional networks for biomedical image segmentation. CoRR abs/1505.04597 (2015). http://arxiv.org/abs/1505.04597

13. Sheagren, C.D., et al.: Motion-compensated T1 mapping in cardiovascular magnetic resonance imaging: a technical review. Front. Cardiovasc. Med. **10** (2023). https://www.frontiersin.org/articles/10.3389/fcvm.2023.1160183

14. Smith, L.N.: Cyclical learning rates for training neural networks. arXiv:1506.01186 (2017)

15. Uecker, M., et al.: ESPIRiT-an eigenvalue approach to autocalibrating parallel MRI: where SENSE meets GRAPPA. Magn. Reson. Med. **71**(3), 990–1001 (2014). https://doi.org/10.1002/mrm.24751

16. Wang, C., et al.: Recommendation for cardiac magnetic resonance imaging-based phenotypic study: imaging part. Phenomics **1**(4), 151–170 (2021). https://doi.org/10.1007/s43657-021-00018-x

17. Zhang, L., Barry, J., Pop, M., Wright, G.A.: High-resolution MR characterization of myocardial infarction using compressed sensing with edge preservation. J. Cardiovasc. Magn. Reson. **18**(1), P303 (2016)

DiffCMR: Fast Cardiac MRI Reconstruction with Diffusion Probabilistic Models

Tianqi Xiang[1], Wenjun Yue[1,2], Yiqun Lin[1], Jiewen Yang[1], Zhenkun Wang[3], and Xiaomeng Li[1(✉)]

[1] The Hong Kong University of Science and Technology, Hong Kong, China
eexmli@ust.hk
[2] HKUST Shenzhen Research Institute, Shenzhen, China
[3] Southern University of Science and Technology, Shenzhen, China

Abstract. Performing magnetic resonance imaging (MRI) reconstruction from under-sampled k-space data can accelerate the procedure to acquire MRI scans and reduce patients' discomfort. The reconstruction problem is usually formulated as a denoising task that removes the noise in under-sampled MRI image slices. Although previous GAN-based methods have achieved good performance in image denoising, they are difficult to train and require careful tuning of hyperparameters. In this paper, we propose a novel MRI denoising framework DiffCMR by leveraging conditional denoising diffusion probabilistic models. Specifically, DiffCMR perceives conditioning signals from the under-sampled MRI image slice and generates its corresponding fully-sampled MRI image slice. During inference, we adopt a multi-round ensembling strategy to stabilize the performance. We validate DiffCMR with cine reconstruction and T1/T2 mapping tasks on MICCAI 2023 Cardiac MRI Reconstruction Challenge (CMRxRecon) dataset. Results show that our method achieves state-of-the-art performance, exceeding previous methods by a significant margin. Code is available at https://github.com/xmed-lab/DiffCMR.

Keywords: Under-sampled MRI · Cardiac MRI · MRI reconstruction · Denoising diffusion probabilistic models

1 Introduction

Magnetic resonance imaging (MRI) is an important non-invasive imaging technique that visualizes internal anatomical structures without radiation doses. However, the acquisition time for MRI is significantly longer than X-ray-based imaging since a series of data points should be collected in the k-space. Particularly, cardiac magnetic resonance imaging (CMR) requires more time for acquisition as the heart beats uncontrollably and the data acquisition period should cover several heartbeat cycles. Long scanning time for MRI usually brings

---

T. Xiang and W. Yue—These authors contributed equally.

© The Author(s), under exclusive license to Springer Nature Switzerland AG 2024
O. Camara et al. (Eds.): STACOM 2023, LNCS 14507, pp. 380–389, 2024.
https://doi.org/10.1007/978-3-031-52448-6_36

discomfort and stress to patients, which may induce artifacts in the MRI recon-struction process. In this work, we study the under-sampled MRI reconstruction that sparsely samples k-space data for image reconstruction, which is one of the ways to accelerate MRI acquisition.

The reconstruction from under-sampled k-space data is an ill-posed problem, which is challenging and has received much attention. In the past, compressive sensing [9] is used to solve an optimization problem that seeks to find the most compressible representation that is consistent with the under-sampled k-space data. With the development of deep learning, the reconstruction problem is usu-ally formulated as a denoising task that removes the artifacts in under-sampled MRI images. For example, encoder-decoder-based methods [10,13,15] are pro-posed to learn a mapping from under-sampled images to fully-sampled images. GAN-based methods [3] introduce a discriminator network that learns to iden-tify the differences between the generated and real images, which can further improve image quality. However, GAN-based methods are difficult to train and require careful tuning of hyperparameters.

Recently, denoising diffusion probabilistic models [5,14] (DDPMs) are pro-posed to use a series of transformations to increase the complexity of the gen-erated output iteratively. Compared with GAN-based methods, DDPMs have been shown to be more stable during training and demonstrate outstanding performance on a variety of computer vision tasks [4], including image synthe-sis [2,5,14], inpainting [8,19], segmentation [1,17], and denoising [18,20,21]. To this end, we leverage conditional DDPMs into under-sampled MRI reconstruc-tion. Specifically, we employ a conditional DDPM that generates fully-sampled MRI slices with a conditioning signal from the input under-sampled MRI slices and further adopt multi-round inference ensembling to stabilize the denoising process.

To summarize, the main contributions of this work include 1.) we propose DiffCMR for fast MRI reconstruction from under-sampled k-space data by lever-aging conditional diffusion models; 2.) extensive experiments are conducted on MICCAI 2023 CMRxRecon dataset, showing DiffCMR's state-of-the-art perfor-mance that outperforms previous methods by a large margin.

2   Methodology

2.1   Problem Definition

In this work, we formulate the reconstruction of high-quality MRI from under-sampled k-space data as a denoising task. Specifically, inverse Fast Fourier Trans-form (iFFT) is performed to transform fully-sampled and under-sampled k-space data into 2D image slices, which are referred to as fully-sampled and under-sampled image slices in subsequent sections. We denote the set of generated image slices as $D^{(k)} = \{(I_{i,k}^u, I_{i,k}^f)_{i=1}^{N_k}\}$, where k refers to three different acceler-ation factors; N_k is the number of samples generated from the raw data with acceleration factor k; $I_{i,k}^u$ indicates the i-th under-sampled image slice with accel-eration factor k and $I_{i,k}^f$ is the corresponding fully-sampled image slice. For each

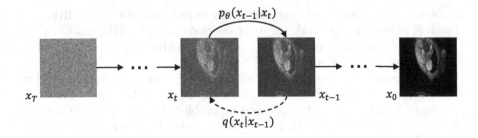

Fig. 1. Forward and backward denoising diffusion processes.

acceleration factor k, we aim to train a denoising model $\phi_\theta^{(k)}$ with the dataset $D^{(k)}$, which has the ability to recover fully-sampled image slice $I_{i,k}^u$ from under-sampled image slice $I_{i,k}^f$.

2.2 Data Preprocess

The original dataset provided by the challenge organizer is composed of single-coil and multi-coil data. Each data is stored in .mat format, accompanied by striped masks of different acceleration factors (e.g., 4, 8, and 10). Under-sampled k-space data can be generated by covering the striped mask on the fully-sampled data. Based on the problem formulation, we first process fully-sampled and under-sampled k-space data to make data pairs (i.e., ground truth and input) for network training. Specifically, for single-coil data, we directly apply iFFT to obtain 2D image slices; for mult-coil data, we first apply RSS [6] to aggregate k-space data from multiple coils and then apply iFFT to obtain 2D image slices. In other words, for a multi-coil input in shape $[t, s, c, h, w]$ (time-frame, slice, coil, height, width) or a single-coil input in shape $[t, s, h, w]$, the preprocessed data will be $t \times s$ 2D image slices in shape $[h, w]$.

We observe that blur noise is introduced by missing frequency information blocked by the striped mask. Hence, for padding these slices to a fixed shape, we choose to add zero padding to the k-space instead of the image space to keep the purify of the source of blur noise because padding zeros in the k-space will not bring new information from the frequency perspective while padding zeros in the image space will introduce unnecessary bias.

2.3 Framework — DiffCMR

We briefly introduce the formulation of the diffusion model proposed in [5,14]. Diffusion models are generative models based on parametrized Markov chain and are comprised of forward and backward processes, as shown in Fig. 1. The forward process iteratively transforms a clean image x_0 into a series of noisier images $\{x_1, x_2, ..., x_T\}$. The following formulation can express the iteration of the forward process:

$$q(x_t|x_{t-1}) = \mathcal{N}(x_t; \sqrt{1 - \beta_t}x_{t-1}, \beta_t I_{n \times n}), \tag{1}$$

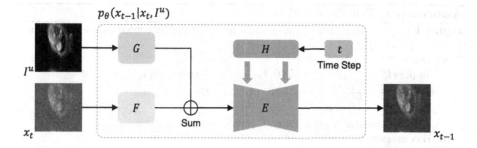

Fig. 2. The figure above illustrates the pipeline of our proposed DiffCMR. F and G encode the noisy signal x_t and the under-sampled image I^u, respectively. H encodes timestamp t to obtain timestamp embeddings. E is a modified U-net that receives both summed features and timestamp embeddings for denoising.

where β_t is a constant to define the ratio of adding Gaussian noise. The reverse process is parametrized by θ and can be simplified as:

$$p_\theta(x_{t-1}|x_t) = \mathcal{N}(x_{t-1}; \epsilon_\theta(x_t, t), \sigma_t^2 I_{n \times n}), \tag{2}$$

where σ_t is a fixed variance, and ϵ_θ is a trained step estimation function.

Conditional generations with diffusion models [1,2,11] are formulated by letting backward process p_θ simultaneously manipulate the current step noisy image x_t and the given condition, which allows the generation of images based on extra conditions without any additional learning. Here we merge the information of the under-sampled image and the current step denoising result by adding the extracted features from their corresponding encoders.

Our proposed DiffCMR, as illustrated in Fig. 2, employs a conditional diffusion model that conditions its step estimation function ϵ_θ on the aggregated information from both the input under-sampled image slice I^u and the current step recovery x_t. In our architecture, the estimation function ϵ_θ is a modified U-Net [12] and can be further expressed as:

$$\epsilon_\theta(x_t, I^u, t) = E(F(x_t) + G(I^u), H(t)), \tag{3}$$

where H encodes the current time step t into timestamp embedding; G and F encode the under-sampled input slice I^u and the current step denoising result x_t, respectively. E is a modified U-Net encoder-decoder structure that receives summed features from F and G, and estimates the noise for the current step. The current time step t is embedded using a learned look-up table H and inserted into layers of both the encoder and the decoder of network E by summation.

2.4   Training and Inference Procedure

The training process for DiffCMR is demonstrated in Algorithm 1. For each step, we sample a random data pair (I^u, I^f) from the dataset D_k, a timestamp

---

Algorithm 1. Training Algorithm

Input: total denoising steps T, under-sampled and fully-sampled image pair dataset
$D_k = \{(I_i^u, I_i^f)\}_{i=1}^{N_k}$
  repeat
    Sample $(I_i^u, I_i^f) \sim D_k$, $\epsilon \sim \mathcal{N}(\mathbf{0}, \mathbf{I}_{n \times n})$, $t \sim \mathbf{Uniform}(\{\mathbf{1}, ..., \mathbf{T}\})$
    $\beta_t = \frac{10^{-4}(T-t)+2\times10^{-2}(t-1)}{T-1}$, $\alpha_t = 1 - \beta_t$, $\overline{\alpha}_t = \prod_{s=0}^t \alpha_s$
    $x_t = \sqrt{\overline{\alpha}_t} I_i^f + \sqrt{1-\overline{\alpha}_t}\epsilon$
    Compute gradient $\nabla_\theta \|\epsilon - \epsilon_\theta(x_t, I_i^u, t)\|$
  until convergence

Algorithm 2. Inference Algorithm

Input: total denoising steps T, under-sampled image I^u, ensemble rounds R
  for $r = 1, 2, ..., R$ **do**
    Sample $x_{T,r} \sim \mathcal{N}(\mathbf{0}, \mathbf{I}_{n \times n})$
    for $t = T, T-1, ..., 1$ **do**
      Sample $z \sim N(\mathbf{0}, \mathbf{I}_{n \times n})$
      $\beta_t = \frac{10^{-4}(T-t)+2\times10^{-2}(t-1)}{T-1}$
      $\alpha_t = 1 - \beta_t$, $\overline{\alpha}_t = \prod_{s=0}^t \alpha_s$, $\widetilde{\beta}_t = \frac{1-\overline{\alpha}_{t-1}}{1-\overline{\alpha}_t}\beta_t$
      $x_{t-1,r} = \alpha_t^{\frac{1}{2}}(x_t - \frac{1-\alpha_t}{\sqrt{1-\overline{\alpha}_t}}\epsilon_\theta(x_t, I^u, t)) + \mathbb{1}_{[t>1]}\widetilde{\beta}_t^{\frac{1}{2}} z$
  return $\sum_{r=1}^R x_{0,r}/R$

$t \in [1, T]$ from a Uniform distribution, and a noise ϵ from a standard Gaussian distribution. We then obtain the current step recovery x_t by reparametrizing Eq. 1:

$$\alpha_t = 1 - \beta_t, \; \overline{\alpha}_t = \prod_{s=0}^t \alpha_s, \; x_t = \sqrt{\overline{\alpha}_t} I^f + \sqrt{1-\overline{\alpha}_t}\epsilon. \tag{4}$$

Then according to our pipeline in Fig. 2, x_t, t, and I^u are sent through the networks F, H, G, and E to obtain $\epsilon_\theta(x_t, I_i^u, t)$. Our training target is to minimize the term:

$$\|\epsilon - \epsilon_\theta(\sqrt{\overline{\alpha}_t} I^f + \sqrt{1-\overline{\alpha}_t}\epsilon, I_i^u, t)\|. \tag{5}$$

Our inference process is described in Algorithm 2. As the procedure to recover x_{t-1} includes an addition with a random sampled $z \sim \mathcal{N}(\mathbf{0}, \mathbf{I}_{n \times n})$, which yields discrete random noise points in the denoising results. We adopt a multi-round ensembling inference strategy to diminish the noise points and stabilize the denoising results. The inference procedure on the same input is carried out for multiple rounds, and the final result is ensembled by taking the average. The effectiveness of this strategy is proved by the experiment results from the ablation study, see Table 4 (b).

Table 1. Training and Validation pairs for both tasks and acceleration factors (Acc-Factor) 4, 8 and 10.

	AccFactor04		AccFactor08		AccFactor10	
	Train	Valid	Train	Valid	Train	Valid
Task 1	14304	1272	14304	1272	14304	1272
Task 2	32904	2904	32904	2904	32904	2904

3 Experiments

3.1 Dataset

The CMRxRecon dataset is released in the MICCAI 2023 CMRxRecon Challenge, comprises cine reconstruction and T1/T2 mapping tasks. Both tasks have two coil types, each with 120 cases for training and 60 cases for validation. The training cases provide fully sampled k-space data and under-sampled k-space data with acceleration factors (AccFactor) 4, 8, and 10. The validation cases only provide under-sampled k-space data with acceleration factors 4, 8, and 10.

As described in Sect. 2.2, we first preprocess the raw data by zero-padding the k-space to size 512×512, and transforming it to 2D images with iFFT. Finally, we resize the image slices to 128×128 to speed up the experiment process. Our local training-validation split is set by assigning images extracted from case P001 to P110 to the training set and those from case P111 to P120 to the validation set. We randomly shuffle the validation set and select the first 240 samples for inference. The detailed numbers of our preprocessed samples for training and validation are listed in Table 1.

3.2 Implementation Details

As for model architectures, we follow the conventions in [1] to build our model. The network G has 10 Residual in Residual Dense Blocks [16] and a depth of six. The number of channels was set to $[C, C, 2C, 2C, 4C, 4C]$ with $C = 128$. The augmentation schemes include horizontal and vertical flips with a probability of 0.5. The training process took place with a batch size of 6 images at size 128×128. We used AdamW [7] optimizer for all experiments. We used 100 diffusion steps for training and 1000 for inference. As there are two different tasks together with 3 acceleration factors, we trained the network 6 times to obtain 6 different sets of weight. The whole training and inference process is carried out on a single NVIDIA GeForce RTX 3090 GPU.

3.3 Results

As the online validation platform limits the daily attempts to 3 trials per task, we perform the validation and report the results on our local split dataset to speed up the upgrading procedure of our method. For the fairness of comparison,

Table 2. Experiment results for Task 1 - Cine Reconstruction

	AccFactor04			AccFactor08			AccFactor10		
	PSNR↑	SSIM↑	NMSE↓	PSNR↑	SSIM↑	NMSE↓	PSNR↑	SSIM↑	NMSE↓
RAW	28.79	0.8150	0.2187	27.94	0.8082	0.2781	27.75	0.8113	0.2883
U-Net [12]	33.11	0.9212	0.0673	33.31	**0.9460**	0.0646	32.71	**0.9391**	0.0740
cGAN [3]	33.85	**0.9435**	0.0573	33.14	0.9449	0.0669	32.56	0.9209	0.0760
DiffCMR	**36.10**	0.9277	**0.0346**	**34.85**	0.9061	**0.0457**	**34.47**	0.9016	**0.0493**

Table 3. Experiment results for Task 2 - T1/T2 Mapping

	AccFactor04			AccFactor08			AccFactor10		
	PSNR↑	SSIM↑	NMSE↓	PSNR↑	SSIM↑	NMSE↓	PSNR↑	SSIM↑	NMSE↓
RAW	28.17	0.8167	0.2003	27.15	0.8041	0.2578	27.17	0.8100	0.2711
U-Net [12]	32.06	0.9340	0.0537	31.05	**0.9286**	0.0695	29.49	0.9106	0.0987
cGAN [3]	32.67	**0.9460**	0.0488	30.65	0.8899	0.0805	31.38	**0.9304**	0.0655
DiffCMR	**34.60**	0.9071	**0.0372**	**33.17**	0.8937	**0.0537**	**33.04**	0.8941	**0.0536**

all the experiments are trained and validated with the same split, input resolution, and the same data augmentation scheme. The evaluation metrics for our experiments are peak signal-to-noise ratio (PSNR), structural similarity index measure (SSIM), and normalized mean square error (NMSE).

We compare our proposed DiffCMR with U-Net [12] and cGAN [3]. Qualitative visualization comparisons are shown in Fig. 3. Quantitative results for Task1 and Task2 are listed in Table 2 and Table 3, where RAW means the direct comparison results between the input under-sampled image slices and the fully-sampled image slices. As can be seen, our method outperforms both baseline methods across most tasks and acceleration factors.

4   Ablation Study

We evaluate two alternatives of hyper-parameters in our DiffCMR method at the inference stage. The first variant determines the number of inference diffusion steps. The second variant determines the number of inference ensembling rounds. These variant experiments were carried out on the data from Task 1 with an acceleration factor equal to 4.

Varying the Number of Inference Diffusion Steps T. In this part, we set the ensembling rounds $R = 4$ and explore the effect of inference diffusion steps on the denoising quality. Quantitative results are shown in Table 4(a). As can be observed, the denoising performance is positively correlated with T and starts to outperform the raw input at around $T = 100$. We choose $T = 1000$ in all other experiments.

Varying the Number of Inference Ensembling Rounds R. In this part, we set the diffusion steps $T = 1000$ and explore the effect of inference ensembling

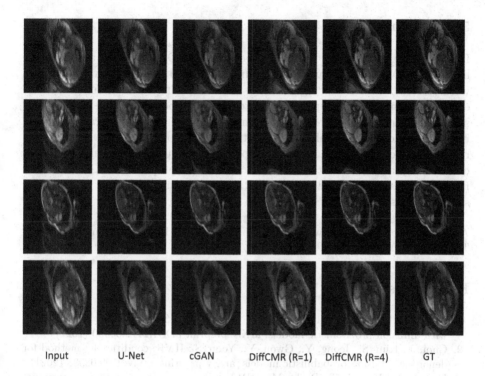

Input   U-Net   cGAN   DiffCMR (R=1)   DiffCMR (R=4)   GT

Fig. 3. Qualitative comparisons between raw input, results of U-Net [12], results of cGAN [3], results of DiffCMR with single-round ensembling inference ($R = 1$), results of DiffCMR with multi-round ensembling inference ($R = 4$), and ground truth.

Table 4. Ablation study on Task 1-AccFactor04. (a) Results with different inference diffusion steps. (b) Results with different inference ensembling rounds.

#Steps T	PSNR↑	SSIM↑	NMSE↓	#Rounds R	PSNR↑	SSIM↑	NMSE↓
Raw	28.79	0.8150	0.2187	Raw	28.79	0.8150	0.2187
20	25.52	0.3664	0.3830	1	33.89	0.8491	0.0561
100	29.86	0.6006	0.1393	2	35.25	0.8994	0.0417
500	35.39	0.9010	0.0399	4	36.10	0.9277	0.0346
1000	**36.10**	**0.9277**	**0.0346**	8	**36.68**	**0.9430**	**0.0305**

(a)   (b)

rounds on the denoising quality. Quantitative results are shown in Table 4(b), and Qualitative comparisons between $R = 1$ and $R = 4$ are visualized in Fig. 3. The results show a positive correlation between denoising effectiveness and R with a diminishing marginal effect when R is large. Therefore, we set $R = 4$ in all other experiments for a reasonable performance to runtime tradeoff.

5 Conclusion

In this paper, we present DiffCMR, a conditional DDPM-based approach for high-quality MRI reconstruction from under-sampled k-space data. Our framework receives conditioning signals from the under-sampled MRI image slice at each denoising diffusion step and generates the corresponding fully-sampled MRI image slice. In addition, we adopt the multi-round ensembling strategy during inference which largely enhances the stableness of our approach. Experiment results show our DiffCMR method outperforms the existing popular denoisers qualitatively and quantitatively. In conclusion, our proposed DiffCMR offers a novel perspective for handling fast MRI reconstruction problems and demonstrates impressive robustness.

Acknowledgement. This work was supported by the Hong Kong Innovation and Technology Fund under Projects PRP/041/22FX.

References

1. Amit, T., Shaharbany, T., Nachmani, E., Wolf, L.: Segdiff: image segmentation with diffusion probabilistic models. arXiv preprint arXiv:2112.00390 (2021)
2. Choi, J., Kim, S., Jeong, Y., Gwon, Y., Yoon, S.: ILVR: conditioning method for denoising diffusion probabilistic models. arXiv preprint arXiv:2108.02938 (2021)
3. Defazio, A., Murrell, T., Recht, M.: MRI banding removal via adversarial training. In: Larochelle, H., Ranzato, M., Hadsell, R., Balcan, M., Lin, H. (eds.) Advances in Neural Information Processing Systems, vol. 33, pp. 7660–7670. Curran Associates, Inc. (2020)
4. Dhariwal, P., Nichol, A.: Diffusion models beat GANs on image synthesis. Adv. Neural. Inf. Process. Syst. **34**, 8780–8794 (2021)
5. Ho, J., Jain, A., Abbeel, P.: Denoising diffusion probabilistic models. Adv. Neural. Inf. Process. Syst. **33**, 6840–6851 (2020)
6. Larsson, E.G., Erdogmus, D., Yan, R., Principe, J.C., Fitzsimmons, J.R.: SNR-optimality of sum-of-squares reconstruction for phased-array magnetic resonance imaging. J. Magn. Reson. **163**(1), 121–123 (2003)
7. Loshchilov, I., Hutter, F.: Decoupled weight decay regularization. arXiv preprint arXiv:1711.05101 (2017)
8. Lugmayr, A., Danelljan, M., Romero, A., Yu, F., Timofte, R., Van Gool, L.: Repaint: inpainting using denoising diffusion probabilistic models. In: Proceedings of the IEEE/CVF Conference on Computer Vision and Pattern Recognition, pp. 11461–11471 (2022)
9. Lustig, M., Donoho, D., Pauly, J.M.: Sparse MRI: the application of compressed sensing for rapid MR imaging. Magn. Reson. Med. Off. J. Int. Soc. Magn. Reson. Med. **58**(6), 1182–1195 (2007)
10. Muckley, M.J., et al.: Training a neural network for Gibbs and noise removal in diffusion MRI. Magn. Reson. Med. **85**(1), 413–428 (2021)
11. Nichol, A.Q., Dhariwal, P.: Improved denoising diffusion probabilistic models. In: International Conference on Machine Learning, pp. 8162–8171. PMLR (2021)

12. Ronneberger, O., Fischer, P., Brox, T.: U-Net: convolutional networks for biomedical image segmentation. In: Navab, N., Hornegger, J., Wells, W.M., Frangi, A.F. (eds.) MICCAI 2015. LNCS, vol. 9351, pp. 234–241. Springer, Cham (2015). https://doi.org/10.1007/978-3-319-24574-4_28

13. Schlemper, J., Caballero, J., Hajnal, J.V., Price, A.N., Rueckert, D.: A deep cascade of convolutional neural networks for dynamic MR image reconstruction. IEEE Trans. Med. Imaging **37**(2), 491–503 (2017)

14. Song, J., Meng, C., Ermon, S.: Denoising diffusion implicit models. arXiv preprint arXiv:2010.02502 (2020)

15. Sriram, A., et al.: End-to-end variational networks for accelerated MRI reconstruction. In: Martel, A.L., et al. (eds.) MICCAI 2020. LNCS, vol. 12262, pp. 64–73. Springer, Cham (2020). https://doi.org/10.1007/978-3-030-59713-9_7

16. Wang, X., et al.: Esrgan: enhanced super-resolution generative adversarial networks. In: Proceedings of the European Conference on Computer Vision (ECCV) Workshops (2018)

17. Wu, J., Fang, H., Zhang, Y., Yang, Y., Xu, Y.: Medsegdiff: medical image segmentation with diffusion probabilistic model. arXiv preprint arXiv:2211.00611 (2022)

18. Xiang, T., Yurt, M., Syed, A.B., Setsompop, K., Chaudhari, A.: DDM^2: self-supervised diffusion MRI denoising with generative diffusion models (2023)

19. Xie, S., Zhang, Z., Lin, Z., Hinz, T., Zhang, K.: Smartbrush: text and shape guided object inpainting with diffusion model. In: Proceedings of the IEEE/CVF Conference on Computer Vision and Pattern Recognition, pp. 22428–22437 (2023)

20. Xie, Y., Yuan, M., Dong, B., Li, Q.: Diffusion model for generative image denoising. arXiv preprint arXiv:2302.02398 (2023)

21. Zhu, Y., et al.: Denoising diffusion models for plug-and-play image restoration. In: Proceedings of the IEEE/CVF Conference on Computer Vision and Pattern Recognition, pp. 1219–1229 (2023)

C³-Net: Complex-Valued Cascading Cross-Domain Convolutional Neural Network for Reconstructing Undersampled CMR Images

Quan Dou$^{(\boxtimes)}$, Kang Yan, Sheng Chen, Zhixing Wang, Xue Feng, and Craig H. Meyer

University of Virginia, Charlottesville, VA 22901, USA
{qd9zb,cmeyer}@virginia.edu

Abstract. Cardiac magnetic resonance (CMR) plays an important role in clinically assessing cardiovascular diseases. However, CMR is inherently slow leading to patient discomfort and degraded image quality. Compared with parallel imaging (PI) and compressed sensing (CS), deep-learning-based methods have demonstrated superior image reconstruction performance, in terms of image quality and substantially reduced reconstruction times from highly undersampled CMR data. In this work, we proposed a novel complex-valued cascading cross-domain convolutional neural network, dubbed "C³-Net", for improved image quality for accelerated CMR. C³-Net outperformed L1-ESPIRiT reconstruction, a baseline U-Net, and a real-valued cascading cross-domain CNN, especially with high acceleration factors (>8). The short-axis results from C³-Net showed reduced residual artifacts and improved temporal fidelity of cardiac motion. In long-axis results, C³-Net excelled in mitigating artifacts surrounding the heart wall and adipose regions in 2-chamber and 4-chamber views, while in a 3-chamber view all the listed methods resulted in suboptimal performance compared to the reference. The quantitative assessment indicated results consistent with assessment of the reconstructed images.

Keywords: cardiac MRI · deep learning · image reconstruction · complex-valued network · cascading network

1 Introduction

Cardiac magnetic resonance (CMR) plays an important role in clinically assessing cardiovascular disease, with advantages including non-invasiveness, excellent soft-tissue contrast, and lack of ionizing radiation [1]. CMR has been recommended by the European Society of Cardiology (ESC) as the gold standard for the left ventricle ejection fraction measurement [2] and is also widely used in detecting diffuse myocardial fibrosis, intracardiac shunts, and other cardiac malfunctions [3].

However, CMR is inherently slow due to the complex workflow (e.g., breath-holding) and slow imaging speed (e.g., retrospective ECG-gating) [4]. The prolonged scan time

---

Q. Dou, K. Yan, S. Chen, and Z. Wang—Contributed equally to this work.

© The Author(s), under exclusive license to Springer Nature Switzerland AG 2024
O. Camara et al. (Eds.): STACOM 2023, LNCS 14507, pp. 390–399, 2024.
https://doi.org/10.1007/978-3-031-52448-6_37

leads to patient discomfort and degrades image quality, which adversely impacts the accuracy in clinical diagnosis.

Parallel imaging (PI) [5, 6] and compressed sensing (CS) [7] are two commonly used methods to accelerate CMR acquisition by exploiting the data redundancy in multi-coil acquisition and in certain sparse domains (e.g., total variation), respectively. Moreover, the combination of PI and CS can be used to further increase the acquisition speed by nearly two-fold over either PI or CS individually. To date, these conventional methods have been investigated for several applications, demonstrating the ability to achieve good image quality with high acceleration rates. Nonetheless, advanced versions of these methods often require off-line image reconstruction taking several minutes, thus limiting clinical practice for low latency real-time imaging.

In recent years, deep-learning-based methods have been studied extensively in MR image reconstruction. In contrast to the PI and CS methods, deep-learning-based approaches have demonstrated superior reconstruction performance, in terms of image quality and significantly reduced reconstruction times from highly undersampled CMR data [3]. For example, a combination of deep-learning and L1-ESPIRiT reconstruction that handles complex-valued data as separate channels was proposed and evaluated on retrospective 12× undersampled datasets [8]; However, the inherent interrelationship between the components of the complex k-space values has not been fully exploited. Thus, an improved complex-valued cascading network architecture has been introduced, and the architectural paradigm exhibited better performance compared to CS-based approaches, as demonstrated on prospective undersampled datasets [9].

In this work, we propose a novel complex-valued, cascading, cross-domain convolutional neural network, abbreviated as C^3-Net, for improved image quality of accelerated CMR. The proposed network was first trained with fully- and undersampled datasets and then compared to the results from other networks and against the fully-sampled images as the reference.

2 Methods

2.1 Network Architecture

Let $\mathbf{x}$ represent a 2D complex-valued cardiac CINE slice. Our purpose is to reconstruct $\mathbf{x}$ from the multi-coil undersampled k-space $\mathbf{y}_u$, such that:

$$\mathbf{y}_u = \mathbf{uFSx} \tag{1}$$

where $\mathbf{u}$ is the binary k-space undersampling mask, $\mathbf{F}$ is the Fourier transform (FT), and $\mathbf{S}$ is the sensitivity maps, which can be derived from the fully sampled calibration region through ESPIRiT [10].

When dealing with data acquired with an undersampling ratio beyond the capability of the coil hardware, Eq. (1) is underdetermined. To tackle the ill-posed inversion problem, we propose a cascading convolutional neural network (CNN), which resembles a classical iterative algorithm with a fixed number of iterations. The proposed C^3-Net alternates between the restoration step and the data consistency (DC) step, as shown in Fig. 1. The restoration part consists of two major components: a k-space subnetwork (K-Net) and an image subnetwork (I-Net). At the i-th iteration, the K-Net generates an estimation of the true k-space:

$$\mathbf{y}_{\text{knet}} = f_{\text{knet}}(\mathbf{y}_{\text{in}}|\boldsymbol{\theta}_{\text{knet}}) \tag{2}$$

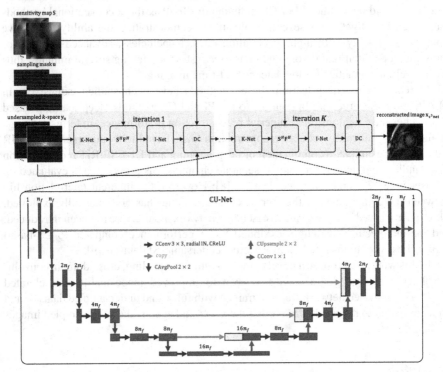

Fig. 1. Illustration of the proposed framework for undersampled CMR image reconstruction.

where f_{knet} is the forward mapping of the K-Net parameterized by $\boldsymbol{\theta}_{\text{knet}}$, and $\mathbf{y}_{\text{in}}$ is the input k-space, of which $\mathbf{y}_{\text{u}}$ is at the first iteration. Then, the output k-space $\mathbf{y}_{\text{knet}}$ is transformed into the image domain and combined using the sensitivity maps $\mathbf{x}_{\text{in}} = \mathbf{S}^H \mathbf{F}^H \mathbf{y}_{\text{knet}}$, where H represents the conjugate transpose. The I-Net generates an estimation of the true image:

$$\mathbf{x}_{\text{inet}} = f_{\text{inet}}(\mathbf{x}_{\text{in}}|\boldsymbol{\theta}_{\text{inet}}) \tag{3}$$

where f_{inet} is the forward mapping of the I-Net parameterized by $\boldsymbol{\theta}_{\text{inet}}$. To incorporate the data consistency, for k-space entries that are initially missing, we use the predicted values from the I-Net; for the k-space entries that are initially sampled, we simply replace the predicted values with the original values:

$$\mathbf{y}_{\text{dc}} = \begin{cases} \mathbf{F}\mathbf{S}\mathbf{x}_{\text{inet}}, & \mathbf{u}_i = 0 \\ \mathbf{y}_{\text{u}}, & \mathbf{u}_i = 1 \end{cases} \tag{4}$$

Both the K-Net and I-Net use a U-Net [11] as the backbone network structure. The complex U-Net (CU-Net) consists of a series of encoding and decoding blocks. Each encoding or decoding block consists of two complex-valued convolutional layers (CConv) with kernel size 3 × 3 [12]. Each convolutional layer is followed by a radial instance normalization (IN) [13] and a complex-valued rectified linear unit (CReLU) [12]. The radial IN scales the magnitude while maintains the phase information, and the CReLU function activates the real and imaginary parts separately. A 2 × 2 average pooling is applied at the end of each encoding block, while a 2 × 2 upsampling is performed at the start of each decoding block. The skip connections between the encoding block and the corresponding decoding block are important to expedite training and avoid vanishing gradients. The final output is generated using a CConv with kernel size 1 × 1.

2.2 Training

The training, validation, and testing datasets used in this work were built from the CMRxRecon challenge dataset (https://cmrxrecon.github.io/). The raw CMRxRecon training dataset includes 120 fully sampled multi-coil CINE MRIs obtained on 3T scanners [14]. Detailed descriptions can be found on the project website. We randomly divided the dataset into training (90 subjects), validation (10 subjects), and testing (20 subjects) subsets. For training, a total of 15192 2D slices were generated by splitting up each 4D sample (matrix size in readout direction, matrix size in phase encoding direction, slice number, time frame). Random flipping along readout and phase encoding directions was employed as training augmentation to further expand the size of training dataset. During training, the undersampling ratio R was randomly selected between 4 to 12 and the equispaced undersampling mask was generated on-the-fly for each 2D slice. The central 24 phase encoding lines were always fully sampled to be used as the autocalibration signal (ACS) region. The sensitivity maps were pre-computed from the time-averaged ACS using ESPIRiT.

The C³-Net was implemented in the open-source machine learning library PyTorch [15]. All subnetworks in the reconstruction pipeline were jointly trained in an end-to-end manner using a mixed L1 and structural similarity index (SSIM) loss [16]:

$$\mathcal{L} = \frac{1}{N} \sum_{i=1}^{N} \left| \mathbf{x}_{c^3\text{net}} - \mathbf{x}_{\text{ref}} \right| + \alpha \left(1 - \text{SSIM}\left(\left| \mathbf{x}_{c^3\text{net}} \right|, \left| \mathbf{x}_{\text{ref}} \right| \right) \right) \tag{5}$$

where $\mathbf{x}_{c^3\text{net}}$ is the complex-valued output of C³-Net, $\mathbf{x}_{\text{ref}}$ is the complex-valued reference image reconstructed from the fully sampled k-space data, N is the total number of training pairs, and α is the weight parameter (empirically set to 1). Training was carried out with an Adam optimizer [17] for 50 epochs with a learning rate of 0.0001, $\beta_1 = 0.9$, $\beta_2 = 0.999$, and $\varepsilon = 10^{-8}$. Due to the limit of GPU memory, the number of filters n_f of the first CConv is set to 8, the number of iterations K is set to 2, and the training batch size was set to 1 in this work. The validation loss was monitored after each epoch to avoid overfitting. Other hyperparameters were optimized through grid search to achieve the best validation loss.

2.3   Evaluation

To evaluate C^3-Net performance compared to other deep learning methods, a baseline U-Net and a real-valued cascading cross-domain CNN, abbreviated as C^2-Net, were trained with the same training setup. The proposed C^3-Net was also compared to the L1-ESPIRiT reconstruction using the code provided by the CMRxRecon organizers. For quantitative assessment, the peak signal-to-noise ratio (PSNR), SSIM, and normalized mean squared error (NMSE) of the magnitudes were calculated.

3   Results

The comparison results of all methods are reported in Table 1 using quantitative metrics (PSNR, SSIM, and NMSE) across different acceleration ratios ($R = 4$, 8, and 10). The proposed C^3-Net yields better image quality than L1-ESPIRiT, U-Net, and C^2-Net in most cases, especially at higher acceleration ratios, as indicated by the bold fonts which show the best reconstruction performance.

The visualization results of the one short-axis and three long-axis of cardiac MR images are depicted in Fig. 2 and Fig. 3, respectively. In Fig. 2, L1-ESPIRiT shows good image quality at $R = 4$ with fewer artifacts, because the acceleration ratio is far smaller than the number of coils used for data acquisition. As the acceleration ratio increases (e.g., 8 and 10), images reconstructed from L1-ESPIRiT degrade, with noticeable residual aliasing artifacts. In contrast, the proposed method presents the best results in both spatial and temporal dimensions, as can be clearly seen from the difference images with much reduced aliasing artifacts and from x-t profiles where the cardiac motion is well preserved in the temporal domain. Figure 3 provides examples of reconstruction results from 2-, 3-, and 4-chamber views at an acceleration ratio of 10. Our model excels, particularly in 2- and 4- chamber datasets, where the artifacts surrounding the heart wall and adipose regions are notably mitigated. The factors underpinning the suboptimal performance observed in 3-chamber images for all deep-learning-based methods will be discussed in the following section.

Table 1. Quantitative assessment of reconstruction performance on the CMRxRecon dataset

SAX ($R = 4$)	Zero-Filling	L1-ESPIRiT	U-Net	C^2-Net	C^3-Net
PSNR [dB]	29.77	38.52	36.28	37.22	**38.53**
SSIM	0.859	0.937	0.957	0.967	**0.972**
NMSE ($\times 10^2$)	8.411	1.407	1.872	1.876	**1.636**
SAX ($R = 8$)	Zero-Filling	L1- ESPIRiT	U-Net	C^2-Net	C^3-Net
PSNR [dB]	29.05	32.59	34.88	35.12	**36.63**
SSIM	0.845	0.849	0.941	0.946	**0.956**

(continued)

Table 1. (*continued*)

NMSE	9.203	3.050	1.996	2.145	**1.729**
SAX ($R = 10$)	Zero-Filling	L1- ESPIRiT	U-Net	C^2-Net	C^3-Net
PSNR [dB]	28.89	31.75	34.49	34.41	**36.19**
SSIM	0.843	0.837	0.937	0.938	**0.951**
NMSE	9.727	3.941	1.969	2.518	**1.661**
LAX ($R = 4$)	Zero-Filling	L1- ESPIRiT	U-Net	C^2-Net	C^3-Net
PSNR [dB]	29.76	**41.55**	34.51	34.66	36.67
SSIM	0.821	**0.949**	0.930	0.896	0.927
NMSE	8.393	**1.514**	4.876	8.205	6.194
LAX ($R = 8$)	Zero-Filling	L1- ESPIRiT	U-Net	C^2-Net	C^3-Net
PSNR [dB]	29.08	32.70	33.09	32.74	**33.12**
SSIM	0.818	0.838	0.920	0.881	**0.924**
NMSE	8.590	**3.574**	4.058	7.706	4.043
LAX ($R = 10$)	Zero-Filling	L1- ESPIRiT	U-Net	C^2-Net	C^3-Net
PSNR [dB]	28.84	31.70	32.60	32.31	**32.39**
SSIM	0.815	0.824	0.915	0.878	**0.919**
NMSE	8.806	4.667	4.135	7.667	**4.273**

4 Discussion

In this study, we designed and implemented a cascading network that operates in both the k-space domain and the image domain. The K-Net estimates the missing k-space values utilizing the information from neighboring sampled points. The I-Net reduces the residual aliasing artifacts and further improves the image quality. The DC layer ensures that the already acquired k-space samples remain unchanged and enhances the output image fidelity. Compared to the U-Net acting in the image domain without the DC layer, the cross-domain cascading networks (C^2-Net and C^3-Net) showed superior performance on the testing dataset and were less prone to generate unrealistic small structures on the output images. Also, all operations used in C^3-Net were complex-valued, enabling the network to fully exploit the complex-valued input data [12]. Compared to C^2-Net which performs operations on the real and imaginary channels separately, C^3-Net achieved better metrics and generated output with less residual artifacts.

For the DC layer, we employed a simple but effective operation in our study: replacing the predicted values with the original values for sampled k-space points. However, if non-negligible noise exists in the acquisition, this method may fail, as shown in the 3-chamber view in Fig. 3. Utilizing a linear combination of the predicted values and the original values weighted by a fixed or trainable parameter [18, 19] is expected to improve the network performance on noisy data.

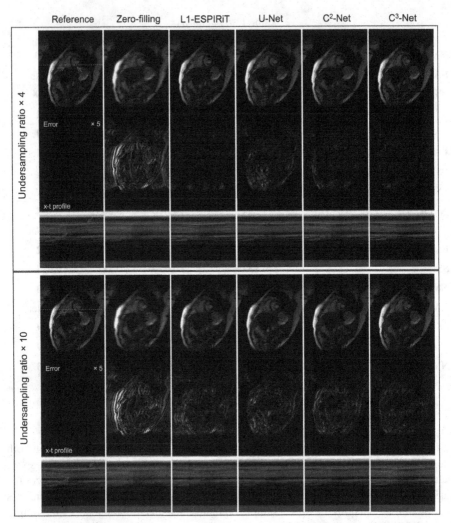

Fig. 2. Comparison of one short -axis slice (end-diastole frame) generated using five reconstruction methods, from datasets retrospectively undersampled with accelerated ratios of 4 and 10. From left to right are images reconstructed from fully-sampled k-space as the reference, zero-filling k-space, L1-ESPIRiT, U-Net, C2-Net and C3-Net, respectively. From top to bottom are magnitude image (top), absolute difference image relative to the fully sampled reference (middle), and x-t plots below the white lines (bottom). The error maps are windowed by scaling the image intensity by a factor of 5. The red arrows point to structures that show fine papillary muscles in ventricles and the blue arrows indicate preserved temporal fidelity. (Color figure online)

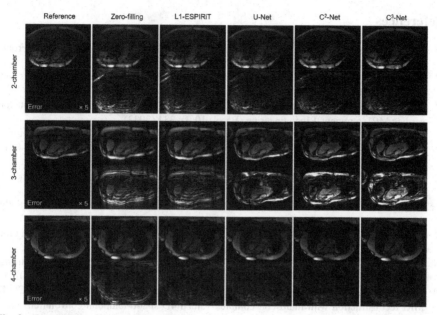

Fig. 3. Comparison of long-axis cardiac images from 2-chamber (top), 3-chamber (middle), and 4-chamber views (bottom), reconstructed using fully-sampled k-space as the reference, zero-filling k-space, L1-ESPIRiT, U-Net, C²-Net, and C³-Net, respectively. Absolute difference images relative to the fully sampled reference are shown for performance comparison, as well. The error maps are windowed by scaling the image intensity by a factor of 5.

The sampling pattern may play an important role in the performance of image reconstruction. Although in this work the same sampling mask was used for retrospective undersampled datasets, using a variable sampling pattern, such as Cartesian Poisson sampling, could potentially improve the de-aliasing performance when using CS- or DL-based approaches. Also, non-Cartesian sampling patterns, such as radial or spiral sampling, have been demonstrated to provide advantages of shorter scan time and higher motion insensitivity [20, 21] than Cartesian acquisitions, which are critical for dynamic imaging. An approach to apply the proposed C³-Net to non-Cartesian scenarios involves pre-gridding the radial or spiral k-space data onto a Cartesian grid [22, 23] prior to inputting it into the K-Net. Subsequently, the training procedure remains consistent with that of the Cartesian case.

5   Conclusion

The proposed C³-Net integrates both the complex-value of MR data and the coupled domain information (k-space domain and image domain) in the CNN model, providing a significant improvement of image quality at high acceleration rates in comparison with the state-of-the-art methods (L1-ESPIRiT, U-Net, and C²-Net).

References

1. Ismail, T.F., Strugnell, W., Coletti, C., et al.: Cardiac MR: from theory to practice. Front. Cardiovasc. Med. **9**, 826283 (2022)
2. McDonagh, T.A., Metra, M., Adamo, M., et al.: 2021 ESC Guidelines for the diagnosis and treatment of acute and chronic heart failure. Eur. Heart J. **42**(36), 3599–3726 (2021)
3. Oscanoa, J.A., Middione, M.J., Alkan, C., et al.: Deep learning-based reconstruction for cardiac MRI: a review. Bioengineering (Basel) **10**(3), 334 (2023)
4. Wang, C., et al.: Recommendation for cardiac magnetic resonance imaging-based phenotypic study: imaging part. Phenomics **1**(4), 151–170 (2021). https://doi.org/10.1007/s43657-021-00018-x
5. Pruessmann, K.P., Weiger, M., Scheidegger, M.B., Boesiger, P.: SENSE: sensitivity encoding for fast MRI. Magn. Reson. Med. **42**(5), 952–962 (1999)
6. Griswold, M.A., Jakob, P.M., Heidemann, R.M., et al.: Generalized autocalibrating partially parallel acquisitions (GRAPPA). Magn. Reson. Med. **47**(6), 1202–1210 (2002)
7. Lustig, M., Donoho, D., Pauly, J.M.: Sparse MRI: the application of compressed sensing for rapid MR imaging. Magn. Reson. Med. **58**(6), 1182–1195 (2007)
8. Sandino, C.M., Lai, P., Vasanawala, S.S., Cheng, J.Y.: Accelerating cardiac cine MRI using a deep learning-based ESPIRiT reconstruction. Magn. Reson. Med. **85**(1), 152–167 (2021)
9. Küstner, T., Fuin, N., Hammernik, K., et al.: CINENet: deep learning-based 3D cardiac CINE MRI reconstruction with multi-coil complex-valued 4D spatio-temporal convolutions. Sci. Rep. **10**(1), 13710 (2020)
10. Uecker, M., Lai, P., Murphy, M.J., et al.: ESPIRiT—an eigenvalue approach to autocalibrating parallel MRI: where SENSE meets GRAPPA. Magn. Reson. Med. **71**(3), 990–1001 (2014)
11. Ronneberger, O., Fischer, P., Brox, T.: U-Net: convolutional networks for biomedical image segmentation. In: Navab, N., Hornegger, J., Wells, W., Frangi, A. (eds.) Medical Image Computing and Computer-Assisted Intervention – MICCAI 2015. MICCAI 2015. LNCS, vol. 9351, pp. 234–241. Springer, Cham (2015). https://doi.org/10.1007/978-3-319-24574-4_28
12. Trabelsi, C., Bilaniuk, O., Zhang, Y., et al.: Deep Complex Networks (2017). arXiv:1705.09792 [cs.NE]
13. Ulyanov, D., Vedaldi, A., Lempitsky, V.: Instance Normalization: The Missing Ingredient for Fast Stylization (2016). arXiv:1607.08022 [cs.CV]
14. Wang, C., Lyu, J., Wang, S., et al.: CMRxRecon: an open cardiac MRI dataset for the competition of accelerated image reconstruction (2023). arXiv:2309.10836 [cs.CV]
15. Paszke, A., Gross, S., Massa, F., et al.: Pytorch: an imperative style, high-performance deep learning library. Adv. Neural Inf. Process. Syst. **32** (2019)
16. Zhao, H., Gallo, O., Frosio, I., Kautz, J.: Loss functions for image restoration with neural networks. IEEE Trans. Comput. Imaging **3**(1), 47–57 (2017)
17. Kingma, D.P., Ba, J.: Adam: A Method for Stochastic Optimization (2014). arXiv:1412.6980 [cs.LG]
18. Schlemper, J., Caballero, J., Hajnal, J.V., Price, A.N., Rueckert, D.: A deep cascade of convolutional neural networks for dynamic MR image reconstruction. IEEE Trans. Med. Imaging **37**(2), 491–503 (2018)
19. Eo, T., Jun, Y., Kim, T., Jang, J., Lee, H.J., Hwang, D.: KIKI-net: cross-domain convolutional neural networks for reconstructing undersampled magnetic resonance images. Magn. Reson. Med. **80**(5), 2188–2201 (2018)
20. Meyer, C.H., Hu, B.S., Nishimura, D.G., Macovski, A.: Fast spiral coronary artery imaging. Magn. Reson. Med. **28**(2), 202–213 (1992)
21. Feng, L., Axel, L., Chandarana, H., Block, K.T., Sodickson, D.K., Otazo, R.: XD-GRASP: golden-angle radial MRI with reconstruction of extra motion-state dimensions using compressed sensing. Magn. Reson. Med. **75**(2), 775–788 (2016)

22. Jackson, J.I., Meyer, C.H., Nishimura, D.G., Macovski, A.: Selection of a convolution function for Fourier inversion using gridding. IEEE Trans. Med. Imaging **10**(3), 473–478 (1991)
23. Seiberlich, N., Breuer, F.A., Blaimer, M., Barkauskas, K., Jakob, P.M., Griswold, M.A.: Non-Cartesian data reconstruction using GRAPPA operator gridding (GROG). Magn. Reson. Med. **58**(6), 1257–1265 (2007)

Space-Time Deformable Attention Parallel Imaging Reconstruction for Highly Accelerated Cardiac MRI

Lifeng Mei[1], Kexin Yang[1], Yi Li[1], Shoujin Huang[1], Yilong Liu[2], and Mengye Lyu[1(✉)]

[1] College of Health Science and Environmental Engineering, Shenzhen Technology University, Shenzhen, China
lvmengye@sztu.edu.cn

[2] Guangdong-Hongkong-Macau Institute of CNS Regeneration, Key Laboratory of CNS Regeneration (Ministry of Education), Jinan University, Guangzhou, China

Abstract. Cardiac magnetic resonance imaging (MRI) provides excellent soft tissue contrast resolution and stands as a pivotal noninvasive modality for assessing cardiac structure and function. However, owing to the intricate balance between spatial and temporal resolution, the reconstruction of cardiac cine MRI sequences dedicated to the heart presents a more complex challenge compared to the swift reconstruction of general magnetic resonance images. While numerous deep learning techniques have emerged to MRI reconstruction, a majority of these endeavors have tended to overlook the dynamic nuances of cardiac motion. In response to this gap, we propose a Space-Time Deformable Attention Parallel Imaging Reconstruction (STDAPIR) framework. This approach is initially refined through the utilization of the Variational Network (VarNet), where the subsequently reconstructed high-frequency data serves as a means to attain enhanced precision in coil sensitivity map estimation. Then, we extend this framework through the integration of Nonlinear Activation Free Network (NAFNet), incorporating the Space-Time Deformable Attention (STDA) module to accommodate spatiotemporal considerations. By introducing these advancements, our methodology aims to elevate the quality of reconstructed images within the cardiac domain. Empirical findings gleaned from our experiments underscore the efficacy of our proposed method, revealing a notable enhancement in both precision and perceptual fidelity of the resulting reconstructed images.

Keywords: Cardiac MRI · Spatio-temporal Information · Parallel Imaging Reconstruction

1 Introduction

1.1 Background

Magnetic Resonance Imaging (MRI) is widely used in human body imaging and disease diagnosis, which makes it possible to perform non-invasive evaluation of

O. Camara et al. (Eds.): STACOM 2023, LNCS 14507, pp. 400–409, 2024.
https://doi.org/10.1007/978-3-031-52448-6_38

the heart. Cardiac Magnetic Resonance (CMR) imaging has emerged as a powerful diagnostic tool in cardiovascular medicine [1], providing detailed insights into the structure, function, and blood flow of the heart. CMR has played a crucial role in the assessment of various cardiac conditions. However, traditional CMR techniques often suffer from lengthy acquisition times, limiting their applicability in certain clinical scenarios, such as critically ill patients or individuals unable to tolerate prolonged scans.

1.2 Challenges and Advances

The demand for fast CMR imaging has grown as clinicians strive to balance the need for accurate diagnosis with patient comfort and clinical workflow efficiency. Long scan times can lead to motion artifacts, decreased patient compliance, and increased costs. However, the development of fast CMR techniques presents significant challenges due to the need for high spatial and temporal resolution, while maintaining image quality and diagnostic accuracy. Technological advancements in both hard-ware and software have enabled researchers to explore novel strategies for accelerating CMR acquisitions. In addition, the current mainstream research of cardiac magnetic resonance lacks a uniform and universal standard. At present, most of the mainstream datasets related to magnetic resonance imaging are static data sets, and most of the relevant researches are conducted on static images. However, video imaging itself is more challenging than static images.

1.3 Fast CMR Reconstruction Techniques Overview

As the field of fast cardiac magnetic resonance imaging continues to evolve, it offers exciting opportunities to overcome the limitations of traditional CMR techniques. One of the hottest topics is CMR image reconstruction from highly undersampled k-space (raw data) [10] in recent years. Parallel imaging [9] techniques, such as sensitivity encoding (SENSE) [11] and generalized auto-calibrating partially parallel acquisitions (GRAPPA) [4], have been pivotal in reducing scan times. These methods leverage multiple receiver coils to under-sampled k-space and subsequently reconstruct high-quality images. Compressed sensing (CS) [8] is another innovative approach that exploits the sparsity of MR images in certain domains, enabling the reconstruction of high-quality images from highly under-sampled data. CS has the potential to significantly shorten CMR scan times while preserving diagnostic information.

Recent advances in machine learning and deep learning [12] have shown promise in accelerating CMR imaging. Neural networks and deep learning architectures can learn complex relationships in the data, allowing for improved image reconstruction from under-sampled k-space data. Techniques such as convolutional neural networks (CNNs) [3] and generative adversarial networks (GANs) [16] have demonstrated the ability to reconstruct high-quality CMR images from limited data, enhancing the feasibility of rapid imaging. Furthermore, the application of Shifted Windows Transformer [5] in MRI reconstruction has also achieved certain results. Its combination with deformable attention can to some extent reduce the complexity of the network and enhance the interpretability of the reconstruction process.

2   Methods

2.1   Problem Formulation of MRI Reconstruction

MRI reconstruction can be viewed as an inverse problem [8]. Given the acquired under-sampled data, the goal is to find the underlying spatial image that best explains the data. However, this problem is ill-posed and often leads to solutions that amplify noise and artifacts. To mitigate this, regularization techniques are employed. Regularization adds constraints to the problem to encourage solutions that are both consistent with the data and follow certain properties, such as smoothness.

2.2   Overall Architecture

The proposed architecture is shown in Fig. 1. The framework is divided into two stages. The first stage is to train a frame-by-frame reconstruction of the Variational Network (VarNet). Taking advantage of multiple iterative expansions, the under-sampled fuzzy MRI image is firstly optimized. The second stage is based on the first stage. Based on the Space-Time Deformable Attention (STDA) module, the multi-frame information reconstruction model is constructed. Since the initial optimization of each frame has been completed in the first stage, we do not need too many iterative reconstructions in the second stage, which improves the computational efficiency of our model. Besides, coil compression is carried out with the Sensitivity Map Estimation (SME) module, and then it is sent to the Nonlinear Activation Free Network (NAFNet). It is worth noting that the STDA block is used here to make better use of spatiotemporal information. After the coil is expanded, the Data Consistency (DC) operation can be performed, and the final result is obtained. The full Model parameter number is 41.5MB. More details will be described in the following sections.

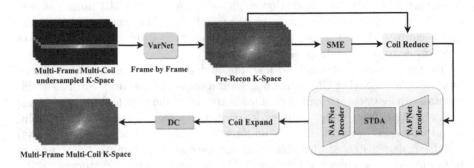

Fig. 1. An illustration of the proposed framework.

2.3   Variational Network

The Variational Network (VarNet) [13] is a deep learning architecture that can accelerate the reconstruction of MRI data. Its primary objective is to enhance the quality of MRI images while simultaneously reducing the time required for data acquisition.

In this work, we employ the VarNet model for optimization. The raw k-space data serves as the primary input to the network. Contrary to some implementations, we do not pre-compute coil sensitivity maps. Instead, within the VarNet's coil estimation module, the central low-frequency information from the k-space is fed into a UNet architecture to estimate the coil sensitivities. Subsequently, the measured raw data and the derived coil sensitivities are used for further processing within VarNet.

2.4   Sensitivity Map Estimation

Consistent with the approach in VarNet [13], we employ the Sensitivity Map Estimation (SME) module to estimate the sensitivity map. As depicted in Fig. 2, the SME module accepts undersampled k-space data as input. Following this, blurred images are produced by applying the inverse Fourier transform (IFT) to this data. We then concurrently execute two operations on the image: RSS, which computes the sum of the square roots of the image, and a processing step using UNet. The final sensitivity map of the image is derived by dividing the output results of these two operations. Unlike the combination of VarNet with SME, where only the Auto-Calibration Signal (ACS) is used, we leverage the entire image optimized by VarNet for coil sensitivity calculation. This allows for capturing more high-frequency details, resulting in a more precise coil estimation.

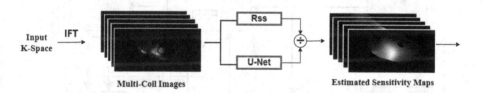

Input
K-Space   IFT → Multi-Coil Images   Rss / U-Net ÷ → Estimated Sensitivity Maps

Fig. 2. Details of the SME module.

2.5   Nonlinear Activation Free Network

For the reconstruction task, we adopted the U-net network architecture for module stacking. For the specific design of the module, we used NAFnet (Non linear Activation Free Network). With the widespread application of Transformer in deep learning, LayerNorm (LN) has also been increasingly used to count the values of all dimensions on each sample, calculate the mean and variance. The application of LN makes the distribution of each sample stable, while also making our training process smoother. NAFnet uses a simplified GELU instead of the GELU activation function in the activation function, which directly divides the features into two parts along the channel dimension and multiplies them, ultimately demonstrating a good deblurring ability in the model. NAFNet encodes and decodes the input information in 2D, sending features from multiple frames into the network as single frame information.

2.6 Space-Time Deformable Attention

Inspired by [6], we introduce the Space-Time Deformable Attention (STDA) block to capture the feature of frame. Details show in the Fig. 3. The entire module completes an operation similar to registration. We first align the features between the target frame and adjacent frames, and then use a multi-head attention mechanism to complete the operation between the target frame and adjacent frames. After cross attention operation, the result is subjected to self attention calculation. It should be noted that the f_{offset} here is a lightweight offset network composed of convolution layers. Finally, we obtained the feature of frame as the output of the module. STDA block is employed to effectively align consecutive frames within a sequence. By utilizing STDA blocks, the output is obtained in the form of distance features that represent the dissimilarity between frames. Additionally, within the STDA framework, we have used a approach for capturing the space-time flow specifically between the previous frame and the current frame.

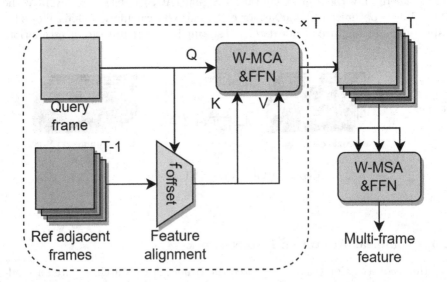

Fig. 3. Details of the STDA block. W-MCA: window multi-head cross attention. W-MSA: window multi-head self attention. FFN: feed-forward network.

2.7 Data Consistency

Preserving data consistency is essential in MRI reconstruction, ensuring that the network does not modify the k-space data inappropriately [2]. The DC layer's role is to integrate parts of the originally sampled k-space into the reconstructed k-space. In more detail, the DC layer takes four specific inputs: the reconstructed image, sub-sampled k-space data, a sub-sampling mask, and the estimated sensitivity maps. The DC layer combines these inputs to maintain data consistency throughout the reconstruction process.

2.8 Loss Function

In our methodology, we utilize the L1 loss function as our primary objective criterion. Specifically, for both the reconstructed k-space, denoted as $\mathbf{K}_{\text{reconstructed}}$, and the image domain, represented by $\mathbf{I}_{\text{reconstructed}}$, we compute the L1 loss relative to their respective ground-truths, $\mathbf{K}_{\text{ground-truth}}$ and $\mathbf{I}_{\text{ground-truth}}$. The overall loss $\mathcal{L}$ for our network is thus formulated as:

$$\mathcal{L} = ||\mathbf{K}_{\text{reconstructed}} - \mathbf{K}_{\text{ground-truth}}||_1 + ||\mathbf{I}_{\text{reconstructed}} - \mathbf{I}_{\text{ground-truth}}||_1$$

3 Experiment Settings

3.1 Cine Dataset and Baselines

The dataset used in this experiment is from the CMRxRecon competition [14]. The acquired images consisted of short-axis (SA) views, as well as two-chamber (2CH), three-chamber (3CH), and four-chamber (4CH) long-axis (LA) views. There are raw k-space data for both SA (multiple slices) and LA (multiple views). The dataset consisting of 120 training data samples and 60 validation data points, with the test data remaining inaccessible to participants. Accordingly, we allocated the existing 180 data points into training, validation, and test sets, maintaining a ratio of 8:1:1. Regarding the cardiac cine data from the challenge, during training, we randomly select one slice and use 6 frames of multi-coil k-space data as input to the model. No additional cropping or data augmentation was performed.

The baselines we compared with are several lightweight and effective techniques for MRI reconstruction, including GRAPPA [4], CineNet [7], VarNet [13], and RecurrentVarnet [15].

3.2 Implementation Details

The AdamW optimizer was used in all the experiments for 100 epochs with a batch size of 1 on 6 Nvidia P40 GPUs. The initial learning rates for all methods were set to 4×10^{-4} with cosine learning rate schedule and 5×10^{-7} minimum learning. The performance are evaluated by PNSR and SSIM for acceleration factors of $4\times$, $8\times$ and $10\times$.

4   Results

4.1   Quantitative Results

The quantitative results are summarized in Table 1. The proposed method achieves exceptional performance under high acceleration factors. In particular, when contrasted with the conventional GRAPPA algorithm and prominent deep learning approaches like CineNet, VarNet has demonstrated promising outcomes. Our approach, in comparison with VarNet, incorporates a richer spatiotemporal context, leading to superior achievements in terms of both SSIM and PSNR metrics. This distinction becomes particularly pronounced as the acceleration factor increases. Still, it is noteworthy that the performance of GRAPPA is not significantly distinguishable from our model's performance at a lower acceleration factor of 4×. This observation may be attributed to our model's training on divergent acceleration data simultaneously. Improved results could potentially be attained if the same acceleration data were employed for training.

Table 1. Quantitative results with acceleration factors of 4×, 8× and 10×, in terms of PSNR and SSIM.

Acceleration factors	4×		8×		10×	
Method	SSIM	PSNR	SSIM	PSNR	SSIM	PSNR
GRAPPA	0.970±0.010	45.05±2.48	0.889±0.023	35.28±1.57	0.884±0.023	34.34±1.50
CineNet	0.944±0.027	37.26±3.58	0.921±0.028	34.96±2.73	0.916±0.029	34.44±2.61
VarNet	0.971±0.009	43.41±2.15	0.952±0.013	39.22±1.85	0.947±0.014	38.29±1.84
RecurrentVarNet	0.967±0.009	42.89±2.08	0.945±0.015	38.87±1.87	0.938±0.017	37.91±1.85
Proposed	0.983±0.006	44.76±2.30	0.967±0.010	40.05±1.90	0.962±0.011	39.04±1.87

4.2   Qualitative Evaluation

As shown in Fig. 4. The acceleration factors considered are 4×, 8×, and 10×, respectively. For each method, error maps are presented using the dataset. These error plots illustrate the magnitude of recovery errors, where more pronounced texture indicates poorer recovery quality. Evidently, the proposed method exhibits the least error in comparison to alternative approaches.

4.3   Ablation Study

We conducted an ablation study using the cine dataset at 8× acceleration. The NAFNet baseline was a simplifed implementation using the U-net network architecture. Upon this baseline, we progressively added the proposed components including STDA for multi frame information reconstruction and Varnet for preliminary optimization of undersampling blurred MRI images. The results of adding modules to the overall architecture is presented in the Table 2.

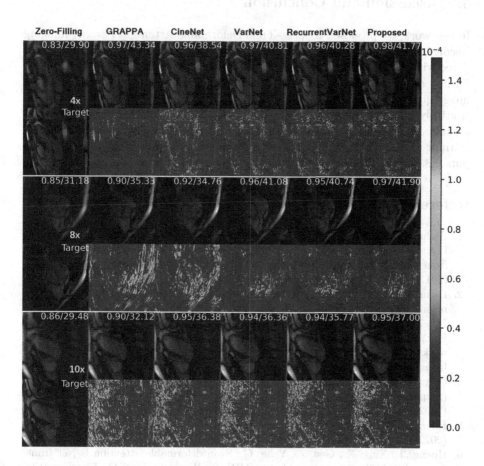

Fig. 4. Visual comparison of reconstruction results and error maps for acceleration factors of $4\times$, $8\times$, and $10\times$. The numbers in each subgraph represent SSIM and PSNR respectively.

Table 2. Ablation study with $8\times$ acceleration.

Variants	SSIM↑	PSNR↑	NMSE↓
NAFNet	0.939	36.46	0.033
NAFNet+STDA	0.952	38.09	0.022
VarNet+NAFNet+STDA	0.967	40.05	0.014

5   Discussion and Conclusion

In this work, we proposed a Space-Time Deformable Attention Parallel Imaging Reconstruction (STDAPIR) method for highly accelerated cardiac MRI. Compared with many advanced methods including frame-by-frame based (GRAPPA, VarNet and Recurrent VarNet) and spatiotemporal based (CineNet) ones, the image quality is substantially improved by our method both visually and quantitatively. Nonetheless, it is imperative to undertake further investigations into a range of loss functions and additional optimizations within the domain of learning. This study offers a promising framework to guide subsequent research pursuits.

References

1. Bulluck, H., Hammond-Haley, M., Weinmann, S., Martinez-Macias, R., Hausenloy, D.J.: Myocardial infarct size by CMR in clinical cardioprotection studies: Insights from randomized controlled trials. JACC: Cardiovasc. Imaging 10(3), 230–240 (2017). https://doi.org/10.1016/j.jcmg.2017.01.008
2. Duan, J., et al.: VS-Net: variable splitting network for accelerated parallel MRI reconstruction. In: Shen, D., et al. (eds.) MICCAI 2019. LNCS, vol. 11767, pp. 713–722. Springer, Cham (2019). https://doi.org/10.1007/978-3-030-32251-9_78
3. Fu, Y., et al.: A novel MRI segmentation method using CNN-based correction network for MRI-guided adaptive radiotherapy. Med. Phys. 45(11), 5129–5137 (2018). https://doi.org/10.1002/mp.13221
4. Griswold, M.A., et al.: Generalized autocalibrating partially parallel acquisitions (grappa). Magn. Reson. Med. Off. J. Int. Soc. Magn. Reson. Med. 47(6), 1202–1210 (2002)
5. Huang, J., et al.: Swin transformer for fast MRI. Neurocomputing 493, 281–304 (2022). https://doi.org/10.1016/j.neucom.2022.04.051
6. Huang, J., Xing, X., Gao, Z., Yang, G.: Swin deformable attention U-Net transformer (SDAUT) for explainable fast MRI. In: Wang, L., Dou, Q., Fletcher, P.T., Speidel, S., Li, S. (eds.) Medical Image Computing and Computer Assisted Intervention – MICCAI 2022. MICCAI 2022. LNCS, vol. 13436, pp. 538–548. Springer, Cham (2022). https://doi.org/10.1007/978-3-031-16446-0_51
7. Kofler, A., Haltmeier, M., Schaeffter, T., Kolbitsch, C.: An end-to-end-trainable iterative network architecture for accelerated radial multi-coil 2d cine MR image reconstruction. Med. Phys. 48(5), 2412–2425 (2021). https://doi.org/10.1002/mp.14809
8. Lustig, M., Donoho, D., Pauly, J.M.: Sparse MRI: the application of compressed sensing for rapid MR imaging. Magn. Reson. Med. Off. J. Int. Soc. Magn. Reson. Med. 58(6), 1182–1195 (2007)
9. Murphy, M., Alley, M., Demmel, J., Keutzer, K., Vasanawala, S., Lustig, M.: Fast ℓ_1-spirit compressed sensing parallel imaging MRI: scalable parallel implementation and clinically feasible runtime. IEEE Trans. Med. Imaging 31(6), 1250–1262 (2012)
10. Oscanoa, J.A., et al.: Deep learning-based reconstruction for cardiac MRI: a review. Bioengineering 10(3), 334 (2023)

11. Pruessmann, K.P., Weiger, M., Scheidegger, M.B., Boesiger, P.: Sense: sensitivity encoding for fast MRI. Magn. Reson. Med. Off. J. Int. Soc. Magn. Reson. Med. **42**(5), 952–962 (1999)

12. Ravishankar, S., Bresler, Y.: MR image reconstruction from highly undersampled k-space data by dictionary learning. IEEE Trans. Med. Imaging **30**(5), 1028–1041 (2010)

13. Sriram, A., et al.: End-to-end variational networks for accelerated MRI reconstruction. In: Martel, A.L., et al. (eds.) MICCAI 2020. LNCS, vol. 12262, pp. 64–73. Springer, Cham (2020). https://doi.org/10.1007/978-3-030-59713-9_7

14. Wang, C., et al.: Cmrxrecon: an open cardiac MRI dataset for the competition of accelerated image reconstruction (2023)

15. Yiasemis, G., Sonke, J.J., Sánchez, C., Teuwen, J.: Recurrent variational network: a deep learning inverse problem solver applied to the task of accelerated MRI reconstruction (2022)

16. Zhan, B., et al.: D2FE-GAN: decoupled dual feature extraction based GAN for MRI image synthesis. Knowl.-Based Syst. **252**, 109362 (2022). https://doi.org/10.1016/j.knosys.2022.109362

Multi-level Temporal Information Sharing Transformer-Based Feature Reuse Network for Cardiac MRI Reconstruction

Guangming Wang[1], Jun Lyu[2(✉)], Fanwen Wang[3,4], Chengyan Wang[5], and Jing Qin[2]

[1] School of Computer and Control Engineering, Yantai University, Yantai, China
[2] Centre for Smart Health, School of Nursing, The Hong Kong Polytechnic University, Hung Hom, Hong Kong
junlyu@polyu.edu.hk
[3] Bioengineering Department and Imperial-X, Imperial College London, London, UK
[4] Cardiovascular Research Centre, Royal Brompton Hospital, London, UK
[5] Human Phenome Institute, Fudan University, Shanghai, China

Abstract. The accurate reconstruction of accelerated Magnetic Resonance Imaging (MRI) brings significant clinical benefits, including improved diagnostic accuracy and reduced examination costs. Traditional cardiac MRI requires repetitive acquisitions over multiple heartbeats, leading to longer acquisition times. Deep learning-based MRI reconstruction methods have made significant progress in accelerating MRI. However, existing methods suffer from the following limitations: (1) Due to the involvement of multiple complex time-series data and image information in the process of heart reconstruction, exploring the nonlinear dependencies between temporal contexts is challenging. (2) Most of the research has neglected weight sharing in iterative frameworks, which prevents better capturing of long-range/non-local information in the data, thus restricting the improvement of model performance. In this paper, we propose a novel Multi-level Temporal Information Sharing Transformer to enhance cardiac MRI reconstruction. Based on the Transformer's multi-level encoder and decoder architecture, we perform multi-level temporal information feature aggregation across multiple adjacent views, establishing nonlinear dependencies between features and effectively learning crucial information between adjacent cardiac temporal frames. We also incorporate cross-view attention for temporal information interaction and fusion to enhance contextual understanding between adjacent views. Additionally, in the reconstruction process, we introduce a training approach of feature reuse to update weights, enhancing feature fusion in important regions and calculating global feature dependencies with fewer computations. Numerous experiments have indicated that this method is significantly superior to state-of-the-art techniques, thereby holding the potential for widespread application in the clinical field.

Keywords: Cardiac MRI reconstruction · Multi-level · Transformer · Temporal Information

O. Camara et al. (Eds.): STACOM 2023, LNCS 14507, pp. 410–420, 2024.
https://doi.org/10.1007/978-3-031-52448-6_39

1 Introduction

Magnetic Resonance Imaging (MRI) is a non-invasive medical imaging technique that plays a pivotal role in clinical diagnoses. Nevertheless, MRI encounters challenges in specific applications, such as cardiac magnetic resonance (CMR) imaging. CMR acquisitions typically demand lengthy scanning times, especially when employing techniques like T1 mapping. To acquire accurate cardiac images, patients are required to hold their breath during the scan (breath-hold), which can cause discomfort and limit patient cooperation. Consequently, the development of accelerated MRI techniques and enhancement of image reconstruction algorithms are imperative.

To shorten the scan time and improve imaging efficiency, researchers have proposed various methods and techniques for accelerated MRI. Among them, traditional methods utilize frequency-domain statistical approaches combined with Inverse Discrete Fourier Transform (IDFT) for interpolation of undersampled frequency-domain data (known as k-space). The k-t FOCUSS algorithm [7] is one of the undersampling techniques derived from Compressed Sensing (CS). Subsequently, other CS-based and low-rank methods have been developed [2,11,17], providing improved MRI reconstruction.

Deep learning-based MRI reconstruction methods have garnered significant attention in recent years. Convolutional Neural Network (CNN)-based approaches have been proposed to address these issues. For example, Schlemper et al. proposed [21] a deep cascaded CNN for reconstructing dynamic sequences of 2D Cardiac Magnetic Resonance (MR) images from undersampled data to accelerate data acquisition. Ramanarayanan et al. introduced DC-WCNN [20], a modified U-Net architecture for recovering fine structures in MR image reconstruction. Murugesan et al. presented Recon-GLGAN [16], a model consisting of a generator and a contextual discriminator, delivering better performance than previous approaches. Similarly, Lyu et al. proposed DuDoCAF [15], a novel dual-domain cross-attention fusion scheme with a recurrent transformer. Most of these studies fall under the category of static MRI reconstruction methods, yet they still exhibit significant limitations. These methods struggle to explore the nonlinear relationships between adjacent temporal frames, a crucial aspect for dynamic MRI reconstruction.

Video MRI reconstruction is a complex task that requires accurate reconstruction of motion structures in three-dimensional space and extracting relevant information from consecutive video frames. Despite efforts made in video reconstruction [4,8,10,12,14,26], these methods still fall short in fully addressing the challenges in dynamic MRI reconstruction. The Transformer [24] is a novel deep-learning model that has attracted attention in recent years and has been applied to video processing tasks. Traditional video processing methods [3,5,22] typically employ Convolutional Neural Networks to extract temporal features from videos. However, there are limitations when dealing with long sequential videos, such as difficulties in capturing long-term dependencies and constraints on input sequence length. Video Transformers address these issues by introducing self-attention mechanisms. By applying self-attention mechanisms in the temporal dimension, Video Transformers achieve modeling and analysis of video sequences. Recent research [18,23,27–29] has further propelled the development of Video Transformers.

To address the challenges in video MRI reconstruction, our team proposes a novel solution called the Multi-level Temporal Information Sharing Transformer (MTIST). The design of MTIST aims to fully leverage the temporal information in video frame sequences and improve the accuracy and stability of dynamic cardiac MRI reconstruction through efficient feature fusion and model optimization. In order to further enhance reconstruction accuracy, we opted for the Feature Reuse Network (FRN), where the number of network parameters does not increase with the growth in the number of reuse instances. In summary, the main contributions of this paper are as follows: 1) We introduce a feature reuse network that significantly improves reconstruction accuracy and reduces degradation caused by the limited receptive field of convolutional kernels. 2) We propose a novel Multi-level Temporal Information Sharing Transformer that achieves multi-level temporal information aggregation for multiple adjacent views through multi-encoders and a single decoder structure. We establish dense connections between features at different levels, utilize residual connections to facilitate information propagation, establish nonlinear dependencies between features, and effectively learn crucial information between adjacent cardiac temporal frames to enhance cardiac MRI reconstruction.

2   Related Work

2.1   Problem Formulation

We have a fully sampled image y, which contains complete spatial domain information. We apply undersampling by using a mask $\mathcal{M}$ and element-wise multiplication $\odot$ with the k-space data obtained after performing the Fourier transform operation ($\mathbb{F}$). Then, we apply the inverse Fourier transform $\mathbb{F}^{-1}$ to obtain the result. The process of obtaining the zero-filled (ZF) image is as follows:

$$y_{zf} = \mathbb{F}^{-1}(\mathbb{F}(y) \odot \mathcal{M} + \varepsilon), \tag{1}$$

where ε represents noise. Through the aforementioned steps, we obtain the ZF image, which is still a complex-valued image. In the subsequent preprocessing, the data are split into two channels: the real part and the imaginary part, which will serve as the input.

2.2   Overall Architecture

Figure 1 illustrates the overall architecture of the proposed network, where a sequence of consecutive cardiac MR images is inputted into the network. As shown in Fig. 1, it consists of three components: 1) Multi-level Encoder, which obtains multi-scale representations of different frames; 2) Multi-temporal aggregation (MTA) module, which explores relevant features between adjacent video

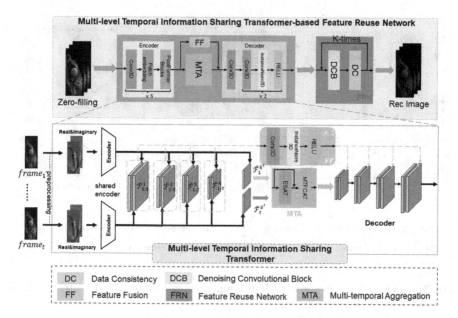

Fig. 1. Our MTIST architecture. Furthermore, the weights are updated using the feature reuse network (FRN) to accomplish the ultimate reconstruction image.

frames; and 3) Multi-level Decoder, responsible for restoring full-resolution feature maps. Specifically, given the adjacent t frames of images ($frame_{1,2,...,t}$), they are first preprocessed by converting them from complex form to a combination of real and imaginary parts. To obtain multi-scale feature maps for contextual matching, feature extraction is performed through t branches. Then, the generated feature maps from t branches are individually subjected to contextual matching at each scale to obtain matching features at different scales. Subsequently, the last layer of matching features is inputted into the MTA module, which guides the aggregation of the target Zero-filling Image at multiple scales. Then, feature restoration is performed using the Multi-level Decoder and Feature Fusion (FF) module. Finally, the reconstructed results are passed through the Feature Reuse Network (FRN) to obtain the final reconstructed target (Rec Image) image.

2.3 Multi-temporal Aggregation

To better extract complementary features between adjacent frames and incorporate a custom number of views fusion, we propose a novel MTA module. The network establishes long-range dependencies between views through two Transformer modules: Enhanced Spatial Attention Transformer (ESAT) and Multi-Temporal Frame Cross-Attention Transformer (MTFCAT), enabling aggregation of spatial attention and multi-temporal frame attention. As shown in Fig. 2, firstly, we obtain deep features $\widehat{\mathcal{F}}_{1:t}^{5} = \{\widehat{\mathcal{F}}_{1}^{5'}, \widehat{\mathcal{F}}_{2}^{5'}, ...\widehat{\mathcal{F}}_{t}^{5'}\}$ from multiple encoders.

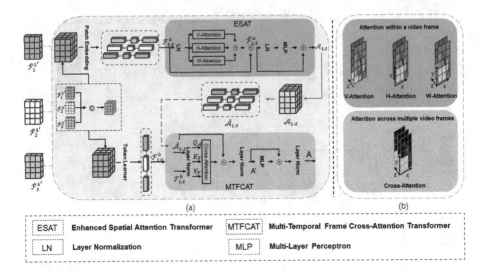

Fig. 2. (a) Overall architecture of Multi-temporal Aggregation. (b) Intra-frame attention and Inter-frame attention.

These features $\widehat{\mathcal{F}}_{1:t}^5$ are then transformed spatially into $\mathcal{F}_{1:t}^a$ and $\mathcal{F}_{1:t}^b$, serving as inputs to ESAT and MTFCAT, respectively. ESAT utilizes self-attention mechanism within each frame to compute long-range correlations between different patches, resulting in spatially enhanced attention features. Subsequently, MTFCAT takes $\mathcal{F}_{1:t}^b$ and the attention features obtained in the previous step as inputs to calculate global relationships between different time frames, achieving feature aggregation across multiple views.

Enhanced Spatial Attention Transformer. Following the inspiration from [6,13], we employ vertical attention (Att_v), horizontal attention (Att_h), and window attention (Att_w) for feature extraction, as shown in Fig. 2. Specifically, Att_v effectively captures long-range dependencies between different positions along the vertical direction. Att_h focuses attention on different regions or features within each frame image to better identify and utilize important local information. Att_w combines the mechanisms of vertical and horizontal attention to effectively capture both global context and local relationships. The calculation formulas are as follows:

$$Att_{all}(x) = Att_v(x) + Att_h(x) + Att_w(x), \tag{2}$$

$$\widetilde{\mathcal{F}}_{1:t}^a = Att_{all}(LNorm(\mathcal{F}_{1:t}^a)) + \mathcal{F}_{1:t}^a, \tag{3}$$

$$\mathcal{A}_{1:t} = MLP(LNorm(\widetilde{\mathcal{F}}_{1:t}^a)) + \widetilde{\mathcal{F}}_{1:t}^a, \tag{4}$$

where $x \equiv LNorm(\mathcal{F}_{1:t}^a)$, the feature x belongs to $\mathbb{R}^{(d \times w \times h) \times c}$, where d, w, h and c represent the depth, width, height, and number of channels of the feature, respectively. $LNorm$ denotes Layer Normalization. Finally, we obtain attention features $\mathcal{A}_{1:t} \in \mathbb{R}^{d \times w \times h \times c}$ through the ESAT module.

Multi-temporal Frame Cross-Attention Transformer. To better extract complementary features between adjacent frames and fuse a customizable number of views, as shown in Fig. 2, we propose a novel module called Multi-Temporal Frame Cross-Attention Transformer (MTFCAT). Specifically, given the high-level features $\mathcal{F}_{1:t}^b$ and the spatial attention features $\mathcal{A}_{1:t}$, we compute the Q, K, and V components of the attention as follows:

$$Q_{\mathcal{A}} = C^q(\mathcal{A}_{1:t}), \quad K_{\mathcal{F}} = C^k(\mathcal{F}_{1:t}^b), \quad V_{\mathcal{F}} = C^v(\mathcal{F}_{1:t}^b), \tag{5}$$

where variables C^q, C^k, and C^v represent correlation calculations. We can further derive the formula for cross-attention as follows:

$$Att_{cross}(Q_{\mathcal{A}}, K_{\mathcal{F}}, V_{\mathcal{F}}) = softmax(\frac{Q_{\mathcal{A}} \otimes (K_{\mathcal{F}})^T}{\sqrt{d_k}}) \otimes V_{\mathcal{F}}, \tag{6}$$

where $\otimes$ represents matrix multiplication, MTFCAT is defined as follows:

$$A' = LNorm(Att_c(LNorm(\widetilde{\mathcal{A}}_{1:t}, \mathcal{F}_{1:t}^b)) + \widetilde{\mathcal{A}}_{1:t}), \tag{7}$$

$$A = LNorm(MLP(A') + A'), \tag{8}$$

where $\mathcal{F}_{1:t}^b$ represents deep features from $\widehat{\mathcal{F}}_{1:t}^5$, $\widetilde{\mathcal{A}}_{1:t}$ is the output from the previous module ESAT, and A' is an intermediate variable in MTFCAT. $A \in \mathbb{R}^{c \times d \times w \times h}$ is the final output of MTFCAT.

2.4   Feature Reuse Network

Feature reuse can reduce redundant computations, saving time and resources, by inputting the features of the previous layer into subsequent layers to enhance network learning and parameter efficiency. We incorporate feature reuse into the training process of the reconstruction model. $F_{1:t} = \{F_1, F_2, \ldots\}, F_i \in \widetilde{F}$, where $1 \leq i \leq t$ and $\widetilde{F}$ represents the output from the Transformer module. For the feature reuse operation, we define the reuse times as K, which can be determined by the following formula:

$$out_{1:t} = \sum_{k=0}^{K-1} \mathcal{DC}(\mathcal{DCB}(\widetilde{F}_k)), \tag{9}$$

where $\mathcal{DCB}$ stands for Denoising Convolutional Block and $\mathcal{DC}$ stands for Data Consistency. In our network, the parameter K is configured as 10. By utilizing this network framework, we are able to obtain K times weight iterations with similar feature propagation and reconstruction performance. In the ablation study, we demonstrate its effectiveness.

Table 1. The comparison results between our reconstruction network and the existing network are given in terms of data metrics, including PSNR, SSIM, and NMSE ($\times 10^{-2}$). The best quantitative metric results are marked in **bold**.

Method	4×			6×			10×		
	PSNR	SSIM	NMSE($\times 10^{-2}$)	PSNR	SSIM	NMSE($\times 10^{-2}$)	PSNR	SSIM	NMSE($\times 10^{-2}$)
ZF	23.32	0.5706	37.52	20.59	0.4787	67.90	18.04	0.3614	117.60
MODL [1]	27.96	0.7996	12.24	26.35	0.7542	18.54	23.85	0.6706	31.16
CRNN [19]	30.24	0.8046	6.71	27.60	0.7515	12.46	24.83	0.6552	22.64
SwinIR [9]	31.50	0.8256	4.54	27.93	0.7684	10.66	25.13	0.6765	21.53
Ours	**32.59**	**0.8663**	**3.890**	**29.09**	**0.7998**	**9.22**	**25.96**	**0.7199**	**20.03**

Table 2. Ablation Study of Various Variants Models under 4× AF.

Variant	Modules			PSNR	SSIM	NMSE ($\times 10^{-2}$)
	MTIFA	FRN	$K = 10$			
w/o MTIST	✗	✓	✓	29.89	0.8398	7.79
w/o FRN	✓	✗	✓	30.97	0.7757	5.74
w/o K	✓	✓	✗	32.03	0.8124	4.12
MTISTFRN	✓	✓	✓	32.59	0.8663	3.89

2.5   Loss Functions

In order to ensure that the model has excellent reconstruction ability and detail recovery ability, we use $L1$ pixel loss and name it $\mathcal{L}_{rec}$:

$$\mathcal{L}_{rec} = \parallel I_{rec} - I_{hr} \parallel_1, \tag{10}$$

where I_{rec} represents the reconstructed MR image and I_{hr} represents the original HR image.

3   Experiments

3.1   Dataset and Baselines

We used the MICCAI Challenge 2023 (T1 mapping) [25] to train our model, which consists of two parts: a training set and a test set. The training set includes 120 healthy volunteers, while the test set contains 60 healthy volunteers. Each volunteer is represented by multiple slices and 9 temporal frames. All complex-valued images were reshaped to a size of 320×128 in the image domain through cropping. Random Cartesian undersampling masks were used along with three different acceleration factors (AF) (4×, 6×, and 10×) to obtain the corresponding ZF cardiac data sets. We compared our method with MODL [1], CRNN [19] and SwinIR [9] under single-coil conditions.

3.2   Implementation Details

The proposed MTISTFRN was implemented using PyTorch on an NVIDIA A100 (40 GB) GPU. We set the learning rate to 1e−4, the number of epochs to 50, and utilized the Adam optimization method. The reconstruction performance was evaluated using peak signal-to-noise ratio (PSNR), structural similarity index (SSIM), and normalized mean squared error (NMSE). More detailed experimental information can be found in Table 3.

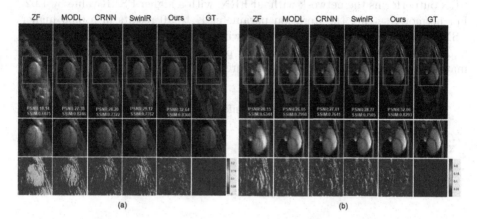

Fig. 3. In the case of $AF = 4$, we compared the cardiac reconstruction results obtained from different methods. The first row displays the overall reconstruction results, while the second row shows the reconstruction results of the local cardiac region. The third row represents the error maps, illustrating the differences between the reconstruction results and the ground truths.

3.3   Qualitative Visualization Result

Fig. 3 illustrates the reconstruction results and corresponding error map of the cardiac image with AF = 4. The darker textures in the error map indicate poorer reconstruction quality. It is evident that our reconstructed MR image surpasses the performance of the MODL [1], CRNN [19] and SwinIR [9] methods, which validates the effectiveness of embedding supplementary information from multiple frames in the reconstruction task.

3.4   Quantitative Metrics Results

Table 1 presents the scores of different datasets under 4x zero-padding in terms of quantitative metrics. It can be observed that our model yields the best results across all metrics. We further computed the results of each method for 6x and 10x zero-padding, where restoring MR images at higher levels of zero-padding is more challenging. However, our method consistently outperforms existing approaches, achieving the highest scores in all metrics.

3.5   Ablation Study

In Table 2, we conducted quantitative comparisons under 4x acceleration with and without the FRN and the MTIST, as well as a sensitivity analysis regarding the iterative parameter K. The results demonstrate significant improvements when employing the FRN and MTIST. Specifically, as shown in Table 2, the network utilizing MTIST achieves a higher PSNR value compared to the network without MTIST, with an increase of 2.70. Similarly, the network using FRN outperforms the network without FRN, with a higher PSNR value by 1.62. Furthermore, setting the iteration parameter K to 10 results in a 0.56 higher PSNR value compared to the network without iteration. In conclusion, these findings indicate the effectiveness of the proposed Multi-level Temporal Information Sharing Transformer-based Feature Reuse Network.

Table 3. Regarding some detailed information about the experiment.

Model Information			
Task of participation	CMR×Recon	Use of pre-training (if applicable, specify the data)	-
University/organization	School of Computer and Control Engineering, Yantai University, Yantai, China	Data augmentation	Yes
Single-channel or multi-channel	Single-channel	Data standardization	Min-Max Normalization
Hardware configuration (GPU, VRAM, CPU, number of cores)	NVIDIA A100, 40 GB, Intel Xeon Gold 5220, 26 cores	Model parameter number	6.528M
Training time	3.49 days	Loss function	L1 Loss
Inference time	77.66 min	Incorporation of a physical model (if applicable, specify the model)	-
Performance on the training set (PSNR, SSIM, NMSE)(4×)	35.49, 0.9031, 0.0194	Use of unrolling	-
Performance on the validation set (PSNR, SSIM, NMSE)(4×)	32.59, 0.8663, 0.0389	Application of k-space fidelity	Yes
Docker submitted?	-0	Model backbone	Transformer; U-net
Use of segmentation labels	-	Amplitude or complex-valued operations	Complex-valued operations

4   Conclusion

In this work, we propose a novel Multi-level Temporal Information Sharing Transformer-based Feature Reuse Network (MTISTFRN) to improve cardiac MRI reconstruction. Specifically, we perform multi-level temporal information feature aggregation across multiple neighboring views to establish nonlinear dependencies between features and effectively capture key information between adjacent cardiac frames. Lastly, we introduce a feature reuse training method to update weights, enhancing feature fusion in important regions while computing global feature dependencies with reduced computational complexity. Extensive experiments validate that MTISTFRN achieves state-of-the-art performance. Furthermore, while our feature reuse training method has proven effective, there is potential for further refinement. Investigating more advanced weight update strategies to enhance feature fusion efficiency and performance represents an avenue for future research. Our future work will be dedicated to addressing these limitations and incorporating multi-coil capabilities for cardiac MRI image reconstruction.

References

1. Aggarwal, H.K., Mani, M.P., Jacob, M.: MoDL: model-based deep learning architecture for inverse problems. IEEE Trans. Med. Imaging **38**(2), 394–405 (2018)
2. Ahmed, A.H., Zhou, R., Yang, Y., Nagpal, P., Salerno, M., Jacob, M.: Free-breathing and ungated dynamic MRI using navigator-less spiral storm. IEEE Trans. Med. Imaging **39**(12), 3933–3943 (2020)
3. Carreira, J., Zisserman, A.: Quo vadis, action recognition? A new model and the kinetics dataset. In: Proceedings of the IEEE Conference on Computer Vision and Pattern Recognition, pp. 6299–6308 (2017)
4. Guo, X., Guo, X., Lu, Y.: SSAN: separable self-attention network for video representation learning. In: Proceedings of the IEEE/CVF Conference on Computer Vision and Pattern Recognition, pp. 12618–12627 (2021)
5. Hara, K., Kataoka, H., Satoh, Y.: Learning spatio-temporal features with 3D residual networks for action recognition. In: Proceedings of the IEEE International Conference on Computer Vision Workshops, pp. 3154–3160 (2017)
6. Ho, J., Kalchbrenner, N., Weissenborn, D., Salimans, T.: Axial attention in multi-dimensional transformers. arXiv preprint arXiv:1912.12180 (2019)
7. Jung, H., Ye, J.C., Kim, E.Y.: Improved k-t blast and k-t sense using FOCUSS. Phys. Med. Biol. **52**(11), 3201 (2007)
8. Liang, J., et al.: VRT: a video restoration transformer. arXiv preprint arXiv:2201.12288 (2022)
9. Liang, J., Cao, J., Sun, G., Zhang, K., Van Gool, L., Timofte, R.: SwinIR: image restoration using swin transformer. In: Proceedings of the IEEE/CVF International Conference on Computer Vision, pp. 1833–1844 (2021)
10. Lin, J., et al.: Flow-guided sparse transformer for video deblurring. arXiv preprint arXiv:2201.01893 (2022)
11. Lingala, S.G., Hu, Y., DiBella, E., Jacob, M.: Accelerated dynamic MRI exploiting sparsity and low-rank structure: kt SLR. IEEE Trans. Med. Imaging **30**(5), 1042–1054 (2011)
12. Liu, R., et al.: FuseFormer: Fusing fine-grained information in transformers for video inpainting. In: Proceedings of the IEEE/CVF International Conference on Computer Vision, pp. 14040–14049 (2021)
13. Liu, Z., et al.: Swin transformer: hierarchical vision transformer using shifted windows. In: Proceedings of the IEEE/CVF International Conference on Computer Vision, pp. 10012–10022 (2021)
14. Lyu, J., et al.: Region-focused multi-view transformer-based generative adversarial network for cardiac cine MRI reconstruction. Med. Image Anal. **85**, 102760 (2023)
15. Lyu, J., Sui, B., Wang, C., Tian, Y., Dou, Q., Qin, J.: DuDoCAF: dual-domain cross-attention fusion with recurrent transformer for fast multi-contrast MR Imaging. In: Wang, L., Dou, Q., Fletcher, P.T., Speidel, S., Li, S. (eds.) MICCAI 2022. LNCS, vol. 13436, pp. 474–484. Springer, Cham (2022). https://doi.org/10.1007/978-3-031-16446-0_45
16. Murugesan, B., Vijaya Raghavan, S., Sarveswaran, K., Ram, K., Sivaprakasam, M.: Recon-GLGAN: a global-local context based generative adversarial network for MRI reconstruction. In: Knoll, F., Maier, A., Rueckert, D., Ye, J.C. (eds.) MLMIR 2019. LNCS, vol. 11905, pp. 3–15. Springer, Cham (2019). https://doi.org/10.1007/978-3-030-33843-5_1
17. Otazo, R., Candes, E., Sodickson, D.K.: Low-rank plus sparse matrix decomposition for accelerated dynamic MRI with separation of background and dynamic components. Magn. Reson. Med. **73**(3), 1125–1136 (2015)

18. Piergiovanni, A., Kuo, W., Angelova, A.: Rethinking video viTs: sparse video tubes for joint image and video learning. In: Proceedings of the IEEE/CVF Conference on Computer Vision and Pattern Recognition, pp. 2214–2224 (2023)
19. Qin, C., Schlemper, J., Caballero, J., Price, A.N., Hajnal, J.V., Rueckert, D.: Convolutional recurrent neural networks for dynamic MR Image reconstruction. IEEE Trans. Med. Imaging **38**(1), 280–290 (2018)
20. Ramanarayanan, S., Murugesan, B., Ram, K., Sivaprakasam, M.: DC-WCNN: a deep cascade of wavelet based convolutional neural networks for MR Image reconstruction. In: 2020 IEEE 17th International Symposium on Biomedical Imaging (ISBI), pp. 1069–1073. IEEE (2020)
21. Schlemper, J., Caballero, J., Hajnal, J.V., Price, A., Rueckert, D.: A deep cascade of convolutional neural networks for MR image reconstruction. In: Niethammer, M., et al. (eds.) IPMI 2017. LNCS, vol. 10265, pp. 647–658. Springer, Cham (2017). https://doi.org/10.1007/978-3-319-59050-9_51
22. Simonyan, K., Zisserman, A.: Two-stream convolutional networks for action recognition in videos. In: Advances in Neural Information Processing Systems, vol. 27 (2014)
23. Touvron, H., Cord, M., Sablayrolles, A., Synnaeve, G., Jégou, H.: Going deeper with image transformers. In: Proceedings of the IEEE/CVF International Conference on Computer Vision, pp. 32–42 (2021)
24. Vaswani, A., et al.: Attention is all you need. In: Advances in Neural Information Processing Systems, vol. 30 (2017)
25. Wang, C., et al.: CMR×Recon: an open cardiac MRI dataset for the competition of accelerated image reconstruction. arXiv preprint arXiv:2309.10836 (2023)
26. Wang, Y., et al.: End-to-end video instance segmentation with transformers. In: Proceedings of the IEEE/CVF Conference on Computer Vision and Pattern Recognition, pp. 8741–8750 (2021)
27. Xing, Z., Yu, L., Wan, L., Han, T., Zhu, L.: NestedFormer: nested modality-aware transformer for brain tumor segmentation. In: Wang, L., Dou, Q., Fletcher, P.T., Speidel, S., Li, S. (eds.) MICCAI 2022. LNCS, vol. 13435, pp. 140–150. Springer, Cham (2022). https://doi.org/10.1007/978-3-031-16443-9_14
28. Xing, Z., Dai, Q., Hu, H., Chen, J., Wu, Z., Jiang, Y.G.: SVFormer: semi-supervised video transformer for action recognition. In: Proceedings of the IEEE/CVF Conference on Computer Vision and Pattern Recognition, pp. 18816–18826 (2023)
29. Yan, S., et al.: Multiview transformers for video recognition. In: Proceedings of the IEEE/CVF Conference on Computer Vision and Pattern Recognition, pp. 3333–3343 (2022)

Cine Cardiac MRI Reconstruction Using a Convolutional Recurrent Network with Refinement

Yuyang Xue[1], Yuning Du[1], Gianluca Carloni[2,3], Eva Pachetti[2,3], Connor Jordan[4(✉)], and Sotirios A. Tsaftaris[1,5]

[1] Institute for Imaging, Data and Communications, University of Edinburgh, Edinburgh, UK
[2] Institute of Information Science and Technologies "Alessandro Faedo", National Research Council of Italy (CNR), Pisa, Italy
[3] Department of Information Engineering, University of Pisa, Pisa, Italy
[4] Institute for Infrastructure and Environment, University of Edinburgh, Edinburgh, UK
c.jordan@ed.ac.uk
[5] The Alan Turing Institute, London, UK

Abstract. Cine Magnetic Resonance Imaging (MRI) allows for understanding of the heart's function and condition in a non-invasive manner. Undersampling of the k-space is employed to reduce the scan duration, thus increasing patient comfort and reducing the risk of motion artefacts, at the cost of reduced image quality. In this challenge paper, we investigate the use of a convolutional recurrent neural network (CRNN) architecture to exploit temporal correlations in supervised cine cardiac MRI reconstruction. This is combined with a single-image super-resolution refinement module to improve single coil reconstruction by 4.4% in structural similarity and 3.9% in normalised mean square error compared to a plain CRNN implementation. We deploy a high-pass filter to our ℓ_1 loss to allow greater emphasis on high-frequency details which are missing in the original data. The proposed model demonstrates considerable enhancements compared to the baseline case and holds promising potential for further improving cardiac MRI reconstruction.

Keywords: Cardiac MRI Reconstruction · MRI Acceleration · MRI Refinement · CRNN

1 Introduction

Cardiac magnetic resonance imaging is a powerful, non-invasive tool to aid visualise the heart's chambers, valves, blood vessels and surrounding tissue. To gain a 3D depiction of the heart, a sequential acquisition process of 2D slices is used, with the scanning duration increasing with the number of slices and temporal

---

Y. Xue, Y. Du, G. Carloni, E. Pachetti, C. Jordan—These authors contributed equally.

© The Author(s), under exclusive license to Springer Nature Switzerland AG 2024
O. Camara et al. (Eds.): STACOM 2023, LNCS 14507, pp. 421–432, 2024.
https://doi.org/10.1007/978-3-031-52448-6_40

resolution desired. Thus for detailed scanning, multiple cardiac cycles must be monitored and the duration of the MRI process can consequently exceed the patients ability to remain steady and hold their breath. By undersampling in the k-space data acquisition process, the scan time can be substantially reduced at the cost of missing information that must be interpolated. Deep learning achieves k-space reconstruction with greater prior knowledge for the regularisation term that covers the missing k-space domain, without requiring an iterative optimisation process and hence greatly accelerating the reconstruction rate.

Various architectures have been explored for MRI reconstruction, including convolutional neural networks (CNNs) and U-Nets [6,7,11,21], variational networks [4] and generative adversarial networks [12,22]. Other deep learning methods that exploit prior knowledge and extend traditional iterative methods include the model-based deep-learning architecture [1] and deep density priors [17]. Enhancing cine MRI through deep learning involves not only capitalising on the spatial relationships acquired from a given dataset but also leveraging temporal correlations. This has been evidenced across various model architectures [9,12,18,25] as well as through registration-based [23] and motion-guided alignment [5] approaches. In [15], data sharing layers were incorporated in a cascaded CNN, whereby adjacent time step k-space data was used to fill the unsampled lines. In [14], recurrent connections are employed across each iteration step as well as bidirectional convolutional recurrent units facilitating knowledge sharing between iterations and input time frames, respectively.

Working within the confines of the challenge, we explored various architectures and found the CRNN block of [14] to perform best within the given limitations in memory and reconstruction time set by the organisers. This was subsequently combined with a lightweight refinement module inspired by single-image super-resolution approaches [3] to perform further de-noising and resolve finer details. The rest of the paper is organised as follows: Sects. 2 and 3 describes the problem, dataset, and methodology, Sect. 4 presents the results of experiments and Sect. 5 and 6 provide discussion and conclusions, respectively.

2   Problem Formulation and Dataset

The objective of MRI reconstruction is to address an ill-posed inverse problem, retrieving image information denoted as $\mathbf{x} \in \mathbb{C}^N$ from acquired undersampled signals $\mathbf{y} \in \mathbb{C}^K$, where $K \ll N$. This procedure can be depicted using a linear forward operator $\mathbf{E}$, which defines the characteristics of the forward problem:

$$\mathbf{y} = \mathbf{E}\mathbf{x} + \epsilon. \tag{1}$$

Equation 1 represents the general form of MRI reconstruction. The goal of reconstruction is to minimise the difference between $\mathbf{x}$ and the ground truth. Therefore, the reconstruction problem can be defined as follows:

$$\tilde{\mathbf{x}} = \underset{\mathbf{x} \in \mathbb{C}^N}{\arg\min} \frac{\lambda}{2} \|\mathbf{E}\mathbf{x} - \mathbf{y}\| + f_\theta(\mathbf{x}). \tag{2}$$

Here, f_θ denotes a neural network for image reconstruction with trainable parameters θ and λ controls the balance between the network and data consistency.

Data. Our model is evaluated on the CMRxRecon Challenge dataset from the 26th International Conference on Medical Image Computing and Computer Assisted Intervention. The dataset includes both short-axis (SA) and long-axis (LA) (two-chamber, three-chamber and four-chamber) views under acceleration rates of 4×, 8×, 10×. The dataset was obtained following recommended protocols and processing [19,24], more details of which can be found in [20]. The 300-patient dataset is split 120:60:120 between challenge training, validation, and testing respectively. Only the challenge training set contained ground truth reference data, hence this was further split 90:20:10 for training, evaluation, and testing respectively for all models.

Data Pre-processing. The unpadded image size varies between widths of 132, 162, 204 & 246 and heights of 448 & 512 pixels. To maintain consistent input size, we apply zero-padding for image sizes of 256 × 512 following the Inverse Fourier Fast Transform, with the outputs cropped after inference.

When using the approach from [9], the computationally intensive conjugate gradient step was a limiting factor due to the GPU's initial 24 GB memory constraint set by the challenge organisers. We thus chose to use the coil combined data rather than the multi-coil format, which allowed use of a simpler data consistency step, at a potential loss of accuracy without using the extra information. Likewise, we used a single channel for the processed image instead of using independent channels for amplitude/phase or real/imaginary components, as adopted by [4,6,11,15].

The SA data are 3-dimensional spatially with an additional time component. It is therefore conceivable that full 4D convolutional kernels could be used to fully utilise spatio-temporal redundancies, but this would result in extremely large memory requirements as discussed in [10]. Furthermore, studies such as [18] have demonstrated that it is preferable to have a larger $2D + t$ network than a smaller 3D-input network with equivalent memory consumption. Therefore, due to the large image size, we choose to use time-series batches of 2D depth slices as per [14] rather than $3D + t$ or 4D for the long-axis images.

3   Methodology

3.1   Model Exploration

The initial limitations for inference imposed by the organisers were 24 GB GPU VRAM and 4 h for the reconstruction of the test dataset, which was later increased after our initial investigations. Pre-trained models or loss functions were not permitted. Denoising diffusion probabilistic models (DDPM) were found to take too long in inference, whilst transformer models have been found to lead to heavily pixelated reconstructions. Hence, more conventional approaches were tested, building upon an existing repository[1].

We compared networks similar to the CineNet [9] and CRNN networks [14]. The number of parameters in each network were maximised such that the full 24GB VRAM memory would be used in training. A 2D U-Net is deployed to serve as an additional baseline to compare all models to. The U-Net is trained on a slice-by-slice basis with 3 cascades and 48 feature map channels. Weight sharing is used when training the model, and λ is set to be learnable with an initialisation of $\log(10^{-1})$. The learning rate is 3×10^{-4}, and the Adam optimiser is deployed to guide the training process.

3.2 Model Architecture

A high-level depiction of the complete architecture of the final model is presented in Fig. 1. The backbone of proposed architecture is based on the CRNN block detailed in [14]. The first step in CRNN is a bidirectional convolutional recurrent unit (BCRNN) with three convolution layers: a standard convolution between layers, one convolution between temporal slices, and one between iterations. This is followed by three convolutional recurrent units (CRNN) which evolve only over iterations before a plain CNN. Finally, residual connections are employed prior to a data consistency term, preserving the information from sampled data.

Aiming to improve performance, we include an additional BCRNN unit to further exploit spatio-temporal dependencies, followed by a refinement module

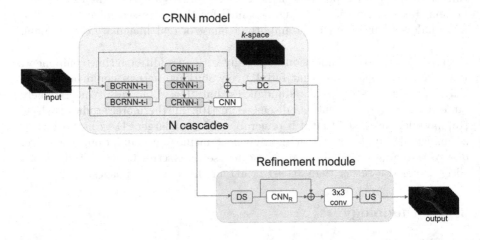

Fig. 1. Final model architecture: BCRNN, CRNN, and CNN units with a data consistency (DC) step from [14] for primary reconstruction. "t" and "i" denote time and iterations, respectively. The low-cost refinement module, inspired by [3], includes downsampling (DS), CNN, and upsampling (US) units.

---

[1] https://github.com/f78bono/deep-cine-cardiac-mri.

to further denoise the output of the CRNN model and refine further details. We deploy a very lightweight single-image super-resolution network, Bicubic++ [3], which maintains short reconstruction times. The refinement module first learns lower resolution features to decrease computational cost and then performs numerous convolutions to denoise the image before a final convolutional filter and upscaling back to the original image size. We test the performance of the network with end-to-end and separate learning for each module.

Loss Function. In the context of image reconstruction, ℓ_1 loss, ℓ_2 loss and SSIM loss are widely used to constrain models for high-quality reconstructed images, but often disregard the complex nature of MRI data. Thus, we investigate a range of losses using an additional loss term, denoted the $\perp$-loss [16]. The $\perp$-loss adds a phase term which can be combined with ℓ_1, ℓ_2 or SSIM losses to address the asymmetry in the magnitude/phase loss landscape. This operates on the polar representation of complex numbers, rather than on two real value channels for magnitude and phase, thus taking advantage of the fact that fully symmetric loss functions can improve task performance [13]. For the separate training of the CRNN and the refinement module, $\perp$-losses are only utilised for the CRNN output, with ℓ_1 and SSIM loss functions deployed for the refinement module. For the end-to-end training, ℓ_1 and SSIM loss are employed to constrain both CRNN and refinement module. We split the ℓ_1 loss by introducing a high-pass frequency filter, allowing us to emphasise the high-frequency content in our reconstructed images to resolve finer details. We denote this as $\ell_{1 split}$.

In training, losses were quantified across the entire image, whereas for the validation leaderboard, assessment was limited to the initial 3 time frames and the central sixth portion of the images. The competition metrics were structural similarity index measure (SSIM), normalised mean square error (NMSE) and peak signal-to-noise ratio (PSNR). Hence, whilst the complete reconstructed images often surpassed SSIM values of 0.98, validation scores only reached 0.85.

3.3   Implementation Details

Implementation details and our code is available at: https://github.com/vios-s/ CMRxRECON_Challenge_EDIPO.

4   Results

Model Choice and Weight Sharing. Figure 2 shows the training losses between models, demonstrating the stronger performance of the CRNN model. The point of convergence for the CRNN with weight-sharing and the CineNet models is similar, but the CRNN networks start at a much lower loss value. This is despite the non-weight-sharing model (1.1M) having over 2× more trainable parameters than the 6-cascade CineNet model (0.5M). Between CRNN models, we see more rapid convergence in the weight-sharing model as there are less parameters to optimise and reduced likelihood of early overfitting. However, the lower number of parameters leads to reduced expressive power and is out-performed by the non-weight-sharing model with insufficient memory gains to justify its use. Using the ⊥-loss only, the weight-sharing model had SSIM of

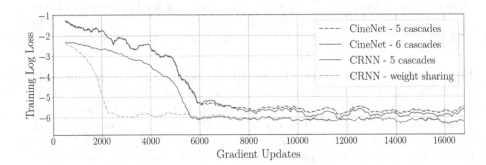

Fig. 2. Log loss during exploratory training of modified CineNet and CRNN (with and without weight-sharing between kernels). Note that the implementation is not identical to the original works.

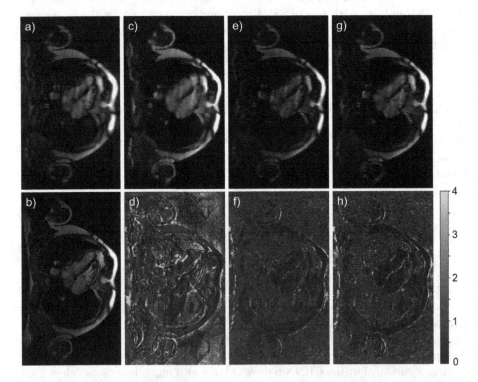

Fig. 3. Reconstruction (top) and associated error maps (bottom) for the initial network investigation. (a) 8 × undersampled LAX input (b) fully sampled ground truth (c,d) CineNet model (6 cascades) (e,f) CRNN model (weight-sharing between cascades) (g,h) CRNN model (no weight-sharing).

0.683, NMSE of 0.123 and PSNR of 23.917, performing notably worse than the non-weight-sharing model, as presented in the next section. Figure 3 shows the reconstruction through the CineNet and the CRNN (with and without weight-sharing) models.

Loss Function Investigation. Table 1 presents the findings of investigations of the loss functions using a low-cost CRNN model. Use of the $\perp$-ℓ_1 loss led to an improvement in SSIM and PSNR compared to ℓ_1 loss alone, but a slightly higher NMSE error. Providing greater emphasis on high-frequency data using a high-pass filter led to improved SSIM but lower NMSE and PSNR performance. Further tuning to improve the ratio of high- to low- frequency led to better results for the higher cascade models. Notably, combining SSIM with the $\perp$-ℓ_1 loss was counter-productive for all metrics and suggests that further tuning of the weighting of each loss component is required.

Table 1. Performance comparisons for different loss function combinations, evaluated on the validation data. A 48-channel 5 cascade CRNN network was used without the refinement module. $\ell_{1\,split}$ denotes the ℓ_1 loss whereby a high-pass filter is used to provide more focus on the high frequency content, in addition to the conventional ℓ_1 loss.

Metric	Loss function				
	$\perp$	ℓ_1	$\perp$-ℓ_1	$\perp$-$\ell_{1\,split}$	$\perp$-SSIM-$\ell_{1\,split}$
SSIM	0.712	0.741	0.752	**0.753**	0.739
NMSE	0.0925	**0.0646**	0.0655	0.0671	0.0719
PSNR	25.143	26.525	**26.535**	26.487	26.067

Introduction of the Refinement Module. Table 2 highlights the improvements in the quality of inference made by introducing the refinement module. Deploying the refinement as a separately trained post-processing module shows notable benefit improving more than adding an additional cascade to the plain CRNN. The end-to-end model results in further improvements upon separate training, of 4.4% in structural similarity and 3.9% in normalised mean square error relative to the plain CRNN, in spite of no longer being able to take advantage of the $\perp$ loss.

Figure 4 shows qualitatively the improvements made by the introduction of the refinement module at full scale. The error is reduced substantially and some finer details are resolved, though there is still scope for improvement at smaller scales. We generally see that the model is incapable of generating details that are completely lost in the undersampling process.

Validation Results. Our final tests prior to submission are presented in Table 3. Across all models, the short-axis reconstruction performs better quantitatively as there is more short-axis data available in training. For both views, the performance reduces with increased undersampling, as more detail is lost.

Table 2. Performance comparisons of various model set-ups. Sequential (separate) and end-to-end training (combined) of the CRNN and refinement module are presented.

Cascades	6	6	6	7
Refinement module	None	Sequential	End-to-end	None
SSIM	0.768	0.792	**0.802**	0.765
NMSE	0.0516	0.0496	**0.0454**	0.0535
PSNR	27.354	27.597	**27.969**	27.351

Table 3. Performance comparisons on evaluation of CMRxRecon cine cardiac MRI coil combined validation data for different acceleration rates (AR).

AR	Metric	Short-axis			Long-axis		
		U-Net Baseline	Plain CRNN	Proposed	U-Net Baseline	Plain CRNN	Proposed
4×	SSIM	0.641	0.824	**0.854**	0.573	0.757	**0.792**
	NMSE	0.180	0.0311	**0.0277**	0.188	0.0485	**0.0433**
	PSNR	23.084	29.842	**30.295**	22.040	26.958	**27.540**
8×	SSIM	0.637	0.796	**0.829**	0.574	0.723	**0.763**
	NMSE	0.201	0.0428	**0.0377**	0.191	0.0687	**0.0586**
	PSNR	22.603	28.364	**29.002**	22.234	25.588	**26.370**
10×	SSIM	0.641	0.788	**0.822**	0.588	0.717	**0.753**
	NMSE	0.210	0.0464	**0.0408**	0.198	0.0724	**0.064**
	PSNR	22.429	28.030	**28.644**	22.065	25.343	**25.965**

5   Discussion

On Model Choice. Without extensive hyperparameterisation, the CRNN architecture demonstrated more promising performance than the CineNet, both qualitatively and quantitatively as shown in Fig. 2 and 3. In this network, the temporal average is subtracted from each slice and then the residuals are transformed into $x-t$ and $y-t$ axes being passed through a 2D U-Net structure. The 2D U-Net is computationally lightweight as the CineNet was originally designed for multi-coil radial acquisition processes, but even with an increased number of cascades fails to resolve finer details as clearly shown in Fig. 3. The recurrent connections exploit the temporal dependencies between slices more effectively than by attempting to capture these relationships through transformation and sparsification.

However, training of the plain CRNN has still resulted in lower perceptual quality than presented in the original implementation, which may have been further improved with more hyperparameterisation tuning. The large image size of up to 246 × 512 proved challenging, and in our implementation limited us to 5 cascades of 48 channels for the original 24GB inference limitation for the CRNN network. In the original work [14], the model consisted of 10 cascades with 64 feature channels, which operated on smaller data with a smaller GPU.

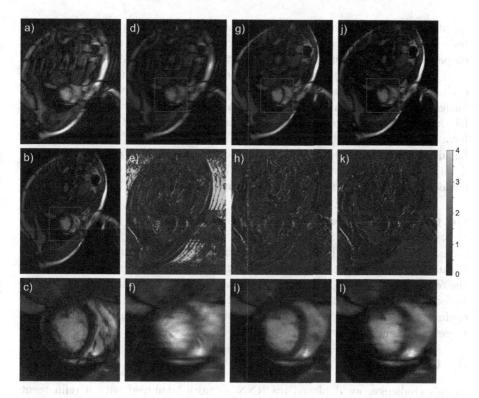

Fig. 4. Reconstruction (top) and associated error maps (middle) for the U-Net baseline and CRNN models. Finer details (bottom) are not resolved by the U-Net, which are partially captured by the plain CRNN model. The refinement module subsequently deblurs the image and provides better resolution at boundaries. (a) 10 × undersampled SAX input (b, c) fully sampled ground truth (d, e, f) U-Net (g, h, i) 6 cascades with combined refinement (j, k, l) 7 cascades, no refinement.

On the Final Model. The plain CRNN implementation substantially outperforms the baseline 2D U-Net, which has over double the number of trainable parameters, demonstrating the importance of exploiting temporal correlations. The reconstruction of the plain CRNN is a considerable improvement upon the 10× undersampled input, however smaller scale details that are resolved are blurry (e.g. Fig. 4l). The introduction of the low-cost refinement module led to better results with further denoising as presented in Fig. 4i. This shows promise for the implementation of lightweight single-image super-resolution models to assist in improving cardiac cine MRI reconstruction, either combined with the main reconstruction model or as a post-processing step. Relative to the ground truth, we still see that finer details are being missed that have been obscured due to the undersampling process. In [2], numerous MRI reconstruction experiments were conducted to test current deep learning reconstruction models. They proposed that "networks must be retrained on any subsampling patterns" for end to end CNN networks. In our approach, we trained all the acceleration rate together to get a stable but averaged reconstruction in a single model, subsequently resolving less finer details. The failure in generating details that have been lost, could be better tackled by a generative model [8]. As such, performance on patient volumes where more aliasing artefacts were present was poorer.

Therefore, to improve the proposed model, there is potential that training exclusively for each view can improve the final details. However, our model performs relatively well on the validation stage leaderboard for high acceleration factors, where finer details are more difficult to resolve, whilst we generally perform worse at lower acceleration factors. This suggests that whilst our model failed to generate some finer details, other architectures also struggled once these details were lost or heavily obscured. There are numerous further modules that could have been implemented, had more time been available.

On the Loss Function. We found that introduction of the $\perp$ loss to the ℓ_1 loss improved both SSIM and PSNR, though at the expense of a slightly reduced ℓ_1 value itself. Treating the weightings of the loss functions as learnable parameters could lead to improved results in all metrics, as anticipated due to the results presented in [16]. Likewise, the introduction of the high-pass filter loss to focus the ℓ_1 loss on higher-frequency information increases the complexity of optimisation but was beneficial after the weightings were improved, though not presented quantitatively here.

6   Conclusions

In this challenge, we deployed a CRNN network combined with a refinement module to perform MRI reconstruction of cardiac cine data. We train the model for a range of acceleration factors and views using a high-pass filter to focus our loss on high-frequency details. From the quantitative analysis of the evaluation data and from direct viewing of the validation portion of the training data, the refinement module provides additional image quality with improvements of around 4% in all metrics relative to the plain CRNN implementation. As is typically found, some finer details at smaller scales remain unresolved that may be improved upon with further hyperparameter tuning and new modules. Nonetheless, the improvement upon the baseline is substantial and our model shows promise for improving cardiac MRI reconstruction.

Acknowledgements. This work was supported in part by National Institutes of Health (NIH) grant 7R01HL148788-03. C. Jordan, Y. Du and Y. Xue thank additional financial support from the School of Engineering, the University of Edinburgh. Sotirios A. Tsaftaris also acknowledges the support of Canon Medical and the Royal Academy of Engineering and the Research Chairs and Senior Research Fellowships scheme (grant RCSRF1819\8\25). The authors would like to thank Dr. Chen and K. Vilouras for inspirational discussions and assistance.

CMRxRecon Summary Information

Task: Cine. **Data used:** Single-channel. **Docker submitted:** Yes.
Final model: CRNN backbone with SISR module (2.3M parameters).

Unrolling: Yes. **Domain:** Complex and amplitude. **k-space fidelity:** Yes.
Pretraining: No. **Augmentation/standardisation:** No.
Trained on: 3×GPU (40GB vRAM). **Loss function:** $\perp$-SSIM-ℓ_1
Training time: 23h (33 epochs). **Inference time:** 1 h 45 min (120 subjects).
Test results: PSNR = 35.582, SSIM = 0.946, NMSE = 0.0374

References

1. Aggarwal, H.K., Mani, M.P., Jacob, M.: MoDL: model-based deep learning architecture for inverse problems. IEEE Trans. Med. Imaging **38**(2), 394–405 (2019). https://doi.org/10.1109/TMI.2018.2865356
2. Antun, V., Renna, F., Poon, C., Adcock, B., Hansen, A.C.: On instabilities of deep learning in image reconstruction and the potential costs of AI. Proc. Natl. Acad. Sci. **117**(48), 30088–30095 (2020)
3. Bilecen, B.B., Ayazoglu, M.: Bicubic++: slim, slimmer, slimmest - designing an industry-grade super-resolution network (2023). https://arxiv.org/abs/2305.02126
4. Hammernik, K., et al.: Learning a variational network for reconstruction of accelerated MRI data. Magn. Reson. Med. **79**(6), 3055–3071 (2018). https://doi.org/10.1002/mrm.26977
5. Han, X., Liu, Y., Lin, Y., Chen, K., Zhang, W., Liu, Q.: MDAMF: reconstruction of cardiac cine MRI under free-breathing using motion-guided deformable alignment and multi-resolution fusion (2023). https://arxiv.org/abs/2303.04968
6. Han, Y., Yoo, J., Kim, H.H., Shin, H.J., Sung, K., Ye, J.C.: Deep learning with domain adaptation for accelerated projection-reconstruction MR. Magn. Reson. Med. **80**(3), 1189–1205 (2018). https://doi.org/10.1002/mrm.27106
7. Hyun, C.M., Kim, H.P., Lee, S.M., Lee, S., Seo, J.K.: Deep learning for undersampled MRI reconstruction. Phys. Med. Biol. **63**(13), 135007 (2018). https://doi.org/10.1088/1361-6560/aac71a
8. Jalal, A., Arvinte, M., Daras, G., Price, E., Dimakis, A.G., Tamir, J.: Robust compressed sensing MRI with deep generative priors. In: Advances in Neural Information Processing Systems, vol. 34, pp. 14938–14954 (2021)
9. Kofler, A., Haltmeier, M., Schaeffter, T., Kolbitsch, C.: An end-to-end-trainable iterative network architecture for accelerated radial multi-coil 2D cine MR image reconstruction. Med. Phys. **48**(5), 2412–2425 (2021). https://doi.org/10.1002/mp.14809
10. Küstner, T., et al.: CINENet: deep learning-based 3D cardiac CINE MRI reconstruction with multi-coil complex-valued 4D spatio-temporal convolutions. Sci. Rep. **10**, 13710 (2020). https://doi.org/10.1038/s41598-020-70551-8
11. Lee, D., Yoo, J., Tak, S., Ye, J.C.: Deep residual learning for accelerated MRI using magnitude and phase networks. IEEE Trans. Biomed. Eng. **65**(9), 1985–1995 (2018). https://doi.org/10.1109/TBME.2018.2821699
12. Lyu, J., et al.: Region-focused multi-view transformer-based generative adversarial network for cardiac cine MRI reconstruction. Med. Image Anal. **85**, 102760 (2023). https://doi.org/10.1016/j.media.2023.102760
13. Patel, D., Sastry, P.S.: Memorization in deep neural networks: does the loss function matter? In: Karlapalem, K., et al. (eds.) PAKDD 2021. LNCS (LNAI), vol. 12713, pp. 131–142. Springer, Cham (2021). https://doi.org/10.1007/978-3-030-75765-6_11

14. Qin, C., Schlemper, J., Caballero, J., Price, A.N., Hajnal, J.V., Rueckert, D.: Convolutional recurrent neural networks for dynamic MR Image reconstruction. IEEE Trans. Med. Imaging **38**(1), 280–290 (2019). https://doi.org/10.1109/TMI.2018.2863670

15. Schlemper, J., Caballero, J., Hajnal, J.V., Price, A.N., Rueckert, D.: A deep cascade of convolutional neural networks for dynamic MR Image reconstruction. IEEE Trans. Med. Imaging **37**(2), 491–503 (2018). https://doi.org/10.1109/TMI.2017.2760978

16. Terpstra, M.L., Maspero, M., Sbrizzi, A., van den Berg, C.A.: ⊥-loss: a symmetric loss function for magnetic resonance imaging reconstruction and image registration with deep learning. Med. Image Anal. **80**, 102509 (2022). https://doi.org/10.1016/j.media.2022.102509

17. Tezcan, K.C., Baumgartner, C.F., Luechinger, R., Pruessmann, K.P., Konukoglu, E.: MR Image reconstruction using deep density priors. IEEE Trans. Med. Imaging **38**(7), 1633–1642 (2019). https://doi.org/10.1109/TMI.2018.2887072

18. Vornehm, M., Wetzl, J., Giese, D., Ahmad, R., Knoll, F.: Spatiotemporal variational neural network for reconstruction of highly accelerated cardiac cine MRI. Eur. Heart J. - Cardiovasc. Imaging **23**(Supplement 2), 34–35 (2022). https://doi.org/10.1093/ehjci/jeac141.018

19. Wang, C., et al.: Recommendation for cardiac magnetic resonance imaging-based phenotypic study: imaging part. Phenomics **1**, 151–170 (2021). https://doi.org/10.1007/s43657-021-00018-x

20. Wang, C., et al.: CMRxRecon: an open cardiac MRI dataset for the competition of accelerated image reconstruction (2023)

21. Wang, S., et al.: Accelerating magnetic resonance imaging via deep learning. In: 2016 IEEE 13th International Symposium on Biomedical Imaging (ISBI), pp. 514–517 (2016). https://doi.org/10.1109/ISBI.2016.7493320

22. Yang, G., et al.: DAGAN: deep de-aliasing generative adversarial networks for fast compressed sensing MRI reconstruction. IEEE Trans. Med. Imaging **37**(6), 1310–1321 (2018). https://doi.org/10.1109/TMI.2017.2785879

23. Yang, J., Küstner, T., Hu, P., Liò, P., Qi, H.: End-to-end deep learning of non-rigid groupwise registration and reconstruction of dynamic MRI. Front. Cardiovasc. Med. **9**, 880186 (2022). https://doi.org/10.3389/fcvm.2022.880186

24. Zhang, T., Pauly, J.M., Vasanawala, S.S., Lustig, M.: Coil compression for accelerated imaging with Cartesian sampling. Magn. Reson. Med. **69**(2), 571–582 (2013). https://doi.org/10.1002/mrm.24267

25. Zhang, Y., Hu, Y.: Dynamic cardiac MRI reconstruction using combined tensor nuclear norm and Casorati matrix nuclear norm regularizations. In: 2022 IEEE 19th International Symposium on Biomedical Imaging (ISBI), pp. 1–4 (2022). https://doi.org/10.1109/ISBI52829.2022.9761409

ReconNext: A Encoder-Decoder Skip Cross Attention Based Approach to Reconstruct Cardiac MRI

Ruiyi Li[1], Hanyuan Zheng[1], Weiya Sun[2], and Rongjun Ge[3(✉)]

[1] School of Computer Science and Technology, Nanjing University of Aeronautics and Astronautics, Nanjing, China
[2] Beijing Institute of Tracking and Telecommunications Technology, Beijing, China
[3] School of Instrument Science and Engineering, Southeast University, Nanjing, China
rongjun_ge@seu.edu.cn

Abstract. Cardiac magnetic resonance imaging (MRI) is an advanced medical imaging technique widely used for the diagnosis and assessment of cardiovascular diseases. However, the acquisition time for cardiac MRI is generally longer, and cardiac motion can easily introduce artifacts that negatively impact image quality. Reconstructing cardiac MRI from under-sampled K-Space data has emerged as a viable approach to effectively reduce cardiac MRI acquisition time. In recent years, methods based on Deep Learning have been employed for image denoising and dehazing, yielding promising results. In this paper, we propose an MRI reconstruction network called ReconNext, based on MedNext and Encoder-Decoder Skip Cross Attention (EDSCA). The backbone of the proposed network is built upon MedNext, which leverages large-kernel convolutions to achieve a balance between local and non-local detail reconstruction. We introduce an Encoder-Decoder Skip Cross Attention structure, akin to self-attention, into the connection between the encoder and decoder. EDSCA incorporates cross-attention inputs from both the encoder and decoder. Compared to traditional skip connection, EDSCA better integrates information from the encoder and decoder, addressing issues of information loss. Experiment results demonstrate that ReconNext outperforms other conventional encoder-decoder architectures in cardiac MRI reconstruction, showcasing superior reconstruction effectiveness.

Keywords: Cardiac MRI · Reconstruction · Cross Attention

1 Introduction

Cardiac Magnetic Resonance Imaging (MRI) is an advanced medical imaging technique that non-invasively provides detailed information about the structure and function of the heart [1]. By harnessing powerful magnetic fields and harmless radio waves, cardiac MRI can generate high-resolution images that vividly depict different tissues and organs of the heart. Cardiac MRI finds widespread applications in the fields of medical diagnosis and research, and can be used to assess cardiac abnormalities, myocardial infarction,

© The Author(s), under exclusive license to Springer Nature Switzerland AG 2024
O. Camara et al. (Eds.): STACOM 2023, LNCS 14507, pp. 433–442, 2024.
https://doi.org/10.1007/978-3-031-52448-6_41

cardiac hypertrophy, valvular function, myocardial wall motion anomalies, and other cardiovascular diseases. Compared to traditional imaging techniques, cardiac MRI not only offers more precise diagnostic information but also obtains images without using ionizing radiation, thereby reducing patients' exposure to radiation risks. However, MRI imaging takes a long time, and there is a possibility of missing the optimal timing for diagnosis due to the extended imaging duration. One method to expedite MRI acquisition is to obtain high-quality MRI images by reconstructing under-sampled k-space data.

Compressed Sensing is employed for MRI reconstruction [2]. However, methods based on Compressed Sensing often require significant computational costs to achieve better reconstruction results, and the substantial time investment makes it challenging to apply in clinical practice. Utilizing CS for reconstruction necessitates the selection of hyperparameters, which need to be readjusted for different application scenarios.

In recent years, methods based on Deep Learning have been continuously maturing, demonstrating impressive performance in image denoising and reconstruction. Convolutional neural networks have proven to yield higher-quality images compared to methods based on Compressed Sensing (CS). Olaf Ronneberger et al. introduced the U-Net, which employs skip connections to supplement input information to the encoder, and it has been demonstrated to be versatile in various medical tasks [3]. Chen Qin et al. proposed CRNN, which utilizes a recurrent structure to capture spatiotemporal correlations in data for reconstruction [4]. Bo Zhou et al. presented DuDoRNet, which fuses multi-domain information from MRI and employs an RNN structure for reconstruction [5]. Networks like E2E-VarNet and KIKI-Net employ cascading structures for reconstruction [6, 7]. Most of the commonly used Deep Learning networks are encoder-decoder architectures, requiring skip connection structures to recover lost information during decoding. However, simple skip connections or concatenation can result in information loss, failing to fully leverage the global information in feature maps.

The Transformer [8] has found widespread application in the field of image processing due to its remarkable ability to model long-range relationships. The exceptional long-range modeling capability of the Transformer arises from its Multi-Head Self-Attention mechanism, which employs input-generated embedding matrices to compute long-range relationships within the image. Alexey Dosovitskiy et al. introduced the Vision Transformer (ViT) [9], applying the Transformer to the visual domain. Networks like Swin Transformer [10] have improved upon ViT's shortcomings and reduced the computational complexity of the Transformer. Jun Lyu et al. [11] proposed a model based on multi-view cross-attention, achieving impressive results in Cine Cardiac MRI reconstruction. Pengfei Guo introduced ReconFormer [12], a model based on a combination of RNN and Transformer, demonstrating strong performance on brain MRI. However, the attention mechanisms in the aforementioned networks utilize embeddings sourced solely from the encoder's outputs, lacking the fusion of information from the decoder's outputs.

MedNext [13], proposed by Saikat Roy et al., adopts a Unet architecture for its overall structure, while employing the ConvNext [14] framework for its internal modules. Inspired by the Transformer, the model employs large kernel convolutions to simulate long-range dependencies, effectively reducing computational costs and enhancing the global modeling performance of the convolutional structure. MedNext demonstrates excellent performance in various medical segmentation tasks, with significant untapped potential. However, MedNext utilizes only simple residual connections between the encoder and decoder, resulting in information loss. It requires the incorporation of appropriate attention mechanisms.

In this paper, we introduce a model based on MedNext and cross-attention mechanisms, which we refer to as ReconNext. The network adopts a MedNext-inspired encoder-decoder architecture, tailored for image reconstruction of fine details. Within the original residual components of MedNext, we modified it to incorporate an Encoder-Decoder Skip Cross Attention (EDSCA) mechanism. EDSCA computes similarity information between the encoder and decoder, leveraging the original encoder information to guide decoder outputs. This approach capitalizes on both deep-level information from intermediate encoder outputs and deep-level information from the decoder, as opposed to a simple skip connection. For Multicoil data, we design Multicoil Aggregation model to extract information from Multicoil feature map. Additionally, we appended a Data-Consistency layer at the end of the network to preserve undistorted information in the reconstructed images. Experimental results demonstrate that ReconNext outperforms other single encoder-decoder structures in terms of reconstruction quality.

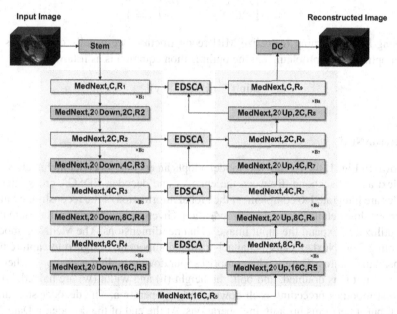

Fig. 1. The general structure of ReconNext. The backbone network is MedNext, which consists of MedNext Blocks and upsampling and downsampling blocks. EDSCA performs information exchange between codecs. At the end of the network, DC blocks are used to match the original k-space data

2 Methodology

2.1 MRI Reconstruction

Under-sampled data is obtained from fully sampled data. We assume that x represents the under-sampled k-space data, and y represents the fully sampled data. Here, F_a stands for the binary k-space matrix mask.

$$x = F_a y \tag{1}$$

Our objective is to reconstruct the image x, denoted as $\tilde{x}$, to be close to the image y. Here, f^{-1} represents the inverse Fourier transform, and f represents the Fourier transform. DC stands for Data Consistency, ensuring that the data retains partial information from the original image. The meaning of Eq. (2) is to retain the original k-space information at positions where mask $= 1$, while using a Deep Learning network F with a parameterized setting to reconstruct at positions where mask $= 0$, and subsequently restore the reconstructed results to obtain k-space information.

$$\tilde{x} = \begin{cases} f(F(f^{-1}(x)|\theta)), & F_a = 0 \\ DC(X), & F_a = 1 \end{cases} \tag{2}$$

Using Deep Learning to realize MRI reconstruction, we usually convert this task into an optimization problem, and the optimization equation is as follows:

$$\min|f^{-1}(\tilde{x}) - f^{-1}(y)| \tag{3}$$

2.2 ReconNext

As shown in Fig. 1, the ReconNext model adopts an encoder-decoder architecture with MedNext as its backbone. Between the encoder and decoder, EDSCA cross-attention modules are integrated to complement the global information of the reconstructed image and recover lost details during deep propagation. Given an input image x, the Stem module is utilized to expand the input image's channel dimensions. The MedNext module, resembling ConvNext, acts as a channel-scaling component, designed to capture high-dimensional features. The encoder comprises four downsampling operations, where the channel count C is doubled, and both the height (h) and width (w) are halved, with B MedNext modules preceding each downsampling operation. The decoder structure is similar, but it performs upsampling operations. At the end of the decoder, a Data Consistency (DC) [15] layer is added to retain vital information from the original K-space data.

2.3 EDSCA

After undergoing deep propagation using the CNN-based MedNext, the model loses some of the global information from the input image, necessitating the supplementation

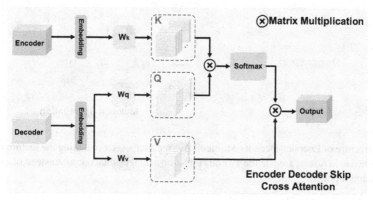

Fig. 2. Structure of Encoder Decoder Skip Cross Attention. Obtaining K from the encoder and Q, and V from the decoder, cross-attention computation results in a fused output from both the encoder and decoder with attention incorporated.

of the original global information from the input image. According to the nature of K-space data, while there is significant information loss in the under-sampled data, some overall information is still partially preserved, and the encoder layers can still retain some global information, which can be used for decoding supplementation. However, traditional residual connections fail to truly capture the internal relationships within the useful encoder feature maps and also overlook the utilization of decoder information, necessitating a new attention mechanism to complement the global information.

As shown in Fig. 2, We design the EDSCA (Encoder-Decoder Cross-Attention) module to simultaneously leverage information from both the encoder and decoder, yielding improved skip information. Unlike the direct use of skip connections in Med-Next, we employ a cross-attention mechanism similar to Transformer, which extracts outputs from corresponding layers of the encoder and decoder and blends deep information. The decoder's output is transformed into embeddings and multiplied by the W_q and W_v matrices to generate the Query Matrix and Value Matrix, respectively. Meanwhile, the encoder's output is multiplied by the W_k matrix to generate the Key Matrix. Here, E_d represents embeddings from the decoder, and E_e represents embeddings from the encoder.

$$Q = E_d W_q, K = E_e W_k, V = E_d W_V \qquad (4)$$

$$Attention = softmax(\frac{QK^T}{\sqrt{d_k}})V \qquad (5)$$

The reason for selecting the decoder's output as Q and V, and the encoder as K, is twofold: a) The encoder output still retains global information, so the encoder output is the one that needs to be queried. $Q*K$ calculates the attention distribution of Q with respect to K using vector dot product, requiring the decoder's output to query the attention distribution of the encoder's output. b) Because the ultimate goal is to supplement the attention distribution onto the decoder's output, the decoder's output serves as V to receive the attention distributions of both Q and K.

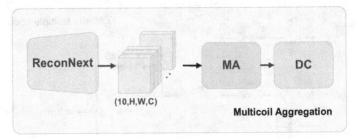

Fig. 3. Structure of Encoder-Decoder Multicoil Aggregation. After obtaining the feature maps of 10 coils, we use MA to aggregate the 10 coils into a one-dimensional output, instead of using the RSS combine function.

2.4 Multicoil Aggregation

Multi-coil data contains richer information compared to single-coil data, but using the traditional RSS method [16] directly for aggregating multi-coil data can lead to information loss. We need to design a new method for aggregating multi-coil information. As shown in Fig. 3, we have designed the Multi-coil Aggregation (MA) module. The MA module consists of two convolutional layers. The first convolutional layer extracts information by dimensionally expanding the input multi-coil data, while the second convolutional layer aggregates the data. After passing through the second layer, the coil dimensions are aggregated into one. Compared to outputting and utilizing single-coil data one by one and using RSS fusion, MA can fully utilize the high-dimensional information in multi-coil data.

2.5 Loss Function

The loss function employs L1 loss and perceptual loss [17]. Perceptual loss leverage the feature-capturing ability of VGG in the perceptual domain, we use VGG to calculate the loss to enhance the visual quality of the image. We compute the loss using the intermediate layer features obtained by feeding the images from the reconstructed image domain, produced by ReconNext, as inputs to VGG.

$$Loss = \frac{1}{N}(\sum_{n=1}^{N}\left|f^{-1}(\tilde{x}) - f^{-1}(y)\right| + \sum_{n=1}^{N}\left|VGG(f^{-1}(\tilde{x})) - VGG(f^{-1}(y))\right|) \quad (6)$$

3 Experiment

Dataset. We use the cardiac MRI Dataset from CMRxRecon2023 for training, validation, and testing [18]. The datasets provided by the competition are undersampled k-space data, fully sampled k-space data and the mask used for undersampling. The undersampled k-space data is used for reconstruction, the fully sampled k-space data is used for supervision, and the mask is used in the Data Consistency layer to maintain data consistency. The data set is divided into T1 and T2 mapping, which is further divided

T104   T108   T110

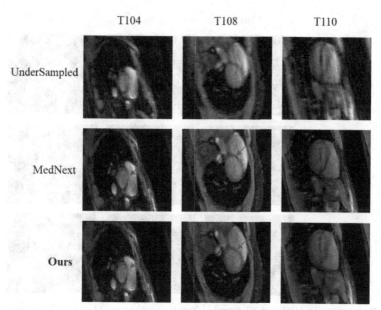

Fig. 4. Reconstruction results on T1 mapping from ReconNext and MedNext. It is obvious that ReconNext's reconstruction results preserve more details. T104 represents T1 mapping AccFactor04, T108 represents T1 mapping AccFactor08, T110 represents T1 mapping AccFactor10

into Multicoil and Singlecoil. According to the Accelerate Factor, Multicoil and Singlecoil are divided into AccFactor04, AccFactor08, and AccFactor10. Each AccFactor has a training set of 120 cases of T1 and T2 mapping data, and a verification set containing 60 cases of data. Multicoil data is collected by multiple coils, and the data has one more coil dimension than singlecoil to represent the data collected by different coils.

Training. We set the learning rate to 0.0002 and decay it by 0.9 every 5 epochs, with a batch size of 8. Regarding the attention module, for training on multi coil data, the patch size is set to 8 and the embeddings length is 256. For training on single coil data, the patch size is set to 4, and the embeddings length is 512. Due to the extensive dataset and time constraints, for multi coil data, T1 is trained for 1 epoch, and T2 is trained for 8 epochs. For single coil data, T1 is trained for 30 epochs, and T2 is trained for 50 epochs.

Validation. We use the validation set provided by the competition for inference and upload it to the designated platform of the competition for validation. After the inference using the trained ReconNext is completed, the generated data will be cropped to 1/6 of the original width and height, and only 2 slices of data will be retained for validation.

Result. As shown in Table 1, ReconNext exhibits a significant performance improvement compared to MedNext. As evident from Fig. 3 and Fig. 4, the reconstructed visual quality has also been notably enhanced, effectively alleviating the issue of unclear myocardial walls observed in MedNext. For example, in the T104 image, ReconNext reconstructed a more complete cardiac wall edge, while in T108, MedNext exhibited

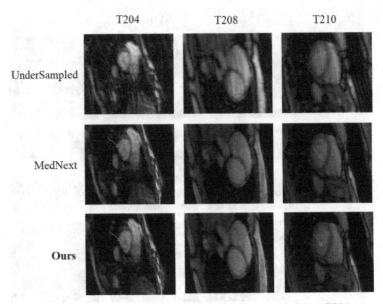

Fig. 5. Reconstruction results on T2 mapping from ReconNext and MedNext. T204 represents T2 mapping AccFactor04, T208 represents T2 mapping AccFactor08, T210 represents T2 mapping AccFactor10

Table 1. Results of ReconNext and MedNext on SingleCoil validation dataset. Evaluate reconstructed images using SSIM, PSNR and NMSE

Method		T1ACC04	T1ACC08	T1ACC10	T2ACC04	T2ACC08	T2ACC10
MedNext	SSIM	0.8453	0.8009	0.7730	0.8275	0.8022	0.8041
	PSNR	31.1484	28.5806	27.2708	28.8198	27.2038	26.8848
	NMSE	0.0283	0.0516	0.0699	0.0236	0.0333	0.0355
Ours	SSIM	**0.8597**	**0.8158**	**0.8022**	**0.8463**	**0.8225**	**0.8171**
	PSNR	**31.9072**	**29.3405**	**28.0833**	**29.5437**	**27.9186**	**27.3210**
	NMSE	**0.0231**	**0.0414**	**0.0583**	**0.0193**	**0.0280**	**0.0321**

irregularities in the cardiac wall shape, which ReconNext did not encounter. In T110, ReconNext produced images with less noise. In T204, ReconNext reconstructed internal details of the heart, and the cardiac wall was also more complete, whereas in T208, MedNext's reconstructed cardiac wall appeared somewhat blurred. In T210, ReconNext provided an overall clearer reconstruction of the cardiac region. As shown in Table 2, ReconNext has achieved excellent results on multi-coil data (Fig. 5).

Table 2. Results of ReconNext on MultiCoil validation dataset. Evaluate reconstructed images using SSIM, PSNR and NMSE

Method		T1ACC04	T1ACC08	T1ACC10	T2ACC04	T2ACC08	T2ACC10
Ours	SSIM	0.9330	0.8843	0.8661	0.9012	0.8805	0.8738
	PSNR	36.4828	32.1382	30.3734	31.8641	29.8087	28.9267
	NMSE	0.0084	0.0230	0.0344	0.0113	0.0191	0.0223

4 Conclusion

In this paper, we propose a model called ReconNext, which is based on an Encoder-Decoder Skip Cross Attention architecture. The model utilizes MedNext as a backbone and supplements the original image information in the encoder using EDSCA. For Multicoil data, we design Multicoil Aggregation model to extract information from Multicoil feature map. Experimental results demonstrate that the model effectively reconstructs cardiac MRI images. There are still several aspects of the model that can be improved, including multi-scale EDSCA fusion structure and the design of model's recurrent units. Future work will focus on enhancing these aspects.

Acknowledgements. This study was supported by the National Natural Science Foundation (No. 62101249 and No. 62136004), the Natural Science Foundation of Jiangsu Province (No. BK20210291), and the China Postdoctoral Science Foundation (No. 2021TQ0149 and No. 2022M721611).

References

1. Wang, C., et al.: Recommendation for cardiac magnetic resonance imaging-based phenotypic study: imaging part. Phenomics 1(4), 151–170 (2021).https://doi.org/10.1007/S43657-021-00018-X
2. Majumdar, A.: Improving synthesis and analysis prior blind compressed sensing with low-rank constraints for dynamic MRI reconstruction. Magn. Reson. Imaging 33(1), 174–179 (2015). https://doi.org/10.1016/J.MRI.2014.08.031
3. Ronneberger, O., Fischer, P., Brox, T.: U-net: convolutional networks for biomedical image segmentation. In: Navab, N., Hornegger, J., Wells, W.M., Frangi, A.F. (eds.) MICCAI 2015. LNCS, vol. 9351, pp. 234–241. Springer, Cham (2015). https://doi.org/10.1007/978-3-319-24574-4_28
4. Qin, C., Schlemper, J., Caballero, J., Price, A.N., Hajnal, J.V., Rueckert, D.: Convolutional recurrent neural networks for dynamic MR image reconstruction. IEEE Trans. Med. Imaging 38(1), 280–290 (2019). https://doi.org/10.1109/TMI.2018.2863670
5. Zhou, B., Kevin Zhou, S.: DuDoRNet: learning a dual-domain recurrent network for fast MRI reconstruction with deep T1 Prior. In: Proceedings of the IEEE Computer Society Conference on Computer Vision and Pattern Recognition, pp. 4272–4281 (2020).https://doi.org/10.1109/CVPR42600.2020.00433
6. Sriram, A., et al.: End-to-end variational networks for accelerated MRI reconstruction. In: Martel, A.L., et al. (eds.) MICCAI 2020. LNCS, vol. 12262, pp. 64–73. Springer, Cham (2020). https://doi.org/10.1007/978-3-030-59713-9_7

7. Eo, T., Jun, Y., Kim, T., Jang, J., Lee, H.J., Hwang, D.: KIKI-net: cross-domain convolutional neural networks for reconstructing undersampled magnetic resonance images. Magn. Reson. Med. **80**(5), 2188–2201 (2018). https://doi.org/10.1002/MRM.27201

8. Vaswani, A., et al.: Attention is all you need. In: Advances in Neural Information Processing Systems, vol. 2017, pp. 5999–6009 (2017). https://arxiv.org/abs/1706.03762v7. Accessed 13 Aug 2023

9. Dosovitskiy, A., et al.: An image is worth 16×16 words: transformers for image recognition at scale. In: ICLR 2021 - 9th International Conference on Learning Representations (2020). https://arxiv.org/abs/2010.11929v2. Accessed 13 Aug 2023

10. Liu, Z., et al.: Swin transformer: hierarchical vision transformer using shifted windows. In: Proceedings of the IEEE International Conference on Computer Vision, pp. 9992–10002 (2021).https://doi.org/10.1109/ICCV48922.2021.00986

11. Lyu, J., et al.: Region-focused multi-view transformer-based generative adversarial network for cardiac cine MRI reconstruction. Med. Image Anal. **85**, 102760. https://doi.org/10.1016/j.media.2023.102760

12. Guo, P., Mei, Y., Zhou, J., Jiang, S., Patel, V.M.: ReconFormer: accelerated MRI reconstruction using recurrent transformer (2022). https://arxiv.org/abs/2201.09376v2. Accessed 10 July 2023

13. Roy, S., et al.: MedNeXt: transformer-driven scaling of ConvNets for medical image segmentation. MICCAI (2023). https://arxiv.org/abs/2303.09975v4. Accessed 25 July 2023

14. Liu, Z., Mao, H., Wu, C.Y., Feichtenhofer, C., Darrell, T., Xie, S.: A ConvNet for the 2020s. In: Proceedings of the IEEE Computer Society Conference on Computer Vision and Pattern Recognition, vol. 2022, pp. 11966–11976 (2022). https://doi.org/10.1109/CVPR52688.2022.01167

15. Schlemper, J., Caballero, J., Hajnal, J.V., Price, A.N., Rueckert, D.: A deep cascade of convolutional neural networks for dynamic MR Image reconstruction. IEEE Trans. Med. Imaging **37**(2), 491–503 (2018). https://doi.org/10.1109/TMI.2017.2760978

16. Kellman, P., McVeigh, E.R.: Image reconstruction in SNR Units: a general method for SNR Measurement. Magn. Reson. Med. **54**(6), 1439 (2005). https://doi.org/10.1002/MRM.20713

17. Fang, Y., Deng, W., Du, J., Hu, J.: Identity-aware CycleGAN for face photo-sketch synthesis and recognition. Pattern Recognit **102**, 107249 (2020). https://doi.org/10.1016/J.PATCOG.2020.107249

18. Wang, C., et al.: CMRxRecon: an open cardiac MRI dataset for the competition of accelerated image reconstruction (2023). https://arxiv.org/abs/2309.10836v1. Accessed 27 Sept 2023

Temporal Super-Resolution for Fast T1 Mapping

Xunkang Zhao[1], Jun Lyu[2]([✉]), Fanwen Wang[3,4], Chengyan Wang[5],
and Jing Qin[2]

[1] School of Computer and Control Engineering, Yantai University, Yantai, China
[2] Centre for Smart Health, School of Nursing, The Hong Kong Polytechnic
University, Hung Hom, Hong Kong
junlyu@polyu.edu.hk
[3] Bioengineering Department and Imperial-X, Imperial College London, London, UK
[4] Cardiovascular Research Centre, Royal Brompton Hospital, London, UK
[5] Human Phenome Institute, Fudan University, Shanghai, China

Abstract. Cardiac T1 mapping can provide important biomarkers for
many cardiovascular diseases. However, the acquisition time of the T1
mapping sequence is relatively long, which not only affects image quality
but also causes discomfort to patients. Existing strategies to accelerate
T1 mapping imaging mostly involve reducing the sampling at each time
point and using reconstruction algorithms to predict complete images
from undersampled k-space data. These in-plane acceleration methods
often neglect the correlation between frames in the time series and may
introduce more artifacts with increasing acceleration factors, without
effectively reducing the overall scan time. We propose a novel inter-plane
acceleration method for T1 mapping, which is completely different from
the in-plane acceleration methods. This method can truly shorten the
scan time. Specifically, instead of using reconstruction algorithms at each
sampling time point, we reduce the number of sampling time points and
employ a time super-resolution algorithm to generate missing time point
images. For this purpose, we design a cross-frame adaptive enhancement
network to achieve temporal super-resolution, which effectively utilizes
input frame features to generate missing frame features and employs
cross-frame deformable attention mechanism to enrich the expression of
missing frame features, enhancing the reconstruction of missing frames
and avoiding artifacts. Experimental results on the CMRxRecon T1 mapping dataset demonstrate that the proposed network predicts missing
frames with clearer details and fewer artifacts compared to the comparative algorithms. This method opens up a new pathway for accelerating
T1 mapping imaging and holds promise for clinical applications.

Keywords: T1 mapping · temporal super-resolution · deformable
attention mechanism

O. Camara et al. (Eds.): STACOM 2023, LNCS 14507, pp. 443–453, 2024.
https://doi.org/10.1007/978-3-031-52448-6_42

1 Introduction

Cardiac magnetic resonancee (CMR) imaging is a non-invasive technique that provides essential information for the diagnosis and treatment of various cardiovascular diseases [19]. T1 mapping, utilizing a preparatory pulse and a series of images to sample the T1 relaxation time of tissues at different time points, is a significant application of CMR. It serves as a crucial biomarker for early detection of many cardiovascular diseases. Common T1 mapping sequences include Modified Look-Locker Inversion Recovery (MOLLI), Inversion Recovery Gradient Echo (IR-GRE), and Phase-Sensitive Inversion Recovery (PSIR) sequences, among others.

Traditional T1 mapping imaging is time-consuming due to the need to acquire complete k-space data and reconstruct images at each time point. [1,4,15] employed deep learning methods to reconstruct highly undersampled k-space data into complete images. Although this in-plane acceleration method enables rapid T1 mapping with good image quality, with increasing acceleration factors, the reconstructed images suffer from more detail loss, and importantly, they do not truly reduce the overall scanning time. In contrast to the in-plane acceleration methods, we propose a novel inter-plane acceleration method for rapid T1 mapping. Specifically, our approach still acquires complete k-space data at each time point, but reduces the number of sampling time points and restores the low temporal resolution fully-sampled image sequence to its original temporal resolution using frame interpolation techniques, ensuring high-quality and efficient T1 mapping. To the best of our knowledge, we are the first to propose the application of inter-plane acceleration techniques to rapid T1 mapping imaging.

Currently, there are numerous studies on time super-resolution. For instance, [5,12,16,18] estimate the optical flow between two adjacent frames and warps the neighboring frames to generate intermediate frames. However, for non-rigid deformations such as cardiac motion, the estimated optical flow field may be inaccurate. In addition to optical flow methods, [2,7,13,14] employ two consecutive frames to predict an adaptive convolutional kernel for each pixel, and estimates the pixel values of the intermediate frame by convolving the predicted kernel with the input frames on a per-pixel basis. However, these kernel-based approaches face a trade-off when dealing with large object displacements: using smaller kernels may lead to blurriness in the intermediate frames, while choosing larger kernels requires more computational resources.

We propose a corresponding solution to address the drawbacks of the above interpolation methods. Considering the drawbacks of optical flow-based methods and the characteristics of large cardiac motion, inspired by [21,23,24], we adopt a multi-scale deformable convolution approach to implicitly align two adjacent frames from coarse to fine in spatial domain and generate preliminary intermediate frames using linear weighting. Additionally, [11] has demonstrated the effectiveness of utilizing temporal information for CMR cine frame interpolation, so we design a refinement mechanism [21] to explore the auxiliary role of the already synthesized frames in predicting the current missing frame. However, the refinement mechanism still struggles to introduce more temporal information into the missing frame features, thus, we aim to integrate information from the

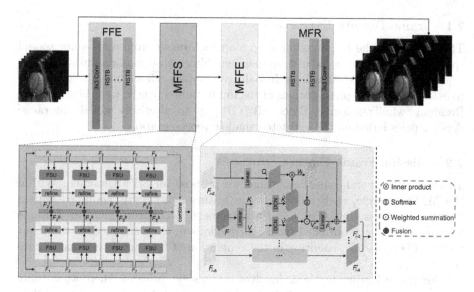

Fig. 1. The architecture of our proposed network. Within the MFFE module, $K = \{K_{j\to i}|j \in 2T-1 \wedge j \neq i\}$, $V = \{V_{j\to i}|j \in 2T-1 \wedge j \neq i\}$, $\hat{K} = \{\hat{K}_{j\to i}|j \in 2T-1 \wedge j \neq i\}$, $\hat{V} = \{\hat{V}_{j\to i} \mid j \in 2T-1 \wedge j \neq i\}$, $W = \{W_{j\to i}|j \in 2T-1 \wedge j \neq i\}$

entire time series to enrich the expression of missing frame features. Inspired by the widespread use of attention mechanisms [10,17,21], we employ a cross-frame deformable attention mechanism to adaptively aggregate temporal context and fully exploit the potential correlations in the time series.

This paper proposes a Cross-Frame Adaptive Enhancement Network (CFAE) for time super-resolution of low temporal resolution MOLLI sequence, achieving, for the first time, rapid cardiac T1 mapping imaging through inter-plane acceleration method. The CFAE network consists of four key modules: Frame Feature Extraction (FFE) module, Missing Frame Feature Synthesis (MFFS) module, Missing Frame Feature Enhancement (MFFE) module, and Missing Frame Reconstruction (MFR) module. Specifically, the FFE module aims to extract frame features, while the MFFS module generates missing frame features by performing multi-scale deformable convolutions in space from coarse to fine. The MFFE module utilizes a cross-frame deformable attention mechanism [21,22] to dynamically capture the temporal context of different frames within the time series, enhancing the feature representation of the synthesized missing frames by the MFFS module. Finally, MFR module aims to reconstruct the missing frames from enhanced features. The model achieved better performance on the CMRxRecon T1 mapping dataset compared to the contrast algorithms.

2 Methods

The architecture diagram of the CFAE network is illustrated in Fig. 1. The input to the network is a low frame-rate sequence $\{I_{2t-1}\}_{t=1}^{T}$ with T frames, and the output is a high frame-rate sequence $\{I_t\}_{t=1}^{2T-1}$, where $T = 5$.

2.1   Frame Feature Extraction

The purpose of the FFE module is to map each image from a low-dimensional space to a high-dimensional feature space, enabling subsequent modules to better capture non-linear relationships. Initially, we employ a 3×3 convolutional layer to extract shallow representations of the input. Subsequently, we utilize multiple Residual Swin Transformer Blocks (RSTB) [8,9] to effectively model long-range dependencies based on the extracted shallow representations.

2.2   Missing Frame Feature Synthesis

To achieve temporal super-resolution, we propose an MFFS module. Within the MFFS module, each missing frame feature synthesis is equipped with a Feature Synthesis Unit (FSU) and a refinement mechanism [21]. The T frame features $\{F_{2t-1}\}_{t=1}^{T}$ obtained from the previous module are used as the input to the MFFS module, resulting in the generation of the complete frame feature sequence $\{F_t\}_{t=1}^{T-1}$.

Our frame feature synthesis involves a two-stage process, where we initially generate preliminary intermediate frame features and then proceed to refine them to achieve a more precise representation.

To generate preliminary missing frame features, we employ the FSU module to align two adjacent frame features and subsequently fuse the alignment results. Considering the complexity of cardiac motion, each FSU module is equipped with two Pyramid Deformable Alignment (PDA) modules [22], enabling multi-scale alignment of the adjacent frame features F_t and F_{t+2}:

$$F_{t+2 \rightarrow t} = \mathrm{PDA}(F_{t+2}, o_{t+2 \rightarrow t}), F_{t \rightarrow t+2} = \mathrm{PDA}(F_t, o_{t \rightarrow t+2}), \qquad (1)$$

$o_{t+2 \rightarrow t}$ and $o_{t \rightarrow t+2}$ represent the backward and forward learnable offsets between frame features F_t and F_{t+2}, respectively. This coarse-to-fine approach in the spatial domain has been demonstrated to improve alignment to sub-pixel accuracys [22]. At the end of the FSU module, we apply pixel-level linear weighting to the warped results $F_{t+2 \rightarrow t}$ and $F_{t \rightarrow t+2}$, resulting in the preliminary missing frame feature $\tilde{F}_{t+1}$.

To address the potential inaccuracies in the synthesis of preliminary missing frame feature $\tilde{F}_{t+1}$ due to insufficient information, we introduce a bidirectional refinement mechanism. This mechanism leverages the previously refined missing frame features from the previous step to further refine the current preliminary missing frame feature, as illustrated in Fig. 1:

$$F_{t+1}^f = Refine(\tilde{F}_{t+1}^f, F_{t-1}^f), F_{t+1}^b = Refine(\tilde{F}_{t+1}^b, F_{t+3}^b), \qquad (2)$$

Refine denotes the refinement function.

The bidirectional refinement mechanism facilitates the incorporation of neighboring synthesized missing frame features into $\tilde{F}_{t+1}$, thereby compensating for the lack of information from adjacent frames and enhancing both the accuracy and temporal continuity of the frame interpolation results. At the end of the MFFS module, the fusion of F_{t+1}^f and F_{t+1}^b results in the final missing frame feature F_{t+1}.

2.3   Missing Frame Feature Enhancement

The purpose of this module is to enhance the feature representation of missing frames for subsequent missing frame reconstruction. Based on the following considerations: firstly, the refinement mechanism only introduces adjacent synthesized missing frame features to assist in the synthesis of the current missing frame feature, without exploring the significance of other frames for the missing frame; secondly, if individually processing consecutive frames $\{F_{2t}\}_{t=1}^{T-1}$ containing similar information, a substantial amount of temporal context would be lost. Therefore, in the MFFE module, we employ a cross-frame deformable attention mechanism to adaptively aggregate the temporal context of the sequence, deeply mining the connections between frames in the time series. Within the MFFE module, we denote the current missing frame feature as F_i and the remaining $2T - 2$ frame features as F, where $F = \{F_j | j \in 2T - 1 \wedge j \neq i\}$, as depicted in Fig. 1.

The implementation of MFFE module involves two steps. First, calculate the correlation between the feature vector of the missing frame feature F_i and the feature vectors of each frame feature F_j in the set F at the corresponding location of the object to obtain the attention map $W_{j \to i}(j \in 2T - 1, j \neq i)$. Secondly, the enhanced missing frame feature F_i^* is obtained through temporal aggregation of frame features.

In order to obtain more accurate correlation weights $W_{j \to i}$, we need to employ deformable convolution (DCN) [25] for inter-frame alignment. The alignment addresses the issue where the feature vectors representation of the missing frame feature F_i and the frame feature F_j at the same position differs due to inter-frame motion. Specifically, after obtaining the embedded features Q_i, K_j, and V_j through linear layers from the missing frame feature F_i and frame feature F_j, we apply DCN to warp K_j and V_j:

$$\hat{K}_{j \to i} = DCN(K_j), \hat{V}_{j \to i} = DCN(V_j), \tag{3}$$

$\hat{K}_{j \to i}$ and $\hat{V}_{j \to i}$ represent the embedding features aligned with Q_i, and they have now been spatially aggregated. Next, we perform an inner product between the aligned embedded feature $\hat{K}_{j \to i}$ and Q_i to obtain a more precise correlation weight $W_{j \to i}$:

$$W_{j \to i} = Q_i \cdot \hat{K}_{j \to i}. \tag{4}$$

The next step is to adaptively aggregate all the embedding features $\hat{V}_{j \to i}(j \in 2T - 1 \wedge j \neq i)$ along the temporal dimension:

$$V_i^* = \sum_{j=1}^{2T-1, j \neq i} softmax(W_{j \to i}) \cdot \hat{V}_{j \to i} = \sum_{j=1}^{2T-1, j \neq i} \frac{e^{W_{j \to i}}}{\sum_{j=1}^{2T-1, j \neq i} e^{W_{j \to i}}} \cdot \hat{V}_{j \to i}, \tag{5}$$

V_i^* represents the enhanced embedded features. In the final stage of the MFFE module, the enhanced missing frame feature F_i^* are obtained by adding the residual feature $\hat{F}_i$ [3], obtained from a linear layer applied to V_i^*, to the frame feature F_i:

$$F_i^* = F_i + \hat{F}_i. \qquad (6)$$

By modeling the feature vectors at the same location with respect to the relative objects, F_i^* adaptively aggregates spatial and temporal context from the other $2T - 2$ frame features.

2.4 Missing Frame Reconstruction

To reconstruct the missing frames from the enhanced missing frame features F_i^*, we employ a combination of RSTBs and convolutional layers. To optimize our network, we utilize the *Charbonnier* function [6] as the reconstruction loss:

$$L_{rec} = \sqrt{\left\| I_t^{rec} - I_t^{GT} \right\|^2 + \varepsilon^2}, \qquad (7)$$

I_t^{rec} represents the prediction of the $t - th$ missing frame, while I_t^{GT} represents the ground truth of the $t - th$ missing frame. Based on empirical knowledge, we set ε to a value of 1×10^{-3}. By utilizing a loss function, our network can be trained end-to-end to generate high-frame-rate sequences from low-frame-rate sequences.

3 Experiments

3.1 Dataset and Implementation

By using the MOLLI technique, a single-coil T1 mapping sequence was acquired, with each sequence consisting of 9 images with different T1-weightings [20]. T1 mapping was performed only in the short-axis (SAX) view, with a typical field-of-view (FOV) of $340 \times 340\,mm^2$, spatial resolution of $1.5 \times 1.5\,mm^2$, 5–6 slices, and slice thickness of $5.0\,mm$. The repetition time (TR) was set to $2.7\,ms$, the echo time (TE) to $1.1\,ms$, partial Fourier factor to $6/8$, and parallel imaging acceleration factor to $R = 2$. A total of 240 MOLLI sequences from 44 healthy volunteers were included, with each image being centrally cropped to a size of 96×96 pixels. Among them, 160 sequences were used as the training set, 40 sequences as the test set, and the remaining as the validation set.

The Cross-Frame Adaptive Enhancement Network was trained on a NVIDIA Tesla A100 GPU using PyTorch implementation. The FFE module and MFR module were configured with 2 and 6 layers of RSTB, respectively. We utilized the Adam optimizer for 150,000 iterations with a learning rate decay from $2e-4$ to $1e-7$ using the cosine annealing strategy. The batch size was set to 1. We evaluated the interpolation quality using the Peak Signal-to-Noise Ratio (PSNR) and the Structural Similarity Index (SSIM) between the predicted images and the ground truth images. Detailed information of the models provided in Table 3.

The proposed algorithm is compared with the modified MRI-CRNN [15], ConvLSTM [11], and SwinIR [8] models for frame interpolation. Quantitative

Table 1. Quantitative results of different models for time super-resolution of T1 mapping sequences.

Dataset	Model	PSNR	SSIM	FLOPs
T1mapping from CMRxRecon	MRI-CRNN	31.38	0.8580	154.740G
	ConvLSTM	29.12	0.7993	348.211G
	SwinIR	29.95	0.8364	147.344G
	Ours	33.45	0.8847	395.504G

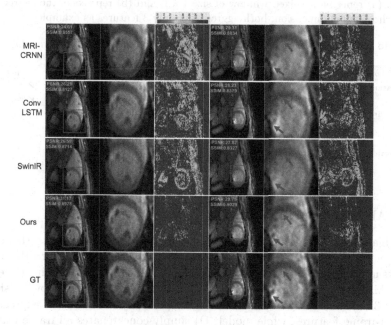

Fig. 2. The imaging results of missing frames generated by different algorithms and the T1 mapping using the time super-resolution enhanced sequences.

results are presented in Table 1. Despite the relatively higher FLOPs of the proposed models, they exhibit the best performance in terms of evaluation metrics. This indicates superior T1 mapping image reconstruction quality. Given the stringent image quality requirements in medical image diagnosis and research, our model stands as an excellent choice, especially in the context of ample computational resources available today. As depicted in Fig. 2, the artifacts in the missing frame generated by our model are minimized, revealing a clear visibility of the cardiac contours, and more intricate image details are predicted. According to the principle that darker regions in the error map correspond to smaller discrepancies between the predicted and the ground truth image, our algorithm achieves heart reconstruction that is closer to the ground truth. As indicated by the arrows, our algorithm demonstrates clearer details and contrast in the ventricular wall and epicardium.

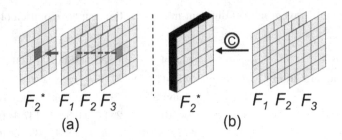

$$F_2^* \quad F_1 \, F_2 \, F_3 \qquad\qquad F_2^* \qquad F_1 \, F_2 \, F_3$$
(a)   (b)

Fig. 3. (a) represents a fixed window of size 1×1, and (b) represents the aggregation along the channel dimension, both aggregating only 3 features as examples.

Table 2. Ablation experiments of the proposed modules under the same dataset.

Method		(A)	(B)	(C)	(D)	Proposed
Feature Interpolation	FSU	✔	✔	✔	✔	✔
	Refinement					✔
Feature Aggregation	MFFE with a 1×1 fixed window		✔			
	MFFE with a deformable window			✔		✔
	channel-wise cat				✔	
Metrics	PSNR	31.48	32.74	32.85	32.23	33.45
	SSIM	0.8544	0.8733	0.8784	0.8718	0.8847

3.2 Ablation Study

The ablation experiment investigated the roles of two key components: the refinement module and the MFFE module. The baseline model (A) directly reconstructs intermediate frame features synthesized using FSU. To examine the effect of the MFFE module, models (B) and (C) utilize a fixed 1×1 window, as shown in Fig. 3 (a), and a deformable window, respectively, for adaptive aggregation of inter-frame features, while model (D) simply concatenates all frame features along the channel dimension, as depicted in Fig. 3 (b). To investigate the effect of the refinement module, a comparison is made between the proposed model and model (C). Evaluation metrics for all models on the same dataset are shown in Table 2.

From Table 2, it can be observed that: 1) Performing feature aggregation before reconstruction contributes to improving reconstruction quality. 2) The deformable attention mechanism in the MFFE module yields significantly better reconstruction results in aggregating frame features compared to simple concatenation. 3) The performance of the MFFE module in aggregating temporal information using a deformable window is better than that of a fixed 1×1 window. 4) By comparing the proposed model with model (C), it can be concluded that incorporating the refinement module helps in synthesizing frame features more accurately, resulting in high-quality reconstruction of missing frames.

Table 3. Model Information.

Model Information	
Task of participation	CMRxRecon
University/organization	School of Computer and Control Engineering, Yantai University, Yantai, China
Single-channel or multi-channel	Single-channel
Hardware configuration (GPU, VRAM, CPU, number of cores)	NVIDIA A100-PCIE, 40GB, Intel(R) Xeon(R) Gold 5320 CPU @ 2.20 GHz, 26
Training time	7 days
Inference time	60 s
Performance on the training set (PSNR, SSIM, NMSE)	42.03, 0.9707, 0.0001
Performance on the validation set (PSNR, SSIM, NMSE)	33.45, 0.8847, 0.0012
Docker submitted?	NO
Use of segmentation labels	NO
Use of pre-training (if applicable, specify the data)	NO
Data augmentation	Crop
Data standardization	$[-1.1]$
Model parameter number	8.29M
Loss function	Charbonnier function
Incorporation of a physical model (if applicable, specify the model)	NO
Use of unrolling	NO
Application of k-space fidelity	NO
Model backbone	Swin Transformer
Amplitude or complex-valued operations	complex-valued operations

4 Conclusion

We propose a novel Cross-Frame Adaptive Enhancement Network (CFAE), which achieves inter-plane acceleration for T1 mapping imaging for the first time. The MFFS module in the network utilizes more temporal information for frame interpolation, while the MFFE module adaptively integrates spatial and temporal context of the entire sequence to enhance missing frame features. Our experiments demonstrate that the network effectively utilizes the input frame information to reconstruct missing frames of high quality, providing favorable images for subsequent T1 mapping imaging. In the future, we will explore using the generated missing frames by CFAE as ground truth for further frame interpolation, achieving unsupervised rapid T1 mapping.

References

1. Beaumont, J., et al.: Multi t1-weighted contrast imaging and t1 mapping with compressed sensing flaws at 3 t. Magnetic Resonance Materials in Physics, Biology and Medicine, pp. 1–14 (2023)
2. Cheng, X., Chen, Z.: Multiple video frame interpolation via enhanced deformable separable convolution. IEEE Trans. Pattern Anal. Mach. Intell. 44(10), 7029–7045 (2021)
3. He, K., Zhang, X., Ren, S., Sun, J.: Deep residual learning for image recognition. In: Proceedings of the IEEE Conference on Computer Vision and Pattern Recognition, pp. 770–778 (2016)
4. Jeelani, H., Yang, Y., Zhou, R., Kramer, C.M., Salerno, M., Weller, D.S.: A myocardial t1-mapping framework with recurrent and u-net convolutional neural networks. In: 2020 IEEE 17th International Symposium on Biomedical Imaging (ISBI), pp. 1941–1944 (2020). https://doi.org/10.1109/ISBI45749.2020.9098459
5. Jiang, H., Sun, D., Jampani, V., Yang, M.H., Learned-Miller, E., Kautz, J.: Super slomo: high quality estimation of multiple intermediate frames for video interpolation. In: Proceedings of the IEEE Conference on Computer Vision and Pattern Recognition, pp. 9000–9008 (2018)
6. Lai, W.S., Huang, J.B., Ahuja, N., Yang, M.H.: Deep Laplacian pyramid networks for fast and accurate super-resolution. In: Proceedings of the IEEE Conference on Computer Vision and Pattern Recognition, pp. 624–632 (2017)
7. Lee, H., Kim, T., Chung, T.Y., Pak, D., Ban, Y., Lee, S.: Adacof: adaptive collaboration of flows for video frame interpolation. In: Proceedings of the IEEE/CVF Conference on Computer Vision and Pattern Recognition, pp. 5316–5325 (2020)
8. Liang, J., Cao, J., Sun, G., Zhang, K., Van Gool, L., Timofte, R.: Swinir: image restoration using swin transformer. In: Proceedings of the IEEE/CVF International Conference on Computer Vision, pp. 1833–1844 (2021)
9. Liu, Z., et al.: Swin transformer: Hierarchical vision transformer using shifted windows. In: Proceedings of the IEEE/CVF International Conference on Computer Vision, pp. 10012–10022 (2021)
10. Lyu, J., Sui, B., Wang, C., Tian, Y., Dou, Q., Qin, J.: Dudocaf: dual-domain cross-attention fusion with recurrent transformer for fast multi-contrast MR imaging. In: Wang, L., Dou, Q., Fletcher, P.T., Speidel, S., Li, S. (eds.) MICCAI 2022. LNCS, vol. 13436, pp. 474–484. Springer, Cham (2022). https://doi.org/10.1007/978-3-031-16446-0_45
11. Lyu, Q., Shan, H., Xie, Y., Li, D., Wang, G.: Cine cardiac MRI motion artifact reduction using a recurrent neural network (2020)
12. Niklaus, S., Liu, F.: Context-aware synthesis for video frame interpolation. In: Proceedings of the IEEE Conference on Computer Vision and Pattern Recognition, pp. 1701–1710 (2018)
13. Niklaus, S., Mai, L., Liu, F.: Video frame interpolation via adaptive convolution. In: Proceedings of the IEEE Conference on Computer Vision and Pattern Recognition, pp. 670–679 (2017)
14. Niklaus, S., Mai, L., Liu, F.: Video frame interpolation via adaptive separable convolution. In: Proceedings of the IEEE International Conference on Computer Vision, pp. 261–270 (2017)
15. Qin, C., Schlemper, J., Caballero, J., Price, A.N., Hajnal, J.V., Rueckert, D.: Convolutional recurrent neural networks for dynamic MR image reconstruction. IEEE Trans. Med. Imaging 38(1), 280–290 (2019). https://doi.org/10.1109/TMI.2018.2863670

16. Ranjan, A., Black, M.J.: Optical flow estimation using a spatial pyramid network. In: Proceedings of the IEEE Conference on Computer Vision and Pattern Recognition, pp. 4161–4170 (2017)
17. Shi, Z., Xu, X., Liu, X., Chen, J., Yang, M.H.: Video frame interpolation transformer. In: Proceedings of the IEEE/CVF Conference on Computer Vision and Pattern Recognition, pp. 17482–17491 (2022)
18. Sun, D., Yang, X., Liu, M.Y., Kautz, J.: PWC-net: CNNs for optical flow using pyramid, warping, and cost volume. In: Proceedings of the IEEE Conference on Computer Vision and Pattern Recognition, pp. 8934–8943 (2018)
19. Wang, C., et al.: Recommendation for cardiac magnetic resonance imaging-based phenotypic study: imaging part. Phenomics 1, 151–170 (2021)
20. Wang, C., et al.: CMRxrecon: an open cardiac MRI dataset for the competition of accelerated image reconstruction. arXiv preprint arXiv:2309.10836 (2023)
21. Wang, H., Xiang, X., Tian, Y., Yang, W., Liao, Q.: STDAN: deformable attention network for space-time video super-resolution. IEEE Trans. Neural Networks Learn. Syst. (2023)
22. Wang, X., Chan, K.C., Yu, K., Dong, C., Change Loy, C.: EDVR: video restoration with enhanced deformable convolutional networks. In: Proceedings of the IEEE/CVF Conference on Computer Vision and Pattern Recognition Workshops, pp. 0–0 (2019)
23. Xiang, X., Tian, Y., Zhang, Y., Fu, Y., Allebach, J.P., Xu, C.: Zooming slow-MO: fast and accurate one-stage space-time video super-resolution. In: Proceedings of the IEEE/CVF Conference on Computer Vision and Pattern Recognition, pp. 3370–3379 (2020)
24. Xu, G., Xu, J., Li, Z., Wang, L., Sun, X., Cheng, M.M.: Temporal modulation network for controllable space-time video super-resolution. In: Proceedings of the IEEE/CVF Conference on Computer Vision and Pattern Recognition, pp. 6388–6397 (2021)
25. Zhu, X., Hu, H., Lin, S., Dai, J.: Deformable convnets v2: more deformable, better results. In: Proceedings of the IEEE/CVF Conference on Computer Vision and Pattern Recognition, pp. 9308–9316 (2019)

NoSENSE: Learned Unrolled Cardiac MRI Reconstruction Without Explicit Sensitivity Maps

Felix Frederik Zimmermann$^{(\boxtimes)}$ and Andreas Kofler

Physikalisch-Technische Bundesanstalt (PTB), Berlin, Germany
felix.zimmermann@ptb.de

Abstract. We present a novel learned image reconstruction method for accelerated cardiac MRI with multiple receiver coils based on deep convolutional neural networks (CNNs) and algorithm unrolling. In contrast to many existing learned MR image reconstruction techniques that necessitate coil-sensitivity map (CSM) estimation as a distinct network component, our proposed approach avoids explicit CSM estimation. Instead, it implicitly captures and learns to exploit the inter-coil relationships of the images. Our method consists of a series of novel learned image and k-space blocks with shared latent information and adaptation to the acquisition parameters by feature-wise modulation (FiLM), as well as coil-wise data-consistency (DC) blocks.

Our method achieved PSNR values of 34.89 and 35.56 and SSIM values of 0.920 and 0.942 in the cine track and mapping track validation leaderboard of the MICCAI STACOM CMRxRecon Challenge, respectively, ranking 4th among different teams at the time of writing.

Code is available at https://github.com/fzimmermann89/CMRx Recon.

Keywords: Calibration-Free MRI · Model-based Deep Learning · Accelerated Cardiac MRI · Recurrent U-Nets

1 Introduction

Deep Learning-based approaches have recently been widely applied in the field of image reconstruction across different imaging modalities [1,2,8,23]. Recently, several image reconstruction challenges have been organized for different image reconstruction problems [10,14,16], and neural networks based on algorithm unrolling [15] are often among the winning teams.

Here, we present a new method developed for the MICCAI STACOM CMRxRecon Challenge, where the task is to reconstruct cardiac cine (and quantitative mapping) MR images from undersampled k-space measurements. In the following, we will mainly focus on the cine reconstruction. Our method is based on algorithm unrolling and utilizes both model-based and learned components. Unlike other recently published methods (e.g., [25,30]), our approach does not explicitly estimate coil sensitivity maps (CSM), and we argue that employing CSMs might potentially introduce model errors for the considered task.

© The Author(s), under exclusive license to Springer Nature Switzerland AG 2024
O. Camara et al. (Eds.): STACOM 2023, LNCS 14507, pp. 454–466, 2024.
https://doi.org/10.1007/978-3-031-52448-6_43

2   Methods

2.1   Theory and Motivation

Let $\mathbf{x}_{\text{true}} \in \mathbb{C}^N$ be the vector representation of the unknown cardiac MR image with $N = N_x \cdot N_y \cdot N_z \cdot N_t$. The forward model considered in multi-coil MRI is given by

$$\mathbf{y}_I := \mathbf{A}_I(\mathbf{C})\mathbf{x}_{\text{true}} + \mathbf{e}, \tag{1}$$

where $\mathbf{A}_I(\mathbf{C}) : \mathbb{C}^N \to \mathbb{C}^M$ denotes the multi-coil MR operator with

$$\mathbf{A}_I(\mathbf{C}) := (\mathbf{I}_{N_c} \otimes \mathbf{F}_I)\mathbf{C}, \quad \text{with} \quad \mathbf{C}^{\mathsf{H}}\mathbf{C} = \mathbf{I}_N \tag{2}$$

where $\mathbf{C} = [\mathbf{C}_1, \ldots, \mathbf{C}_{N_c}]^{\mathsf{T}} \in \mathbb{C}^{(N_c \cdot N) \times N}$ with $\mathbf{C}_c = \text{diag}(\mathbf{s}_c) \in \mathbb{C}^{N \times N}$ denotes the operator of the N_c stacked CSMs, which are represented by diagonal operators. Here, $\otimes$ represents the Kronecker product, $\mathbf{F}_I := \mathbf{S}_I \mathbf{F}$ with $\mathbf{F}$ being a frame-wise 2D Fourier operator, $\mathbf{S}_I$ a binary sub-sampling mask that collects a subset $I \subset J = \{1, \ldots, N_x \cdot N_y\}$ of the k-space coefficients of a 2D image, $\mathbf{I}_{N_c}$ and $\mathbf{I}_N$ are the N_c- and N-dimensional identity operators, respectively, and $\mathbf{e}$ is complex-valued Gaussian noise. Additionally, let $\mathbf{A}$ be defined as in (1) but with $I = J$, i.e. it models a full acquisition with no acceleration, and thus, $\mathbf{S}_I = \mathbf{I}_N$. The operator $\mathbf{A}_I$ depends on $\mathbf{C}$ which is unknown and must be estimated with appropriate methods [26]. Once $\mathbf{C}$ is fixed, one can consider variational problems of the form

$$\min_{\mathbf{x}} \mathcal{F}_\lambda(\mathbf{x}), \quad \mathcal{F}_\lambda(\mathbf{x}) := \frac{1}{2}\|\mathbf{A}_I(\mathbf{C})\mathbf{x} - \mathbf{y}_I\|_2^2 + \lambda \mathcal{R}(\mathbf{x}), \tag{3}$$

where $\mathcal{R}(\mathbf{x})$ encodes the regularizing properties that can be learned from data and $\lambda > 0$. From (3), different reconstruction methods based on algorithm unrolling [15] can be derived. For example, by identifying the learned components as learned proximal operators [4], learned sparsifying transforms [11], learned filter transforms with learned potential functions [8] or learned denoisers/artifact reduction methods [2,23]. Generally, these methods alternate between the application of learned blocks and model-based blocks, which use information on the forward and adjoint operators $\mathbf{A}_I$ and $\mathbf{A}_I^{\mathsf{H}}$. For multi-coil acquisitions, the latter blocks often require solving inner minimization problems [2,5,23]. Last, instead of first estimating the operator $\mathbf{C}$ and fixing it when deriving a reconstruction network from (3), recent methods often estimate CSMs using special blocks within a learned reconstruction [25,30].

In this work, we adopt a strategy based on the structure of the specific problem formulation and task posed by the CMRxRecon Challenge. Unlike other methods, our approach does not require a prior estimate of the CSMs or their refinement within the network architecture. We illustrate that unrolled methods derived from a variational problem as in (3) may introduce systematic errors that the learned components must compensate for, rendering them sub-optimal.

2.2   CMRxRecon Dataset

We briefly describe the given dataset (see [27] for more details) and the associated challenge task motivating our proposed approach. Let the multi-coil k-space data acquired according to (1) be of the form $\mathbf{y}_I = [\mathbf{y}_I^1, \ldots, \mathbf{y}_I^{N_c}]^\mathsf{T}$. The provided dataset consists of undersampled k-space data of short- and long-axis (SAX/LAX) images, retrospectively undersampled with different acceleration factors $R \in \{4, 8, 10\}$ (not counting the 24 central ACS lines that are always sampled) from fully-sampled data and root sum of squares (RSS) reconstructions obtained from the latter, i.e. $\mathcal{D} = \{(\mathbf{y}_I, \mathbf{S}_I, \mathbf{y}, \mathbf{x}_{\mathrm{RSS}})\}$, where

$$\mathbf{y}_I = \mathbf{S}_I \mathbf{y}, \quad \mathbf{y} := \mathbf{A}(\mathbf{C}_{\mathrm{true}})\mathbf{x}_{\mathrm{true}} + \mathbf{e}, \quad \mathbf{x}_{\mathrm{RSS}} = \Big(\sum_{c=1}^{N_c} |\mathbf{F}_I^\mathsf{H} \mathbf{y}_I^c|^2 \Big)^{1/2}. \quad (4)$$

Importantly, we note that, first, we do not have access to the pairs $(\mathbf{C}_{\mathrm{true}}, \mathbf{x}_{\mathrm{true}})$ that generate the complete k space data $\mathbf{y}$ according to the forward model as in (1). Secondly, due to the retrospective undersampling, the undersampled data as well as the targets are subject to the same realization of the random noise $\mathbf{e}$. Finally, the task of the challenge is to obtain an estimate of $\mathbf{x}_{\mathrm{RSS}}$ from $\mathbf{y}_I$, rather than finding an estimate of $\mathbf{x}_{\mathrm{true}}$. This has an important consequence as illustrated in the following.

Let $\mathbf{B} : V \to W$ be a linear operator between Hilbert spaces V and W (possibly the forward model in (2)), and $\{\mathbf{R}_\lambda\}_{\lambda>0}$ with $\mathbf{R}_\lambda : W \to V$ be a family of continuous operators. Then, $\{\mathbf{R}_\lambda\}_{\lambda>0}$ is a regularization of the Moore-Penrose generalized inverse $\mathbf{B}^\dagger$, if $\mathbf{R}_\lambda \to \mathbf{B}^\dagger$ for $\lambda \to 0$ pointwise on the domain of $\mathbf{B}^\dagger$. In particular, for the fully-sampled case of $\mathbf{A}$ in (1), we have

$$\mathbf{x}^* := \arg\min_{\mathbf{x}} \mathcal{F}_\lambda(\mathbf{x}) \xrightarrow{\lambda \to 0} \mathbf{A}^\dagger \mathbf{y} = \mathbf{A}^\mathsf{H} \mathbf{y} = \sum_{c=1}^{N_c} \mathbf{F}_I^\mathsf{H} \mathbf{C}_c^\mathsf{H} \mathbf{y}_I^c, \quad (5)$$

where we used that $\mathbf{A}^\dagger := (\mathbf{A}^\mathsf{H}\mathbf{A})^{-1}\mathbf{A}^\mathsf{H} = \mathbf{A}^\mathsf{H}$ due the normalization of $\mathbf{C}$ in (2). From (5), we observe that even in the fully-sampled case, $|\mathbf{x}^*| \xrightarrow{\lambda \to 0} |\mathbf{x}_{\mathrm{RSS}}|$ due to the CSMs. Specifically, this implies that any reconstruction network designed to approximate a solution to problem (3) inherently introduces an error that the learned component must counteract. This observation motivates the construction of a network that addresses a different problem than stated in (3). Rather than learning to obtain an estimation of the noise-free complex-valued ground-truth image (and subsequently utilizing its magnitude as an estimate for the RSS-reconstruction), it seems a more promising approach to learn to estimate fully-sampled (potentially noisy) k-space data, obtain the respective coil-images by a simple Fourier transform, and finally perform an RSS-reconstruction. Indeed, there have been methods proposed that estimate k-space data instead of the underlying image [30]; nevertheless, these methods lack an explanation for the rationale behind this preference.

2.3 Proposed Approach

We propose a learned method that, instead of estimating $\mathbf{x}_{\text{true}}$ by deriving a reconstruction scheme from (3), directly estimates the fully-sampled multi-coil k-space data $\tilde{\mathbf{y}} := [\tilde{\mathbf{y}}^1, \ldots, \tilde{\mathbf{y}}^{N_c}]^\mathsf{T}$. Given $\tilde{\mathbf{y}}$, an RSS can be performed to obtain an estimate of $\mathbf{x}_{\text{RSS}}$. Hence, a suitable problem formulation is

$$\min_{\mathbf{y}=(\mathbf{y}^1,\ldots,\mathbf{y}^{N_c})} \mathcal{G}(\mathbf{y}), \quad \mathcal{G}(\mathbf{y}) := \mathcal{S}_\theta(\mathbf{y}) \text{ such that } \mathbf{S}_I \mathbf{y} = \mathbf{y}_I, \tag{6}$$

where $\mathcal{S}_\theta(\mathbf{y})$ encodes the regularization imposed on the sought k-space data. Comparing (6) and (3), we observe two key differences: first, the problem is now formulated in k-space rather than in image space. Second, instead of only requiring approximate data-consistency between the image and the acquired k-space data, it might be desirable to obtain hard data consistency, as suggested by the constraint in (6) in the absence of noise.

In line with related approaches [23,25], we propose to construct a cascade that alternates between the application of CNN-blocks as well as data-consistency layers. Instead of relying on CSM estimation, we work directly with the coil-weighted k-space/ image data, i.e. each vector in the following consists of N_c components. Let $\mathbf{E} := (\mathbf{I}_{N_c} \otimes \mathbf{F})$ denote the operator that performs a frame-wise 2D Fourier transformation on each coil-image, and $\mathbf{E}^\mathsf{H}$ its adjoint. We unroll T iterations of the following scheme for $k = 0, \ldots, T-1$:

$$\tilde{\mathbf{z}}_k, \tilde{\mathbf{h}}_{k+1} := \mathrm{X}_\theta\left(\mathbf{E}^\mathsf{H}\mathbf{y}_k, \tilde{\mathbf{h}}_k, \mathbf{c}_k\right), \tag{7}$$

$$\mathbf{z}_k, \mathbf{h}_{k+1} := \mathrm{Y}_\theta\left(\mathbf{E}\,\tilde{\mathbf{z}}_k, \mathbf{h}_k, \mathbf{y}_I, \mathbf{c}_k\right), \tag{8}$$

$$\mathbf{y}_{k+1} := \left[\arg\min_{\mathbf{y}^1} \mathcal{G}_{\lambda_k^1, \mathbf{z}_k^1}(\mathbf{y}^1), \ldots, \arg\min_{\mathbf{y}^{N_c}} \mathcal{G}_{\lambda_k^{N_c}, \mathbf{z}_k^{N_c}}(\mathbf{y}^{N_c})\right]^\mathsf{T}, \tag{9}$$

$$\tilde{\mathbf{x}}_{\text{RSS}} := \mathrm{RSS}(\mathbf{y}_T) \tag{10}$$

where X_θ and Y_θ denote U-Net [22] blocks operating in image- and k-space, respectively [6]. In both, information from the latent space is passed as hidden variables to the next iteration as $\mathbf{h}_{k+1}$ and $\tilde{\mathbf{h}}_{k+1}$, respectively, inspired by the recent success of recurrent networks [30] for MR reconstruction. Moreover, the blocks use information about the acquisition, i.e., the axis (SAX or LAX), the slice/view, the current iteration k, and the acceleration factor R, which are encoded in one hot and transformed into $\mathbf{c}_k$ by a fully connected network (MLP). In addition, Y_θ utilizes the acquired undersampled $\mathbf{y}_I$ data as additional input in each iteration.

The functionals in (9) take the form

$$\mathcal{G}_{\lambda_k^c, \mathbf{z}_k^c}(\mathbf{y}^c) = \frac{1}{2}\left\|\mathbf{S}_I\mathbf{y}^c - \mathbf{y}_I^c\right\|_2^2 + \frac{\lambda_k^c}{2}\left\|\mathbf{y}^c - \mathbf{z}_k^c\right\|_2^2 \tag{11}$$

and separately regularize each coil-weighted k-space data. Note that the CNNs in (7), in contrast, jointly utilize all coil information. For each DC block and each coil c, we use a learned linear function with bias from $\mathbf{c}_k$ to λ_k^c. In contrast

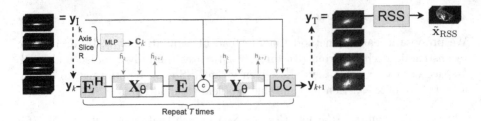

Fig. 1. Schematic overview of the proposed method. It applies U-Net blocks to the multi-coil image and k-space data and uses intermediate (coil-wise) data-consistency blocks and a final RSS reconstruction. Note the information sharing between different iterations via the hidden latent variables h_k (k-space) and $\tilde{h}_k$ (image-space). Both U-Nets are conditioned on $\mathbf{c}_k \in \mathbb{R}^{192}$, in which we encode information about the axis, slice, iteration, and acceleration factor via a MLP (details in the main text). In each iteration, the DC block performs (12) with λ_k^c obtained from $\mathbf{c}_k$ by a learned affine mapping.

to (6), no hard data-consistency is enforced in (11). In preliminary experiments, we experimented with hard data-consistency constraints by projecting the intermediate k-space outputs onto the kernel of the sub-sampling operator $\mathbf{S}_I$ in the spirit of [24]. We observed a more stable training when employing a soft data-consistency in (9), i.e., solving (11) as [21]

$$\mathbf{y}_{k+1}^c = \mathbf{S}_I \left(\frac{1}{1+\lambda_k^c}\mathbf{y}_I^c + \frac{\lambda_k^c}{1+\lambda_k^c}\mathbf{z}_k^c \right) + \overline{\mathbf{S}}_I \mathbf{z}_k^c , \tag{12}$$

where $\overline{\mathbf{S}}_I = \mathbf{I}_N - \mathbf{S}_I$. In our network, we obtain λ_k^c from $\mathbf{c}_k$ by a learned affine transformation, possibly allowing the resulting λ_k^c to become zero. This in turn allows the network to enforce hard data-consistency.

Figure 1 shows a schematic overview of our proposed approach. The U-Net blocks X_θ and Y_θ within our network are depicted in Fig. 2. These blocks include encoders/decoders comprising separate SiLU-activated spatial convolutions in the (xy)-dimensions and time-wise convolutions [20,32]. We employ group normalization, FiLM-based conditioning [17,19] on $\mathbf{c}_k$, and, in the encoders, Coord-Convs [12]. Notably, we introduce novel *LatentGU* blocks after each encoder except the first. These blocks, inspired by the minimal gated unit [31], consist of $1 \times 1 \times 1$ convolutions. They are used to FiLM-modulate the encoded features, depending on the incoming hidden state $\mathbf{h}_k$. Additionally, they form $\mathbf{h}_{k+1}$ for the next iteration based on the current encoded features to share information across the T iterations. The MLP used to obtain the 192-dimensional $\mathbf{c}_k$ from the iteration counter k, the one-hot encoded imaging axis, slice position, and acceleration factor consists of two SiLU-activated hidden layers with 192 features each. Overall, our network has 7 million trainable parameters.

We used the AdamW optimizer [13] with a cosine annealed learning rate and linear warmup [7] and gradient accumulation over 4 samples The training was performed slice/view-wise on the full training dataset provided by the challenge

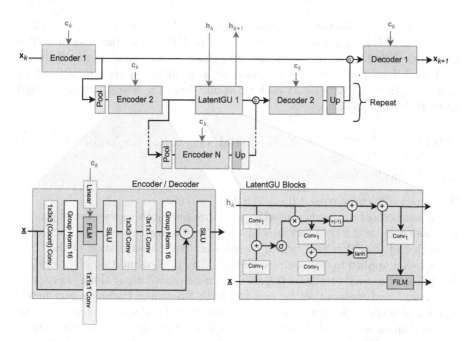

Fig. 2. Detailed view of one of the U-Nets used within our reconstruction network as X_θ (at 4 resolution scales) and Y_θ (3 resolution scales). The residual blocks employed as encoder/decoder apply FiLM conditioning [19] and separate spatial/temporal convolutions. We introduced MGU-inspired [31] *LatentGU* blocks in the skip connections at different resolutions for sharing information between different iterations.

except for one subject which we held back and used as an additional validation sample for hyper-parameter tuning. As data augmentation techniques, we employed spatial and temporal flips as well as shuffling of the different coils. We used oversampling of the fewer LAX views compared to the SAX slices.

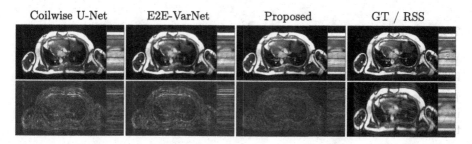

Fig. 3. An example of an LAX image reconstructed with the reported methods of comparison and our proposed approach at $R = 8$ and corresponding amplified pointwise absolute error images (bottom) as well as the fully-sampled ground truth and the zero-filled RSS reconstruction. The example is in the official *train* split of the CMRxRecon challenge, but has not been used for training.

The training objective $\mathcal{L}(\theta)$ was a linear combination of coil-wise L_1-loss, as well as SSIM- and L_2-loss between the final RSS and the reference RSS-reconstruction. Due to the particular evaluation metrics used in the challenge, we included an additional term penalizing a deviation of the intensity of the brightest pixel in each slice at each timepoint from the corresponding ground-truth RSS's brightest pixel,

$$\mathcal{L}(\theta) = \frac{1}{N}\left(\frac{1}{N_c}\|\mathbf{E^H y} - \mathbf{E^H \tilde{y}}\|_1 - \alpha_1 \text{SSIM}(\mathbf{x}_{\text{RSS}}, \mathbf{\tilde{x}}_{\text{RSS}}) + \alpha_2\|\mathbf{x}_{\text{RSS}} - \mathbf{\tilde{x}}_{\text{RSS}}\|_2^2\right)$$
$$+ \alpha_3\|\max \mathbf{x}_{\text{RSS}}, \max \mathbf{\tilde{x}}_{\text{RSS}}\|_2^2 . \tag{13}$$

We empirically chose $\alpha_1 = 1.0$, $\alpha_2 = 0.2$, $\alpha_3 = 0.05$.

As methods of comparison used for guidance during the development process, we chose a basic *2.5 D* U-Net as well as the well-known end-to-end variational network [25] (E2E VarNet), which includes an explicit estimation of the CSMs and a U-Net applied to coil-combined images. Our implementation of the *2.5D* U-Net follows the construction of [32] with separate convolutions along (xy)-spatial dimensions and temporal dimension [20] with 3.5 million trainable parameters. The input of the U-Net consists of a concatenation of real and imaginary parts of each coil image and the RSS reconstruction. The output is combined with a residual for each coil before an RSS reconstruction is performed. Our implementation of E2E VarNet follows the one in the original work [25] and extends it for the application to cardiac cine MR data. Based on preliminary experiments, we also employed 2.5 D U-Nets instead of computationally more demanding 3 D U-Nets [9] for the image- and CSMs-refinement modules. This approach has a total of 8 million trainable parameters. Both methods were trained with AdamW, a cosine learning rate schedule with warmup, and a combination of L_1, L_2, and SSIM as loss functions with weightings and maximum learning rate determined by a grid search. An implementation of the proposed method with the final hyperparameters used in our challenge submission, trained weights, and all training code is available at https://github.com/fzimmermann89/CMRxRecon.

2.4 Ablations

We performed the following ablations to investigate the importance of specific parts our proposed reconstruction network.

K-Space U-Net: We removed the k-space U-Net (8) from the network, thus leaving only the learned image space refinement and data-consistency, to investigate the influence of Y_θ.

Image-Space U-Net: Similarly, to investigate the influence of the image-space U-Net, X_θ (7), we performed an experiment with only the learned image space refinement and data-consistency layer in each iteration of the cascade.

LatentGU Blocks: We removed the LatentGU blocks from both U-Nets, thus ablating the information sharing between different iterations.

Conditioning on the Meta-Information: We removed all conditioning on meta-information such as acceleration factor, axis, and current iteration k embedding in c_k. Specifically, we remove the MLP as well as the FiLM blocks inside the U-Nets. Also, we replaced the mapping from c_k to λ_k^c inside of the DC layer by $T \cdot N_c$ learned constant parameters.

Iterations of the Cascade: Finally, we set $T = 1$, thus performing only a single iteration of (7), (8) and (9) each to investigate whether the cascading approach is beneficial in our specific network.

2.5   Extension to Quantitative Mapping

Although we mainly focus on cine imaging, our proposed architecture can also be applied to the CMRxRecon mapping track. Similarly to the cine track, it involves reconstructing qualitative images for evaluation. Thus, approaches that integrate image reconstruction and the parameter estimation task [32] within a single reconstruction network are not readily applicable. However, our *NoSENSE* approach can easily be adapted. The main difference between these two tasks in the CMRxRecon challenge lies in the effect of the time-dimension. Whereas in the cine task, between neighboring time points the images change mainly in anatomy due to cardiac motion, in the mapping track the changes are mainly in the form of changes in the contrast due to the signal preparation [29]. We slightly modify our architecture by including a learned data-dependent normalization before each image-space U-Net [18]. Here, for each pixel, the RSS intensities at time points are fed as channels into a simple two-layer, 128 feature, window size 3 CNN, which outputs two scaling factors for each time point for each pixel to be applied before and after each image space U-Net, respectively.

3   Results

We report the results of the CMRxRecon cine reconstruction, T1/T2-mapping reconstruction task, and ablation experiments for the cine task as reported by the validation leaderboard. Please note, that for all metrics, the official leaderboard normalizes the ground truth by its maximum value and the prediction by its maximum value, respectively.

3.1   Cine Reconstruction

Figure 3 shows an example of cine images reconstructed with the methods reported. We see that our proposed approach accurately removes undersampling artifacts and clearly outperforms baseline comparison methods. Table 1 lists the metrics on the (cropped) CMRxRecon validation dataset as reported by the public leaderboard.

Table 1. Results of our method compared to RSS reconstruction, a coil-wise U-Net baseline, and an E2E-VarNet with explicit sensitivity map estimation as reported by the CMRxRecon multi-coil validation leaderboard.

			RSS	U-Net	E2E VarNet	Proposed
LAX	$R = 4$	**PSNR**	22.75	29.14	30.60	**36.43**
		NMSE	0.166	0.030	0.025	**0.006**
		SSIM	0.640	0.828	0.862	**0.942**
	$R = 8$	**PSNR**	22.90	26.96	28.24	**32.08**
		NMSE	0.168	0.052	0.041	**0.015**
		SSIM	0.639	0.785	0.816	**0.891**
	$R = 10$	**PSNR**	22.77	26.47	27.67	**31.30**
		NMSE	0.175	0.059	0.045	**0.018**
		SSIM	0.634	0.773	0.805	**0.880**
SAX	$R = 4$	**PSNR**	24.43	32.69	33.35	**39.92**
		NMSE	0.136	0.016	0.018	**0.003**
		SSIM	0.718	0.888	0.907	**0.965**
	$R = 8$	**PSNR**	23.70	30.22	30.90	**35.42**
		NMSE	0.163	0.028	0.027	**0.008**
		SSIM	0.692	0.848	0.869	**0.929**
	$R = 10$	**PSNR**	23.40	29.50	30.15	**34.22**
		NMSE	0.174	0.034	0.031	**0.011**
		SSIM	0.685	0.835	0.856	**0.915**

Table 2. Results of the ablation experiments on the CMRxRecon cine validation leaderboard, highlighting the importance of the different parts of our proposed network. Shown are average values over different acceleration factors.

Ablation	Multi-Coil SAX		Multi-Coil LAX	
	PSNR	SSIM	PSNR	SSIM
No k-space U-Net	34.88	0.921	31.87	0.887
No Image-space U-Net	28.62	0.811	26.24	0.760
No LatentGU	35.11	0.923	32.06	0.888
No Conditioning	34.64	0.916	31.58	0.872
Single Iteration ($T=1$)	33.69	0.902	30.69	0.860
Full Proposed Network	**36.46**	**0.936**	**33.22**	**0.904**

3.2 Ablations

In Table 2 we report the results of the different ablations, evaluated on the (cropped) cine validation data of the CMRxRecon challenge in terms of SSIM and PSNR compared to our full network as proposed.

3.3 Quantitative Mapping Image Reconstruction

Application of our network trained on the cine task to the reconstruction of the images used in parameter mapping without any additional training resulted in 0.900 SSIM and 32.35 PSNR on the CMRxRecon mapping track validation leaderboard. Using the same architecture and re-training on the mapping task training dataset resulted in 0.934 SSIM/34.92 PSNR. By incorporating the learned data-dependent normalization, this improved further to 0.942 SSIM/35.57 PSNR. As a comparison, a simple zero-filled RSS reconstruction achieves 0.701 SSIM/23.06 PSNR on the same validation dataset.

4 Discussion and Conclusion

We have presented a novel approach for the reconstruction of undersampled k-space cardiac MR data inspired by unrolled dual-domain networks [1,6] with recurrent blocks [30] and adaption via feature modulation [19]. A fundamental aspect of our method is its direct estimation of multi-coil k-space data as opposed to the coil-combined image, as well as the usage of coil-wise image data within the convolutional blocks. Therefore, it is custom-tailored to the specific CMRxRecon challenge task. This design choice was also influenced in part by the unexpectedly strong performance of a single U-Net when applied to non-coil-combined images in contrast to the typically more potent E2E VarNet [25].

A performance gap between LAX and SAX is evident in Table 1, which might be attributed to our unsophisticated oversampling approach not completely mitigating the impact of the fewer LAX images present in the training dataset. So far, we trained the network only in a supervised manner. However, the challenge organizers have provided a relatively large validation dataset across various acceleration factors. While this dataset lacks target fully sampled RSS-reconstructions, it can be harnessed for training using self-supervised learning techniques [28]. Also, the hyperparameters of the loss and the U-Net blocks might be further optimized outside the time constraint of a challenge.

In our coarse ablations, we were able to show the great importance of the image-space U-Net and the incremental improvements by including a k-space U-Net, the latent space information sharing between different iterations, and the conditioning on the meta-information. Finally, multiple iterations, i.e., $T > 1$, are required for good performance. A more detailed investigation of the effect of the different parts of our architecture will be explored in future works.

An inherent limitation of the proposed approach is its substantial GPU memory requirements, even when compared to other unrolled methods, especially during training. This is due to the increased number of features necessary for the highest resolution within the U-Nets. We successfully mitigated the issue by activation checkpointing of these less computationally dense but memory-intensive blocks [3], enabling single GPU training.

The straight transfer of the network trained on cine data to quantitative mapping data already resulted in surprisingly good performance. By proper re-training and incorporating a learned normalization layer we further improved the result. A more sophisticated normalization approach can be further explored in future works.

Acknowledgments. This project has received funding from the European Partnership on Metrology, co-financed from the European Union's Horizon Europe Research and Innovation Programme, and by the Participating States. This work was supported in part by the Metrology for Artificial Intelligence in Medicine (M4AIM) project, which is funded by the German Federal Ministry for Economic Affairs and Climate Action (BMWi) as part of the QIDigital initiative.

Supplement

See the Table 3.

Table 3. Challenge Participation Information

Challenge Track	Cine and Mapping
University/organization	Physikalisch Technische Bundesanstalt (PTB), Braunschweig & Berlin, Germany
Single-channel or multi-channel	Multi-Coil
Hardware configuration	Nvidia A6000 (48 GB), Intel Xeon 6326 (16 Core)
Training time	approx. 48 h (Cine), 12 h (Mapping)
Inference time	approx. 1 h (full validation set)
Validation Performance (Cine)	PSNR 34.89, SSIM 0.920, NMSE 0.010
Validation Performance (Mapping)	PSNR 35.57, SSIM 0.942, NMSE 0.008
Docker submitted	Yes
Use of segmentation labels	No
Use of pre-training	No, not allowed by the challenge
Data augmentation	Spatial and temporal flips, shuffling of the different coils
Data standardization	Mean 0, Standard Deviation 1
Model parameter number	approx. 7 million trainable parameters
Loss function	Combination of coil-wise L_1-loss, SSIM and L_2-loss
Incorporation of a physical model	Yes, frame-wise 2D FFT
Use of unrolling	Yes
Application of k-space fidelity	Yes, approximate data-fidelity as a solution of a minimization problem
Model backbone	2.5 D U-Nets with LatentGU and FilM, MLP
Complex-valued operations	complex-valued images represented by real/imag. channels in U-Nets

References

1. Adler, J., Öktem, O.: Learned primal-dual reconstruction. IEEE Trans. Med. Imaging **37**(6), 1322–1332 (2018). https://doi.org/10.1109/TMI.2018.2799231
2. Aggarwal, H.K., Mani, M.P., Jacob, M.: MoDL: model-based deep learning architecture for inverse problems. IEEE Trans. Med. Imaging **38**(2), 394–405 (2019). doi.org/10/gg2nb6
3. Chen, T., Xu, B., Zhang, C., Guestrin, C.: Training Deep Nets with Sublinear Memory Cost (2016). https://doi.org/10.48550/arXiv.1604.06174
4. Cheng, J., Wang, H., Ying, L., Liang, D.: Model learning: primal dual networks for fast MR imaging. In: Shen, D., et al. (eds.) MICCAI 2019. LNCS, vol. 11766, pp. 21–29. Springer, Cham (2019). https://doi.org/10.1007/978-3-030-32248-9_3
5. Duan, J., et al.: Vs-net: variable splitting network for accelerated parallel MRI reconstruction. In: Shen, D., et al. (eds.) MICCAI 2019. LNCS, vol. 11767, pp. 713–722. Springer, Cham (2019). https://doi.org/10.1007/978-3-030-32251-9_78
6. Eo, T., Jun, Y., Kim, T., Jang, J., Lee, H.J., Hwang, D.: KIKI-net: cross-domain convolutional neural networks for reconstructing undersampled magnetic resonance images. MRM **80**(5), 2188–2201 (2018). doi.org/10/gdbpc4
7. Gotmare, A., Shirish Keskar, N., Xiong, C., Socher, R.: A closer look at deep learning heuristics: learning rate restarts, warmup and distillation. In: ICLR (2019). https://doi.org/10.48550/arXiv.1810.13243
8. Hammernik, K., et al.: Learning a variational network for reconstruction of accelerated MRI data. MRM **79**(6), 3055–3071 (2018). https://doi.org/10.1002/mrm.26977
9. Hauptmann, A., Arridge, S., Lucka, F., Muthurangu, V., Steeden, J.A.: Real-time cardiovascular MR with spatio-temporal artifact suppression using deep learning. MRM **81**(2), 1143–1156 (2019). doi.org/10/ggcg23
10. Knoll, F., et al.: Overview of the 2019 fastMRI challenge. MRM **84**(6), 3054–3070 (2020). doi.org/10/gsmqdj
11. Kofler, A., Wald, C., Schaeffter, T., Haltmeier, M., Kolbitsch, C.: Convolutional dictionary learning by end-to-end training of iterative neural networks. In: European Signal Processing Conference. vol. 2022-August, pp. 1213–1217. IEEE (2022). doi.org/10/gsmqdf
12. Liu, R., et al.: An intriguing failing of convolutional neural networks and the Coord-Conv solution. In: NeurIPS, pp. 9605–9616 (2018). https://doi.org/10.48550/arXiv.1807.03247
13. Loshchilov, I., Hutter, F.: Decoupled weight decay regularization. In: ICLR (2019). https://doi.org/10.48550/arXiv.1711.05101
14. McCollough, C.H., et al.: Results of the 2016 Low Dose CT Grand Challenge. Med. Phys. **44**(10), e339–e352 (2017). doi.org/10/gcggv5
15. Monga, V., Li, Y., Eldar, Y.C.: Algorithm unrolling: interpretable, efficient deep learning for signal and image processing. IEEE Signal Process Mag. **38**(2), 18–44 (2021). doi.org/10/gh5z3t
16. Muckley, M.J., Riemenschneider, B., Radmanesh, A.E.A.: Results of the 2020 fastMRI challenge for machine learning MR image reconstruction. IEEE Trans. Med. Imaging **40**(9), 2306–2317 (2021). doi.org/10/gj24fq
17. Nichol, A., Dhariwal, P.: Improved denoising diffusion probabilistic models. Proc. Mach. Learn. Res. **139**, 8162–8171 (2021). https://doi.org/10.48550/arXiv.2102.09672

18. Park, T., Liu, M.Y., Wang, T.C., Zhu, J.Y.: Semantic image synthesis with spatially-adaptive normalization. In: 2019 IEEE/CVF CVPR, pp. 2332–2341 (2019). https://doi.org/10.1109/CVPR.2019.00244
19. Perez, E., Strub, F., De Vries, H., Dumoulin, V., Courville, A.: FiLM: visual reasoning with a general conditioning layer. In: 32nd AAAI Conference on Artificial Intelligence, pp. 3942–3951 (2018). doi.org/10/gsk6mb
20. Qiu, Z., Yao, T., Mei, T.: Learning spatio-temporal representation with pseudo-3D residual networks. In: Proceedings of the IEEE ICCV, vol. 2017-October, pp. 5534–5542 (2017). doi.org/10/ggz7r7
21. Ravishankar, S., Bresler, Y.: MR image reconstruction from highly undersampled k-space data by dictionary learning. IEEE Trans. Med. Imaging **30**(5), 1028–1041 (2011). doi.org/10/c9dqs4
22. Ronneberger, O., Fischer, P., Brox, T.: U-net: convolutional networks for biomedical image segmentation. In: Navab, N., Hornegger, J., Wells, W., Frangi, A. (eds.) MICCAI 2015. LNCS, vol. 9351, pp. 234–241. Springer, Cham (2015). https://doi.org/10.1007/978-3-319-24574-4_28
23. Schlemper, J., Caballero, J., Hajnal, J.V., Price, A.N., Rueckert, D.: A deep cascade of convolutional neural networks for dynamic MR image reconstruction. IEEE Trans. Med. Imaging **37**(2), 491–503 (2018). doi.org/10/ggbv8j
24. Schwab, J., Antholzer, S., Haltmeier, M.: Deep null space learning for inverse problems: convergence analysis and rates. Inverse Prob. **35**(2), 25008 (2019). doi.org/10/gfvm7t
25. Sriram, A., et al.: End-to-End variational networks for accelerated MRI reconstruction. In: Martel, A.L., et al. (eds.) MICCAI 2020. LNCS, vol. 12262, pp. 64–73. Springer, Cham (2020). https://doi.org/10.1007/978-3-030-59713-9_7
26. Uecker, M., et al.: ESPIRiT - an eigenvalue approach to autocalibrating parallel MRI: Where SENSE meets GRAPPA. MRM **71**(3), 990–1001 (2014). doi.org/10/gfvjn3
27. Wang, C., et al.: CMRxRecon: an open cardiac MRI dataset for the competition of accelerated image reconstruction (2023). https://doi.org/10.48550/arXiv.2309.10836
28. Yaman, B., Hosseini, S.A.H., Moeller, S., Ellermann, J., Uğurbil, K., Akçakaya, M.: Self-supervised learning of physics-guided reconstruction neural networks without fully sampled reference data. MRM **84**(6), 3172–3191 (2020). doi.org/10/gj5thf
29. Yang, C., Zhao, Y., Huang, L., Xia, L., Tao, Q.: DisQ: disentangling quantitative MRI mapping of the heart. In: Wang, L., Dou, Q., Fletcher, P.T., Speidel, S., Li, S. (eds.) MICCAI 2022. LNCS, vol. 13436, pp. 291–300. Springer, Cham (2022). https://doi.org/10.1007/978-3-031-16446-0_28
30. Yiasemis, G., Sonke, J.J., Sanchez, C., Teuwen, J.: Recurrent variational network: a deep learning inverse problem solver applied to the task of accelerated MRI reconstruction. In: Proceedings of the IEEE CVPR, vol. 2022-June, pp. 722–731 (2022). doi.org/10/gq8r55
31. Zhou, G.B., Wu, J., Zhang, C.L., Zhou, Z.H.: Minimal gated unit for recurrent neural networks **13**(3), 226–234 (2016). doi.org/10/gftp4q
32. Zimmermann, F.F., Kolbitsch, C., Schuenke, P., Kofler, A.: PINQI: an end-to-end physics-informed approach to learned quantitative MRI reconstruction, pp. 1–20 (2023). https://doi.org/10.48550/arXiv.2306.11023

CineJENSE: Simultaneous Cine MRI Image Reconstruction and Sensitivity Map Estimation Using Neural Representations

Ziad Al-Haj Hemidi[1]([✉])[ID], Nora Vogt[2][ID], Lucile Quillien[2][ID], Christian Weihsbach[1][ID], Mattias P. Heinrich[1][ID], and Julien Oster[2,3][ID]

[1] Institute of Medical Informatics, Universität zu Lübeck, Lübeck, Germany
z.alhajhemidi@uni-luebeck.de
[2] IADI U1254, INSERM, Université de Lorraine, Nancy, France
nora.vogt@inserm.fr
[3] CIC-IT 1433, INSERM, CHRU Nancy, Université de Lorraine, Nancy, France
https://github.com/MDL-UzL/CineJENSE

Abstract. Parallel imaging (PI) techniques have enabled accelerated magnetic resonance imaging (MRI). However, real-time imaging is still hindered by the fact that PI methods typically show insufficient reconstruction performance for high acceleration factors and limited autocalibration signals. In this study, we introduce CineJENSE, an unsupervised implicit neural representation (INR) network designed for simultaneous image reconstruction and sensitivity map estimation in cardiac 2D+t cine MRI. Expanding upon the recently proposed IMJENSE network for 2D data, our model simultaneously processes cine frames of the entire cardiac cycle. It does not only surpass the 2D IMJENSE model in terms of computational efficiency but also enhances reconstruction quality for the cardiac MRI reconstruction challenge (CMRxRecon) dataset. We hypothesize that this enhancement can be explained by the effective learning of robust data priors from spatiotemporal redundancies in undersampled raw data, which generalize well to unseen k-space regions. To prevent the propagation of errors from pre-computed coil sensitivities, the proposed network learns sensitivity maps in an end-to-end manner, utilizing low-resolution hash grid encodings to ensure the generation of smooth estimates, while maintaining low computation times.

Keywords: Cardiac MRI · Parallel Imaging · Neural Representations · Instance optimization · Sensitivity map estimation

1 Introduction

Magnetic resonance imaging (MRI) plays a pivotal role in cardiac diagnosis; yet, its potential is affected by prolonged acquisition times. To address this

Z. A.-H. Hemidi and N. Vogt—These authors contributed equally to this work

© The Author(s), under exclusive license to Springer Nature Switzerland AG 2024
O. Camara et al. (Eds.): STACOM 2023, LNCS 14507, pp. 467–478, 2024.
https://doi.org/10.1007/978-3-031-52448-6_44

limitation, accelerated reconstruction techniques have been developed to recover high-quality images from undersampled k-space acquisitions. Parallel imaging (PI) involves the simultaneous acquisition of undersampled k-space data by multiple receiver coils, exploiting coil sensitivities to derive artifact-free reconstructions from sets of aliased data. Deep learning-based reconstruction approaches are emerging and surpassing the performance of widely adopted PI methods [14] like SENSitiviy Encoding (SENSE) [16] and GeneRalized Autocalibrating Partially Parallel Acquisitions (GRAPPA) [5]. Leading performances have lately been observed for unrolled algorithms, such as variational networks [6,21], which perform image processing with interleaved data consistency evaluation in the k-space domain [12]. A comparably novel research direction explores self-supervised implicit neural representations (INR) [3,4,7], which learn efficient data priors from spatiotemporal redundancies inherent in undersampled data, overcoming the requirement for fully sampled reference data. While numerous reconstruction approaches suffer from the drawback of depending on pre-computed coil sensitivities, the INR model IMplicit representation for Joint coil sENSitivity and image Estimation (IMJENSE) [4] jointly optimizes image reconstruction and sensitivity estimation to mitigate the risk of error propagation.

In this work, we propose CineJENSE, an efficient extension of the IMJENSE model tailored to dynamic data. In contrast to the 2D IMJENSE model, which optimizes 2D sensitivity maps through polynomial function learning, we introduce a 2D+t sensitivity estimation network that learns smooth, low-frequency representations from a multi-resolution spatiotemporal hash grid. We evaluate the performance of CineJENSE using the Cardiac MRI Reconstruction Challenge (CMRxRecon) [23] cine dataset and conduct a performance comparison to state-of-the-art PI approaches.

2   Related Work

PI approaches can be broadly categorized into three categories: k-space domain-based, image domain-based, and hybrid methods. Popular k-space domain-based algorithms are GRAPPA [5] and Robust Artificial neural networks for K-space Interpolation (RAKI) [1,26]. These approaches leverage k-space interpolation to infer missing k-space lines, thus facilitating the reconstruction of unaliased images from densely sampled k-spaces. In general, k-space interpolation involves the estimation of convolution kernels from an auto-calibration signal (ACS), a densely sampled region in the center of the k-space. During inference, the estimated kernels are then applied across the entire k-space to yield dense predictions. While demonstrating promising performance at low acceleration rates, these techniques rely on a sufficient number of ACS lines to mitigate model overfitting [26].

Recently, there has been a growing interest in the application of implicit neural representations for k-space interpolation [3,4,7,10]. These approaches involve the use of coordinate-based multi-layer perceptrons (MLPs) that learn a mapping from 2D spatial (or 2D+t spatiotemporal) coordinates to the complex k-space observations of the undersampled raw data for a given scan. INRs demonstrated the ability to generate efficient data representations without the necessity of fully sampled k-space data or artifact-free reference images. Adhering to a self-supervised subject-specific training paradigm (instance optimization) offers a potential solution to address the generalization challenges encountered by deep learning-based supervised PI methods, which tend to generate hallucinations, blurring, and artifacts when dealing with out-of-distribution test samples [8,20]. Furthermore, self-supervised MLP architectures overcome the large computational requirements of convolutional network models, which typically involve multiple days of training [12], while also eliminating the necessity to share sensitive patient data for network training. Although the application of instance optimization comes with prolonged inference times, ongoing developments in neural representation learning, such as the integration of multi-resolution hash encodings [13], hold promise for speeding up the optimization process.

Another emerging trend is the joint optimization of coil sensitivities and reconstructions [20]. The benefits of this strategy were recently demonstrated by Sriram et al. [21], who reported improved reconstruction performances for high acceleration factors when replacing pre-computed sensitivity maps with end-to-end optimized estimations (E2E-VarNet). In the field of implicit learning, Feng et al. introduced IMJENSE [4], which addresses the reliance on pre-computed sensitivity maps of related INRs [3,7]. In the scope of this study, our objective is to enhance the efficiency of the IMJENSE model in the context of processing dynamic cine data. To achieve lightweight, time-dependent coil sensitivity estimations, we propose replacing the polynomial function representation of the original IMJENSE architecture with a shallow MLP that processes low-resolution hash grid encodings of spatiotemporal locations.

3   Materials and Methods

3.1   Problem Formulation

MR image reconstruction aims to retrieve a high-quality image from measured k-space data. This mapping can be expressed using a forward operator $\mathbf{F}$ as follows:

$$K = \mathbf{F}I + \epsilon. \tag{1}$$

In Eq. (1), $I \in \mathbb{C}^{N_x \times N_y}$ represents the desired MR image, $K \in \mathbb{C}^{N_x \times N_y \times N_c}$ the measured multi-coil k-space data, $N_x \times N_y$ the matrix size of the image, N_c the number of coils, ϵ a complex additive noise, and $\mathbf{F}$ the Fourier encoding operator incorporating the coil sensitivities S, a Fourier transform $\mathcal{F}$, and a k-space sampling mask M. Solving Eq. (1) is challenging in the presence of undersampled k-space data as the problem becomes ill-posed. Therefore, Eq. (1) can be solved by minimizing a least squares problem introducing a regularization term $R(I)$, given by:

$$\min_I \|\mathbf{F}I - K\|_2^2 + \lambda_I R(I). \tag{2}$$

The regularization term $R(I)$ promotes certain image properties, while λ_I controls the regularization strength. In most PI-based methods, image

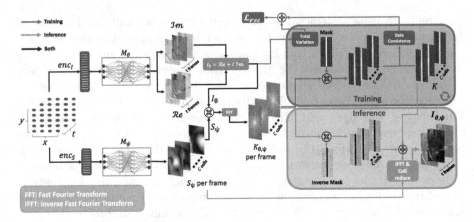

Fig. 1. CineJENSE consists of two networks, an image MLP and a coil MLP, that take hash grid encoded 2D+t coordinates as input and predict complex image reconstructions and sensitivity maps, respectively. The outputs of both networks are multiplied and Fourier-transformed to obtain a coil-expanded k-space prediction. In the training phase, the loss is calculated between the predicted k-space and the ground truth k-space and TV regularization is applied to the image. During inference, the predicted k-space is multiplied by the inverse sampling mask and the measured k-space lines are added to the masked prediction. The final reconstruction is then obtained by coil reduction using the estimated sensitivity maps.

reconstruction from multi-coil k-space data K requires coil sensitivity maps S. Such sensitivity maps can be obtained from separate calibration scans or be derived from the ACS regions using methods like ESPIRiT [22]. In the case of fewer ACS lines or higher accelerations, this problem is particularly challenging and joint image and coil sensitivity map updates were proposed to improve reconstruction performances [24]. This can be achieved through the introduction of an extra sensitivity map regularization term $R(S)$ in the reconstruction process, extending Eq. (2) to:

$$\min_{I,S} \|\mathbf{F}I - K\|_2^2 + \lambda_I R(I) + \lambda_S R(S), \tag{3}$$

with λ_I and λ_S representing the regularization strengths for I and S, respectively.

3.2   CineJENSE

Based on JSENSE [24] and further developing the IMJENSE [4] method, we explore a joint image and sensitivity map reconstruction for 2D + time cine MRI data. IMJENSE extends JSENSE by optimizing an INR for intensity prediction and a continuous polynomial function for sensitivity map estimation. IMJENSE operates only on 2D spatial coordinates. To extend IMJENSE to 2D+t cine data, we propose a spatiotemporal model for simultaneous image reconstruction and sensitivity map reconstruction based on 2D+t coordinates (CineJENSE).

CineJENSE takes 2D+t multi-resolution hash encoded [13] coordinates enabling the use of shallower MLPs and faster convergence. The encoding uses L independent hash grid levels, each consisting of $T \times F$ learnable hash table parameters serving as lookup tables for querying F-dimensional feature vectors for voxel grid vertices, defined at L resolutions (i.e., $N_{\min}, b \times N_{\min}, \ldots, b(L - 1) \times N_{\min}$, where $N_{\min}$ and b are the finest resolution and the growth factor, respectively). For a given input coordinate, feature vectors are derived for corresponding voxel corner nodes and linearly interpolated according to the relative position of the input within the corresponding grid voxel at a level $l \in L$. The final input encoding is obtained by concatenating the interpolated F-dimensional features of the L levels. As mentioned in [13], the five aforementioned hyperparameters can be adjusted to better suit a wide range of tasks: $N_{\min}$ and b govern the increase in resolution among different hash grids, while L, T, and F play crucial roles in balancing performance, memory usage, and quality.

An intensity image I_θ is predicted by MLP $M_I(enc_I; \theta)$ and J coil sensitivity maps S_ψ by MLP $M_S(enc_S; \psi)$, where θ and ψ are the learnable parameters of the MLPs. For the image hash encoding enc_I, we choose a finer resolution than the sensitivity map hash encoding enc_S since sensitivity maps are expected to be smooth. With this aim, we choose fewer levels L and a lower resolution N_{min} than for the image encoding but a higher number of features F to maintain sufficient representation complexity. The outputs of the MLPs are multiplied and Fourier transformed to obtain a coil-expanded k-space prediction:

$$K_{\theta,\psi} = \mathcal{F} I_\theta \odot S_\psi, \tag{4}$$

which is masked and compared to the acquired k-space signal to enforce data consistency. The considered reconstruction loss is a combination of a data-consistency loss $\mathcal{L}_{DC}$ and a total variation loss $\mathcal{L}_{TV}$ loss and is defined as:

$$\mathcal{L}_{rec} = \mathcal{L}_{DC} + \lambda \mathcal{L}_{TV}(I_\theta) = \|M \odot K_{\theta,\psi} - K\|_2^2 + \lambda \|\nabla I_\theta\|_1, \tag{5}$$

where λ is a hyperparameter that weights the TV regularization strength and ∇ is a gradient operator. Minimizing the loss in Eq. (5) optimizes the reconstruction of the image and sensitivity maps updating the MLP parameters θ and ψ. During inference, $K_{\theta,\psi}$ is obtained using Eq. (4), and the final image $I_{\theta,\psi}$ is reconstructed as follows:

$$I_{\theta,\psi} = \Sigma_{j=0}^J \mathcal{F}^{-1}(K_{\theta,\psi}^j \odot (1 - M) + K^j) \odot \overline{S}_\psi^j, \tag{6}$$

where $K_{\theta,\psi}^j$, K^j and $\overline{S}_\psi^j$ are the j-th coil of the predicted k-space, the measured k-space, and the complex conjugate predicted sensitivity maps, respectively. The whole pipeline is visualized in Fig. 1.

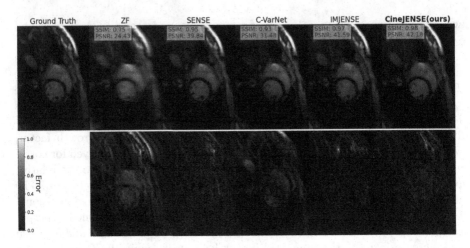

Fig. 2. Results for cine short axis views for an acceleration factor of R=4.

4   Experimental Results

4.1   Dataset

The CMRxRecon [23] challenge dataset consists of 300 cine MRI scans of healthy subjects and is split into training/validation/test sets with 120/60/120 samples, respectively. Fully sampled k-spaces were only available for the training set. We trained and tested using the training set with a 100/20 split, including cine multi-coil short-axis (SAX) and long-axis (LAX) views of different acceleration rates (R=4, R=8, and R=10). The k-space data was acquired using a Cartesian trajectory and compressed to 10 virtual coils. While the SAX views consisted of $5-10$ slices with $12-25$ frames, the LAX views (2-chamber, 3-chamber, and 4-chamber) comprised a single 2D+t slice. Typical imaging properties were spatial resolutions of 2.0×2.0 mm^2, slice thicknesses of 8.00 mm, slice gaps of 4.0 mm, and temporal resolutions of 50 ms.

4.2   Experimental Setup

Implementation Details. All experiments were implemented using *Python* version 3.9 and *PyTorch* version 2.0.0. For evaluation, we considered the peak signal-to-noise ratio (PSNR) and the structural similarity index (SSIM). For hash encoding and efficient MLP implementation, we used the tiny-cuda-nn [13] library, while general k-space operations were implemented using fastMRI [25]. The specifications of the hardware are a 32-core AMD CPU @ 3.78 GHz and an Nvidia® A100 40 GB GPU running with CUDA® driver version 11.6.

Cine MRI Reconstruction. We compared the proposed method to the following baselines: Zero-filled reconstruction, SENSE reconstruction, C-VarNet

Table 1. Performance comparison for acceleration factors R=4. The arrows indicate the direction of improvement.

View	Method	SSIM↑	PSNR (db)↑	Time (minutes)↓
SAX	Zero Filled	0.70 ± 0.05	23.96 ± 2.36	–
	SENSE	0.91 ± 0.05	35.76 ± 4.21	–
	C-VarNet	0.85 ± 0.03	26.71 ± 2.59	–
	IMJENSE	0.95 ± 0.03	37.65 ± 3.22	1.88 ± 0.62
	CineJENSE (ours)	**0.96 ± 0.02**	**39.70 ± 2.84**	**0.66 ± 0.11**
LAX	Zero Filled	0.64 ± 0.06	22.90 ± 1.62	–
	SENSE	0.84 ± 0.07	31.72 ± 2.50	–
	IMJENSE	**0.93 ± 0.03**	35.36 ± 2.34	1.55 ± 0.68
	CineJENSE (ours)	**0.93 ± 0.03**	**35.79 ± 1.96**	**0.46 ± 0.09**

(explored for SAX view only), and IMJENSE. The zero-filled reconstruction was performed by applying the inverse Fourier transform to the undersampled k-space. SENSE was performed using the CMRxRecon SENSE implementation in the base configuration. The C-VarNet is based on the E2E-VarNet [21] and conditioned [11] on the acceleration rates. It was trained on SAX data using the Adam optimizer with a learning rate of $4e^{-5}$ and a batch size of 1 for 100 epochs until convergence. It consisted of 12 cascaded U-Nets [17] with 4 downsampling layers. The training was performed on 100 subjects and evaluated on 20. The iterations for IMJENSE were set to 300 with Adam as the optimizer using learning rates of $3e^{-3}$ and 1 for the MLPs and polynomial coefficients, respectively. The MLP consisted of 2 layers with 64 channels and ReLU activations. Hash encodings were configured as recommended in [13]. The TV regularization weight λ was set to 5 and the maximum polynomial order to 15. The CineJENSE iterations were set to 200 using the Adam optimizer with learning rates of $1e^{-2}$ for both MLPs and λ was set to 5. While using the same network and image encoding configurations as in the IMJENSE experiments, we introduced an additional coil encoding with $L = 4$, $F = 8$, $N_{min} = 2$, and $b = 1.1$.

4.3   Results and Discussion

As shown in Tables 1, 2, and 3, the proposed CineJENSE outperformed the compared approaches in terms of SSIM and PSNR for all acceleration factors. It is worth noting that while there was only a marginal performance difference between IMJENSE and CineJENSE, the latter enabled a faster reconstruction by simultaneously processing multiple time frames. The SENSE algorithm yielded satisfactory reconstructions for the acceleration factor $R = 4$, but performances dropped significantly for $R = 8$ and $R = 10$. On the other hand, the C-VarNet algorithm was the least impacted by the acceleration factor. Yet it produced overly smooth reconstructions for all configurations, resulting in relatively low PSNR scores, as depicted in Figs. 2, 3, and 4. When comparing the noise patterns

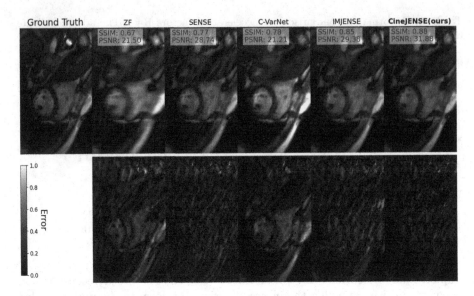

Fig. 3. Results for cine short axis views for an acceleration factor of R=8.

visible in the error maps of these reconstructions (see Figs. 2, 3, and 4), the CineJENSE method appeared favorable for clinical diagnosis as it was the least affected by structured noise.

The higher the acceleration factor, the more pronounced the need for adequate regularization priors becomes. We observed that, without explicit regularization, the networks exhibited stronger overfitting on the sampled k-space lines. The high dynamic range loss [7] was introduced to better balance the impact of k-space frequencies, reducing the influence of high-magnitude lines, and improving generalization. Further research might involve exploring low-rank and sparsity regularization constraints [15], which could facilitate the learning of more robust and disentangled representations of static background and moving structures. Performance and robustness could be improved through the use of 3D+t variants and the incorporation of explicit deformation field modeling. We acknowledge that despite the demonstrated efficiency improvement over the baseline model, training INRs from scratch remains time-consuming. Nevertheless, recent advancements in meta-learning and hyper-networks [9] promise great potential for addressing both inference times and generalization abilities of INRs. While these advancements often involve learning data-driven priors from large databases, promising results were also recently obtained for one-shot priors derived from single reference samples [19]. Improved convergence times and motion artifact reduction are moreover expected for non-Cartesian k-space sampling trajectories (e.g. radial or spiral), which provide efficient k-space coverage for cardiac imaging [18].

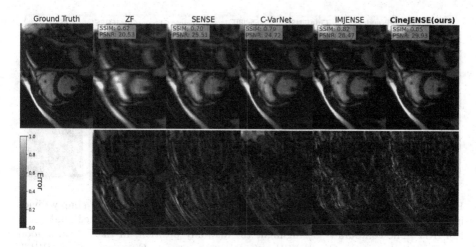

Fig. 4. Results for cine short axis views for an acceleration factor of R=10.

Table 2. Performance comparison for acceleration factors R=8. The arrows indicate the direction of improvement.

View	Method	SSIM↑	PSNR (db)↑	Time (minutes)↓
SAX	Zero Filled	0.67 ± 0.06	23.10 ± 2.27	–
	SENSE	0.72 ± 0.09	27.32 ± 3.41	–
	C-VarNet	0.82 ± 0.04	25.67 ± 2.79	–
	IMJENSE	0.86 ± 0.04	31.53 ± 2.38	1.87 ± 0.71
	CineJENSE (ours)	$\mathbf{0.88 \pm 0.03}$	$\mathbf{32.32 \pm 2.45}$	$\mathbf{0.52 \pm 0.10}$
LAX	Zero Filled	0.63 ± 0.06	22.92 ± 1.53	–
	SENSE	0.65 ± 0.06	24.50 ± 1.33	–
	IMJENSE	$\mathbf{0.84 \pm 0.03}$	$\mathbf{29.24 \pm 1.92}$	1.11 ± 0.17
	CineJENSE (ours)	0.83 ± 0.03	28.92 ± 2.10	$\mathbf{0.41 \pm 0.08}$

Table 3. Performance comparison for acceleration factors R=10. The arrows indicate the direction of improvement.

View	Method	SSIM↑	PSNR (db)↑	Time (minutes)↓
SAX	Zero Filled	0.66 ± 0.06	22.81 ± 2.20	–
	SENSE	0.69 ± 0.08	26.02 ± 3.20	–
	C-VarNet	0.80 ± 0.04	25.11 ± 2.68	–
	IMJENSE	0.84 ± 0.04	30.18 ± 2.30	1.48 ± 0.19
	CineJENSE (ours)	$\mathbf{0.85 \pm 0.04}$	$\mathbf{30.59 \pm 2.58}$	$\mathbf{0.53 \pm 0.12}$
LAX	Zero Filled	0.64 ± 0.06	22.90 ± 1.62	–
	SENSE	0.84 ± 0.07	31.72 ± 2.50	–
	IMJENSE	$\mathbf{0.80 \pm 0.03}$	$\mathbf{27.79 \pm 1.35}$	1.10 ± 0.18
	CineJENSE (ours)	$\mathbf{0.80 \pm 0.03}$	27.61 ± 1.26	$\mathbf{0.42 \pm 0.08}$

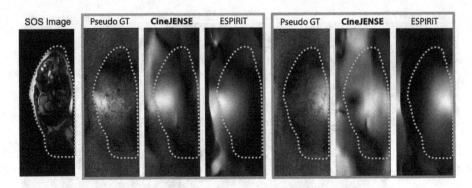

Fig. 5. Comparison of coil sensitivity maps from one frame provided by simply dividing coil images by the sum-of-square (SOS) image (pseudo GT), CineJENSE, and ESPIRiT. The contours (▪ ▪ ▪) highlight the anatomy position. In the □ box, a good agreement between the CineJENSE estimation, the pseudo GT maps, and ESPIRiT can be observed. While in the □ box, our model demonstrated a higher discrepancy. (Color figure online)

In this work, we demonstrated that coil sensitivities could be successfully modeled by a shallow MLP network with low-resolution hash grid encodings. Contrary to our expectations that estimated sensitivity maps would remain constant over time, we found a slight performance improvement when defining a 2D+t model as opposed to a 2D model. Overall, the coil compression of the CMRxRecon pre-processing pipeline posed several challenges in assessing the quality of the estimated maps (see Fig. 5), as the compression to 10 virtual coils seemed to introduce coil entanglements, leading to unexpected violations of smoothness constraints. The proposed joint optimization strategy seems to compensate for suboptimal coil representations, but the impact of imperfect coil sensitivity estimations needs further investigation.

The proposed model showed promise for cardiac mapping (CMRxRecon task 2 - not shown here) by successfully reconstructing 2D+TI and 2D+TE sequences without requiring any architecture modifications, likely due to linear relations in these sequences. Future work will focus on the development of dedicated priors for the mapping task, such as exploiting locally-low-rank constraints [2].

5   Conclusion

In this work, we introduced CineJENSE, an implicit network designed for the joint reconstruction of 2D+t cine MRI images and sensitivity maps. The proposed model leverages spatiotemporal correlations in undersampled, multi-coil k-space data and employs low-resolution hash grid encodings to generate smooth coil sensitivity maps. We assessed the reconstruction performance on the CMRxRecon challenge dataset and demonstrated that our model achieves competitive image quality and reconstruction times.

Acknowledgements. This work was supported by a grant from the Joint French-German ANR-DFR Call on Artificial Intelligence, MEDICARE (ANR-21–FAI1-0007, BMBF funding code: 01IS21094).

References

1. Akçakaya, M., Moeller, S., Weingärtner, S., Uğurbil, K.: Scan-specific robust artificial-neural-networks for k-space interpolation (RAKI) reconstruction: database-free deep learning for fast imaging. Magn. Reson. Med. **81**(1), 439–453 (2019)
2. Demirel, Ö.B., Weingärtner, S., Moeller, S., Akçakaya, M.: Improved regularized reconstruction for simultaneous multi-slice cardiac MRI T1 mapping. In: 2019 27th European Signal Processing Conference (EUSIPCO), pp. 1–5. IEEE (2019)
3. Feng, J., Feng, R., Wu, Q., Zhang, Z., Zhang, Y., Wei, H.: Spatiotemporal implicit neural representation for unsupervised dynamic MRI reconstruction. arXiv preprint arXiv:2301.00127 (2022)
4. Feng, R., et al.: IMJENSE: scan-specific implicit representation for joint coil sensitivity and image estimation in parallel MRI. IEEE Trans. Med. Imaging (2023). https://doi.org/10.1109/TMI.2023.3342156
5. Griswold, M.A., et al.: Generalized autocalibrating partially parallel acquisitions (GRAPPA). Magn. Reson. Med. **47**(6), 1202–1210 (2002)
6. Hammernik, K., et al.: Learning a variational network for reconstruction of accelerated MRI data. Magn. Reson. Med. **79**(6), 3055–3071 (2018)
7. Huang, W., Li, H.B., Pan, J., Cruz, G., Rueckert, D., Hammernik, K.: Neural implicit k-space for binning-free non-cartesian cardiac MR imaging. In: Frangi, A., de Bruijne, M., Wassermann, D., Navab, N. (eds.) Information Processing in Medical Imaging. IPMI 2023. Lecture Notes in Computer Science, vol. 13939, pp. 548–560. Springer, Cham (2023). https://doi.org/10.1007/978-3-031-34048-2_42
8. Johnson, P.M., et al.: Evaluation of the robustness of learned MR image reconstruction to systematic deviations between training and test data for the models from the fastMRI challenge. In: Haq, N., Johnson, P., Maier, A., Würfl, T., Yoo, J. (eds.) MLMIR 2021. LNCS, vol. 12964, pp. 25–34. Springer, Cham (2021). https://doi.org/10.1007/978-3-030-88552-6_3
9. Kim, C., Lee, D., Kim, S., Cho, M., Han, W.S.: Generalizable implicit neural representations via instance pattern composers. In: Proceedings of the IEEE/CVF Conference on Computer Vision and Pattern Recognition, pp. 11808–11817 (2023)
10. Kunz, J.F., Ruschke, S., Heckel, R.: Implicit neural networks with fourier-feature inputs for free-breathing cardiac MRI reconstruction. arXiv preprint arXiv:2305.06822 (2023)
11. Meseguer-Brocal, G., Peeters, G.: Conditioned-U-Net: introducing a control mechanism in the U-Net for multiple source separations. arXiv:1907.01277 (2019)
12. Muckley, M.J., et al.: Results of the 2020 fastMRI challenge for machine learning MR image reconstruction. IEEE Trans. Med. Imaging **40**(9), 2306–2317 (2021)
13. Müller, T., Evans, A., Schied, C., Keller, A.: Instant neural graphics primitives with a multiresolution hash encoding. ACM Trans. Graph. (ToG) **41**(4), 1–15 (2022)
14. Oscanoa, J.A., et al.: Deep learning-based reconstruction for cardiac MRI: a review. Bioengineering **10**(3), 334 (2023)
15. Otazo, R., Candes, E., Sodickson, D.K.: Low-rank plus sparse matrix decomposition for accelerated dynamic MRI with separation of background and dynamic components. Magn. Reson. Med. **73**(3), 1125–1136 (2015)

16. Pruessmann, K.P., Weiger, M., Scheidegger, M.B., Boesiger, P.: SENSE: sensitivity encoding for fast MRI. Magn. Reson. Med. **42**(5), 952–962 (1999)
17. Ronneberger, O., Fischer, P., Brox, T.: U-Net: convolutional networks for biomedical image segmentation. In: Navab, N., Hornegger, J., Wells, W.M., Frangi, A.F. (eds.) MICCAI 2015. LNCS, vol. 9351, pp. 234–241. Springer, Cham (2015). https://doi.org/10.1007/978-3-319-24574-4_28
18. Seiberlich, N., Ehses, P., Duerk, J., Gilkeson, R., Griswold, M.: Improved radial GRAPPA calibration for real-time free-breathing cardiac imaging. Magn. Reson. Med. **65**(2), 492–505 (2011)
19. Shen, L., Pauly, J., Xing, L.: NeRP: implicit neural representation learning with prior embedding for sparsely sampled image reconstruction. IEEE Trans. Neural Networks Learn. Syst. (2022)
20. Singh, D., Monga, A., de Moura, H.L., Zhang, X., Zibetti, M.V., Regatte, R.R.: Emerging trends in fast MRI using deep-learning reconstruction on undersampled k-space data: a systematic review. Bioengineering **10**(9), 1012 (2023)
21. Sriram, A., et al.: End-to-end variational networks for accelerated MRI reconstruction. In: Martel, A.L., et al. (eds.) MICCAI 2020. LNCS, vol. 12262, pp. 64–73. Springer, Cham (2020). https://doi.org/10.1007/978-3-030-59713-9_7
22. Uecker, M., et al.: ESPIRiT - an eigenvalue approach to autocalibrating parallel MRI: where SENSE meets GRAPPA. Magn. Reson. Med. **71**(3), 990–1001 (2014)
23. Wang, C., et al.: CMRxRecon: an open cardiac MRI dataset for the competition of accelerated image reconstruction. arXiv:2309.10836 (2023)
24. Ying, L., Sheng, J.: Joint image reconstruction and sensitivity estimation in SENSE (JSENSE). Magn. Reson. Med. **57**(6), 1196–1202 (2007)
25. Zbontar, J., et al.: fastMRI: an open dataset and benchmarks for accelerated MRI. arXiv preprint arXiv:1811.08839 (2018)
26. Zhang, C., Moeller, S., Demirel, O.B., Uğurbil, K., Akçakaya, M.: Residual RAKI: a hybrid linear and non-linear approach for scan-specific k-space deep learning. Neuroimage **256**, 119248 (2022)

Deep Cardiac MRI Reconstruction
with ADMM

George Yiasemis[1,2]([✉]), Nikita Moriakov[1,2], Jan-Jakob Sonke[1,2],
and Jonas Teuwen[1,2,3]

[1] Netherlands Cancer Institute, Amsterdam, The Netherlands
{g.yiasemis,n.moriakov,j.sonke,j.teuwen}@nki.nl
[2] University of Amsterdam, Amsterdam, The Netherlands
[3] Radboud University Medical Center, Nijmegen, The Netherlands

Abstract. Cardiac magnetic resonance imaging (CMR) is a valuable
non-invasive tool for identifying cardiovascular diseases. For instance,
Cine MRI is the benchmark modality for assessing the cardiac function and anatomy. On the other hand, multi-contrast (T1 and T2) mapping has the potential to assess pathologies and abnormalities in the
myocardium and interstitium. However, voluntary breath-holding and
often arrhythmia, in combination with MRI's slow imaging speed, can
lead to motion artifacts, hindering real-time acquisition image quality. Although performing accelerated acquisitions can facilitate dynamic
imaging, it induces aliasing, causing low reconstructed image quality in
Cine MRI and inaccurate T1 and T2 mapping estimation. In this work,
inspired by related work in accelerated MRI reconstruction, we present
a deep learning-based method for accelerated cine and multi-contrast
reconstruction in the context of dynamic cardiac imaging. We formulate the reconstruction problem as a least squares regularized optimization task, and employ vSHARP, a state-of-the-art Deep Learning-based
inverse problem solver, which incorporates half-quadratic variable splitting and the alternating direction method of multipliers (ADMM) with
neural networks. We treat the problem in two setups; a 2D reconstruction and a 2D dynamic reconstruction task, and employ 2D and 3D deep
learning networks, respectively. Our method optimizes in both the image
and k-space domains, allowing for high reconstruction fidelity. Although
the target data is undersampled with a Cartesian equispaced scheme,
we train our deep neural network using both Cartesian and simulated
non-Cartesian undersampling schemes to enhance generalization of the
model to unseen data, a key ingredient of our method. Furthermore, our
model adopts a deep neural network to learn and refine the sensitivity
maps of multi-coil k-space data. Lastly, our method is jointly trained on
both, undersampled cine and multi-contrast data.

Keywords: Accelerated Cardiac MRI · Deep MRI Reconstruction ·
Dynamic Cardiac MRI Reconstruction · Cine Reconstruction · T1-T2
Reconstruction

O. Camara et al. (Eds.): STACOM 2023, LNCS 14507, pp. 479–490, 2024.
https://doi.org/10.1007/978-3-031-52448-6_45

1   Introduction

Cardiac magnetic resonance (CMR) stands as a vital clinical tool for assessing cardiovascular diseases due to its non-invasive and radiation-free nature, enabling a comprehensive evaluation of cardiovascular aspects, such as structure, function, flow, perfusion, viability, tissue characterization, as well as the assessment of myocardial fibrosis and other pathologies [1–3]. Key CMR applications include cine MR imaging and T1/T2 mapping.

However, CMR faces inherent physical challenges, primarily the time consuming MRI acquisition process. The requirement for increased spatiotemporal resolution in cardiac imaging further amplifies this challenge. To mitigate prolonged scan times, accelerated MRI acquisitions are utilized by obtaining undersampled k-space data, though this approach violates the Nyquist-Shannon sampling criterion [4].

In the broader MRI domain, conventional techniques such as Parallel Imaging (PI) [5,6] and Compressed Sensing (CS) [7,8] have been employed to accelerate MRI data acquisition. These approaches leverage spatial sensitivity information from multiple receiver coil arrays and exploit the sparsity or compressibility of MRI data. However, these methods have limitations, such as noise amplification in PI, and assumptions of sparsity that may not hold for all MRI data in CS, whilst finding optimal parameters for CS methods might be computationally and time consuming.

In the last decade, Deep Learning (DL) has revolutionized MRI image reconstruction, exhibiting superior performance compared to traditional methods, especially in accelerated MRI reconstruction tasks [9]. DL-based algorithms can learn complex image representations directly from available datasets, enabling enhanced image reconstruction from undersampled k-space measurements, often in supervised learning [10–13], or self-supervised settings [14]. This advancement holds significant potential to impact CMR by elevating the image quality of reconstructed highly undersampled data while concurrently reducing breath-hold duration.

In this work, motivated by the need for reducing acquisition times and breath-hold durations further during CMR, we employ vSHARP [15] (variable Splitting Half-quadratic ADMM algorithm for Reconstruction of inverse-Problems), a DL-based inverse problem solver, previously applied on brain and prostate MR imaging exhibiting state-of-the-art performance. We particularize vSHARP for accelerated Cardiac MRI Reconstruction and introduce in Sect. 3.1 two variants by treating the problem at hand as a 2D reconstruction task (2D model) or as a 2D dynamic reconstruction task (3D model). Additionally, in Sect. 3.2 we propose various training techniques to boost model training and generalizability across unseen cardiac (cine and T1/T2) MRI data. In Sect. 5, we experimentally compare our two approaches, highlighting that our 2D dynamic implementation outperforms traditional 2D reconstruction and we further compare our models with current state-of-the-art approaches.

2   Theory and Problem Formulation

2.1   Accelerated MRI Reconstruction

Recovering a two-dimensional image $\mathbf{x}^* \in \mathbb{C}^N$ from undersampled multi-coil (assume N_c coils) k-space measurements $\tilde{\mathbf{y}} \in \mathbb{C}^{N \times N_c}$ can be formulated as a minimisation problem as follows:

$$\mathbf{x}^* = \underset{\mathbf{x} \in \mathbb{C}^N}{\operatorname{argmin}} \frac{1}{2} \sum_{k=1}^{N_c} \left\| \mathcal{A}^k(\mathbf{x}) - \tilde{\mathbf{y}}^k \right\|_2^2 + \mathcal{R}(\mathbf{x}), \quad \mathcal{A}^k = \mathbf{U}\mathcal{F}\mathbf{S}^k, \tag{1}$$

where $\mathcal{A}^k$ represents the forward or corruption operator per coil. It involves mapping the image to an individual coil image using a known sensitivity map $\mathbf{S}^k$, transforming it to the k-space domain via the Fast Fourier Transform (FFT) $\mathcal{F}$, and undersampling with $\mathbf{U}$. The function $\mathcal{R} : \mathbb{C}^N \to \mathbb{R}$ denotes a regularization functional, which is assumed to impose prior knowledge about the image.

In the context of cardiac magnetic resonance, acquisitions are typically dynamic and synchronized with electrocardiography (ECG)-derived cardiac cine. In dynamic acquisitions, multiple undersampled k-space data $\tilde{\mathbf{y}} \in \mathbb{C}^{N \times N_c \times N_f}$ are obtained at N_f time frames. Consequently, Eq. 1 is adapted as follows:

$$\mathbf{x}_{\mathrm{d}}^* = \underset{\mathbf{x} \in \mathbb{C}^{N \times N_f}}{\operatorname{argmin}} \frac{1}{2} \sum_{t=1}^{N_f} \sum_{k=1}^{N_c} \left\| \mathcal{A}^k(\mathbf{x}_{\cdot,t}) - \tilde{\mathbf{y}}_{\cdot,t}^k \right\|_2^2 + \mathcal{R}(\mathbf{x}), \quad \mathcal{A}^k = \mathbf{U}\mathcal{F}\mathbf{S}^k. \tag{2}$$

In dynamic acquisitions, it is often assumed that knowledge can be shared across time frames or that the motion pattern is known, thereby requiring the selection of an appropriate prior $\mathcal{R} : \mathbb{C}^{N \times N_f} \to \mathbb{R}$ that incorporates this information [16].

3   Methods

3.1   Deep Learning Framework

Sensitivity Map Prediction. In conventional settings, sensitivity maps are estimated from the autocalibration signal (ACS) data, often incorporating a portion of the center of the k-space. Advanced techniques for refining these estimated sensitivities include ESPIRiT or GRAPPA [5,17]. However, these approaches can impose computational constraints. To overcome the need for such computationally expensive algorithms, we employ a two-dimensional deep learning module, specifically a 2D U-Net [18]. This model takes ACS-estimated sensitivity maps as input and produces refined versions of them as output. The predicted sensitivity maps $\{\mathbf{S}^k\}_{k=1}^{N_c}$ are used for downstream reconstruction tasks, and the sensitivity module is trained in an end-to-end manner along with the reconstruction model.

Reconstruction via ADMM Unrolled Optimization. Our approach utilizes vSHARP [15], a DL-based inverse problem solver, to address Eq. 1. vSHARP employs the half-quadratic variable splitting method [19] to transform the optimization problem in Eq. 1 by introducing an intermediate variable $\mathbf{w}$. It then unrolls the optimization process over T iterations using the alternating direction method of multipliers algorithm (ADMM) [20], as follows:

$$\mathbf{w}^{(j+1)} = \underset{\mathbf{w} \in \mathbb{C}^N}{\text{argmin}} \, \mathcal{R}(\mathbf{w}) + \frac{\lambda}{2} ||\mathbf{x}^{(j)} - \mathbf{w} + \frac{\mathbf{m}^{(j)}}{\lambda}||_2^2, \tag{3a}$$

$$\mathbf{x}^{(j+1)} = \underset{\mathbf{x} \in \mathbb{C}^N}{\text{argmin}} \, \frac{1}{2} \sum_{k=1}^{N_c} ||\mathcal{A}^k(\mathbf{x}) - \tilde{\mathbf{y}}^k||_2^2 + \frac{\lambda}{2} ||\mathbf{x} - \mathbf{w}^{(j+1)} + \frac{\mathbf{m}^{(j)}}{\lambda}||_2^2, \tag{3b}$$

$$\mathbf{m}^{(j+1)} = \mathbf{m}^{(j)} + \lambda(\mathbf{x}^{(j+1)} - \mathbf{w}^{(j+1)}), \quad j = 0, \cdots, T-1. \tag{3c}$$

Our method incorporates U-Nets to replace the need for manually selecting a prior functional $\mathcal{R}$ in Eq. 3a and learn the solution from data directly, namely the denoising step. Next, data consistency is enforced by solving Eq. 3b via an unrolled (differentiable) gradient descent scheme. Our approach initializes $\mathbf{w}^{(0)}$ and $\mathbf{x}^{(0)}$ using a zero-filled reconstruction with $\tilde{\mathbf{y}}$ and the predicted coil sensitivity maps: $\mathbf{w}^{(0)} = \mathbf{x}^{(0)} := \sum_{k=1}^{N_c} \mathbf{S}^{k*} \mathcal{F}^{-1}(\tilde{\mathbf{y}}^k)$. Additionally, a learned initializer, adapted from [13], is used to determine an initialization for the Lagrange Multipliers $\mathbf{m}^{(0)}$. For dynamic reconstruction as in Eq. 2, Eq. 3 is replaced by:

$$\mathbf{w}^{(j+1)} = \underset{\mathbf{w} \in \mathbb{C}^{N \times N_f}}{\text{argmin}} \, \mathcal{R}(\mathbf{w}) + \frac{\lambda}{2} ||\mathbf{x}^{(j)} - \mathbf{w} + \frac{\mathbf{m}^{(j)}}{\lambda}||_2^2, \tag{4a}$$

$$\mathbf{x}^{(j+1)} = \underset{\mathbf{x} \in \mathbb{C}^{N \times N_f}}{\text{argmin}} \, \frac{1}{2} \sum_{t=1}^{N_f} \sum_{k=1}^{N_c} ||\mathcal{A}^k(\mathbf{x}_{\cdot,t}) - \tilde{\mathbf{y}}_{\cdot,t}^k||_2^2 + \frac{\lambda}{2} ||\mathbf{x} - \mathbf{w}^{(j+1)} + \frac{\mathbf{m}^{(j)}}{\lambda}||_2^2, \tag{4b}$$

$$\mathbf{m}^{(j+1)} = \mathbf{m}^{(j)} + \lambda(\mathbf{x}^{(j+1)} - \mathbf{w}^{(j+1)}), \quad j = 0, \cdots, T-1. \tag{4c}$$

3.2 Model Training Techniques

In this section, we outline the various additional techniques employed in our paper to enhance the performance of our models.

Joint Modality Training. During the training of our DL-based approach, we jointly trained it using all available data at our disposal (see Sect. 4.3). This approach served a dual purpose; Firstly, instead of training separate models for each modality, our joint modality training aimed to utilize a larger dataset promoting more effective learning and generalization. Moreover, by integrating cine and T1/T2-weighted MRI data, we aimed to harness the complementarity between these modalities. This approach enabled the model to exploit the shared features and correlations, potentially improving the reconstruction quality for both modalities.

Random k-space Cropping. To optimize computational efficiency during training, we utilized random cropping on the fully-sampled multi-coil k-space data. Since direct cropping of the k-space would be inappropriate, we first applied the inverse Fast Fourier Transform (FFT) to reconstruct it into fully-sampled multi-coil images. Subsequently, random cropping was performed on this reconstructed image, and the resulting cropped image was transformed back to the k-space domain (via FFT). The k-space data was then undersampled and used as input to our model. This approach not only offered computational benefits but also allowed our model to gain exposure to different parts of the reconstructed data, including background noise and the regions of interest, without compromising overall reconstruction quality as compared to using non-cropped data. Figure 1 illustrates examples of cropped images before the transformation back to the k-space domain. It's important to note that for dynamic data, the same cropping process was applied to all time frames.

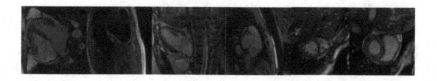

Fig. 1. Randomly cropped (in the image domain) examples of cine and T1/T2-weighted MRI images from the dataset. These images are then transformed to the k-space domain, followed by retrospective undersampling, and are subsequently utilized for training.

Multi-scheme Undersampling. Undersampling for the target (validation) data comprised Cartesian rectilinear equispaced undersampling masks, with 24 fully-sampled ACS (central) lines, and with acceleration factors of $R = 4$, 8 and 10. Inspired by previous work [21], which demonstrated enhanced model generalizability in reconstructing Cartesian rectilinear data, we employed a multi-scheme undersampling setup during training. Alongside the provided undersampling pattern, we used the following undersampling schemes: Equispaced and Random Cartesian rectilinear, Gaussian 2D Cartesian, and pseudo-Radial and pseudo-Spiral schemes. These undersampling schemes are visualized in Fig. 2. Note that for dynamic data, the same undersampling scheme was applied on all time frames.

Dual Domain Loss. To train our models we designed a dual-domain loss:

$$\mathcal{L}_\phi = \mathcal{L}_\phi^{img} + \mathcal{L}_\phi^{freq}, \tag{5}$$

where $\mathcal{L}_\phi^{img}$ and $\mathcal{L}_\phi^{freq}$ represent losses computed in the image and frequency domain, respectively.

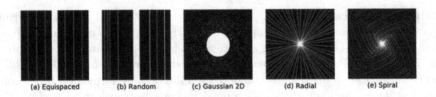

Fig. 2. Undersampling Schemes during training.

Image Domain Loss. The image domain loss, $\mathcal{L}_\phi^{img}$, is computed between the ground truth RSS image $\mathbf{x}$ and the magnitude of the model-predicted image $\hat{\mathbf{x}}_\phi$. This loss comprises several components:

$$\mathcal{L}_\phi^{img} = \lambda_{\text{SSIM}}\mathcal{L}_{\text{SSIM}}\left(\mathbf{x}, \hat{\mathbf{x}}_\phi\right) + \lambda_1 \mathcal{L}_1\left(\mathbf{x}, \hat{\mathbf{x}}_\phi\right) + \lambda_{\text{HFEN}_1}\mathcal{L}_{\text{HFEN}_1}\left(\mathbf{x}, \hat{\mathbf{x}}_\phi\right) \quad (6)$$

which are defined as follows:

$$\mathcal{L}_{\text{SSIM}}(\mathbf{u}, \mathbf{v}) = 1 - \text{SSIM}(\mathbf{u}, \mathbf{v}), \quad \mathcal{L}_1(\mathbf{u}, \mathbf{v}) = ||\mathbf{u} - \mathbf{v}||_1, \\ \text{and,} \quad \mathcal{L}_{\text{HFEN}_1}(\mathbf{u}, \mathbf{v}) = \text{HFEN}_1(\mathbf{u}, \mathbf{v}). \quad (7)$$

In Eq. 7, SSIM denotes the Structural Similarity Index Measure, computed over W windows, each of size 7×7 pixels extracted from images $\mathbf{u}$ and $\mathbf{v}$. It is defined as:

$$\text{SSIM}(\mathbf{u}, \mathbf{v}) = \frac{1}{W} \sum_{i=1}^{W} \frac{(2\mu_{\mathbf{u}_i}\mu_{\mathbf{v}_i} + 0.01)(2\sigma_{\mathbf{u}_i\mathbf{v}_i} + 0.03)}{(\mu_{\mathbf{u}_i}^2 + \mu_{\mathbf{v}_i}^2 + 0.01)(\sigma_{\mathbf{u}_i}^2 + \sigma_{\mathbf{v}_i}^2 + 0.03)}. \quad (8)$$

Here, $\mu_{\mathbf{u}_i}$, $\mu_{\mathbf{v}_i}$, $\sigma_{\mathbf{u}_i}$ and $\sigma_{\mathbf{v}_i}$ represent the means and standard deviations of each window, while $\sigma_{\mathbf{u}_i\mathbf{v}_i}$ signified the covariance between $\mathbf{u}_i$ and $\mathbf{v}_i$. HFEN_1 represents the High-Frequency Error Norm, and is defined as follows:

$$\text{HFEN}_1(\mathbf{u}, \mathbf{v}) = \frac{||G(\mathbf{u}) - G(\mathbf{v})||_1}{||G(\mathbf{u})||_1}, \quad (9)$$

where G denotes a 15×15 Laplacian of Gaussian filter with a standard deviation of 2.5.

SSIM and HFEN are computed per single 2D slice/time frame. For dynamic reconstruction experiments, we also incorporated $\lambda_{\text{SSIM3D}}\mathcal{L}_{\text{SSIM3D}}$, which computes the SSIM metric for volumes using windows of voxel-size $7 \times 7 \times 7$.

Frequency Domain Loss. The frequency domain loss, $\mathcal{L}_\phi^{freq}$, was computed between the ground truth multi-coil k-space $\mathbf{y}$ and the k-space transformation of the model predicted image $\hat{\mathbf{y}}_\phi$:

$$\mathcal{L}_\phi^{freq} = \lambda_{\text{NMAE}}\mathcal{L}_{\text{NMAE}}\left(\mathbf{y}, \hat{\mathbf{y}}_\phi\right), \quad \text{where } \mathcal{L}_{\text{NMAE}}(\mathbf{u}, \mathbf{v}) = \frac{||\mathbf{u} - \mathbf{v}||_1}{||\mathbf{u}||_1}. \quad (10)$$

The choice of the weighting factors λ_{SSIM}, λ_{SSIM3D}, λ_1, λ_{HFEN_1}, $\lambda_{\text{NMAE}} \geq 0$ are hyperparameters that determine the influence of each loss component in the overall optimization process.

4 Experimental Setup

We conducted two sets of experiments, addressing the reconstruction task from two perspectives: a 2D reconstruction problem and a 2D dynamic reconstruction problem involving spatial dimensions and time.

4.1 2D Reconstruction

In this setup, our goal was to solve Eq. 3. We utilized 2D U-Nets with four scales as denoisers, each featuring 32 filters in the initial scale. The optimization process involved 16 steps (T = 16). Data consistency in Eq. 3b was ensured through 14 gradient descent iterations. For the sensitivity model, we employed a 2D U-Net with four scales and 32 filters for the first scale. This configuration focused on reconstructing 2D images. The input consisted of undersampled multi-coil k-space data from single slices or frames, and the output comprised 2D images.

4.2 2D Dynamic Reconstruction

In this configuration, we approached the reconstruction challenge dynamically, utilizing the formulation presented in Eq. 4. Our model took as input a sequential series of time frames featuring 2D undersampled multi-coil k-space data. Our objective was to generate a corresponding sequential series of time-frame images as the output. In contrast to the previous setup, we employed 3D U-Nets, incorporating four scales and 32 filters in the initial scale. However, to accommodate GPU memory constraints, we limited the optimization steps to T = 10 and conducted 8 gradient descent iterations for data consistency. Similarly to the 2D reconstruction setup, for the sensitivity model we utilized a 2D U-Net with four scales and 32 filters in the initial scale.

4.3 Dataset

We conducted our experiments using the CMRxRecon dataset [22], containing 4D multi-coil Cine and multi-contrast k-space data acquired on a 3T MRI scanner with protocols outlined in [23]. The Cine MRI data included short-axis (SAX) and long-axis (LAX) views, while the multi-contrast data encompassed T1 and T2-weighted MRI data. For training, we had access to a total of 203 cine and 240 multi-contrast 4D volumes of fully-sampled k-spaces. The validation dataset comprised 111 cine and 118 multi-contrast 4D volumes of undersampled k-spaces at acceleration factors of 4, 8, and 10.

4.4 Training and Optimization Details

Our models were implemented and optimized using PyTorch [24]. The Deep Image Reconstruction Toolkit (DIRECT) [25] facilitated our pipeline tools. We employed Adam as the model parameter optimizer, with $\epsilon = 10^{-8}$ and $(\beta_1, \beta_2) =$

(0.9, 0.999). Training was conducted on four NVIDIA A100 80GB GPUs with a batch size of 1 and 2 on each GPU, for dynamic and non-dynamic tasks, respectively.

For both experimental setups, the loss computation used these weighting parameters: $\lambda_{SSIM} = \lambda_1 = \lambda_{HFEN_1} = 1.0$, and $\lambda_{NMAE} = 3.0$. For 2D dynamic reconstruction (Sect. 4.2), we employed both versions of the SSIM loss, computed per 2D slice and across the entire sequence, and we set $\lambda_{SSIM3D} = 1.0$.

4.5 Comparisons

To evaluate our proposed methods, we compared them against two state-of-the-art 2D MRI reconstruction approaches, the Recurrent Variational Network (RecurrentVarNet) [13], wining method in the MultiCoil MRI Reconstruction Challenge [10] and the End-to-end Variational Network (E2EVarNet), one of the top-performing solutions in the fastMRI challenge [12]. Both approaches were trained using the same settings and techniques as used for our proposed methods.

4.6 Evaluation Metrics

Metrics used for evaluation were the structural similarity index measure (SSIM), the normalized mean-squared-error (NMSE), and the peak signal-to-noise ratio (PSNR).

5 Results

Table 1. Average evaluation metrics on the validation set for each modality.

Experimental Setup	Acceleration Factor	Cine						Multi-Contrast					
		LAX			SAX			T1-weighted			T2-weighted		
		SSIM	NMSE	PSNR	SSIM	NMSE	PSNR	SSIM	NMSE	PSNR	SSIM	NMSE	PSNR
RecurrentVarNet	4	0.8696	0.0192	31.07	0.9170	0.0118	34.14	0.9016	0.0175	33.21	0.8995	0.0125	31.34
	8	0.7871	0.0505	26.99	0.8499	0.0272	30.42	0.8360	0.0424	29.46	0.8534	0.0266	28.08
	10	0.7763	0.0592	26.46	0.8295	0.0362	29.24	0.8034	0.0601	27.79	0.8451	0.0340	27.04
E2EVarNet	4	0.9521	0.0048	37.43	0.9693	0.0033	40.67	0.9715	0.0038	41.34	0.9543	0.0042	36.64
	8	0.8871	0.0174	31.79	0.9262	0.0095	35.24	0.9354	0.0107	35.63	0.9261	0.0093	33.08
	10	0.8727	0.0209	30.79	0.9112	0.0126	33.91	0.9202	0.0190	33.21	0.9205	0.0114	32.02
2D Reconstruction (2D vSHARP)	4	0.9584	0.0034	38.74	0.9739	0.0025	41.54	0.9766	0.0026	42.16	0.9573	0.0038	36.94
	8	0.9072	0.0111	33.50	0.9410	0.0069	36.75	0.9521	0.0063	37.87	0.9369	0.0069	34.31
	10	0.8944	0.0138	32.48	0.9284	0.0091	35.50	0.9442	0.0092	36.50	0.9334	0.0083	33.57
2D Dynamic Reconstruction (3D vSHARP)	4	0.9658	0.0028	39.57	0.9783	0.0020	42.39	0.9814	0.0021	42.24	0.9655	0.0029	38.23
	8	0.9229	0.0087	34.54	0.9522	0.0055	37.81	0.9609	0.0055	38.80	0.9479	0.0054	35.47
	10	0.9112	0.0111	33.44	0.9407	0.0079	36.48	0.9544	0.0080	37.33	0.9460	0.0063	34.74

In Fig. 3 we present sample reconstructions and in Table 1 are presented the reconstruction evaluation results on the validation dataset, from both of our experimental setups. Additionally, we include results from the two methods employed for comparison: the RecurrentVarNet and the E2EVarNet. We can observe that both, 2D reconstruction and 2D dynamic reconstruction with vSHARP, yielded superior results in terms of quantitative metrics, surpassing both the RecurrentVarNet and the E2EVarNet. However, the 2D dynamic reconstruction setup outperforms the 2D reconstruction for both Cine and Multi-Contrast tasks.

Additionally, in Table 2, we present the time required for volume reconstruction in seconds across the two experimental setups detailed in this work. From Table 2 is evident that in overall, the 2D dynamic reconstruction surpasses the 2D reconstruction in both Cine and Multi-Contrast scenarios.

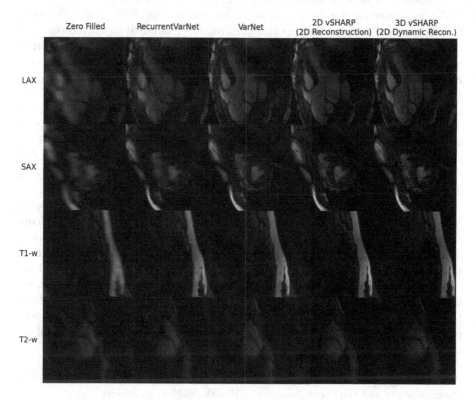

Fig. 3. Sample reconstructions from the 10× undersampled validation set.

Table 2. Time for reconstruction per volume (in seconds).

2D Reconstruction	Cine		Multi-Contrast		2D Dynamic Reconstruction	Cine		Multi-Contrast	
	LAX	SAX	T1-w	T2-w		LAX	SAX	T1-w	T2-w
	8.57	96.49	12.60	2.26		3.63	15.71	5.46	2.72

6   Conclusion and Discussion

In this work we employed the variable Splitting Half-quadratic ADMM algorithm for Reconstruction of inverse-Problems (vSHARP) network, a state-of-the-art DL-based method, to the task of reconstructing undersampled Cardiac MRI data. We adapted vSHARP under two settings, one that considers the reconstruction problem as a 2D reconstruction task, i.e., each image at a specific time frame is treated individually, and one that it considers it as a dynamic task by operating on all time frame data within a given sequence.

Upon reviewing the Table 1, it becomes evident that both of our proposed methods have demonstrated superior performance compared to the alternatives. In addition, as anticipated and demonstrated in other works [26], our empirical findings confirm that 2D dynamic reconstruction outperforms the traditional 2D reconstruction. This improved performance of the 2D dynamic model can be attributed to its ability to leverage shared information across data points within the same time sequence.

Another aspect worth considering is that, in our dynamic setup, we employed all time frames per slice as input. This introduced GPU memory limitations, thereby constraining the parameter count in the reconstruction model (3D vSHARP). However, by utilizing only a subset of the time sequence data (e.g., 2–3 adjacent time frames), it would be feasible to construct a larger model.

Furthermore, Table 2 shows that the 2D dynamic reconstruction setup requires less inference time. This can be attributed to the fact that the 2D reconstruction process involves loading individual slices or time frames into memory and subsequently performing a forward pass through the model. This leads to relatively longer reconstruction times, as evidenced by the higher values for both the Cine and Multi-Contrast datasets. Conversely, in the 2D dynamic reconstruction setup, sequences of data are loaded collectively and processed in a single forward pass through the 2D dynamic model, resulting in significantly reduced reconstruction times. This observation could indeed play a pivotal role in selecting an appropriate reconstruction model for real-time clinical scenarios.

Acknowledgements. This work was funded by an institutional grant from the Dutch Cancer Society and the Dutch Ministry of Health, Welfare and Sport.

References

1. Arai, A.E.: The cardiac magnetic resonance (CMR) approach to assessing myocardial viability. J. Nucl. Cardiol. **18**(6), 1095–1102 (2011). https://doi.org/10.1007/s12350-011-9441-5

2. Kim, P.K., et al.: Myocardial T1 and T2 mapping: techniques and clinical applications. Korean J. Radiol. **18**(1), 113 (2017). https://doi.org/10.3348/kjr.2017.18.1.113

3. Larose, E., Rodés-Cabau, J., Delarochelliere, R., Barbeau, G., Noel, B., Bertrand, O.: Cardiovascular magnetic resonance for the clinical cardiologist. Can. J. Cardiol. **23**, 84B–88B (2007). https://doi.org/10.1016/s0828-282x(07)71017-6

4. Shannon, C.: Communication in the presence of noise. Proc. IRE **37**(1), 10–21 (1949)
5. Griswold, M.A., et al.: Generalized auto calibrating partially parallel acquisitions (GRAPPA). Magn. Reson. Med. **47**(6), 1202–1210 (2002). https://doi.org/10.1002/mrm.10171
6. Niendorf, T., Sodickson, D.K.: Parallel imaging in cardiovascular MRI: methods and applications. NMR in Biomed. **19**(3), 325–341 (2006). https://doi.org/10.1002/nbm.1051
7. Geethanath, S., et al.: Compressed sensing MRI: A review. Crit. Rev. Biomed. Eng. **41**(3), 183–204 (2013). https://doi.org/10.1615/critrevbiomedeng.2014008058
8. Kido, T., et al.: Compressed sensing real-time cine cardiovascular magnetic resonance: accurate assessment of left ventricular function in a single-breath-hold. J. Cardiovasc. Magn. Reson. **18**(1), 50 (2016). https://doi.org/10.1186/s12968-016-0271-0
9. Pal, A., Rathi, Y.: A review and experimental evaluation of deep learning methods for MRI reconstruction (2022)
10. Beauferris, Y., et al.: Multi-coil MRI reconstruction challenge-assessing brain MRI reconstruction models and their generalizability to varying coil configurations. Front. Neurosci. **16**, 919186 (2022). https://www.frontiersin.org/articles/10.3389/fnins.2022.919186
11. Küstner, T., et al.: CINENet: deep learning-based 3D cardiac CINE MRI reconstruction with multi-coil complex-valued 4D spatio-temporal convolutions. In: Scientific Reports, vol. 10, no. 1 (2020). https://doi.org/10.1038/s41598-020-70551-8
12. Sriram, A., et al.: End-to-end variational networks for accelerated MRI reconstruction. In: Martel, A.L., et al. (eds.) MICCAI 2020. LNCS, vol. 12262, pp. 64–73. Springer, Cham (2020). https://doi.org/10.1007/978-3-030-59713-9_7
13. Yiasemis, G., Sonke, J.-J., Sánchez, C., Teuwen, J.: Recurrent variational network: a deep learning inverse problem solver applied to the task of accelerated MRI reconstruction. In: Proceedings of the IEEE/CVF Conference on Computer Vision and Pattern Recognition (CVPR), pp. 732–741 (2022)
14. Hamilton, J.I.: A self-supervised deep learning reconstruction for shortening the breathhold and acquisition window in cardiac magnetic resonance fingerprinting. Front. Cardiovasc. Med. **9**, 928546 (2022). https://doi.org/10.3389/fcvm.2022.928546
15. Yiasemis, G., Moriakov, N., Sonke, J.-J., Teuwen, J.: vSHARP: variable splitting half-quadratic admm algorithm for reconstruction of inverse-problems. arXiv.org (2023). arXiv:2309.09954 [eess.IV], https://doi.org/10.48550/arXiv.2309.09954
16. Ye, J.C.: Compressed sensing MRI: a review from signal processing perspective. BMC Biomed. Eng. **1**(1), 8 (2019). https://doi.org/10.1186/s42490-019-0006-z
17. Uecker, M., et al.: ESPIRiT-an eigenvalue approach to autocalibrating parallel MRI: where SENSE meets GRAPPA. Magn. Reson. Med. **71**(3), 990–1001, (2013). https://doi.org/10.1002/mrm.24751
18. Ronneberger, O., Fischer, P., Brox, T.: U-Net: Convolutional Networks for Biomedical Image Segmentation. In: Navab, N., Hornegger, J., Wells, W.M., Frangi, A.F. (eds.) MICCAI 2015. LNCS, vol. 9351, pp. 234–241. Springer, Cham (2015). https://doi.org/10.1007/978-3-319-24574-4_28
19. Li, R., Luo, L., Zhang, Y.: Convolutional neural network combined with half-quadratic splitting method for image restoration. J. Sens. **2020**, 1–12 (2020). https://doi.org/10.1155/2020/8813413

20. Boyd, S., Parikh, N., Chu, E., Peleato, B., Eckstein, J., et al.: Distributed optimization and statistical learning via the alternating direction method of multipliers. Found. Trends® Mach. Learn. **3**(1), 1–122 (2011)
21. Yiasemis, G., Sánchez, C.I., Sonke, J.-J., Teuwen, J.: On retrospective k-space subsampling schemes for deep MRI reconstruction. Mag. Reson. Imaging S0730725X23002199 (2024). https://doi.org/10.1016/j.mri.2023.12.012
22. Wang, C., et al.: CMR x Recon: An open cardiac MRI dataset for the competition of accelerated image reconstruction (2023)
23. Wang, C., et al.: Recommendation for cardiac magnetic resonance imaging-based phenotypic study: imaging part. Phenomics **1**(4), 151–170 (2021). https://doi.org/10.1007/s43657-021-00018-x
24. Paszke, A., et al.: Automatic differentiation in pytorch (2017)
25. Yiasemis, G., Moriakov, N., Karkalousos, D., Caan, M., Teuwen, J.: Direct: deep image reconstruction toolkit. J. Open Source Softw. **7**(73), 4278 (2022). https://doi.org/10.21105/joss.04278
26. Zhang, C., Caan, M.W., Navest, R., Teuwen, J., Sonke, J.-J.: Radial-rim: accelerated radial 4D MRI using the recurrent inference machine. In: Proceedings of International Society for Magnetic Resonance in Medicine, vol. 31 (2023)

Author Index

A

Agrawal, Shaleka 88, 220
Al-Haj Hemidi, Ziad 467
Alvarez-Florez, Laura 25
Amir-Khalili, Alborz 240
Arri, Satpal 108
Ashby, Joseph 220
Aslanidi, Oleg 55, 174
Axel, Leon 261

B

Bai, Jeiyun 220
Bai, Jieyun 88
Bai, Wenjia 293
Banerjee, Abhirup 163, 209
Beetz, Marcel 163
Berruezo, Antonio 35
Bieging, Erik 230
Bourfiss, Mimount 25
Boyett, Mark 220
Brüning, Jan 140

C

Cai, Xue J. 220
Camara, Oscar 35
Carlhäll, Carl-Johan 88
Carloni, Gianluca 421
Cassady, Nathan 240
Chen, Min 163
Chen, Sheng 390
Chen, Weitian 314
Chiew, Mark 369
Choudhury, Robin P. 209
Corno, Antonio F. 220
Creamer, Stephen A. 98, 108
Curran, Kathleen M. 326

D

De Vecchi, Adelaide 55, 186
Di Folco, Maxime 66

DiBella, Ed 230
Dietlmeier, Julia 326
Dillon, Joshua R. 98, 119
Ding, Wenzhe 339
Dobrzynski, Halina 220
Dou, Quan 390
Doughty, Robert N. 98, 108, 119
Dreger, Henryk 140
Du, Yuning 421
Dylov, Dmitry V. 274

E

Edwards, Nicola C. 98, 108, 119
Elhabian, Shireen 230

F

Falk, Volkmar 140
Feng, Fan 88, 220, 250
Feng, Xue 390

G

Gamage, Thiranja P. Babarenda 98, 108, 119
Garcia-Cabrera, Carles 326
Garcia-Fernandez, Ignacio 35
Ge, Rongjun 433
Gerardo-Giorda, Luca 152
Geven, Bram W. M. 98
Gil, Miriam 35
Goubergrits, Leonid 140
Grau, Vicente 163, 209
Guerra, Jose M. 152

H

Hart, George 220
Hasaballa, Abdallah I. 108
Hashmi, Anam 326
Hassan, Ahmed 186
He, Haorui 209
He, Jin 283
Heinrich, Mattias P. 467

Hennemuth, Anja 140
Hermida, Uxio 186
Huang, Liqin 359
Huang, Shoujin 400
Hutcheon, Robert C. 220

I
Išgum, Ivana 25
Ivantsits, Matthias 140

J
Jacob, Athira J. 44
Jarvis, Jonathan 220
Jones, Caroline B. 220
Jordan, Connor 421

K
Kadota, Brenden T. 369
Kholmovski, Eugene 230
King, Andrew P. 174
Kofler, Andreas 454
Kong, Fanwei 196
Krüger, Nina 140
Kühne, Titus 140
Kunze, Karl P 130
Kwan, Eugene 230

L
Lamata, Pablo 186
Legget, Malcolm E. 98, 108, 119
Leoni, Massimiliano 152
Li, Hongsheng 15
Li, Ruiyi 433
Li, Wei 359
Li, Xiaomeng 380
Li, Xinqi 77
Li, Yansong 303
Li, Yi 400
Lin, Yiqun 380
Liu, Weizhou 283
Liu, Xiaohan 339
Liu, Yilong 400
Liu, Yiming 339
Locas, John 186
Lockhart, Lisette 240
Logantha, Sunil J. R. J. 220
Longobardi, Stefano 55
Lowe, Boris S. 108, 119

Lozano, Miguel 35
Lundberg, Peter 88
Lyu, Jun 410, 443
Lyu, Mengye 400

M
Marsden, Alison L. 196
Mei, Lifeng 400
Melidoro, Paolo 55
Metaxas, Dimitris N. 261
Metaxas, Dimitris 15
Meyer, Craig H. 390
Moriakov, Nikita 479
Morris, Alan 230
Muffoletto, Marica 130

N
Nagy, Michael 186
Nash, Martyn P. 98, 108, 119
Neji, Radhouene 130
Ni, Ziyu 15
Niederer, Steven A 130
Nielles-Vallespin, Sonia 293
Nunn, Alexandra 240

O
O'Connor, Noel E. 326
Obada, George 174
Ogbomo-Harmitt, Shaheim 174
Orkild, Benjamin 230
Oster, Julien 467

P
Pachetti, Eva 421
Pang, Yanwei 339
Patel, Jaykumar H. 369
Paton, Julian F. R. 108
Penela, Diego 35
Peng, Jiachuan 163
Petras, Argyrios 152

Q
Qiao, Mengyun 293
Qin, Jing 410, 443
Quill, Gina M. 98, 108, 119
Quillien, Lucile 467
Qureshi, Ahmed 55

R

Ranjan, Ravi   230
Razumov, Artem   274
Rodrigo, Miguel   35
Romero, Pau   35
Ruckert, Daniel   44
Rueckert, Daniel   130, 293
Ruygrok, Peter N.   108

S

Sabry, Malak   186
Samuel, Irini   186
Sander, Jörg   25
Schnabel, Julia A.   66
Sebastian, Rafael   35
Serra, Dolors   35
Sharma, Puneet   44
Sheagren, Calder D.   369
Smaill, Bruce   220
Smine, Zineb   55
Sonke, Jan-Jakob   479
Stein, Josh   66
Stephenson, Robert S.   220
Stojanovski, David   186
Sultan, K M Arefeen   230
Sun, Weiya   433
Sun, Yong   339
Sutton, Timothy M.   119
Swingen, Cory   240

T

Talou, Gonzalo D. Maso   108
Tan, Yongyao   88, 250
Tang, Mengshi   359
Tänzer, Michael   293
Tao, Qian   77, 349
Teuwen, Jonas   479
Tian, Yun   283, 303
Tjong, Fleur V. Y.   25
Trew, Mark   88
Tsaftaris, Sotirios A.   421

V

van Gemert, Jan   77
Vandersickel, Nele   174
Velthuis, Birgitta K.   25
Vogt, Nora   467
Vohra, Akbar   220

W

Walczak, Lars   140
Wang, Chengyan   410, 443
Wang, Fanwen   293, 410, 443
Wang, Guangming   410
Wang, Vicky Y.   108, 119
Wang, Zhenkun   380
Wang, Zhixing   390
Wei, Linda   15
Weidmann, Zoraida Moreno   152
Weihsbach, Christian   467
Williams, Michelle C   130
Williams, Steven E.   55, 130
Wright, Graham A.   369

X

Xia, Qing   15
Xiang, Tianqi   380
Xin, Bingyu   261
Xu, Hao   130
Xu, Lijian   15
Xu, Yiyang   130
Xue, Yuyang   421

Y

Yacoub, Magdi H.   186
Yan, Kang   390
Yang, Guang   293
Yang, Jiewen   380
Yang, John Zhiyong   88
Yang, Kexin   400
Yanni, Joseph   220
Ye, Meng   261
Ye, Zhiyu   3
Yi, Xin   240
Yiasemis, George   479
Young, Alistair A.   98, 108, 119
Young, Alistair A   130
Yue, Wenjun   380

Z

Zhang, Liping   314
Zhang, Shaoting   15
Zhang, Tong   3
Zhang, Weihua   359
Zhang, Yi   77, 349
Zhao, Debbie   98, 108, 119
Zhao, Jichao   88, 220, 250
Zhao, Lulu   303

Zhao, Shifeng   283, 303
Zhao, Xunkang   443
Zhao, Yidong   77, 349

Zheng, Hairong   3
Zheng, Hanyuan   433
Zimmermann, Felix Frederik   454

Printed in the United States
by Baker & Taylor Publisher Services